W0107529

INERTIAL COORDINATE SYSTEM ON THE SKY

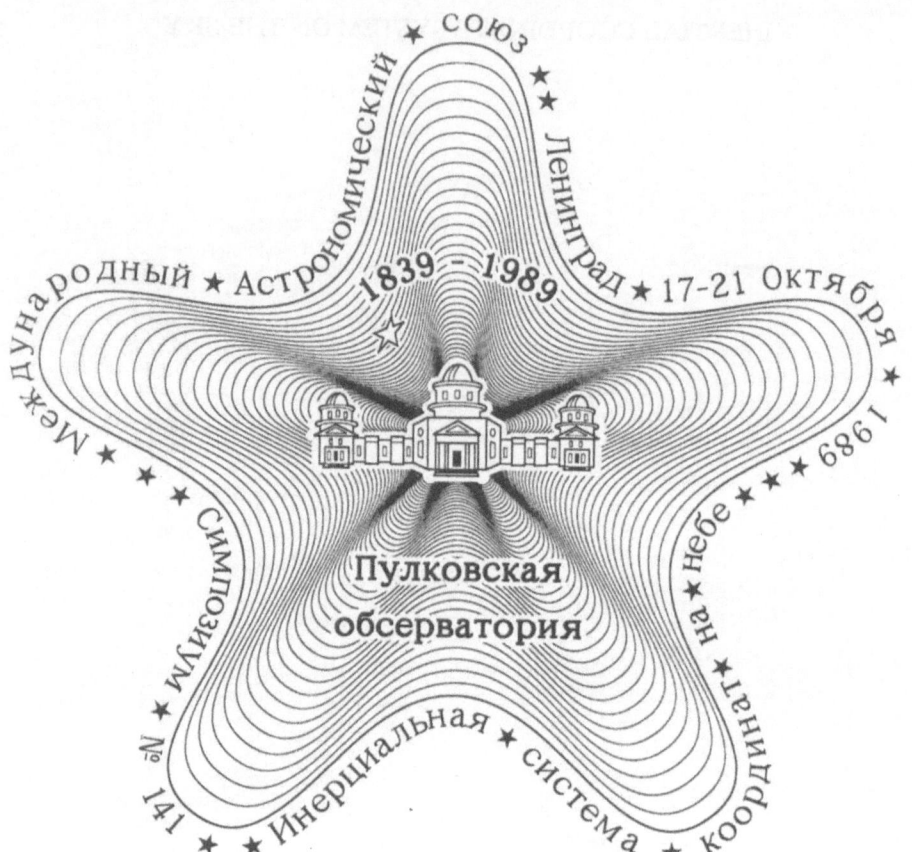

Международный ★ Астрономический ★ союз ★ Ленинград ★ 17-21 Октября 1989

Симпозиум № 141

1839 - 1989

Пулковская обсерватория

Инерциальная ★ система ★ координат ★ на ★ небе

INTERNATIONAL ASTRONOMICAL UNION

UNION ASTRONOMIQUE INTERNATIONALE

INERTIAL COORDINATE SYSTEM ON THE SKY

PROCEEDINGS OF THE 141ST SYMPOSIUM OF THE
INTERNATIONAL ASTRONOMICAL UNION
HELD IN LENINGRAD, U.S.S.R., OCTOBER 17–21, 1989

EDITED BY

JAY H. LIESKE

*Jet Propulsion Laboratory, California Institute of Technology,
Pasadena, California, U.S.A.*

and

VICTOR K. ABALAKIN

*Central Astronomical Observatory Pulkovo, USSR Academy of Sciences,
Leningrad, U.S.S.R.*

KLUWER ACADEMIC PUBLISHERS
DORDRECHT / BOSTON / LONDON

Library of Congress Cataloging in Publication Data

International Astronomical Union. Symposium (141st : 1989 :
  Leningrad, R.S.F.S.R.)
    Inertial coordinate system on the sky : proceedings of the 141st
  Symposium of the International Astronomical Union, held in
  Leningrad, USSR, 17-21 October 1989 / edited by Jay H. Lieske and
  Viktor K. Abalakin.
       p.   cm.
    At head of title: International Astronomical Union=Union
  astronomique internationale.
    Includes bibliographical references.
    ISBN 0-7923-0786-0 (U.S. : alk. paper). -- ISBN 0-7923-0787-9
  (U.S. : pbk. : alk. paper)
    1. Celestial reference systems--Congresses.   I. Lieske, Jay H.
  II. Abalakin, Viktor Kuz'mich.   III. Title.
  QB633.I58   1989
  525--dc20
                                                          90-4538

ISBN-13: 978-0-7923-0787-7   e-ISBN-13: 978-94-009-0613-6
DOI: 10.1007/ 978-94-009-0613-6

Published on behalf of
the International Astronomical Union
by
Kluwer Academic Publishers, P.O. Box 17, 3300 AA Dordrecht, The Netherlands.

Kluwer Academic Publishers incorporates
the publishing programmes of
D. Reidel, Martinus Nijhoff, Dr W. Junk and MTP Press.

Sold and distributed in the U.S.A. and Canada
by Kluwer Academic Publishers,
101 Philip Drive, Norwell, MA 02061, U.S.A.

In all other countries, sold and distributed
by Kluwer Academic Publishers Group,
P.O. Box 322, 3300 AH Dordrecht, The Netherlands.

Printed on acid-free paper

All Rights Reserved
© 1990 by the International Astronomical Union
Softcover reprint of the hardcover 1st edition 1990
No part of the material protected by this copyright notice may be reproduced or utilized in any form or by any means, electronic or mechanical including photocopying, recording or by any information storage and retrieval system, without written permission from the publisher.

TABLE OF CONTENTS

Part 1
The legacy of Pulkovo for inertial systems and reference frames

Part 2
Pulkovo today

Part 3
Concepts, Definitions, Models

Part 4
Realization and comparison of reference frames

PREFACE

IAU Symposium Number 141 *"Inertial Coordinate System on the Sky"* was held in Leningrad, USSR from 17–21 October 1989. The symposium also commemorated the 150th anniversary of the founding of Pulkovo Observatory. The scientific program was presented in ten half-day sessions. Most sessions were held at the Pulkovskaya Hotel, but one session which highlighted Pulkovo's current programs was held at Pulkovo Observatory. The sessions were organized into general categories pertaining to the legacy of Pulkovo for inertial systems; current programs at Pulkovo Observatory; concepts, definitions and models; and the realization and comparision of reference frames. More than 140 scientific papers were presented, either orally or in poster form. Extensive use was made of electronic mail and computer-readable communications, and more than two-thirds of the authors made use of the opportunity to submit papers for formatting by the editors.

The meeting was truly a symposium in the Greek sense of the word—a free-flowing exchange of ideas and opinions. The final two papers presented at the symposium by Wilkins and by Westerhout are presented at an early stage in the published proceedings, in order to help focus the reader's attention on the concepts and problems explored in subsequent papers. As pointed out by G. Westerhout in his summary, the suggestions for terminology proposed by Wilkins (p. 39) and by Han, Huang and Xu (p. 99) appeared late in the program. Indeed, if one followed all of the suggestions, the title of the symposium might have been quite different. Relativists would have removed the word "Inertial" from the title; Wilkins and Xu would have removed "Coordinate System"; and so we are left with simply a symposium "On the Sky", which probably is quite appropriate.

Three resolutions (see page xxxiii) were adopted by the participants in the symposium concerning HIPPARCOS, VLBI, and future astrometry.

We are indebted to the members of the Scientific Organizing Committee and to co-chairman Yatskiv, for their valuable guidance and assistance in planning the symposium on rather short notice, and we appreciate the support from the Local Organizing Committee in planning the local arrangements. Their names are given on the following page.

We are indebted to several organizations and agencies for financial and other support of the symposium. The IAU provided travel grants for the participation of 16 astronomers. The Astronomical Council of the USSR Academy of Sciences and Pulkovo Observatory also generously provided assistance. The Jet Propulsion Laboratory of the California Institute of Technology provided support in organizing the symposium and Kluwer Academic Publishers generously donated some copies of the proceedings to the Scientific Organizing Committee. We are also grateful to Stephan Lieske for his assistance in developing the logo for the symposium, and to A. Sergeyevsky and P. Halamek of JPL for some assistance in translations.

Finally, we are indebted to the many astronomers from convertible-currency countries who pre-paid their registration fee in convertible currency—which was matched by donations from Pulkovo Observatory in local currency to the Local Organizing Committee—and who thus provided the means for Pulkovo Observatory to purchase copies of the proceedings for astronomers from countries without convertible currency.

February 1990   *The Editors.*

Scientific Organizing Committee

Ya.S. Yatskiv (USSR) co-chairman
J.H. Lieske (USA) co-chairman

J.A. Hughes (USA)
B. Kołaczek (Poland)
M. Miyamoto (Japan)
B. Morando (France)
P. Pâquet (Belgium)
Y. Réquième (France)
P.K. Seidelmann (USA)
W.F. van Altena (USA)
R. Wielen (FRG)
G.A. Wilkins (UK)
Ye Shu-Hua (China)

Local Organizing Committee

V.K. Abalakin, chairman
I.I. Kanayev, vice-chairman
A.S. Zherbina, secretary

H.I. Potter
V.A. Fomin
V.A. Naumov
V.F. Mikhailov
N.R. Persiyaninova
E.V. Polyakov
L.V. Rykhlova
Yu.Ya. Vitinsky

List of Participants, IAU Symposium No. 141

Victor K. Abalakin
Central Astronomical Observatory
Pulkovo
196140 Leningrad, USSR

Harold D. Ables
U.S. Naval Observatory
Flagstaff Station
P.O. Box 1149
Flagstaff, Arizona 86001 USA

Praskovya M. Afanas'eva
Central Astronomical Observatory
Pulkovo
196140 Leningrad, USSR

A.N. Alexandrov
Observatory Kiev Univ.
Observatorna St 3
252053 Kiev, USSR

Claudio A. Anguita
Obs. Astronomico Nacional
Universidad de Chile
Casilla 36-D
Santiago, CHILE

J.P. Anosova
Leningrad Univ. Astron. Observatory
Bibliotechnaya Pl 2
198904 Leningrad, USSR

Vadim A. Antonov
Leningrad Univ. Astron. Observatory
199178 Leningrad, USSR

A. Noel Argue
Institute of Astronomy
Madingley Road
Cambridge CB3 0HA, U.K.

E. Felicitas Arias
Central Bureau of IERS
Bureau des Longitudes
77 Ave Denfert-Rochereau
F-75014 Paris, FRANCE

H.A. Avanesov
Space Research Institute
Profsoyuznaya 84 /32
117810 Moscow, USSR

O.V. Avvakumov
Kazan State University
Lenin Str 18
Kazan, Tatar ASSR, USSR

Bronislav K. Bagil'dinskij
Central Astronomical Observatory
Pulkovo
196140 Leningrad, USSR

A.T. Bajkova
Institute of Applied Astronomy
8 Zhdanov Street
197042 Leningrad, USSR

Yu.V. Barkin
N.E. Baumann Technical University
2nd Baumann Str 5
Moscow, USSR

Katalin Barlai
Konkoly Observatory
Box 67
H-1525 Budapest XII, HUNGARY

Ulrich Bastian
Astronomisches Rechen-Institut
Mönchhofstrasse 12-14
D-6900 Heidelberg, GERMANY, Fed. Rep.

Yu.V. Batrakov
Institute for Theoretical Astronomy
10 Kutuzov Quay
191187 Leningrad, USSR

Galina D. Baturina
Central Astronomical Observatory
Pulkovo
196140 Leningrad, USSR

Alan H. Batten
Dominion Astrophys. Observatory
5071 West Saanich Road
Victoria, BC V8X 4M6, CANADA

Annick Bec-Borsenberger
Bureau des Longitudes
77 Ave Denfert-Rochereau
F-75014 Paris, FRANCE

M.V. Belikov
Institute for Theoretical Astronomy
10 Kutuzov Quay
191187 Leningrad, USSR

Ashok Kumar Bhatnagar
Positional Astronomy Center
P 546, Block N, 1st Floor
New Alipore, Calcutta 700053, INDIA

V.V. Bobylev
Central Astronomical Observatory
Pulkovo
196140 Leningrad, USSR

V.I. Bogdanov
Central Astronomical Observatory
Pulkovo
196140 Leningrad, USSR

Eric Bois
CERGA
Ave Copernic
F-06130 Grasse, FRANCE

Tatiana V. Bordovitsyna
Tomsk State University
Inst. Appl. Mechanics and Math.
634029 Tomsk GSP-14, USSR

Victoria N. Boyko
Institute for Theoretical Astronomy
10 Kutuzov Quay
191187 Leningrad, USSR

Nina M. Bronnikova
Central Astronomical Observatory
Pulkovo
196140 Leningrad, USSR

Peter Brosche
Observatorium Hoher List
Univ Sternwarte Bonn
D-5568 Daun, GERMANY, Fed. Rep.

Eugene V. Brumberg
Institute of Applied Astronomy
8 Zhdanov Street
197042 Leningrad, USSR

Viktor A. Brumberg
Institute of Applied Astronomy
8 Zhdanov Street
197042 Leningrad, USSR

V.A. Bykova
Schmidt Institute for Earth Physics
Bol. Gruzinskaya 10
123810 Moscow D-242, USSR

E.N. Bystrov
Central Astronomical Observatory
Pulkovo
196140 Leningrad, USSR

John F. Chandler
Harvard-Smithsonian Center for Astrophysics
60 Garden St
Cambridge, Massachusetts 02138 USA

V.O. Chechet
Institute for Theoretical Astronomy
10 Kutuzov Quay
191187 Leningrad, USSR

A.M. Cherepaschuk
Sternberg Astronomical Institute
Universitetskij Prospekt 13
119899 Moscow, USSR

Yu.A. Chernetenko
Institute for Theoretical Astronomy
10 Kutuzov Quay
191187 Leningrad, USSR

M.S. Chubey
Central Astronomical Observatory
Pulkovo
196140 Leningrad, USSR

Carl S. Cole
U.S. Naval Observatory
34th and Massachusetts Ave, NW
Washington, DC 20392 USA

Thomas E. Corbin
U.S. Naval Observatory
34th and Massachusetts Ave. NW
Washington, DC 20392 USA

L.A. Dautov
Engelhardt Astronomical Observatory
Observatory Station
422526 Kazan, Tatar ASSR, USSR

Suzanne V. Débarbat
Observatoire de Paris
61 Ave de l'Observatoire
F-75014 Paris, FRANCE

A.B. Demitchev
GOS STANDART
State Committee f or National Standards
Leninskij prospekt 9
117049 Moscow, USSR

A.V. Devyatkin
Central Astronomical Observatory
Pulkovo
196140 Leningrad, USSR

Wolfgang R. Dick
Zentralinstitut für Astrophysik
Rosa-Luxemburg-Strasse 17a
DDR-1591 Potsdam, GERMANY, Dem. Rep.

Steven J. Dick
U.S. Naval Observatory
34th and Massachusetts Ave. NW
Washington, DC 20392 USA

Smiliana Dikova
Department of Astronomy
Boulevard Lenin 72
1784 Sofia, BULGARIA

M.A. Dirikis
Latvia State University Observatory
Janis Rainis Boulevard 18
226088 Riga, Latvia, USSR

O.V. Doroshenko
P.N. Lebedev Physical Institute
Lenininskij Prospekt 53
117924 Moscow, USSR

V.N. Dudinov
Kharkov State University Observatory
Sumskaya Str 35
310022 Kharkov, USSR

Dmitrij P. Duma
Main Astronomical Observatory
Ukrainian Academy of Sciences
252127 Kiev, USSR

Raynor L. Duncombe
Dept of Aerospace Engineering
University of Texas
Austin, Texas 78712 USA

S.V. Dyakonov
Schmidt Institute of Earth Physics
Bol. Gruzinskaya 10
123810 Moscow D-242, USSR

N.G. Eliseeva
Institute of Applied Astronomy
8 Zhdanov Street
197042 Leningrad, USSR

Nikolaj V. Emelyanov
Sternberg Astronomical Institute
Universitetskij Prospekt 13
119899 Moscow, USSR

G.I. Eroshkin
Institute for Theoretical Astronomy
10 Kutuzov Quay
191187 Leningrad, USSR

Claus V. Fabricius
Copenhagen University Observatory
Brorfelde
DK-4340 Tølløse, DENMARK

Laurent Fairhead
601 Campbell Hall
Department of Astronomy
University of California
Berkeley, California 94720 USA

P.M. Fedij
Main Astronomical Observatory
Ukrainian Academy of Sciences
252127 Kiev, USSR

V.P. Fedotov
Space Research Institute
Profsoyuznaya 84 /32
117810 Moscow, USSR

Martine Feissel
Observatoire de Paris
61 Ave de l'Observatoire
F-75014 Paris, FRANCE

A.M. Finkelstein
Institute of Applied Astronomy
8 Zhdanov Street
197042 Leningrad, USSR

Valery A. Fomin
Central Astronomical Observatory
Pulkovo
196140 Leningrad, USSR

Alexander M. Fominov
Institute for Theoretical Astronomy
10 Kutuzov Quay
191187 Leningrad, USSR

N.V. Frolova
Urals State Univ. Astron. Observatory
Prospekt Lenina 51
620083 Sverdlovsk, USSR

M.A. Fursenko
Institute for Theoretical Astronomy
10 Kutuzov Quay
191187 Leningrad, USSR

I.S. Gayazov
Institute for Theoretical Astronomy
10 Kutuzov Quay
191187 Leningrad, USSR

A. Gedrovits
Latvia State University Observatory
Janis Rainis Boulevard 18
226088 Riga, Latvia, USSR

N.I. Glebova
Central Astronomical Observatory
Pulkovo
196140 Leningrad, USSR

K.G. Gnevsheva
Central Astronomical Observatory
Pulkovo
196140 Leningrad, USSR

B.O. Gorbachev
Institute for Research Inst. Design
Herzen Str 51
190000 Leningrad, USSR

V.L. Gorshkov
Central Astronomical Observatory
Pulkovo
196140 Leningrad, USSR

A.Ya. Gregul'
Kiev University Observatory
Observatornaya 3
252053 Kiev, USSR

L.P. Gribko
Sternberg Astronomical Institute
Universitetskij Prospekt 13
119899 Moscow, USSR

Erwin Groten
Institut Physikalische Geod.
Petersentrasse 13
D-6100 Darmstadt, GERMANY, Fed. Rep.

N.D. Grushnitskij
Sternberg Astronomical Institute
Universitetskij Prospekt 13
119899 Moscow, USSR

Vadim S. Gubanov
Institute of Applied Astronomy
8 Zhdanov Street
197042 Leningrad, USSR

A.P. Gulyaev
Sternberg Astronomical Institute
Universitetskij Prospekt 13
119899 Moscow, USSR

Rustam I. Gumerov
Engelhardt Astronomical Observatory
Observatory Station
422526 Kazan, Tatar ASSR, USSR

A.A. Gurshtein
Institute History of Science & Tech
Staropansky Lane 1 / 5
103012 Moscow, USSR

I.S. Guseva
Central Astronomical Observatory
Pulkovo
196140 Leningrad, USSR

Han Yanben
Beijing Astronomical Observatory
Chinese Academy of Sciences
100080 Beijing, CHINA, P.R.

He Miao-Fu
Shanghai Observatory
Academia Sinica
Shanghai, CHINA, P.R.

V.N. Hejfets
Space Research Institute
Profsoyuznaya 84 /32
117810 Moscow, USSR

Sonja Hirte
Zentralinstitut für Astrophysik
Rosa Luxemburg Strasse 17A
DDR-1591 Potsdam, GERMANY, Dem. Rep.

Erik Høg
University Observatory
Øster Voldgade 3
DK-1350 Copenhagen K, DENMARK

Hu Xiaochun
Yunnan Observatory
P.O. Box 110, Kunming
Yunnan Province, CHINA, P.R.

James A. Hughes
U.S. Naval Observatory
34th and Massachusetts Ave. NW
Washington, DC 20392 USA

Hieronim Hurnik
Astronomical Observatory
A. Mickiewicz University
Sloneczna 36
60-286 Poznan, POLAND

Yu.P. Ilyasov
P.N. Lebedev Physical Institute
Lenininskij Prospekt 53
117924 Moscow, USSR

G.I. Ivanov
Leningr. Optical-Mechanical Factory
Chugunnaya Str 10
194100 Leningrad, USSR

Tamara V Ivanova
Institute of Applied Astronomy
8 Zhdanov Street
197042 Leningrad, USSR

Violeta Ivanova
Department of Astronomy
72 Lenin Blvd
1784 Sofia, BULGARIA

V.V. Ivashkin
M.V. Keldysh Institute Applied Math.
Miusskaya Ploshchad 4
125047 Moscow, USSR

S.P. Izmailov
NPO Metrology
Dzierzynski Str 40
310022 Kharkov 22, USSR

Carme Jordi Nebot
Department de Astronomia
Universitat de Barcelona
Avda Diagonal 647
E-08028 Barcelona, SPAIN

O.A. Kalinichenko
Central Astronomical Observatory
Pulkovo
196140 Leningrad, USSR

I.A. Kalinin
Institute for Research Inst. Design
Herzen Str 51
190000 Leningrad, USSR

I.M. Kalinina
Sternberg Astronomical Institute
Universitetskij Prospekt 13
119899 Moscow, USSR

K.K. Kamensky
SCTB Institute Applied Problems
of Mechanics and Mathematics
Ul Lermontova 15
290005 Lvov, USSR

Ivan I. Kanaev
Central Astronomical Observatory
Pulkovo
196140 Leningrad, USSR

V.B. Kapkov
Engelhardt Astronomical Observatory
Observatory Station
422526 Kazan, Tatar ASSR, USSR

George H. Kaplan
U.S. Naval Observatory
34th and Massachusetts Ave. NW
Washington, DC 20392 USA

N.V. Kharchenko
Main Astronomical Observatory
Ukrainian Academy of Sciences
252127 Kiev, USSR

A.S. Kharin
Main Astronomical Observatory
Goloseevo
252127 Kiev, USSR

V.V. Khokhlov
188350 Gatchina, USSR

E.Z. Khotimskaya
Institute of Applied Astronomy
8 Zhdanov Street
197042 Leningrad, USSR

G.S. Khromov
All-Union Astron. & Geod. Society
Sadovo-Kudrinskaya Str 24
103001 Moscow K-1, USSR

E.V. Khrutskaya
Central Astronomical Observatory
Pulkovo
196140 Leningrad, USSR

V.M. Kirpatovskij
Kharkov State University Observatory
Sumskaya Str 35
310022 Kharkov, USSR

Tamara P. Kiseleva
Central Astronomical Observatory
Pulkovo M-140
196140 Leningrad, USSR

Alexej A. Kiselyov
Central Astronomical Observatory
Pulkovo
196140 Leningrad, USSR

Vitalij S. Kislyuk
Main Astronomical Observatory
Goloseevo
252127 Kiev, USSR

O.V. Kiyaeva
Central Astronomical Observatory
Pulkovo
196140 Leningrad, USSR

Arnold R. Klemola
Lick Observatory
University of California
Santa Cruz, California 95064 USA

S.A. Klioner
Institute of Applied Astronomy ·
8 Zhdanov Street
197042 Leningrad, USSR

A.Yu. Kogan
Space Research Institute
Profsoyuznaya 84 /32
117810 Moscow, USSR

Barbara Kolaczek
Planetary Geodesy Department
Polish Academy of Sciences
Ul. Bartycka 18
00-716 Warsaw, POLAND

Yu.B. Kolesnik
Astronomical Council
USSR Academy of Sciences
Pyatnitskaya 48
109017 Moscow, USSR

Yu.F. Kolyuka
Space Flight Control Centre
141070 Kaliningrad, Moscow Region, USSR

S.M. Kopejkin
Sternberg Astronomical Institute
Universitetskij Prospekt 13
119899 Moscow, USSR

A.A. Korsun'
Main Astronomical Observatory
Goloseevo
252127 Kiev, USSR

Gennadij S. Kosin
Central Astronomical Observatory
Pulkovo
196140 Leningrad, USSR

Lidija D. Kostina
Central Astronomical Observatory
Pulkovo
196140 Leningrad, USSR

O.V. Kotreleva
Central Astronomical Observatory
Pulkovo M-140
196140 Leningrad, USSR

V.A. Krasikov
Space Research Institute
Profsoyuznaya 84 /32
117810 Moscow, USSR

George A. Krasinskij
Institute of Applied Astronomy
8 Zhdanov Street
197042 Leningrad, USSR

Kevin Krisciunas
Joint Astronomy Centre
665 Komohana Street
Hilo, HI 96720 USA

N.B. Krivova
Institute of Applied Astronomy
8 Zhdanov Street
197042 Leningrad, USSR

A.A. Krivtsov
Central Astronomical Observatory
Pulkovo
196140 Leningrad, USSR

S.M. Kudr'avtsev
Space Flight Control Centre
141070 Kaliningrad, Moscow Region, USSR

Irina I. Kumkova
Institute of Applied Astronomy
8 Zhdanov Street
197042 Leningrad, USSR

Krystyna Kurzynska
Astronomical Observatory of Poznan
Sloneczna 36
60-286 Poznan, POLAND

A.V. Latyshev
Irkutsk State University
Sovetskaya Str 119a
664009 Irkutsk, USSR

Marek Lehmann
Astronomical Latitude Obs.
Borowiec
62-035 Kornik, N. Poznan, POLAND

E.P. Levitan
"Zemlya i Vselennaya"
("Earth and Universe") Magazine
Maronovsky Lane 26
117089 Moscow, USSR

D Lhagvasuren
Astronomical Observatory
P.O.B. 788
Ulan Bator, MONGOL P. R.

Li Zhi-gang
Shaanxi Astronomical Observatory
P.O. Box 18
Lintong, Xian, CHINA, P.R.

Li Zheng-xin
Shanghai Observatory
Academia Sinica
Shanghai, CHINA, P.R.

Jay H. Lieske
Mail Stop 301-150
Jet Propulsion Laboratory
4800 Oak Grove Drive
Pasadena, California 91109 USA

N.G. Litkevich
Astronomical Observatory
Sumskaya ul 35
310022 Kharkov 22, USSR

Alvaro Lopez
Obs. Astronomico Uni de Valencia
Avenide Blasco Ibañez 13
46010 Valencia, SPAIN

Patricio Loyola
Obs. Astronomico Nacional
Universidad de Chile
Casilla 36-D
Santiago, CHILE

Lu Chun-Lin
Purple Mountain Observatory
21008 Nanjing, CHINA, P.R.

V.N. L'vov
Institute for Theoretical Astronomy
10 Kutuzov Quay
191187 Leningrad, USSR

N.S. Lyadovoj
Nikolayev Branch, Pulkovo Observ.
Observatory Str 1
327030 Nikolaev, USSR

Chopo Ma
Mail Code 621.9
Goddard Space Flight Center
Greenbelt, Maryland 20771 USA

S.P. Major
Main Astronomical Observatory
Ukrainian Academy of Sciences
252127 Kiev, USSR

V.V. Makarov
Central Astronomical Observatory
Pulkovo
196140 Leningrad, USSR

E.N. Makarova
Institute for Theoretical Astronomy
10 Kutuzov Quay
191187 Leningrad, USSR

E.I. Malakhov
Institute of Applied Astronomy
8 Zhdanov Street
197042 Leningrad, USSR

Z.M. Malkin
Central Astronomical Observatory
Pulkovo
196140 Leningrad, USSR

A.L. Malkov
Institute for Theoretical Astronomy
10 Kutuzov Quay
191187 Leningrad, USSR

K.S. Mansurova
Irkutsk State University
Sovetskaya Str 119a
664009 Irkutsk, USSR

Giuseppe Massone
Osservatorio Astronomico di Torino
Strada Osservatorio 20
10025 Pino Torinese, ITALY

N.N. Matveev
Inst. Astrophysics, Tadzhik SSR
Sviridenko Str 22
734042 Dushanbe 42, USSR

Dennis D. McCarthy
U.S. Naval Observatory
34th and Massachusetts Ave. NW
Washington, DC 20392 USA

Yu.D. Medvedev
Institute for Theoretical Astronomy
10 Kutuzov Quay
191187 Leningrad, USSR

Mikhail M. Medvedsky
Main Astronomical Observatory
Ukrainian Academy of Sciences
252127 Kiev, USSR

Manfred Meinig
Zentralinstitut für Physik der Erde
Telegrafenberg A 17
DDR-1561 Potsdam, GERMANY, Dem. Rep.

François Mignard
CERGA
Avenue Copernic
F-06130 Grasse, FRANCE

N.F. Minyajlo
Main Astronomical Observatory
Ukrainian Academy of Sciences
252127 Kiev, USSR

Masanori Miyamoto
National Astronomical Observatory
Osawa Mitaka
Tokyo 181, JAPAN

S.M. Molodenskij
Schmidt Institute of Earth Physics
Bol. Gruzinskaya 10
123810 Moscow D-242, USSR

Horst Montag
Zentralinstitut für Physik der Erde
Telegrafenberg A17
DDR-1500 Potsdam, GERMANY, Dem. Rep.

Bruno L. Morando
Bureau des Longitudes
77 Ave. Denfert-Rochereau
F-75014 Paris, FRANCE

Leslie V. Morrison
Royal Greenwich Observatory
181a Huntingdon Road
Cambridge CB3 0DJ, U.K.

C. Andrew Murray
12 Derwent Road
Eastbourne BN20 7PH, U.K.

A.A. Myullyari
Petrozavodsk State University
Lenin Str 33
185640 Petrozavodsk, USSR

Vitalij A. Naumov
Central Astronomical Observatory
Pulkovo M-140
196140 Leningrad, USSR

Yu.A. Nefed'ev
Engelhardt Astronomical Observatory
Observatory Station
422526 Kazan, Tatar ASSR, USSR

Antonina I. Nefed'eva
Engelhardt Astronomical Observatory
Observatory Station
422526 Kazan, Tatar ASSR, USSR

Andrej A. Nemiro
Central Astronomical Observatory
Pulkovo
196140 Leningrad, USSR

V.V. Nesterov
Sternberg Astronomical Institute
Universitetskij Prospekt 13
119899 Moscow, USSR

V.V. Orlov
Leningrad Univ. Astron. Observatory
Bibliotechnaya Pl 2
198904 Leningrad, USSR

A.A. Ovchinnikov
Sternberg Astronomical Institute
Universitetskij Prospekt 13
119899 Moscow, USSR

L.K. Pakulyak
Main Astronomical Observatory
Ukrainian Academy of Sciences
252127 Kiev, USSR

N.N. Parkhomenko
NPO Space Instrumentation
Aviamotornaya Str
111250 Moscow, USSR

L.S. Pavlenko
Kharkov State University Observatory
Sumskaya Str 35
310022 Kharkov, USSR

I.K. Pavlov
Leningr. Optical-Mechanical Factory
Chugunnaya Str 10
194100 Leningrad, USSR

N.R. Persiyaninova
Central Astronomical Observatory
Pulkovo
196140 Leningrad, USSR

S.S. Peruansky
Kazan State University
Lenin Str 18
Kazan, Tatar ASSR, USSR

G.M. Petrov
Nikolayev Branch, Pulkovo Observ.
Observatory Str 1
327030 Nikolaev, USSR

M.S. Petrovskaya
Institute for Theoretical Astronomy
10 Kutuzov Quay
191187 Leningrad, USSR

D. Picca
Dipartimento di Fisica
Universita Degli Studi di Bari
Via G Amendola 173
I-70100 Bari, ITALY

Gennadij I. Pinigin
Nikolayev Branch, Pulkovo Observ.
Observatory Str 1
327030 Nikolaev, USSR

E.V. Pit'eva
Institute of Applied Astronomy
8 Zhdanov Street
197042 Leningrad, USSR

V.V. Podobed
Sternberg Astronomical Institute
Universitetskij Prospekt 13
119899 Moscow, USSR

A.D. Polozhentsev
Leningrad Univ. Astron. Observatory
Bibliotechnaya Pl 2
198904 Leningrad, USSR

Dimitrij D. Polozhentsev, Jr
Central Astronomical Observatory
Pulkovo
196140 Leningrad, USSR

E.V. Polyakov
Central Astronomical Observatory
Pulkovo
196140 Leningrad, USSR

E.N. Polyakhova
Leningrad Univ. Astron. Observatory
Bibliotechnaya Pl 2
198904 Leningrad, USSR

A.A. Popov
Central Astronomical Observatory
Pulkovo
196140 Leningrad, USSR

Heino I. Potter
Central Astronomical Observatory
Pulkovo
196140 Leningrad, USSR

A.A. Pozhalov
Nikolayev Branch, Pulkovo Observ.
Observatory Str 1
327030 Nikolaev, USSR

Yu.I. Prodan
Sternberg Astronomical Institute
Universitetskij Prospekt 13
119899 Moscow, USSR

Vojislava Protitch-Benishek
Astronomical Observatory
Volgina 7
YU-11050 Belgrade, YUGOSLAVIA

E.Ya. Prudnikova
Central Astronomical Observatory
Pulkovo
196140 Leningrad, USSR

Qi Guan-Rong
Shaanxi Observatory
P.O. Box 18
Lintong, Xian, CHINA, P.R.

Luis Quijano
Instituto y Observat. de Marina
11110 San Fernando (Cadiz), SPAIN

A.G. Rakhimov
Uzbek Acad. Sci. Astronomical Inst.
Astronomy 33
700052 Tashkent GSP, USSR

Kavan U. Ratnatunga
Space Data and Computing Div
NASA Goddard Space Flight Ctr
Code 630
Greenbelt, Maryland 20771 USA

Yves Réquième
Observatoire de Bordeaux
B.P. 21
F-33270 Floirac, FRANCE

Naufal G. Rizvanov
Engelhardt Astronomical Observatory
Observatory Station
422526 Kazan, Tatar ASSR, USSR

G.V. Romanova
Sternberg Astronomical Institute
Universitetskij Prospekt 13
119899 Moscow, USSR

Siegfried Röser
Astronomisches Rechen-Institut
Mönchhofstrasse 12-14
D-6900 Heidelberg, GERMANY, Fed. Rep.

L.F. Roze
Latvia State University Observatory
Janis Rainis Boulevard 18
226088 Riga, Latvia, USSR

L.I. Rumyantseva
Institute for Theoretical Astronomy
10 Kutuzov Quay
191187 Leningrad, USSR

Jane L. Russell
Code 4130R
U.S. Naval Research Lab
Washington, DC 20375 USA

S.P. Rybka
Main Astronomical Observatory
Ukrainian Academy of Sciences
252127 Kiev, USSR

Lidija V. Rykhlova
Astronomical Council
USSR Academy of Sciences
Pyatnitskaya Str 48
109017 Moscow, USSR

Sofija Sadzakov
Astronomical Observatory
Volgina 7
11050 Beograd, YUGOSLAVIA

Yu.I. Safronov
Main Astronomical Observatory
Ukrainian Academy of Sciences
252127 Kiev, USSR

T.I. Samusenko
Uzbek Acad. Sci. Astronomical Inst.
Astronomy 33
700052 Tashkent GSP, USSR

Elena Schilbach
Zentralinstitut für Astrophysik
Rosa Luxemburg Strasse 17A
DDR-1591 Potsdam, GERMANY, Dem. Rep.

Ralf-Dieter Scholz
Zentralinstitut für Astrophysik
Rosa-Luxemburg-Strasse 17a
DDR-1591 Potsdam, GERMANY, Dem. Rep.

Heiner Schwan
Astronomisches Rechen-Institut
Mönchhofstrasse 12-14
D-6900 Heidelberg, GERMANY, Fed. Rep.

A.I. Sedyukov
Space Flight Control Centre
141070 Kaliningrad, Moscow Region, USSR

P. Kenneth Seidelmann
Nautical Almanac Office
U.S. Naval Observatory
34th and Massachusetts Ave. NW
Washington, DC 20392 USA

A.V. Sergeev
Main Astronomical Observatory
Ukrainian Academy of Sciences
252127 Kiev, USSR

T.P. Sergeeva
Main Astronomical Observatory
Ukrainian Academy of Sciences
252127 Kiev, USSR

D.V. Sergeyev
Leningr. Optical-Mechanical Factory
Chugunnaya Str 10
194100 Leningrad, USSR

V.I. Sergienko
NPO Etalon
Borodina 57
664018 Irkutsk, USSR

S.V. Serova
Institute for Theoretical Astronomy
10 Kutuzov Quay
191187 Leningrad, USSR

N.A. Shakht
Central Astronomical Observatory
Pulkovo
196140 Leningrad, USSR

N.V. Shcherbakova
Central Astronomical Observatory
Pulkovo
196140 Leningrad, USSR

Vladimir A. Shefer
Tomsk State University
Inst. Appl. Mechanics and Math.
634029 Tomsk GSP-14, USSR

E.K. Sheffer
Sternberg Astronomical Institute
Universitetskij Prospekt 13
119899 Moscow, USSR

Alexander A Shiryaev
Institute for Theoretical Astronomy
Naverezhnaya Kutuzova 10
191187 Leningrad, USSR

Vladimir G. Shkodrov
Department of Astronomy
Bulgarian Academy of Sciences
72 Lenin Blvd
1784 Sofia, BULGARIA

V.D. Shkutov
Central Astronomical Observatory
Pulkovo
196140 Leningrad, USSR

N.A. Shkutova
Central Astronomical Observatory
Pulkovo
196140 Leningrad, USSR

Viktor A. Shor
Institute for Theoretical Astronomy
10 Kutuzov Quay
191187 Leningrad, USSR

O.E. Shornikov
Nikolayev Branch, Pulkovo Observ.
Observatory Str 1
327030 Nikolaev, USSR

A.V. Shul'ga
Nikolayev Branch, Pulkovo Observ.
Observatory Str 1
327030 Nikolaev, USSR

V.P. Sibilev
Nikolayev Branch, Pulkovo Observ.
Observatory Str 1
327030 Nikolaev, USSR

V.D. Simonenko
Kharkov State University Observatory
Sumskaya Str 35
310022 Kharkov, USSR

L.A. Sinenko
VNIIFTRI
Institute for Physical, Technical
 and Radio Measurements
Solnechnogorskij District
141570 Moscow, USSR

Vladimir I. Skripnichenko
Institute of Applied Astronomy
8 Zhdanov Street
197042 Leningrad, USSR

R.I. Smekhacheva
Institute for Theoretical Astronomy
10 Kutuzov Quay
191187 Leningrad, USSR

Clayton A. Smith Jr
U.S. Naval Observatory
34th and Massachusetts Ave. NW
Washington, DC 20392 USA

Alla S. Sochilina
Institute for Theoretical Astronomy
10 Kutuzov Quay
191187 Leningrad, USSR

A.V. Sokolov
Central Astronomical Observatory
Pulkovo
196140 Leningrad, USSR

A.G. Sokol'skij
Institute for Theoretical Astronomy
10 Kutuzov Quay
191187 Leningrad, USSR

Natalia I. Solina
Institute of Applied Astronomy
8 Zhdanov Street
197042 Leningrad, USSR

Mitsuru Sôma
National Astronomical Observatory
Osawa Mitaka
Tokyo 181, JAPAN

Jean Souchay
National Astronomical Observatory
Osawa Mitaka
Tokyo 181, JAPAN

Ojars J. Sovers
Mail Stop 238–700
Jet Propulsion Laboratory
4800 Oak Grove Drive
Pasadena, California 91109 USA

E. Myles Standish
Mail Stop 301–150
Jet Propulsion Laboratory
4800 Oak Grove Drive
Pasadena, California 91109 USA

Klaus-Günter Steinert
Technische Universität Dresden
Mommsen-Strasse 13
DDR-8027 Dresden, GERMANY, Dem. Rep.

Ronald C. Stone
U.S. Naval Observatory
Flagstaff Station
P.O. Box 1149
Flagstaff, Arizona 86001 USA

Kaj Aa. Strand
3200 Rowland Pl., NW
Washington, DC 20008 USA

Yu.S. Streletskij
Central Astronomical Observatory
Pulkovo
196140 Leningrad, USSR

Peter Stumpff
MPI für Radioastronomie
Auf dem Hügel 69
D-5300 Bonn, GERMANY, Fed. Rep.

I.S. Sudnik
Institute for Theoretical Astronomy
10 Kutuzov Quay
191187 Leningrad, USSR

M.L. Sveshnikov
Institute for Theoretical Astronomy
10 Kutuzov Quay
191187 Leningrad, USSR

E.S. Sveshnikova
Institute for Theoretical Astronomy
10 Kutuzov Quay
191187 Leningrad, USSR

Vladimir K. Tarady
Main Astronomical Observatory
Ukrainian Academy of Sciences
252127 Kiev, USSR

S.V. Tarasevich
Institute of Applied Astronomy
8 Zhdanov Street
197042 Leningrad, USSR

V.V. Tel'nyuk-Adamchuk
Astron. Observ. Kiev University
Observatorna Street 3
252035 Kiev, USSR

S.V. Tolbin
Central Astronomical Observatory
Pulkovo
196140 Leningrad, USSR

S. A. Tolchelnikova-Murri
Central Astronomical Observatory
Pulkovo
196140 Leningrad, USSR

Tong Fu
Purple Mountain Observatory
21008 Nanjing, CHINA, P.R.

Jorge Torra
Dept Fisica i Astronomia
Avda. Diagonal 647
E-08028 Barcelona, SPAIN

Robert N. Treuhaft
Mail Stop 238–700
Jet Propulsion Laboratory
4800 Oak Grove Drive
Pasadena, California 91109 USA

A.A. Trubitsyna
Institute for Theoretical Astronomy
10 Kutuzov Quay
191187 Leningrad, USSR

E.V. Trushin
USSR Acad. Sci. Gen. Phys & Astron.
Leninskij Prospekt 14, D-4
117901 GSP-1 Moscow, USSR

R.E. Tselishchev
Engelhardt Astronomical Observatory
Observatory Station
422526 Kazan, Tatar ASSR, USSR

V.I. Turenko
NPO Metrology
Dzierzynski Str 40
310022 Kharkov 22, USSR

V.G. Turyshev
Moscow State University
Faculty of Physics
117234 Moscow, USSR

K. Umarova
Uzbek Acad. Sci. Astronomical Inst.
Astronomy 33
700052 Tashkent GSP, USSR

K.I. Usovich
Institute for Theoretical Astronomy
10 Kutuzov Quay
191187 Leningrad, USSR

V.I. Valyaev
Institute for Theoretical Astronomy
10 Kutuzov Quay
191187 Leningrad, USSR

William F. van Altena
Yale University Observatory
P.O. Box 6666
New Haven, Connecticut 06511 USA

A.S. Vashevich
Institute for Theoretical Astronomy
10 Kutuzov Quay
191187 Leningrad, USSR

N.N. Vasil'ev
Institute for Theoretical Astronomy
10 Kutuzov Quay
191187 Leningrad, USSR

E.A. Vershinskij
Leningr. Optical-Mechanical Factory
Chugunnaya Str 10
194100 Leningrad, USSR

V.V. Vityazev
Leningrad Univ. Astron. Observatory
Bibliotechnaya Pl 2
198904 Leningrad, USSR

B.I. Vlasov
VNIIFTRI
Institute for Physical, Technical
   and Radio Measurements
Solnechnogorskij District
141570 Moscow, USSR

Alexander V. Voinov
Institute of Applied Astronomy
8 Zhdanov Street
197042 Leningrad, USSR

M.Yu. Volyanskaya
Odessa State University
Shevchenko Park
270014 Odessa, USSR

John Wahr
Dept of Physics and CIRES
University of Colorado
Boulder, Colorado 80309 USA

Hans-Georg Walter
Astronomisches Rechen-Institut
Mönchhofstrasse 12–14
D-6900 Heidelberg, GERMANY, Fed. Rep.

Wayne H. Warren Jr
National Space Science Data Ctr
NASA Goddard Space Flight Ctr
Code 633
Greenbelt, Maaryland 20771 USA

Gart Westerhout
U.S. Naval Observatory
34th and Massachusetts Ave. NW
Washington, DC 20392 USA

Roland Wielen
Astronomisches Rechen-Institut
Mönchhofstrasse 12-14
D-6900 Heidelberg, GERMANY, Fed, Rep.

George A. Wilkins
Royal Greenwich Observatory
Herstmonceux Castle
Hailsham BN27 1RP, U.K.

Wu Shou-xian
Shaanxi Astronomical Observatory
No. 3 Xiaozhai East Road
Xian, Shaanxi, CHINA, P.R.

Xu Bang-Xin
Department of Astronomy
Nanjing University
Nanjing, CHINA, P.R.

L.I. Yagudin
Central Astronomical Observatory
Pulkovo
196140 Leningrad, USSR

E.I. Yagudina
Institute of Applied Astronomy
8 Zhdanov Street
197042 Leningrad, USSR

A.Yu. Yatsenko
Engelhardt Astronomical Observatory
Observatory Station
422526 Kazan, Tatar ASSR, USSR

Ya. S. Yatskiv
Main Astronomical Observatory
Ukrainian Academy of Sciences
252127 Kiev, USSR

Ye Shu-Hua
Shanghai Observatory
Shanghai, CHINA, P.R.

V.N. Yershov
Central Astronomical Observatory
Pulkovo
196140 Leningrad, USSR

F.D. Zablotskiy
Lvov Polytechnical Institute
Mira Str 12
290646 Lvov 13, USSR

R.V. Zagretdinov
Kazan State University
Lenin Str 18
Kazan, Tatar ASSR, USSR

V.I. Zhdanovskaya
Observatory Kiev Univ.
Observatorna 3
252053 Kiev, USSR

Mitrofan S. Zverev
Central Astronomical Observatory
Pulkovo
196140 Leningrad, USSR

RESOLUTIONS ADOPTED BY PARTICIPANTS AT
IAU SYMPOSIUM NUMBER 141

Resolution concerning HIPPARCOS

IAU Symposium No. 141, "Inertial Coordinate System on the Sky", meeting in Leningrad, USSR, on 17–21 October 1989:

NOTES with great concern the current critical state of the HIPPARCOS mission,

STRESSES the extreme importance of HIPPARCOS to astronomy and the expectations of the world astronomical community,

CONGRATULATES the operational teams of the European Space Agency (ESA) on their great efforts to operate the satellite in its current highly elliptical orbit, and

URGES that the maximum possible scientific output be obtained from the present satellite, and

IF the objectives of the project cannot adequately be fulfilled by the current mission,

RECOMMENDS that ESA give consideration to implement a second mission with the minimum delay.

Resolution concerning Very Long Baseline Interferometry (VLBI)

The participants of IAU Symposium Number 141, "Inertial Coordinate System on the Sky",

NOTING
1)   the capabilities offered by the observation of extragalactic sources;
2)   the current status of active programs and the potential for future observing programs;

SUPPORT the
1)   development of international cooperation of VLBI programs;
2)   development of observing capabilities for VLBI networks in many countries;
3)   cooperative application of the scientific data for the determination of reference systems.

Resolution concerning Future Astrometry

At the time of the launch of HIPPARCOS and the anticipated launch of the Hubble Space Telescope and the 150th anniversary of the Central Astronomical Observatory of the USSR Academy of Sciences at Pulkovo, the participants in IAU Symposium 141 "Inertial Coordinate System on the Sky"

RECOGNIZE the
1) potential high accuracy, resolution and coverage that can be achieved by space techniques;
2) proposed new projects to be realized at different epochs which offer increased capabilities over space instruments currently available;
3) the value of international cooperation and the need for complementary techniques to achieve the best results;

URGE that every effort be made to
1) develop new satellites for high accuracy astrometric space observations;
2) develop stellar and radio catalogues and reference frames at the milliarcsecond or better level;
3) develop international cooperation in both planning and operations to ensure the achievement of high effectiveness at the lowest cost;
4) obtain observations of star positions at milliarcsecond to microarcsecond levels;
5) encourage the development of complementary high accuracy ground-based techniques.

Part 1
The legacy of Pulkovo
for inertial systems and reference frames

Plate I. The original Pulkovo Observatory viewed from the southwest in winter. Photograph provided by Pulkovo Observatory.

Plate II. The reconstructed Pulkovo Observatory viewed from the southwest. Photograph provided by Pulkovo Observatory.

INTRODUCTORY REMARKS

J.H. LIESKE
co-chairman Scientific Organizing Committee

Jet Propulsion Laboratory
California Institute of Technology
4800 Oak Grove Drive
Pasadena, California 91109 USA

IAU Symposium Number 141, which is titled "Inertial Coordinate System on the Sky," is sponsored by the International Astronomical Union and is co-sponsored by the Astronomical Council of the USSR Academy of Sciences. The sponsoring IAU commissions are Commission 8 (*Positional Astronomy*), Commission 4 (*Ephemerides*), Commission 7 (*Celestial Mechanics*), Commission 19 (*Rotation of the Earth*), Commission 24 (*Photographic Astrometry*), and Commission 31 (*Time*).

We are here to discuss the scientific problems which pertain to defining and to realizing reference frames so that for astronomical work we can establish an "Inertial Coordinate System on the Sky." Our meeting is also being held on the occasion of the 150th anniversary of the founding of Pulkovo Observatory which has been called the "Astronomical capital of the world."

At the time of Pulkovo's 100th anniversary in 1939 the world climate was not conducive to celebrating Pulkovo's centennial. Indeed, the observatory was destroyed a few years later during the siege of Leningrad. We are grateful that today we can commemorate the 150th anniversary of the founding of Pulkovo Observatory amid a reduced state of international tension in the world.

In the history of any individual, or of any organization, there are moments which one recalls with great satisfaction and some with a certain amount of sadness. The history of Pulkovo Observatory is no different. We will commemorate the many instances in which Pulkovo and its astronomers made lasting contributions to our field. We also remember those astronomers, such as B.P. Gerasimovich and B.V. Numerov, whose contributions were ended in the turmoil of the 30s.

From its very beginning, Pulkovo Observatory was at the forefront of scientific research for the astronomical understanding of the universe. We recall with great satisfaction, for example, the work of the early Pulkovo astronomers such as the Struves and C.A.F. Peters who were very influential and very active in the observation of double stars, in the determination of the precessional constant, in the determination of the constant of nutation, in the determination of aberration, in the determination of parallax, and in promoting international cooperation among astronomers.

During the symposium we will briefly review the past history of Pulkovo Observatory, as well as learn more about the current programs at Pulkovo as we discuss the legacy of Pulkovo as it pertains to the development of inertial systems. In other sessions we will discuss the concepts and definitions involved in setting up various models of inertial frames. We will discuss how reference frames are realized: both optical and radio, involving galactic as well as extra-galactic objects. Finally, we will consider how the various reference frames are tied to one another.

J. H. Lieske and V. K. Abalakin (eds.), Inertial Coordinate System on the Sky, 3.
© 1990 *IAU. Printed in the Netherlands.*

Plate III. Portrait of Wilhelm Struve, painted in 1841–42 by C.A. Jensen. Copy provided by Pulkovo Observatory via A.H. Batten.

Concluding Remarks

Gart Westerhout
U.S. Naval Observatory
Washington, DC 20392
USA

We started this meeting with a discussion of Vassily Yakovlevich Struve. I would like to mention again this great founder of the Pulkovo Observatory. Vassily Yakovlevich directed the determination of a constant of precession which is almost identical to the modern value determined by VLBI. How did he and his son Otto do that? They observed and reduced their own data, thereby avoiding the pitfalls of using data compiled by others without a full understanding of what went into such compilations. Can we still do this? I think this excellent symposium has taught us that, with difficulty, we can; that there are many who do not recognize the pitfalls; but that the reward is unprecedented accuracy and worldwide recognition.

What else did we learn? There is a need for a common approach to reference frames. Several speakers discussing relativistic frames tried to get the audience to understand their approach and pointed out inconsistencies. We must have our house in order before the high accuracy data come in. Here is an example: the mean place calculation controversy has now raged for six years. It was made crystal clear during this meeting by Soma and Aoki that Murray is erroneous and by Murray that Soma and Aoki are erroneous! People are reporting "errors" in the FK5 and we don't even know yet how to transform from B1950 to J2000! And by-the-way, the basic FK5 is now in print, a very happy occurrence!

We heard about HIPPARCOS, and the encouraging news that some very important reference frame data may emerge. Even with only six months of operation HIPPARCOS can yield considerable optical reference frame improvements. And we encourage the ESA to construct a second HIPPARCOS with minimum delay. The old ground-based catalogs are now even more important than they were before, as they will carry great weight when used with HIPPARCOS data to provide very accurate proper motions.

The Hubble Space Telescope, hopefully to be launched next year, promises galactic-extragalactic connections (quasars-HIPPARCOS Stars) and other relative astrometry with great precision.

We heard about three proposed USSR astrometric satellites, AIST, Regatta, and Lomonosov, and one proposed USA satellite, POINTS, all projects aimed at achieving very precise global reference frame data in the next ten to fifteen years. We hope that at least some of these projects will be brought to completion in this highly competitive field of space research, as they will bring a major advance to astrometry.

An example of the wide-ranging importance of the field of astrometry to other parts of science is the advance in the area of nutation. The accuracy of the astrometric (VLBI) data now available is well beyond what can be achieved by current theory and provides an entirely new set of data for geophysics.

5

J. H. Lieske and V. K. Abalakin (eds.), Inertial Coordinate System on the Sky, 5–6.
© 1990 *IAU. Printed in the Netherlands.*

In VLBI we now have 1-2 milliarcsecond agreement between reference frames based on entirely independent sets of observations, with the promise of better accuracy using general relativity refinements. This is fifty times better than the current optical reference frame. In local frames, i.e., fields 10 degrees or smaller, accuracies down to a stunning 0.2 milliarcseconds have been achieved. And in VLBI relative astrometry even higher precisions are anticipated with the ambitious plans for the USSR Space VLBI project QUASAT.

VLBI catalogs and the FK5 are representations of reference frames from entirely different techniques. We need to understand these techniques better before we try to improve the optical reference frame by connecting it to the radio frame. A very large amount of effort is going into tying the optical and radio frames, both here in the USSR and abroad. But a warning needs to be sounded. The radio frame contains only a few hundred objects, one for every ten FK5 stars. Current work in connecting the frames is certainly useful, but we have to await new techniques, such as ground-based optical interferometry and the space-based astrometric systems, to get the optical frame to the same accuracy as the radio frame.

Please note the comment that the dynamical frame is defined by the ephemerides, which are only as good as the observations to which the physical models are tied.

George Wilkins presented a list of definitions. I have to admire the wisdom and courage of the organizers of this symposium for placing the definitions at the end, so there would be no cluttering up the discussions with facts! The definitions are complex. I saw several heads nodding vigorously during Wilkins' talk, both YES (agree) and NO (disagree) at the same time. Clearly, there is a need for some major decisions in reference frame building, and the reference frame meeting organized by the USNO next year in Virginia Beach (IAU Colloquium 127: Reference Frames, October 14-20,1990) comes none too early. We must iron out our differences, and a lot of preparation and correspondence is still necessary if we want to achieve a consensus at IAU Colloquium 127.

There is a great amount of optimism for reaching unparalleled accuracies both on the ground and from space. Only with optimism can we succeed. We can learn from HIPPARCOS. Without its optimistic predictions it would not have been built, but even if it cannot now reach its ambitious design goals, it will still provide a major step forward. The same holds for the many astrographic photographic surveys now being undertaken.

Vassily Yakovlevich was a master in international cooperation, using a web of personal friendships and open sharing of data necessary to further international goals. Let us continue in his footsteps and continue the close international cooperation that has always characterized astrometry, so that duplications do not occur and the field advances at the fastest pace possible. This symposium has contributed immeasurably to this goal.

THE LEGACY OF THE STRUVES FOR THE ESTABLISHMENT OF AN INERTIAL COORDINATE SYSTEM ON THE SKY

ALAN H. BATTEN
Dominion Astrophysical Observatory
Herzberg Institute of Astrophysics
Victoria, B.C.
Canada

ABSTRACT. Contributions made by Vassily Yakovlevich Struve and his son Otto Vassilyevich to the eventual establishment of an inertial frame of reference on the sky are discussed under five heads: (i) the foundation of Pulkovo Observatory, (ii) instrumentation, (iii) Vassily Yakovlevich's determination of the constant of aberration, (iv) Otto Vassilyevich's determination of the coefficient of precession and (v) international cooperation.

1. Prologue

Just over 125 years ago, Pulkovo Observatory celebrated its first great jubilee — its twenty-fifth anniversary. Astronomers came from many countries on that occasion, as they have done today, and they assembled in the main rotunda of the Observatory. To inaugurate the proceedings, a frail old man came into the rotunda, supported on the arm of his eldest surviving son. The two walked slowly round the circle of distinguished visitors, greeting each one by name, for they were all personal friends. The old man, of course, was the founding director of the Observatory, usually known in other countries as Wilhelm Struve, but often called in this, his adopted country, Vassily Yakovlevich. This little vignette was drawn for us, not by his eldest son, Otto Vassilyevich, who was also his successor as director, but, rather, by Vassily Yakovlevich's youngest son, Nikolai, who wrote a family history (Struve 1915).

We cannot be privileged on this occasion to be greeted by the founder of this Observatory, but we are in many ways more privileged than those predecessors of ours. Despite our constant complaints about funding, we can bring resources to the study of astronomy beyond their wildest dreams. Travel to gatherings such as this — a rare privilege for most of them — many of us assume to be almost a right. In real terms, travel is probably still cheaper for us than it was for them; it is certainly faster and more convenient. In 1864, the railway from Berlin to St. Petersburg was still incomplete; everyone present at that earlier celebration had made at least part of the journey either by sea or by horse-drawn vehicles. Recalling this fact can help us to feel the poignancy of the occasion. Six years earlier Vassily Yakovlevich had been ravaged by an illness that left him both physically and mentally frail. Friends and colleagues who had looked to him as one of the foremost of their number must have been saddened to see his reduced state. They could not guess that in a little over three months' time he would be dead, but they surely knew that most of them would never see him again.

J. H. Lieske and V. K. Abalakin (eds.), Inertial Coordinate System on the Sky, 7–13.
© 1990 *IAU. Printed in the Netherlands.*

Vassily Yakovlevich and his son dominated Pulkovo for one-third the total span that we are *now* celebrating. They were masters of fundamental astronomy and, in one sense, all their work was a legacy for the establishment of an inertial coordinate system. I am tempted to range widely, or perhaps to concentrate on that aspect of their work that I understand best — the study of double stars. Either course would take too long, however, and would in a large measure repeat what I have recently written (Batten 1988). I shall concentrate, therefore, on five topics (as identified in the Abstract) and, even on these, I must be brief and omit much detail.

2. The founding of Pulkovo

By 1830, it was obvious that the old observatory of the Academy of Sciences, built in the centre of St. Petersburg towards the end of Peter the Great's reign, was no longer adequate. The Academy talked of replacing the old building and instruments, and had even engaged in inconclusive negotiations for a site north of the city, but they lacked both funds and an astronomer of sufficient calibre and dynamism to carry out the project. At the end of 1830, Tsar Nicholas I received Vassily Yakovlevich in audience and, as a result of this meeting, both the Academy's wants were supplied.

Nicholas was not an easy man to love, either in his life or viewed in the retrospective of history, but given the system of autocracy by which imperial Russia was ruled, the building of a new national observatory would have been impossible without the support of the Tsar, and Nicholas appears to have given that support whole-heartedly. Without either of these men, Nicholas or Vassily Yakovlevich, a new observatory probably would have been built sometime in the middle of the nineteenth century, but both of them together were essential for the creation of the "astronomical capital of the world". One provided the money and the political will, the other the scientific abilities and vision. According to Vassily Yakovlevich (Struve 1845), Nicholas himself suggested Pulkovo as the site. Cleveland Abbe (1867), however, who visited Pulkovo shortly after Vassily Yakovlevich's death and talked often with Otto, reports that the elder Struve had a premonition as early as 1828 that the new St. Petersburg Observatory would one day be built there. Perhaps Pulkovo was a rather obvious site, or perhaps Vassily Yakovlevich knew well that the best method of getting your own way is to let your superiors think that it is theirs! It seems likely that there was a mutual respect between the two men, despite the gulf that then separated autocrat and subject, and this was one of the secrets of the successful creation of the new Observatory.

Nicholas, for his part, had two motives for the foundation of Pulkovo. First, the vast lands of the Russian empire needed to be surveyed: the new Observatory was to meet precisely the same sort of practical need as those founded earlier by the world's maritime and colonial powers. Vassily Yakovlevich's success in surveying the arc of the meridian through Dorpat undoubtedly was one reason why he was entrusted with the building of Pulkovo. Nicholas's other motive was quite simply national prestige. At that 1830 audience, the Tsar declared (Struve 1845) that "... l'honneur du pays Lui paraissait réclamer la fondation, près de la capitale, d'un nouvel Observatoire astronomique, conforme à la hauteur actuelle de la science et propre à contribuer à son avancement ultérieur." Those were the motives that loosened the purse strings, but it was the genius and energy of Vassily Yakovlevich that transformed Pulkovo into an observatory destined to make major contributions to the establishment of an inertial coordinate system.

3. The instruments

Instruments are the most important part of an observatory. No matter how good the location, how fine the buildings or how skilled the astronomers, without good instruments all would be wasted. Precisely in the provision of instruments, Nicholas was most ready to trust Vassily Yakovlevich. Cost was to be no consideration, provided only that the instruments were the best available. The trust was well founded. As a youth, Vassily Yakovlevich had installed, without help or supervision, a Dollond transit instrument that had lain several years in its case. Later, when what was then the world's largest refracting telescope was delivered in 22 cases to Dorpat, he discovered that "the maker had forgotten to send the direction" for assembling it (Struve 1826). The maker himself (Fraunhofer) later expressed surprise at the success with which Vassily Yakovlevich assembled the telescope without detailed instructions.

The Dorpat refractor had been bought from Fraunhofer because that was what he had available. Fortified by more than a decade of experience with that refractor, and by the Tsar's *carte blanche*, Vassily Yakovlevich determined that this time *he* would tell the instrument makers what he wanted. It has been said that he told them not even a single screw in the designs was to be altered without his consent. Certainly he had a very large say even in details of design. This can be illustrated by the prime-vertical circle, built by the Repsold brothers, and in some respects Vassily Yakovlevich's favourite among the Pulkovo instruments. He credits his friend Bessel with first pointing out that prime-vertical instruments could be used most advantageously for observations near the zenith. His own use of such an instrument in Dorpat, however, convinced Vassily Yakovlevich that it could be the most powerful means of studying aberration, nutation and annual parallax, provided certain changes were made. "Il fallait placer le tube à l'extremité de l'axe de rotation, pour avoir le niveau dans un emplacement permanent sur l'axe....Le second point essentiel était de joindre à l'instrument un appareil propre pour un renversement aussi prompte que possible..." (Struve 1845). The Repsolds succeeded in realizing these ideas. The prime-vertical instrument, we are told, could be reversed in 16 seconds and the astronomer could resume observing on either side of the pier, one minute and twenty seconds after he had finished on the other. Not everyone thought the instrument perfect; Airy (1848) has left his criticisms. Another early visitor to Pulkovo, the American astronomer G.P. Bond, wrote in his diary, however, "Tenths of a second of arc take the position here that seconds have hitherto done elsewhere." (Holden 1897).

Quantum leaps in precision, such as Bond recorded, often lead to great progress in astronomy. For example, the development of the photomultiplier tube was what gave photoelectric photometry its precision and therefore its value for astrophysical research (Wood 1989). Similarly, new techniques for the determination of radial-velocities (Campbell 1983) are now leading to a significant increase in the precision with which at least some velocities can be determined and make it look likely that, if we ever do detect other planetary systems, it will be from their radial, rather than transverse, motions. We are on the threshold of another quantum leap in astrometric precision, that we hope will come from observations made above the Earth's atmosphere. I am sure we shall hear much about this in the next few days. The Struves, however unconsciously, set us on that road by what they did here in Pulkovo; in this, as in so much else, we are their heirs.

4. The constant of aberration

Vassily Yakovlevich began to work on the determination of all the important astronomical constants — aberration, nutation, precession, solar motion etc. — before ever he came to Pulkovo. As we have just seen, his early experience led him to design the prime-vertical instrument explicitly for the determination of the constants of aberration and nutation. Most of the astronomers initially employed at Pulkovo came from Dorpat and were already engaged in the Observatory programme for the determination of these important constants. At Pulkovo, Vassily Yakovlevich took responsibility himself for the determination of the constant of aberration, assigning that of the coefficient of precession to his son, and of the constant of nutation to C.A.F. Peters. The results eventually appeared in three major memoirs (Struve 1841, Struve 1843a, Peters 1844) of which Vassily Yakovlevich wrote: "Les trois mémoires, celui de mon fils sur la précession et le mouvement du système solaire, celui de M. Peters, sur la nutation et le mien sur l'aberration, forment un corps entier, qui nous a fourni des fondements nouveaux et incomparablement plus exacts pour les réductions des observations astronomiques, fondements qui étaient indispensables pour faire marcher les travaux de l'observatoire centrale d'une manière digne de ses moyens supérieurs." (Struve 1843b).

Vassily Yakovlevich chose seven stars for observation with the new prime-vertical instrument, primarily for the determination of the constant of aberration, although the observations were continued and later used by Nyrén (1883) for revisions of both that constant and the constant of nutation. All seven stars culminated very close to the zenith. In fact, one can almost deduce the latitude of Pulkovo from the stars chosen. Precessed back to 1850.0, the mean of the declinations of the seven stars is $59°08!9$, while the value that Vassily Yakovlevich himself adopted for the latitude was $59°46!3$. For the aberration constant he deduced $20''4451$, with a probable error of about $0''01$. This is within 0.3 percent of the currently accepted I.A.U. (1976) value (Astronomical Almanac 1989). It was long accepted as standard, although Nyrén's higher value of $20''517$ tended to replace it towards the end of the century. S.C. Chandler (1891b), however, wrote "... I desire to state my strong impression that, although the tendency at present among astronomers is to supplant Struve's constant of aberration by Nyrén's, the lower value will very likely prove to be nearer the true one." In fact, Nyrén's value *was* an improvement, but also an over-correction.

I have quoted elsewhere Airy's (1848) appreciation of Vassily Yakovlevich's observing skills. Though the style may strike us as pompous, I believe that the tribute was sincere: "I had the pleasure twice of witnessing complete observations made by him [Struve]; and I trust that he will not be offended by the testimony of one who, though a younger man, is not without experience as an observer, to the caution, the delicacy, the steadily waiting to the proper time, the promptitude at the proper time, which distinguish Mr. Struve's observations." A greater tribute, however, is perhaps that these same observations later revealed an effect of which Vassily Yakovlevich had no inkling — the polar wobble detected by Chandler (1891a), with its approximate period of 427 days. Chandler (1893a) later found that if the wobble was taken into account in the analysis of Vassily Yakovlevich's observations, the constant of aberration was $20''481$ (plus some small indeterminate terms) and could not, in Chandler's opinion, exceed $20''5$. Nyrén's determination, on the other hand, was hardly affected when polar wobble was taken into account (Chandler 1893b). Chandler (1891c) also found evidence for the wobble in Bradley's discovery observations. Vassily Yakovlevich, who seems to have regarded Bradley as a model, would, I believe, have been especially gratified by this evidence that his own work was on a par with Bradley's.

5. The coefficient of precession

The coefficient of precession was determined by Otto Vassilyevich (Struve 1841) and, nine years later, his work won him the Gold Medal of the Royal Astronomical Society. I cannot resist quoting Airy's (1850) own explanation of the delay: "This has arisen partly from the delay which usually occurs in the printing and distribution of foreign memoirs; partly from the time which is necessary for thoroughly reading a paper of such length; but principally from the occupation of the minds of astronomers, as well within as without the Society, by the remarkable planetary discoveries made in several years past."

The determination of the coeffcient of precession did not intimately involve one particular instrument in the same way that that of the constant of aberration had. Otto Vassilyevich's method was to compare modern positions (many determined at Dorpat by his father) of 400 stars with the positions determined by Bradley some 70 years earlier. It was natural, indeed inevitable, that, along with the constant of precession, Otto Vassilyevich should also attempt to derive the magnitude and direction of the solar motion. His treatment of the individual proper motions was painstakingly meticulous and he derived a value of 50."235 per annum for the coefficient of general precession. This differs from the I.A.U. 1976 value by little more than 0.1 percent. I am indebted to one of our co-chairmen, Jay Lieske, for pointing out to me that so simple and direct a comparison is wrong in principle and actually *under*estimates the quality of Otto Vassilyevich's result. When allowance is made for the different epochs and different time units (tropical and Julian centuries) used, the two values are within 0.02 percent. If Peters' 1844 revision is used, the agreement is even better. If present indications from VLBI results that the I.A.U. 1976 value is too high are confirmed, the agreement between the result obtained by Struve and Peters and the best available modern value is almost perfect. That is a measure of the quality of work in Pulkovo: 150 years ago the astronomers here could achieve the same accuracy that our best modern technology can provide. The characteristics of the solar motion were nothing like so well determined; partly because the still limited knowledge of stellar distances led Otto Vassilyevich to an erroneous estimate of the Sun's linear speed and partly because, as he showed himself, determination of the direction of solar motion is very sensitive to quite small systematic errors in the first-epoch observations. Even Bradley's excellent data could not be assumed to be free from such errors. It was indeed the analysis of errors in the memoir on precession that particularly impressed Airy and probably played a role in the decision to award the Gold Medal to Otto Vassilyevich.

Excellent though the determination of the coefficient of precession was, it never quite attained the status of the elder Struve's work on aberration. Newcomb (1895) referred to Otto Vassilyevich's work on precession (and also to Bessel's) as "classic", but he actually favoured the result derived by one of Otto's two astronomer sons, Ludwig (Struve 1887). Perhaps he assumed that the longer time-base available to Ludwig would guarantee the greater reliability of the result. From our perspective, it seems that no improvement was necessary to the earlier value.

6. International cooperation

Astronomy depends on international cooperation even more than do other branches of science; and the field of astronomy that we have come here to discuss perhaps needs international cooperation most of all. For Vassily Yakovlevich, cooperation was mainly informal, through his personal friendships; although the geodetic work of his early career gave him many official contacts, too. Prominent amongst his personal friends were, of course, F.W. Bessel in Germany, G.B. Airy in England and U.J.J. Le Verrier in France. International meetings were much rarer then than now, but Vassily Yakovlevich had a wide freedom to travel and the opportunities this gave him to maintain and to expand his circle of friends undoubtedly contributed to the success with which he enriched the reputation of Russian science.

Otto Vassilyevich inherited many of his father's friends, particularly Le Verrier and Airy, with whom he remained on excellent terms until they each in turn died. Later he made his own friends, of whom the outstanding ones were David Gill in South Africa and Simon Newcomb in Washington. Such informal contacts remained important throughout his life, but the more formal kind of meeting, to which we are accustomed, began to be important too. Thus Otto Vassilyevich was active both in the *Astronomische Gesellschaft* and the *Carte du Ciel*, the first major organized international project in astronomy.

A few observatories thoughout the world are distinguished from all others by the combination of the length and the quality of their traditions. We might differ about just how many there are, but I think we would all include Paris, Greenwich, the Cape of Good Hope, the U.S. Naval Observatory and, of course, our host Observatory of Pulkovo, even though one of these no longer exists as a separate entity and another is undergoing radical change. I think we shall hear more about these observatories from Steven Dick (1989); I wish only to point out that the web of personal friendships constructed by the two Struves, connected Pulkovo with each of the other major observatories of their time. This, again, enriched the science of their homeland.

7. Epilogue

As I began by trying to recreate a scene from the lives of Vassily Yakovlevich and his son, permit me to close by entertaining the thought that they are with us in spirit today and, unseen and unheard by us, they can watch what we do and listen to what we say. How would they react? Astronomy has been transformed since their deaths, although perhaps Vassily Yakovlevich foresaw a few faint hints of what was to come when he wrote *Etudes d'astronomie stellaire* (Struve 1847). They would, I am sure, be gratified that we are still concerned with the problems to which they devoted so much effort; they would be amazed and excited at the prospect of making astrometric observations from above the Earth's atmosphere. They would be thankful that their great observatory has risen, phoenix-like, from the ashes to which it was reduced, soon after its first century, by a grim and horrible struggle. They would be flattered that we still speak of them and their work. If they could, they would certainly come and greet us each by name, as they did their guests of 125 years ago. I venture to suggest, however, that one of the things that would give them most pleasure would be that we have seen fit to celebrate this jubilee with a meeting sponsored by the International Astronomical Union — a body that realizes, however, imperfectly, the vital international cooperation that is as much a part of the Struves' legacy to us as was their scientific work. Although even Otto Vassilyevich died well before there was an I.A.U., his other astronomer son, Hermann, was active in the precursor International Union for Cooperation in Solar Research, and his grandson and namesake, the second Otto Struve, was one of our Presidents. Within the I.A.U., each one of us here has built a web of personal friendships of his or her own, and many of us are rightly proud of such friendships maintained across the barriers of language, culture and different ideas of political and economic justice, even at times of tension in the world. That pride is justified, because friendships of that kind are in the best tradition of the Struves, of Pulkovo Observatory and of science.

I dedicate this paper to the memory of a great-grandson of Vassily Yakovlevich Struve, Dr. Nils Lindhagen, who died earlier this year. In his home in Sweden, I dined with cutlery engraved with the initials "W.S." and I drank wine from glasses owned and used in Pulkovo by Magnus Nyrén. Lindhagen's enthusiasm and practical help did much to ensure whatever success my biography of his ancestors may enjoy and is, therefore, directly responsible for my having been invited to speak at this symposium.

References

Abbe, C. 1867, *Dorpat and Pulkovo*, Report of the Smithsonian Institute, Appendix, p.370.

Airy, G.B. 1848, *Astron. Nachr.* **26**, 353.

Airy, G.B. 1850, *Mem. Roy. Astron. Soc.* **19**, 271.

Astronomical Almanac 1989, U.S. Government Printing Office and Her Majesty's Stationery Office.

Batten, A.H. 1988, *Resolute and Undertaking Characters: The Lives of Wilhelm and Otto Struve*. D. Reidel, Dordrecht, The Netherlands.

Campbell, B. 1983, *Publ. Astron. Soc. Pacific* **95**, 977.

Chandler, S.C. 1891a, *Astron.J.* **11**, 59.

Chandler, S.C. 1891b, *Astron.J.* **11**, 65.

Chandler, S.C. 1891c, *Astron.J.* **11**, 83.

Chandler, S.C. 1893a, *Astron.J.* **13**, 57.

Chandler, S.C. 1893b, *Astron.J.* **13**, 65.

Dick, S. 1989, this symposium.

Holden, E.S. (editor) 1897, *Memoirs of W.C. and G.P. Bond*, C.A. Murdock and Co., San Francisco and Lemke and Buechner, New York City. Reprinted 1980 Arno Press, New York. p.100.

Newcomb, S. 1895, *The Elements of the Four Inner Planets and the Fundamental Constants of Astronomy*, Government Printing Office, Washington, U.S.A.

Nyrén, M. 1883, *Mém. de l' Acad. Impériale des Sciences de Saint-Pétersbourg*, 7ème série, **31**, No. 9, p.47.

Peters, C.A.F. 1844, *Mém. de l' Acad. Imperiale des Sciences de Saint-Pétersbourg*, 6ème série, **3**, p.125.

Struve, F.G.W. (V.Ya.) 1826, *Mem. Roy. Astron. Soc.* **2**, 93.

Struve, F.G.W. 1843a, *Mém. de l' Acad. Impériale des Sciences de Saint-Pétersbourg*, 6ème série, **3**, p.229.

Struve, F.G.W. 1843b, *Astron. Nachr.* **21**, 57.

Struve, F.G.W. 1845, *Description de l' observatoire astronomique centrale de Poulkova*, Imprimerie de l'académie impériale des sciences, Saint-Pétersbourg.

Struve, F.G.W. 1847, *Etudes d' astronomie stellaire*, Imprimerie de l'académie impériale des sciences, Saint-Pétersbourg.

Struve, L. 1887, *Mém. de l' Acad. Impériale des Sciences de Saint-Pétersbourg*, 7ème série, **35**, No.3.

Struve, N. 1915, *Deutsche Monatschrift für Russland*, **57**.

Struve, O.W. (O.V.) 1841, *Mém. de l' Acad. Impériale des Sciences de Saint-Pétersbourg*, 6ème série, **3**.

Wood, F.B. 1989, in *Algols*, (ed. A.H. Batten, *Space Science Reviews* **50**), Kluwer, Dordrecht, p. xix.

Pulkovo Observatory's Status in 19th Century Positional Astronomy

Kevin Krisciunas

Joint Astronomy Centre
665 Komohana Street
Hilo, Hawaii 96720 USA

Abstract

The most significant basis of the reputation of a scientific organization is the accuracy and influence of its work. From a comparison of values of certain astronomical constants and stellar parallaxes obtained at Pulkovo and elsewhere, the extremely high weight given to the Pulkovo values (e.g. by Newcomb) is further justified in retrospect given the small deviations between Pulkovo values and the "modern" values accepted a century later.

Introduction

Most observatories, among them Greenwich and Harvard, the U. S. Naval Observatory, and even Mauna Kea, had very uncertain and modest beginnings. Occasionally, however, it was planned to establish an observatory which would have, from the very start, a number of world class instruments. For a few this aim was, to a large extent, achieved. Tycho Brahe's island observatory was one such institution, where upwards of 17 instruments were built and used (Krisciunas 1988, pp. 49 and 53).

Another institution that can be included in this category is Pulkovo Observatory, whose 150th birthday we celebrate this year. The first and foremost purpose of the observatory was "to provide uninterrupted observations as accurate as possible which serve the progress of astronomy." With the means provided for its construction (600,000 silver rubles = 2,100,000 *assignat* [paper] rubles), it was clearly intended that Pulkovo be a major scientific institution. Given that positional astronomy was so highly regarded in the 19th century, it is no wonder that a number of telescopes were ordered: an Ertel transit instrument (for the determination of absolute right ascensions), an Ertel vertical circle (for the determination of absolute declinations), a Repsold meridian circle (for the determination of differential stellar coordinates) and prime vertical transit (for the determination of aberration and nutation), a Merz & Mahler heliometer, and the 15-inch Merz & Mahler refractor, then the largest in the world, to be used for the discovery and measurement of double stars (Krisciunas 1988, chapter 5, and references therein).

15

J. H. Lieske and V. K. Abalakin (eds.), Inertial Coordinate System on the Sky, 15–24.
© 1990 *IAU. Printed in the Netherlands.*

Expectations

W. Struve's elaborate 1845 *Description de L'Observatoire Astronomique central de Poulkova* gives a breakdown of the 2.1 million rubles allocated for the establishment of the observatory, including 40,000 for the relocation of the 2000 inhabitants of the village, 1,700,000 for construction (including the piers for the instruments), 299,000 for the instruments, 28,000 for the library, and 17,500 for the publication of the *Description* itself (F. G. W. Struve 1845, p. 53). According to Otto Wilhelm Struve, his father was given *carte blanche* by the Tsar to order the "most perfect instruments the technology of the time could provide" (Otto Struve 1895, p. 49, translation Batten).

Of course it is one thing to spend a lot of money, and another to achieve great science. Already in the fall of 1838 W. Struve had a feeling for how good the instruments were to be. He journeyed to Hamburg to visit the Repsolds and to Munich to visit the firms of Ertel and Merz & Mahler. He reports that he looked at celestial and terrestrial targets with the 15-inch objective, remarking on the excellent achromatic image of α Aquilae. He observed an artificial double star situated 2700 feet away at the Church of St. Petri. The two components of this artificial double consisted of two "stars" 0.24 and 0.42 arcsec in diameter, separated by 1.24 arcsec, which he observed easily with a magnification of 1600. Clearly, the optical characteristics of the great refractor were excellent (W. Struve 1839).

One of the fundamental means of operation that was to be put into practice at Pulkovo Observatory we may label "Struve's method", which can be considered a carryover from W. Struve's Dorpat years. It amounts to an attempt to eliminate systematic and random errors by relying to the maximum degree on the data of one's own observatory.

The State of the Art

What kind of progress was being made in positional astronomy prior to 1800? In Fig. 1 we show the improvement in accuracy of determining the coordinates of a star. By 1700 a resolution of 15 arcsec was achieved, 8 arcsec by 1725, and 0.5 arcsec by 1800 (Chapman 1983, Hughes 1984).

As always, actual probable errors are larger than the instrumental resolution. In the *Positiones Mediae* (Struve 1852, pp. 48-50), based on observations at Dorpat, for 109 primary stars we find median probable errors of ±0.022 seconds of time in right ascension and ±0″.19 in declination. This can be compared to the modern Carlsberg meridian observations (L. Morrison 1989, elsewhere in this volume), which are good to ±0″.12 (mean error).

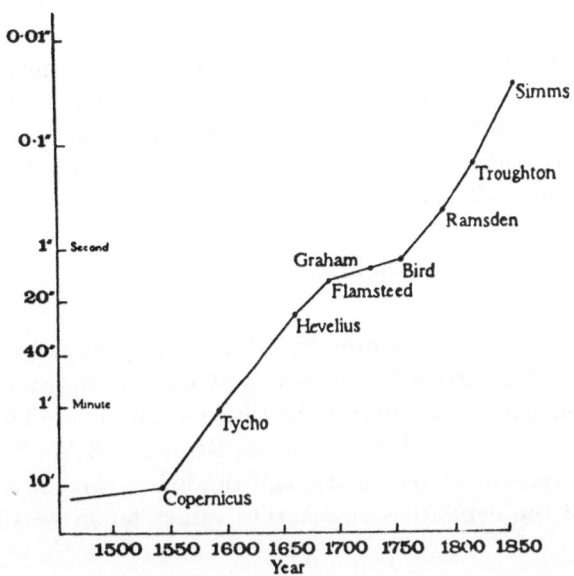

Fig. 1 – The resolution of graduated scales (from Chapman 1983).

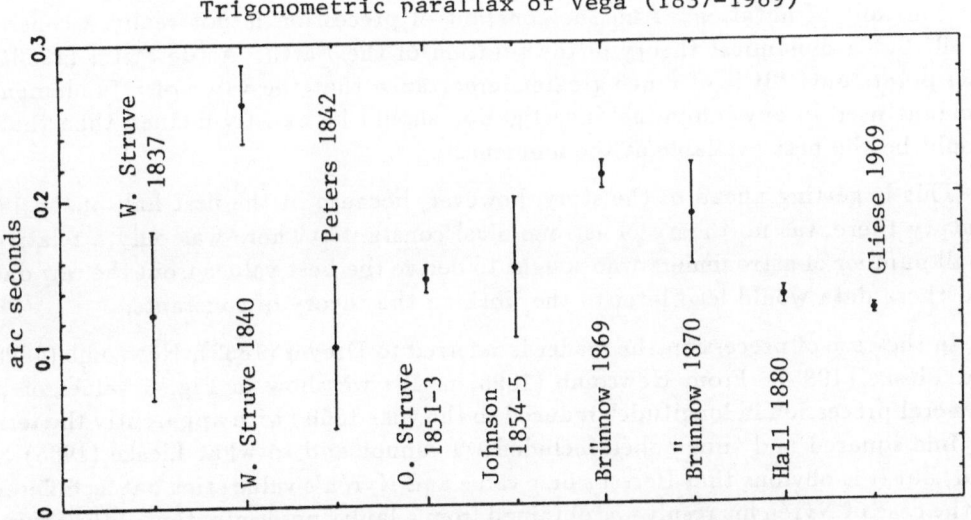

Fig. 2 – Various values of the measured parallax of α Lyrae (Vega), taken from Hall (1888, p. 12), but also including W. Struve's preliminary value (1837) and the modern value from Gliese (1969). Error bars are *probable* errors.

Repeated measurements of the same star yield smaller internal errors. As an example consider measurements of the parallax of Vega (Fig. 2). Wilhelm Struve's preliminary 1837 value of $0\overset{''}{.}125 \pm 0\overset{''}{.}055$, based on 17 micrometrical measures with the Dorpat 9.6-inch refractor (see Batten 1988, p. 121) was later revised to $0\overset{''}{.}261 \pm 0\overset{''}{.}025$ (W. Struve 1840) based on a total of 96 measures. Other values obtained throughout the 19th century show some scatter, but most of the values are within one standard deviation of the modern value ($0\overset{''}{.}124 \pm 0\overset{''}{.}005$, Gliese 1969).

Astronomical Constants

If the underlying goal of Pulkovo Observatory was the production of the most accurate star catalogues, this involved not only well built and carefully operated instruments, but also the use of the best reduction parameters. This required a theory of errors, such as that elaborated by Gauss, Bessel, and W. Struve, and an analysis of the cause and remedy of systematic and random errors. More specifically for astronomy it involved the derivation of accurate values for precession, nutation, and aberration.

The concept of fundamental constants has undergone some major changes since the foundation of Pulkovo Observatory. The solar parallax (or distance to the Sun) was once a primary constant, but now is a derived constant. The same is true of the constant of nutation. And the constant of precession is not really a constant at all, but a dynamical theory of the rotation of the Earth. As de Sitter (1938, p. 213) points out: "It is of much greater importance that the value of a fundamental constant used in any individual investigation should be exactly defined than that it should be the best available at the moment."

This is getting ahead of the story, however, because in the first half of the 19th century there was no theory of astronomical constants. There was only a relatively small number of astronomers who sought to derive the best values from the raw data, and these data would lead later to the work on the theory of constants.

In the case of precession the reader is referred to Dreyer (1882), Newcomb (1898), and Lieske (1985). From Newcomb (1898, p. 10) we show in Fig. 3 values of the "general precession in longitude" reduced to the year 1850 (with apparently the terms in time squared and time cubed included). I cannot add to what Lieske (1985) has said, but it is obvious that Bessel's first value and Nyrén's value stick out as different. In the case of Nyrén his result was obtained from a faulty determination of the equinox of 1865 (Dreyer 1918, on p. 348). Values obtained by Otto Wilhelm Struve and C. A. F. Peters are consistent with the values adopted by Dreyer and Newcomb. Yet, Newcomb (1895, pp. 125-6) states:

The [value of precession] which seems entitled to most weight is that of Ludwig Struve ... This work was suggested by the completion of Auwers' re-reduction of Bradley's Observations, and of the Pulkowa standard catalogues for 1845, 1855, and 1865. It depends entirely on the Bradley stars, and the result, when reduced to the most probable equinox, may be regarded as the best now derivable from those stars, or, at least, as not susceptible of any large correction.

It was Newcomb's value published in 1898 (known as N96, however) that was adopted for *The Nautical Almanac* and in *The American Ephemeris* in 1901 and also by the IAU in 1964 (5025.64 arcsec per tropical century for the year 1900; see *Explanatory Supplement* 1961, pp. 192-3).

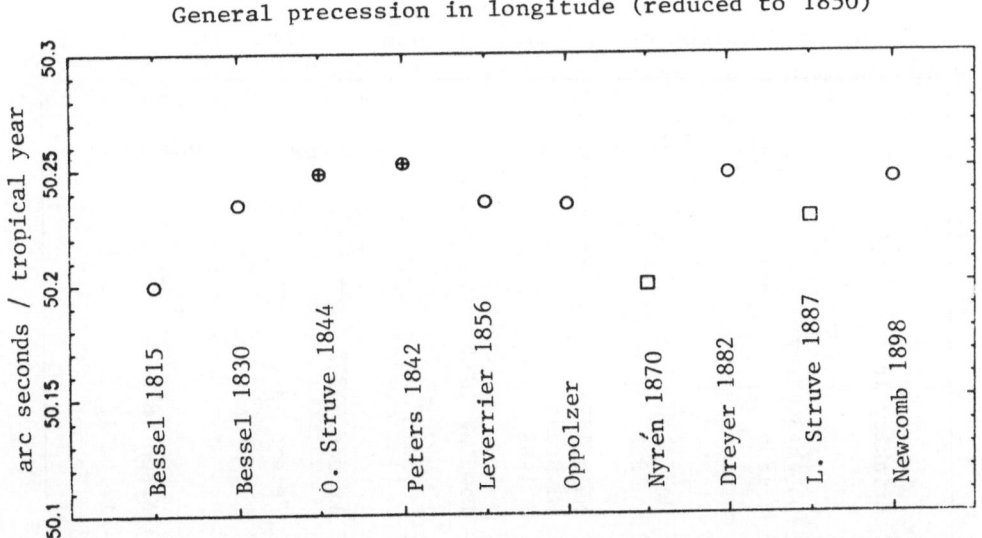

General precession in longitude (reduced to 1850)

Fig. 3 – Various values of the general precession in longitude, reduced to 1850 (from Newcomb 1898, p. 10). Circles with crosses represent values based in part on Dorpat data. Squares represent values based in part on Pulkovo data. The value for 1850 based on modern theory is 50.25656 arcsec per tropical year (Lieske *et al.* 1977, p. 14), and is indicated by the horizontal line.

Derived values for nutation are in close agreement with each other and overlap the modern accepted value (9''.2025). In Fig. 4 we show data compiled by Harkness (1891, p. 25). While there is some scatter in the Greenwich values of 1838-1882, the

Dorpat and Pulkovo results are more self consistent. It is not surprising that the nutation values are on the whole correct. As Newcomb (1895, p. 129) points out:

> The determination of this constant from observations is extremely satisfactory, owing to the completeness with which systematic errors may be eliminated. If, with a meridian instrument, regular observations are made through a draconitic period, on a uniform plan, upon stars equally distributed through the circle of Right Ascension, the observations being made daily through more than 12 hours of Right Ascension, all systematic errors in the determination of the nadir point and all having a diurnal or annual period may be completely eliminated from the constant in question.

Derived values of constant of nutation (1821-1885)

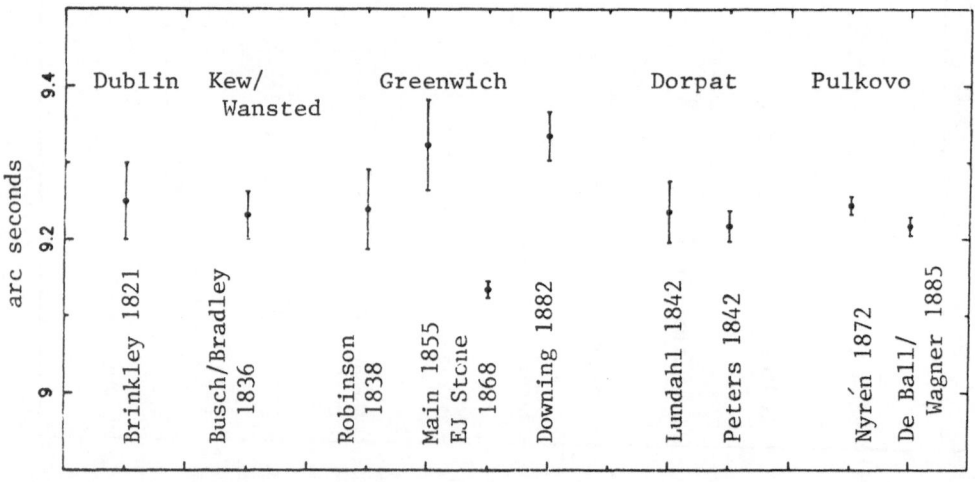

Fig. 4 – Derived values of nutation, given by Harkness (1891, p. 25) and grouped by observatory of origin. The modern value is 9.2025 arcsec (1989 *Astronomical Almanac*, p. K6), indicated by the horizontal line. Error bars are *probable* errors.

Newcomb (1895, p. 130) lists 27 determinations of the constant of nutation, including nine from Pulkovo data and 12 Greenwich values. The highest weight he assigns to observations of southern stars with the Washington transit circle (weight 6). Pulkovo values are generally weighted 3 and Greenwich weights average 2 1/3.

Newcomb's weighted result is 9″.210±0″.008 , the value accepted at the 1896 conference on astronomical constants (*Procès-Verbaux* 1896, p. D54).

Regarding aberration Newcomb (1895, p. 133) states: "This constant is itself the one of which the determination is most likely to be affected by systematic errors. In this respect it is at the opposite extreme from the constant of nutation."

In Fig. 5 we give some representative values of aberration, taken from Harkness (1891, pp. 25-6). Two preliminary values of W. Struve and his "final" value are given (W. Struve 1841, 1842, 1844). If we consider the results observatory by observatory, the Pulkovo results are clearly the most self consistent, and even have some of the smallest error bars.

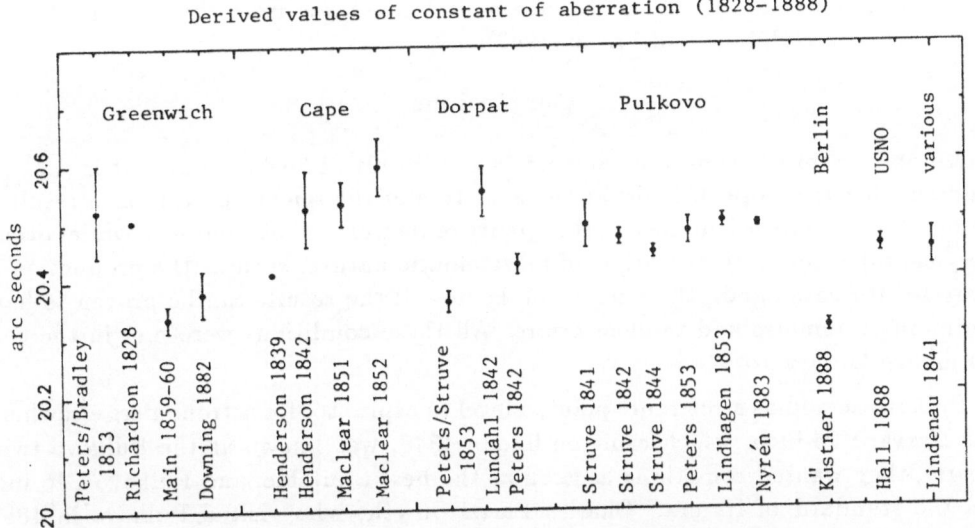

Fig. 5 – Derived values for constant of aberration, given by Harkness (1891, pp. 25-6) and grouped by observatory of origin. For two values no error bars were listed. The preliminary values of W. Struve (1841 and 1842) are shown, as well as his final value. The modern value is 20.49552 arcsec (1989 *Astronomical Almanac*, p. K6), indicated by the horizontal line. Error bars are *probable* errors.

From the self-consistency of the Pulkovo data Newcomb (1895, pp. 137-8) is justified in separating Pulkovo results from all others. For ten adjusted Pulkovo values Newcomb obtains 20″.493 ± 0″.011. The other 24 values from other observatories give 20″.463±0″.013. (The value adopted at the 1896 conference was 20″.47 (*Procès-Verbaux* 1896, p. D54). The modern value is 20″.49552 .)

Later in Newcomb's 1895 treatise (pp. 157, 166) he gives values of the solar parallax. From the Pulkovo constant of aberration he obtains $8''.793 \pm 0''.0046$. (The modern value is $8''.794148$.) There are nine values that Newcomb considers. He assigns a weight of 40 to the solar parallax value based on Pulkovo data. For the others he makes "liberal allowance for the more or less probable sources of systematic error". The other weights range from 1 to 10, so the Pulkovo result is weighted 4 to 40 times higher than any of the other solar parallax values.

In Newcomb's estimation a century ago, or in our estimation with a century of hindsight, Pulkovo values of the 19th century for fundamental constants warranted very high regard. The best example to consider from an historical point of view is that of aberration, because of the possible sources of error, and it is this example that highlights the quality of positional work done at Pulkovo.

For a compilation of values of fundamental constants from the first half of the 20th century, see Böhme and Fricke (1963).

Conclusions

An observatory becomes famous partly as a result of how grandiose it is (i.e. big buildings, big telescopes). It draws the attention of the scientific world as a result of the reputation of its scientists and the quality of its work. If obvious discoveries occur, everyone takes note. If its work is of a systematic nature, such as the production of accurate star catalogues, the scientists take note if the results can be proven to have minimum systematic and random errors. All these conditions were met in the case of Pulkovo Observatory.

Pulkovo served as a reference point, a level to aspire to, for astronomers elsewhere. The Harvard 15-inch, which came on line in 1846, was known as the Pulkovo twin. Clearly, they wanted something as good as the best available, and Pulkovo's 15-inch was the standard of its era. Charles Piazzi Smyth, who visited Pulkovo in 1859, described another observatory on another trip of his as "like Pulkova perfectionised" (Brück and Brück 1988, p. 199). Pulkovo established a tradition of exactitude based on "Struve's method" and produced such fine results that B. A. Gould called it the "astronomical capital of the world." When Simon Newcomb analyzed determinations of astronomical constants, he weighted Pulkovo's results on aberration (a constant whose derivation is susceptible to systematic errors) far higher than all the other results. The Astronomer Royal, George Biddell Airy, stated in 1847 that no modern astrometrist could consider himself sufficiently trained until he had learned the methods developed at Pulkovo, since "a single observation made at Pulkovo is at least as valuable as two made elsewhere" (quoted by Otto Struve 1951).

In his *History of Physical Astronomy*, Robert Grant (18[52], p. 505) writes: "In short, the Observatory of Pulkowa may be regarded as one of the most complete in existence, of those institutions that have been founded for promoting the advancement of astronomical science." Another opinion is given by Agnes Clerke (1904, p. 44), who writes: "The institution created by [Struve] was acknowledged to surpass all others of its kind in splendour, efficiency, and completeness." "

As we celebrate the 150th birthday of Pulkovo Observatory let us keep "Struve's method" in mind and proceed to evaluate the state of the art of positional astronomy in the late 20th century and see what it will become by the beginning of the 21st.

References

The Astronomical Almanac for the year 1989, Washington: U. S. Government Printing Office and London: Her Majesty's Stationery Office.

Batten, Alan H. 1988, *Resolute and Undertaking Characters: the Lives of Wilhelm and Otto Struve*, Dordrecht: D. Reidel.

Böhme, S. and Fricke, W. 1963, in *The System of Astronomical Constants*, J. Kovalevsky ed., Paris: Gauthier-Villars (IAU Symposium No. 21), 269-293.

Brück, H. A. and Brück, M. T. 1988, *The Peripatetic Astronomer: The Life of Charles Piazzi Smyth*, Bristol and Philadelphia: Adam Hilger.

Chapman, A. 1983, *J. Hist. Astr.* **14**, 133-137.

Clerke, Agnes M. 1904, *A Popular History of Astronomy during the Nineteenth Century*, London: Adam and Charles Black, 4th ed.

de Sitter, W. 1938, *Bull. Astron. Insts. Neth.* **8**, 213-231. (This paper was completed by D. Brouwer.)

Dreyer, J. L. E. 1882, *Copernicus* (Dublin) **2**, 135-155.

Dreyer, J. L. E. 1918, *M. N. R. A. S.* **78**, 343-349.

Explanatory Supplement to *The Astronomical Ephemeris* and *The American Ephemeris and Nautical Almanac*, London: Her Majesty's Stationery Office, 1961, in particular pp. 168-174, 180-193.

Gliese, W. 1969, *Catalogue of Nearby Stars*, Veröffentlichungen des Astron. Rechen-Inst. Heidelberg, No. 22.

Grant, Robert 18[52], *History of Physical Astronomy from the Earliest Ages to the Middle of the Nineteenth Century*, London: Henry G. Bohn.

Hall, A. 1888, *Astron. J.* **8**, 1-13.

Harkness, W. 1891, "The solar parallax and its related constants, including the figure and density of the Earth," *Washington Observations for 1885*, Appendix III.

24

Hughes, D. 1984, *Nature* **307**, 15-16.

Krisciunas, Kevin 1988, *Astronomical Centers of the World*, Cambridge: Cambridge Univ. Press.

Lieske, J. H. 1985, *Celestial Mechanics* **37**, 209-238.

Lieske, J. H., Lederle, T., Fricke, W., and Morando, B. 1977, *Astron. Astrophys.* **58**, 1-16.

Newcomb, Simon 1895, *The Elements of the Four Inner Planets and the Fundamental Constants of Astronomy*, Washington: Government Printing Office.

Newcomb, Simon 1898, "A new determination of the precessional constant with the resulting precessional motions," *Astronomical Papers prepared for the use of the American Ephemeris and Nautical Almanac* **8**, Washington: Government Printing Office.

Procès-Verbaux of the *Conférence Internationale des Étoiles Fondamentales de 1896*, *Annales du Bureau des Longitudes* **3**, 18[96], pp. D1-D90.

Struve, Otto [Wilhelm] 1895, *Zur Erinnerung an den Vater den Geschwistern dargebracht*, Karlsruhe: G. Braun'schen Hofbuchdruckerei.

Struve, Otto 1951, *Navigation* **2**, 302-305.

Struve, W. 1839, *Astron. Nachrichten* **16**, No. 371, 163-166.

Struve, W. 1840, *ibid.* **17**, No. 396, 177-180.

Struve, W. 1841, *ibid.* **18**, No. 426, 289-294.

Struve, W. 1842, "Notice sur l'instrument des passages de Repsold, établi a L'Observatoire de Poulkova dans le premier vertical, sur les résultats que cet instrument a donnés pour l'evaluation de la constante de l'aberration, *Bulletin scientifique publié par l'Académie imp. des Sciences de St.-Pétersbourg* **10**, No. 14, 15, 16.

Struve, W. 1844, "Sur le coefficient constant dans l'aberration des étoiles fixes déduit des observations que ont été exécutées à l'observatoire de Poulkova par l'instrument des passages de Repsold établi dans le premier vertical," *Mém. de l'Acad. Imp. des Sci. de St.-Pétersbourg, Sci. Math. et Phys.* **3**, 229-285.

Struve, [F. G.] W. 1845, *Description de L'Observatoire Astronomique central de Poulkova*, St. Petersburg: Imprimerie de L'Académie Imperiale des Sciences.

Struve, [F. G.] W. 1852, *Stellarum Fixarum imprimis Duplicium et multiplicium Positiones Mediae pro epocha 1830.0 deductae ex observationibus meridianis annis 1822 ad 1843 in specula Dorpatensi*, St. Petersburg.

L'ARC GEODESIQUE LE PLUS LONG :
DELISLE , LES STRUVE ET L'OBSERVATOIRE DE PULKOVO

Suzanne DEBARBAT
Observatoire de Paris, URA 1125/CNRS
61, avenue de l'Observatoire
75014 Paris
France

ABSTRACT. The longest geodetic arc ever mesured by classical goedesy is long of 25° from the Baltic to the Black Sea through the Dorprat meridian. This arc is based upon measurements made, from 1816 to 1855, by a russian general, a norvegian, a swedish and F.G.W. Struve, the founder of the Pulkovo Observatory ; its lenght is 2° more than the arc Delisle, then at the Petersburg Observatory, intended to determine along the meridian of this observatory in the year 1737. Russia, at that time, became part of the european triangulation, a prelude to circumterrestrial modern campaigns.

L'arc de plus grande longueur jamais mesuré en Europe par la géodésie classique s'étend sur plus de 25°, de la Baltique à la mer Noire. Issu des opérations géodésiques menées de 1816 à 1855 par un Général russe (C. de Tenner), un Norvégien (Chr. Hansten), un Suédois (N.H. Selander) et F.G.W. Struve fondateur de l'Observatoire de Pulkovo, il a donné lieu à d'importants volumes publiés à Saint-Pétersbourg peu après l'achèvement des opérations. Leur ampleur (28x36cm, 9cm d'épaisseur dont un Atlas de 26 planches, au prix de 10 roubles d'argent) est à l'image de la tâche accomplie par l'ensemble des géodésiens de nombreuses nationalités attelés aux opérations : 258 triangles, 10 bases mesurées, 13 points astronomiques sur le méridien de Dorpat, correspondant à une distance des points extrêmes de près de 3000 kilomètres.

Dans son "Coup d'oeil général" par lequel débute l'introduction de ces ouvrages, W. Struve précise que cet arc le plus long est aussi celui qui se place sensiblement à la longitude moyenne de l'Europe, entre le Cap du Finistère et la ville d'Iekaterinbourg. Doublement européen, en longitude et en latitude, l'arc mesuré représente à la fois le fondement géodésique de la cartographie moderne de l'Europe, le raccordement de deux continents, et l'aboutissement d'idées, de recherches, de développements largement tributaires du 17ème et du 18ème siècles. Certes les idées sur la grandeur et la forme de la Terre ont évolué depuis la publication des "Principia" (1687), la

J. H. Lieske and V. K. Abalakin (eds.), Inertial Coordinate System on the Sky, 25–28.
© 1990 IAU. Printed in the Netherlands.

controverse Cassini-Newton et la victoire de ce dernier assurée par les expéditions du Pérou et de Laponie sous l'égide de l'Académie des sciences de Paris. Mais en 1805 des mesures nouvelles, par Svanberg reprenant celles de Laponie, ne s'accordent pas complètement avec ces dernières.

Lindenau (directeur de l'Observatoire du Seeberg) d'un côté, W. Struve de l'autre forment le projet, vers 1810, d'un arc de méridien le plus long possible. Les projets prendront corps et sept arcs seront mesurés, du Danube à la mer Glaciale comme la dénomme W. Struve. En 1844, l'ensemble des arcs s'étend déjà du Dnestr au sud à Ténéa au nord, dans la zone mesurée un siècle plus tôt par Maupertuis, le grand aplatisseur de Voltaire. Venu à Paris à cette époque, Otto Struve (1819-1905), le fils de Wilhelm (1793-1864), rencontre Arago et trouve à la Bibliothèque de l'Observatoire les manuscrits établis par Delisle, manuscrits dont une partie importante concerne précisément un projet d'arc de méridien en Russie conçu un siècle plus tôt.

Invité par Pierre le Grand (venu à Paris, et à l'Observatoire les 12 et 17 mai 1717), puis par Catherine 1ère, Delisle était parti en Russie, pour quelques années, en 1726. Dix ans plus tard il est toujours là. La controverse sur la forme de la Terre bat son plein; il propose alors - dans un mémoire lu le 21 janvier 1737 à l'Académie des sciences de Russie - la mesure d'un arc de 23° en terrain favorable, aligné sur le méridien du nouvel Observatoire de Saint-Pétersbourg.

Gravure extraite des manuscrits de Delisle
conservés à la Bibliothèque de l'Observatoire de Paris

Et s'il y a une montagne, qu'à cela ne tienne : M. Euler (il est alors en Russie) trouve "que l'attraction d'une montagne supposée sphérique d'une demi-lieue de diamètre posée sur la superficie de la terre doit déranger la direction du perpendicule d'environ deux minutes ce qui mérite bien d'être examiné très exactement par l'observation". Delisle entreprend aussitôt des mesures, notamment celle d'une base qu'à l'instar de Maupertuis il mesure sur la glace entre Doubky et Petershoff. Petershoff qui se voit de l'Observatoire impérial est sur le continent, tandis que sur l'île de Rétusari se trouve la nouvelle église de Cronstadt. La base a 20 wersts (orthographe de Delisle), et un triangle pourra être formé entre ces deux points et la montagne de Douderhof.

Bientôt Maupertuis, Clairaut, l'abbé Outhier, Celsius et d'autres rendent caduques les recherches sur la figure de la Terre ; mais restent les nécessités de la cartographie de la Russie et les besoins de la navigation. Pourtant Delisle ne poursuit pas l'exécution de son projet. Il part en Sibérie à la poursuite de passages de Mercure sur le Soleil, puis est occupé à d'autres tâches. Il rentre en France - après une absence de 21 ans - en 1747. Une partie de ses manuscrits sera prêtée à W. Struve en 1847, après un voyage de celui-ci à Oxford où il eut "pour voisin de logis M. Le Verrier" qui "l'invita à l'accompagner à Paris". Quelques années auparavant ont eu lieu des opérations chronométriques : en 1843 entre Pulkovo et Altona, en 1844 entre Altona et Greenwich. Schumacher (alors au Danemark) avait donné dès 1821 la différence de longitude entre Hambourg et Copenhague à une seconde d'heure. O. Struve va bénéficier des derniers progrès de la technique: une "vitesse de transport encore plus merveilleuse" |....| "le concours de plusieurs horloges transportables" |....| ; les bateaux à vapeur, le chemin de fer seront les artisans du succès des opérations.

De 1844 à 1851, l'arc de méridien est prolongé, vers le nord et vers le sud ; puis, après 1851, sont effectués des travaux supplémentaires permettant d'assurer le nivellement trigonométrique des rattachements avec d'autres pays d'Europe. La Carte des triangulations exécutées en Russie englobe, avec la méridienne de Dorpat, les opérations tranversales avec le nom des responsables qui les ont menées. En 1854, le raccordement en longitude est effectué entre Pulkovo et Dorpat.

Les instruments ont été nombreux et variés (émanant de constructeurs anglais, allemands et français) : une dizaine pour les visées géodésiques, une demi-douzaine pour celles de caractère astronomique, sans compter les chronomètres, les instruments météorologiques,... il y a également les instruments pour les mesures linéaires, par exemple une échelle de fer de 7 pieds anglais (1 sagène), avec la référence : la toise à bout de Fortin de l'Observatoire de Dorpat. Cette dernière a été comparée en 1821, par Arago, à la toise du Pérou, celle qui a servi pour la définition du mètre (3 pieds 11.44 lignes pour le mètre provisoire de 1793, 3 pieds 11.296 de cette même toise pour le mètre définitif de 1799). Par ailleurs, W. Struve lors de son voyage à Oxford a pris contact avec G. Everest pour effectuer un raccordement des étalons russes avec celui ayant servi aux mesures géodésiques indiennes menées sous la direction de ce dernier de 1823 à 1843. L'ensemble de ces raccordements permet à Struve de donner indifféremment des mesures en toises, en verstes, en kilomètres, en pieds anglais,...

Ainsi dès le milieu du 19ème siècle, grâce aux opérations conduites par l'un ou l'autre des Struve, leurs collaborateurs et les autres chefs d'expéditions, était assuré un extraordinaire raccordement des pays qui avaient mené leurs opérations géodésiques. La Pologne, la Prusse, la Silésie se joignaient à la France, l'Angleterre, l'Autriche, les pays scandinaves, prélude aux opérations circumterrestres de notre époque. D'emblée l'Observatoire de Pulkovo créé en 1839, s'inscrivait dans la modernité. Wilhelm et Otto Struve, ses deux premiers directeurs, ont simultanément développé les recherches qui, à côté de l'astronomie géodésique - créée par Picard à la fin du 17ème siècle - en ont fait le "leader" de l'astronomie fondamentale du 19ème siècle. Côté français ils ont noué des relations avec l'Observatoire de Paris, ses directeurs de l'époque, notamment Le Verrier. Ce n'est sans doute pas un hasard si O. Struve est présent à l'inauguration de la statue de Le Verrier le 27 juin 1889, il y a à peine plus d'un siècle, s'il figure dans un projet de plafond pour une salle de l'Observatoire de Paris, relatif à la découverte de Neptune, et si la famille de Le Verrier a remis à l'Observatoire de Pulkovo, un portrait en pied de ce dernier par Duverdoing exposé au Salon de Paris de 1847.

Poulkova. Le nom de Struve, successivement illustré par son père et par lui, restera après celui de Le Verrier, comme le plus célèbre et le plus respecté parmi les astronomes du 19e Siècle.

**Extrait de la minute d'une lettre
de l'Amiral Mouchez à son ministre le 15 juin 1887**

Le réseau géodésique européen comme la détermination de l'ellipsoïde terrestre (par Tenner, par Struve, et d'autres) à partir du grand arc de méridien du centre de l'Europe, raccordant Pulkovo aux autres points du réseau, portent témoignage de la valeur des mesures qui ont mené géodésiens et astronomes de Tflis et Azoff à Odessa, Nikolaïeff et Niznyi-Nowgorod, aussi bien à Varsovie, Kieff, Smolensk qu'à Tornéa, Helsingfors, Pulkova et Riga, sans oublier Saint-Pétersbourg. L'hommage qui leur est rendu ici s'appuie sur les importantes publications des Struve et de leurs collaborateurs, sur les manuscrits et documents, ceux de Delisle notamment, que conserve la Bibliothèque de l'Observatoire de Paris.

PULKOVO OBSERVATORY AND THE NATIONAL OBSERVATORY MOVEMENT: AN HISTORICAL OVERVIEW

STEVEN J. DICK
U.S. Naval Observatory
34th and Massachusetts Ave., NW
Washington, D. C. 20392
USA

ABSTRACT: The patronage of national governments has played an important role in the history of astronomy, classically in the form of National Observatories. In this paper we 1) argue that the last three centuries have seen what we may call a "national observatory movement," in that national governments during this period increasingly supported astronomical observatories, and in that such institutions share certain common properties of origin, purpose and evolution; 2) demonstrate the important role that Pulkovo Observatory has played in this movement; and 3) compare certain aspects of the Pulkovo Observatory and the United States Naval Observatory as exemplars of this species founded within a decade of each other under very different political conditions.

1. The National Observatory Movement

The phrase "national observatory movement" implies an ongoing process with definite goals shared by members of the national observatory community. The term has been applied before to attempts to form national observatories within particular countries, such as the United States (Paullin, 1923) and India (Ansari, 1985), but we apply it here in a more encompassing global sense, where the national observatories of particular nations are seen as members of the movement. We may consider that this broader process began, after the important but abortive founding of Tycho Brahe's observatory by Frederick II of Denmark in 1576, with the Paris Observatory (1667) and the Royal Observatory at Greenwich (1675). But the need for national observatories did not end in France and England. As Table 1 shows, Germany and Imperial Russia added two important examples in the early 18th century, the rate of new institutions actually increased in the 19th century, and new members have continued to be added in the 20th century. It is clear that all national observatories are not listed in Table I, particularly those of the Far East. A more comprehensive list would be an interesting and major task, but the sample given here is large enough to demonstrate several points and the overall benefits of this institutional approach to the history of astronomy. It should also be clear that there are many other "astronomical centers of the world" (Krisciunas, 1988) that may receive partial government funding but are not considered national observatories.

We may distinguish three eras in this movement: the *first era*, in which the prototype Paris, Greenwich, Berlin and St. Petersburg observatories were founded; the *second era*, characterized by offshoots from previous national observatories (Royal Observatory Cape), by new observatories of younger nations (USNO), and by the rise of astrophysical observatories; and the *third era*, post World

J. H. Lieske and V. K. Abalakin (eds.), Inertial Coordinate System on the Sky, 29–38.
© 1990 *IAU. Printed in the Netherlands.*

War II, characterized by national or international consortia, large budgets relative to the previous eras, and the study of old and new wavelength regions with increasingly sophisticated telescopes, detectors and spacecraft. National Observatories were the original "big science" of their time, but in this third era, which encompasses both the Computer Age and the Space Age, the movement has benefited from the general trend toward even bigger science seen in the national laboratories of many disciplines (Price, 1963; Weinberg, 1967).

Table 1. Some Important National Observatories and Their Patrons

	Institution	founded *	patron
1st Era	Uraniborg (Tycho Brahe)	1576 (abortive)	Frederick II
	Paris Observatory	1667	Louis XIV
	Royal Observatory, Greenwich	1675	Charles II
	Berlin Observatory	1701	Frederick I
	St. Petersburg Observatory	1725	Peter the Great
2nd Era	Royal Observatory, Cape	1820	Britain
	U. S. Naval Observatory	1830	U.S. Navy
	Pulkovo Observatory	1839	Nicholas I
	Chilean National Observatory	1852	Chile
	Argentine National Observatory	1870	Argentina
	Potsdam Astrophysical	1874	German Acad. Science
	Smithsonian Astrophysical	1891	Smithsonian/ U.S.
	Dominion Observatory	1903	Canada
	Dominion Astrophysical	1918	Canada
3rd Era	NRAO (U. S.)	1956	NSF/AUI
	Kitt Peak	1957	NSF/AURA
	NRAO (Australia)	1959	CSIRO
	Cerro-Tololo Inter-American	1963	NSF/AURA/Chile
	European Southern Observatory	1964	5 countries (now 8)
	Anglo-Australian (Siding Spring)	1967	Britain/Australia
	Space Telescope Science Institute	1981	NASA/AURA

* A number of criteria can be used for founding dates. The majority of dates here indicate when funding was assured. Abbreviations: AUI (Associated Universities, Inc.), AURA (Association of Universities for Research in Astronomy), CSIRO (Commonwealth Scientific and Industrial Research Organization), NASA (National Aeronautics and Space Administration), NSF (National Science Foundation), NRAO (National Radio Astronomy Observatory).

Aside from the striking association of all the early institutions with their national scientific societies (whether Academies of Science or the Royal Society), the common property that stands out in the first era is the largely practical nature of the work for which the first national observatories were founded. Whether for the improvement of navigation, geographic and geodetic work, or calendar reform, all these institutions were founded to meet a national need. In meeting these national needs the precise determinations of celestial positions formed the backbone of much of their work; for example, the method of lunar distances for navigation required precise ephemerides of the Moon and precise positions of stars as the reference frame, a task that was brought to fruition only with Maskelyne's publication of the British *Nautical Almanac* in 1766 (Sadler, 1976). Byproducts of this practical work were the great star catalogues of Flamsteed and others, Bradley's discovery of aberration of light, the determination of proper motions, and many other results also of interest to pure astronomy.

The early institutions of the second era were also founded for similar purposes, but now with the determination of longitude by chronometers as the most promising method of navigation at sea, and an important method for determining geographical positions as well. With accuracies now on the order of tenths of arcseconds *vs.* about 15 arcseconds for Flamsteed (Chapman, 1983) greatly improved star catalogues were the byproducts of the Cape, U. S. Naval, Pulkovo, Chilean and Argentine observatories, whereas in this era of improved instruments stellar parallaxes too were at last possible. With the rise of the astrophysical national observatories at the end of the 19th century, results beyond any practical need were the goal, and the question of how far public money should support such research became increasingly important. The very existence of the third era gives an answer to that question, for the research of these observatories has gone beyond anything that the public would consider practical. The longevity of the older institutions, as well as the occasional addition of new ones, attests to a continuing national necessity, though one expanded to include the benefits of pure research, whether to national prestige or the advancement of astronomy.

A second common characteristic of national observatories, rooted in the first property of common purpose in the service of a perceived national necessity, is that most of them at the time of their founding undertook similar programs in order to carry out their tasks. This led to substantial interactions and cooperation among national observatories, not only in international programs such as the transits of Venus, the *Carte du Ciel*, the Eros campaign and long-term astrometric projects such as the AGK3R and Southern Reference Star systems, but also in many colorful episodes in the history of astronomy. A notable example of the latter is the involvement of the observatories of Paris, Greenwich and Berlin in the discovery of Neptune. Of particular interest to this meeting, the work of determining fundamental reference frames has largely fallen to these national observatories, for few other observatories can undertake such long-term programs.

National Observatories have also faced in turn many similar problems and issues: the proper balance between practical and pure research, the relative roles of astrometry and astrophysics, the most efficient mode of administration, and patterns of funding, to name only a few. It is true that all astronomical institutions share to some extent these problems, but government institutions, particularly in regard to a mission to be accomplished, share them in peculiar and related forms.

Finally, it is notable that national observatories have historically been perceived as a distinct group of astronomical institutions, both by the institutions themselves and externally. Aside from cooperative programs, the younger national observatories such as Pulkovo and USNO often consulted their predecessors regarding buildings, instrumentation, and programs; visits and comparisons were often made among the national observatories to improve efficiency; and moral support was given in times of crisis such as threats of retrenchment or even abolishment. For all their similarities, each observatory has developed its own character, as it has met these problems and challenges in different ways. Even with obvious differences reflecting national styles, a global view of the history of astronomy shows that national observatories, with their common purpose, common programs and common problems, form a distinct group of institutions and comprise a worldwide national observatory movement.

2. The Place of Pulkovo Observatory

Let me proceed to the second point, namely the role of Pulkovo Observatory in this movement. We note first from Table 1 that Pulkovo Observatory was founded in what we have called the second era of national observatories. By the 1830s the older national observatories were in many ways in their prime. In 1835 G.B. Airy began his tenure of almost a half century as Astronomer Royal at RGO (Meadows, 1975). At Paris Observatory the reign of Delambre was over, Alexis Bouvard had been Director since 1822, and Arago and Leverrier were coming on stage (Debarbat, Grillot and Levy, 1984). In Berlin J. F. Encke was well into his long tenure at a revitalized observatory newly constructed in 1828 (Dick, 1951).

In Russia itself, the St. Petersburg Observatory was now in a run-down condition, and its demise figured directly in the founding of Pulkovo Observatory. Like the Berlin Observatory a few years earlier Pulkovo thus began with new buildings and instruments on a new site, but unlike Berlin it also began with a new head, Wilhelm Struve, who as the head of the Dorpat Observatory in Tartu had completely outdone the St. Petersburg Observatory. Struve relates how in December, 1830 he frankly told Tsar Nicholas I of the sad condition of the St. Petersburg observatory. The result was the Tsar's decision to build an observatory near the capital, to be located on the hill of Pulkovo, outside the city. The details of Pulkovo's founding, and the life and work of the Struves, are by now well-known, thanks to Struve's 1845 *Description*, and more recently the work of Daedev (1972) and Sokolovskaya (1976) in Russia, and Krisciunas (1978, 1988) and Batten (1988) in the U. S. and Canada. For the record I show in Table 2 some of the more important dates in the founding of Pulkovo, taken from Struve.

Table 2. Important Dates in Pulkovo's Early History

1830	December. Struve meets with Nicholas I.
1833	October 28. Nicholas orders purchase of instruments
1834	February 24. Architects present their plans to the Commission
1834	April 15/3. Nicholas names Struve Director
1834	June-October. Struve's trip abroad to purchase instruments
1835	3 July/21 June. Foundation stone laid.
1839	August 19/7. Official inauguration
1845	Struve's *Description* published.

Source: Struve (1845). Double dates represent New Style/Old Style.

Pulkovo was one of the few observatories to have so many interests so early in its career. The official statute for the observatory specified not only the perfection of navigation and geographical positions, but also the advancement of astronomy beyond any practical need. While it is clear that

Struve took seriously the practical aspects of his job, his contributions to the advancement of astronomy are even better known. The breadth of interest and the development of the work at Pulkovo in its first 50 years may be seen from Table 3, according to categories established by Struve himself.

We see from this table that the work of the Pulkovo Observatory during its first 50 years was dominated by sidereal and solar system studies, some of which contrasted with the more purely practical work such as the measure of the arc of the meridian. The stellar work included double stars, parallaxes, positional catalogues and the constants of precession, aberration and nutation; and the planetary work consisted largely of comets, satellites and eclipses. As a member of the second era of national observatories, its practical problems of navigation and geodesy were more refined than the problems that faced its older predecessors at their founding. The chronometer method, rather than the method of lunar distances, was on the ascendancy, with all that implied not only for navigation but also for geographic positions by chronometer expeditions, the latter for which Pulkovo is famous. We also note from this table the early entry of Pulkovo into the realm of astrophysics; from 1864 to 1888 10% of the publications dealt with astrophysics, while the number of geodesic and geographic publications declined by half relative to the first 25 years.

Table 3. Research Publications at Pulkovo, 1839–88

Subject	1839–64		1864–89		Differences
sidereal	49	30%	63	27%	−3
solar system	42	26%	54	23%	−3
practical astronomy	14	9%	21	9%	0
geodesy and geographical positions	36	22%	22	9%	−13
history, bibliography & physical experiments	22	14%	33	14%	0
astrophysics	0		23	10%	+10
mathematics, physics & theoretical studies			21	9	+ 9

Sources: Struve (1865) and Pulkovo Observatory (1889)

Pulkovo Observatory quickly became legendary among national observatories. Early visitors to the observatory, such as the Astronomer Royal George B. Airy, and the American astronomers B.A. Gould (1849) and C. Abbe (1868), left fascinating accounts of their visits. On his return from St. Petersburg in 1847, Airy typically remarked that "... no astronomer can feel himself perfectly acquainted with modern observing astronomy in its most highly cultivated form, whether as regards the personal establishment, the preparation of the buildings, the selection and construction of the instruments, or the delicacy of using them, who has not well studied the Observatory of Pulkowa," (Airy, 1848). This is high praise indeed coming from the head of another national observatory, especially one with the personality of George Airy. In Russian history of science Pulkovo represents a revitalization of government support for astronomy in that country; in the international sense it represents the addition to the community of national observatories that quickly became a model for the others.

3. Pulkovo Observatory and the United States Naval Observatory

The detailed comparative history of astronomical institutions, especially those in a similar class such as national observatories, is an important task. It is, however, a very large one that cannot be carried out in this paper, and that indeed cannot be carried out fully until the individual histories of these institutions are written. But I would like to touch here on a more manageable task — the comparison of USNO and Pulkovo in just a few areas during their first 50 years, as two exemplars founded very close together in time, though very far in political systems. This comparison will serve to illustrate some of the similarities and differences, some of the variations on the theme of national observatories.

We may briefly compare the two institutions in four areas during the first half century of their lives: Origin, administration, instruments, and programs. I will conclude with some remarks on 20th century political effects on these institutions.

3.1. Origins

Whereas as we have seen in Russia, Nicholas I was convinced of the need for a new observatory, and his wishes were of course carried out, in the United States President John Quincy Adams had called for such an observatory in 1825, but it was expressly forbidden by Congress as a waste of money. There were advantages to having a czar if he happened to be on your side. Thus in November, 1830 it was left for a lowly Navy Lieutenant to recommend to the Secretary of the Navy that a Depot of Charts and Instruments be founded in Washington, D.C. This direct forerunner to the Naval Observatory was established in December, 1830, the very same month that Struve had his audience with Nicholas I. It was founded to care for navigational instruments and to rate chronometers by astronomical observation, and here was the opening to astronomy. It thus had only very modest astronomical instruments, and was not yet an observatory in the true sense of the word.

The founding of Pulkovo Observatory in 1839 did not go unnoticed in the United States. John Quincy Adams, now an ex-president, pointed out to Congress a few months later "Here is the sovereign of the mightiest empire and the most absolute government upon earth, ruling over a land of serfs, gathering a radiance of glory around his throne by founding and endowing the most costly and most complete establishment for astronomical observation on the face of the earth... The committee of the House [of Representatives] ... in casting their eyes around over the whole length and breadth of their native land, must blush to acknowledge that not a single edifice deserving the name of an astronomical observatory is to be seen" (Rhees, 1879).

Though Adams' arguments were to no avail, another Navy Lieutenant, James M. Gilliss, officer-in-charge of the Depot of Charts and Instruments, pushed through Congress a bill for a new Depot in 1842, which became a national observatory in fact if not yet in name. The purposes of founding for both Pulkovo and USNO were thus similar, but the purpose of the USNO was more limited at first, with the determination of geographical positions being left to the Army and the Coast Survey and a role for less practical aspects of astronomy only gradually evolving (Dick, 1980).

3.2. ADMINISTRATION

Whereas Pulkovo was founded by Royal decree and came under its country's Academy of Sciences, the U. S. Naval Observatory was totally under the control of the Navy, and headed at first by Navy Lieutenants and later Navy Captains or Admirals. While the distinguished astronomer Struve brought instant fame to Pulkovo Observatory, the Naval Observatory was at first better known for Matthew Maury's oceanographic wind and current charts than for its astronomy. This illustrates an important point: that even though national observatories are bound to carry out certain specific duties necessary for national needs, the character of the institution can be largely shaped by the personality and interests of its Director and senior staff, in addition to the overriding importance of its original charter and administrative sponsor. In the United States it is perhaps not surprising that oceanography rather than astronomy dominated the observatory's first 20 years; there were as yet no distinguished astronomers in the country, indeed no tradition of astronomy to draw on as there was in Russia. Only in the post-civil war era, with the names of Newcomb, Hall, Harkness and G.W. Hill, did the Naval Observatory achieve fame in astronomy.

3.3. INSTRUMENTS

In 1834 Struve journeyed to Europe to obtain instruments, and in 1842 Lt. Gilliss made a similar journey.

Table 4. Early Instruments at Pulkovo and USNO

Pulkovo (1839)	USNO (1844)
15-inch achromatic refractor (Merz and Mahler, Munich)	9.6-inch achromatic refractor (Merz and Mahler, Munich)
heliometer (Merz and Mahler)	
small parallactic refractor	
two comet seekers (Merz and Mahler)	3.9-inch comet seeker (Merz and Mahler)
large meridian telescope (Ertel)	5.5-inch transit instrument (Ertel)
large vertical circle (Ertel)	
large meridian circle (Repsold)	4-inch mural circle (Troughton & Simms)
prime vertical transit (Repsold)	5-inch prime vertical transit (Pistor & Martins

As Table 4 shows, both turned to the expert instrument makers of Europe, and their choices are an interesting testimony to the rise of the German astronomical technology as opposed to the English. Both obtained transit instruments from the German makers Ertel (successor to Reichenbach at Munich). Both obtained achromatic refractors and comet seekers from Merz and Mahler (successor to Fraunhofer at Munich). Struve opted for a Repsold prime vertical transit instrument and Gilliss for a Pistor and Martins from Berlin. Whereas Merz and Mahler made a 15-inch refractor for Struve, Gilliss could only afford a 9.6 inch, exactly similar to the Dorpat telescope Struve had previously used. Only with the mural circle did Gilliss show some faith in English instrument makers; Struve showed none.

Though the U. S. Naval Observatory began with a relatively small achromatic refractor, it is notable that by 1873, American technology had advanced to the extent that the Naval Observatory had the Alvan Clark 26-inch refractor, the largest in the world. In another example of direct interaction between the two observatories, Otto and Hermann Struve visited the U. S. in 1879, where Newcomb persuaded them to purchase from Alvan Clark the 30-inch lens for their new refractor. This turnabout over a period of 30 years is testimony the rapid rise of astronomy, and astronomical technology, in America.

3.4. PROGRAMS

Like Pulkovo the Naval Observatory carried on long-term programs for star catalogues, double stars and specialized observations, leading in some cases to direct cooperation, most recently in the form of the Southern Reference Star (SRS) catalogue. If we were to categorize publications at the USNO by subject over its first 50 years, similar to Table 3, a larger proportion of practical work would be evident at the Naval Observatory, again due largely to Maury.

Unlike Pulkovo, where we have seen in Table 3 a substantial amount of research in astrophysics by 1888, the Naval Observatory did not undertake a sustained program in astrophysics until John Hall's work on interstellar polarization beginning in 1948. The difference lies partly in a more restricted concept of mission at the USNO, but also in individual differences — whereas Struve pushed astrophysics at Pulkovo beginning in 1866, Newcomb in the 1890s was still questioning its importance. With the founding of the Smithsonian Astrophysical Observatory in the 1890s, the USNO left the subject to such specialized observatories, or to university or privately endowed observatories such as Lick, at least until in the 20th century the close relationship between astrometry and astrophysics became evident.

3.5 POLITICAL EVENTS

It is clear that both Pulkovo and the U.S. Naval Observatory have been subject to political and economic events in their own countries. The 1930s was a particular time of crisis for both observatories. In the United States in 1932 President Herbert Hoover, as part of his policy to counter the Great Depression, recommended that the Naval Observatory be transferred from the Navy to the Commerce Department, a proposal seriously considered by Congress, but not carried out. Hardly had this crisis been weathered when President Roosevelt came into office and proposed abolishing the Naval Observatory completely in a money-saving move, a proposal obviously also not carried out.

In the Soviet Union the situation for Pulkovo Observatory during the same decade was even more serious. Not only were the threats more serious, they were followed by action, as Pulkovo Observatory was caught up in Stalin's purge of Soviet astronomers in 1936-37. Throughout the Soviet Union 29 astronomers were arrested, many never heard from again. At Pulkovo alone 13 scientists, almost half the staff, disappeared during these years. This tragic event has recently been studied in detail by Robert McCutcheon (1989). Nor did this end the problems of Pulkovo Observatory ; it was completely destroyed in 1942 during the war. With all of its scientific achievements, we should not forget that the very existence of the Pulkovo Observatory today is a testimony to endurance and hope.

4. Summary and Conclusion

Many more comparisons could be made, but these examples suffice to indicate some of the interesting problems that arise in the comparative history of that subset of astronomical institutions known as "national observatories".

In summary, looking back on the last three centuries we may conclude that national observatories as a group do indeed represent a related process responding to the needs of each nation; that Pulkovo arose suddenly like a bright star in this movement due largely to the Struves, who not only filled a national need but also greatly advanced astronomy; and that despite their similarities, national observatories may evolve in quite different ways, as a result of political systems, science organization, and individual enterprise, as shown in the case of Pulkovo and its American counterpart.

It is clear that over the centuries national observatories have achieved many of their goals, perhaps too well. Navigation and other practical needs are no longer the driving force for many of them, and most have changed in ways their founders could have hardly forseen. The Berlin Observatory has gone through many reincarnations, Cape Observatory was amalgamated with others, and Greenwich Observatory is at present undergoing radical change.

Despite the appearance of new evolutionary forms in the movement, such as the multinational European Southern Observatory and the Space Telescope, these developments warn us that the future of the national observatory movement is not clear. Among astronomical institutions, national observatories especially face the problem of justifying their relevance to national needs. They must surely change, not only in advancing new techniques and instrumentation, but also in reshaping their goals to meet modern requirements. Just as surely, they are bound to be challenged about the relevance of any new goals to national needs, and affected by national priorities and economies much more directly than their colleagues at private institutions or universities. Perhaps the comparative study of the histories of these institutions will help to illuminate future directions, as well as past patterns. The roads taken will determine the fate of this venerable movement, now well into its fourth century, that has contributed so much to astronomy.

5. References

Abbe, C. (1868). "Dorpat and Poulkova," in *Annual Report of the Board of Regents of the Smithsonian Institution*, Government Printing Office, Washington. 370-390.

Airy, G. B. (1848). "Schreiben des Herrn Airy, Konigl. Astronomen in Greenwich, an den Herausgeber. Royal Observatory Greenwich, 1847 November 26," in *Astronomische Nachrichten*, 26, 353-360.

Ansari, S. M. R. (1985). "The Observatories Movement in India During the 17 - 18th Centuries," *Vistas in Astronomy*, 28, 379-85.

Batten, A. H. (1988). *Resolute and Undertaking Characters: The Lives of Wilhelm and Otto Struve*, D. Reidel, Dordrecht.

Chapman, C. (1983). "The Accuracy of Angular Measuring Instruments used in Astronomy between 1500 and 1850," *Journal for the History of Astronomy*, 14, 133-37.

Daedev, A. N. (1972). *Pulkovo Observatory: An Essay on its History and Scientific Activity*, Izdatel'stvo Nauka, Leningrad, trans. K. Krisciunas (1977).

Debarbat, S., Grillot, S. and Levy, J. (1984). *L'Observatoire de Paris: Son Histoire (1667-1963)*, Observatoire de Paris, Paris.

38

Dick, J. (1951). "The 250th Anniversary of the Berlin Observatory," *Popular Astronomy*, 59, 524-535.

Dick, S. J.(1980). "How the U. S. Naval Observatory Began, 1830-1865," *Sky and Telescope*, 60, 466-471; reprinted with corrections and notes in S. J. Dick and L. E. Doggett, *Sky with Ocean Joined: Proceedings of the Sesquicentennial Symposia of the U.S. Naval Observatory*, U. S. Naval Observatory, Washington, D. C., 1983, 167-181.

Gould, B. A. (1849). "An Account of the Observatory at Pulkowa," reprinted from *North American Review*, July, 1849, 1-20.

Krisciunas, K. (1978). "A Short History of Pulkovo Observatory," *Vistas in Astronomy*, 22, 27-37.

Krisciunas, K. (1988). *Astronomical Centers of the World*, Cambridge University Press, Cambridge, England.

Meadows, A. J. (1975). *Greenwich Observatory: Recent History (1836-1975)*, volume 2 of the Greenwich Tercentenary history, Taylor and Francis, London.

McCutcheon, Robert. "Stalin's Purge of Soviet Astronomers," *Sky and Telescope* (October, 1989), 352-57.

Paullin, Charles O. (1923). "Early Movements for a National Observatory, 1802-1842, *Records of the Columbia Historical Society*, 25, 36-56.

Price, Derek J. de Solla (1963). *Little Science, Big Science*, Columbia University Press, New York.

Pulkovo Observatory. (1889). *Zum 50-Jahrigen Bestehn der Nicolai-Hauptsternwarte, Kaiserlichen Akademie der Wissenschaften*, St. Petersburg.

Rhees, W. J. (1879). "Congressional Proceedings, Twenty-Sixth Congress, 1839-41," in *The Smithsonian Institution: Documents Relative to its Origin and History*, Smithsonian Institution, Washington, 200-246, esp. 220-222.

Sadler, D. H. (1976). "Lunar Distances and the Nautical Almanac," in *The Origins, Achievement and Influence of the Royal Observatory, Greenwich: 1675-1975*, *Vistas in Astronomy*, 20, 113-121.

Sokolovskaya, Z. K. (1976). "Struve" in *Dictionary of Scientific Biography*, 13, 108-121.

Struve, F. G. W. (1845), *Description de L'Observatoire Astronomique Central de Poulkova*, L'Academie Imperiale des Sciences, St. Petersburg.

Struve, O. (1865) *Ubersicht der Thatigkeit der Nicolai-Hauptsternwarte wahrend der ersten 25 Jahre ihres Bestehens*, Kaiserlichen Akademie der Wissenschaften, St. Petersburg.

Weinberg, Alvin (1967). *Reflections on Big Science*, The MIT Press, Cambridge, Mass.

Discussion

QUIJANO: The "Real Observatorio de la Armada en San Fernando" (Spain) was founded in 1753 as a naval observatory by King Fernando VI. The "Observatorio de Madrid" in Spain was founded as a national observatory at the end of the 18th century.

S. DICK: It is clear that I have not listed all national observatories in Table 1. This is only a sample, but I think a good sample, to demonstrate some patterns. Further research is needed to see how other national observatories fit into this scheme.

THE PAST, PRESENT AND FUTURE OF REFERENCE SYSTEMS FOR ASTRONOMY AND GEODESY

G.A. WILKINS
c/o Royal Greenwich Observatory
Madingley Road
Cambridge
England CB3 0EZ

ABSTRACT. Standard values of astronomical and geodetic constants were adopted in the late nineteenth century for use in the determination of terrestrial and celestial coordinates, and their values were changed subsequently as better values became known. The great improvements in the precision of observation during the past decade has made it necessary to develop reference systems that also include the models and procedures that are used for the analysis of observations by particular techniques. The accurate detemination of the connections between the corresponding reference frames and the adoption of unified standards are major tasks for the future.

1. Introduction

This paper is based on the informal review that I gave at the end of IAU Symposium 141 on "The inertial coordinate system on the sky". My principal aim was to draw attention to the importance of developing a coherent system of standards that encompasses the specification and use of the terrestrial reference frame as well as of the the celestial frame. I included some material from the introductory paper on the connections between reference frames that I had expected to present at the beginning of the conference. It also appeared to be necessary to discuss terminology since, in particular, the terms *reference frames* and *reference systems* had not been properly distinguished in papers given earlier in the conference. I have discussed these topics in further detail in earlier papers (Wilkins 1989a and 1989b) and many relevant papers will be found in the recent monograph on *Reference Frames* (Kovalevsky *et al.* 1989). The review papers by Mueller (1988) and Groten (1989) contain much additional information and extensive lists of references.

2. The Development of the Current Systems of Constants

2.1 ASTRONOMICAL CONSTANTS

The desirability of the use of standard values for certain parameters used in the computation of the principal ephemerides of the Sun, Moon and planets was recognised at a conference held in Paris in 1896. The recommended values were adopted in the ephemerides published in the *American Ephemeris* and in the (British) *Nautical Almanac* from 1901 onwards. A more comprehensive

J. H. Lieske and V. K. Abalakin (eds.), Inertial Coordinate System on the Sky, 39–46.
© 1990 *IAU. Printed in the Netherlands.*

"system of astronomical constants" was developed by De Sitter (1927), but this was not formally adopted. An international conference held in Paris in 1950 (Danjon 1950) decided to make no changes in the constants then in use, but it did recommend the introduction of a new system of *ephemeris time* to replace mean solar time (universal time, UT) as the argument of the ephemerides. Improved ephemerides using the new timescale were introduced in the principal almanacs in 1960; the development of the system was reviewed in the *Explanatory Supplement to the Astronomical Ephemeris* (NAO 1961).

By this time the inaccuracies of the values and the inconsistencies with the theoretical relations between some of the constants were so significant that the conference held in Paris in 1963 (Kovalevsky 1965) decided to set up a working group to prepare a new system of constants. The report of the group (Fricke *et al.* 1966) was adopted by the IAU in 1964, and the new system was introduced in the almanacs in 1968 (NAO 1967). Improvements in the techniques of observation and computation were so rapid that by 1970 it was clear that new ephemerides based on the relativistic theories would be needed by 1980. The development of a new system of constants and of new ephemerides was, however, slower than had been hoped, and they were not introduced until 1984 (NAO 1983). The new ephemerides used a new standard celestial coordinate frame, but the corresponding fundamental star catalogue FK5 is not yet published in full (Fricke *et al.* 1988). Moreover, it was clear that some of the constants adopted in 1976 (Duncombe *et al.* 1977) were not appropriate for use in 1984 and other values were used in some almanacs (*e.g.*, see NAO 1983, pp. S11–S12).

2.2 GEODETIC CONSTANTS

The constants that specify the size and shape of a spheroid (ellipsoid of revolution) that approximates to the actual surface of the Earth were usually chosen to give a best fit over a limited region of the surface, so that many sets of such constants have been used for mapping (for example, see the list in the *Astronomical Almanac* for 1987 onwards). The inclusion of geodetic constants in the IAU (1964) system of astronomical constants prompted the International Association of Geodesy (IAG) to introduce the IAG (1967) Geodetic Reference System, although the formal specification did not relate the axis of figure to any particular reference frame (IAG 1971). It was later stated that the axis is defined by Conventional International Origin of the coordinates of the pole of rotation. Formally the zero of terrestrial longitude was that adopted in 1884 by the international conference that was held in Washington, D.C., in 1884 for the purpose of fixing a prime meridian and a universal day, but in practice it was determined by the adopted longitudes of the observatories that contributed to the determination of universal time. The system was revised in 1980 (Moritz 1980).

The development of new techniques that were potentially able to determine the variations in the rotation of the Earth with higher precision than the technique of optical astrometry led to the setting up in 1978 of a working group that organised Project MERIT to Monitor Earth Rotation and Intercompare the Techniques of observation and analysis (Wilkins 1980). The group recognised the importance of using the same constants and reference frames for all of the six techniques in the project, and it developed a set of MERIT Standards that included many more astronomical and geodetic quantities than the currently-adopted systems (Melbourne 1983). At that time it was not possible to adopt standard reference frames, but the determination of the differences between the reference frames for the different techniques became a major objective of the project. In particular, the MERIT group worked closely with another group that had been set up in 1981 to develop a new COnventional TErrestrial reference System (COTES).

2.3 THE CURRENT POSITION

Following the success of the MERIT Main Campaign, the IAU and IUGG decided to accept the recommendations of the MERIT/COTES Working Group (Wilkins and Mueller 1986) and to set up a new International Earth Rotation Service (IERS). This is responsible for the establishment and maintenance of terrestrial and celestial reference frames whose relative orientation is monitored continuously. These frames have been specified in the IERS Annual Report for 1988, and new IERS standards to replace the MERIT standards have just been published (McCarthy 1989). The IERS standard celestial reference frame is based on the positions of radio sources determined by the VLBI technique, whereas the IAU standard frame is based on the FK5 catalogue of the positions of stars determined by optical techniques.

3. A Digression on Terminology and Concepts

3.1 FRAMES AND SYSTEMS

In this context it is useful to make distinctions between the following terms:

coordinate frame	coordinate system
reference frame	reference system

For this purpose it is possible to use the Newtonian concepts of space and time; the same distinctions may be made when relativistic concepts must be used. The following definitions appear to be the most appropriate, although it must be recognised that current usage does not always conform to them:

Coordinate frame: a triad of rectangular coordinate axes or other geometrical construction with respect to which a direction or the position of a point may be specified by a set of coordinates.

Coordinate system: a method of specifying the position of a point with respect to a particular coordinate frame. The most common coordinate systems are rectangular coordinates and spherical polar coordinates, but geodetic coordinates defined with respect to a spheroid are also important in geodesy.

Reference frame: a catalogue of the adopted coordinates of a set of reference objects that serves to define, or realize, a particular coordinate frame. It is usually necessary to adopt expressions that give the coordinates as functions of time; the timescale is then part of the frame. A reference frame may be intended for general use or for use with a particular technique; the former type are sometimes said to be *conventional*. In the case of dynamical reference frames the catalogues are replaced by ephemerides of the geodetic satellites, planets or other celestial objects that serve to define the frames.

Reference system: the totality of procedures, models and constants that are required for the use of one or more reference frames. An equivalent statement is that a reference system is the combination of a reference frame (represented by a catalogue of the positions of reference objects) and a set of theories and parameters that can be used to derive the positions of other objects at measured times from observations of particular types.

3.2 IDEAL AND STANDARD FRAMES

It is also useful to make a distinction between an *ideal* coordinate frame that is appropriate for explanatory purposes and for use in theoretical studies and a *standard* reference frame that is appropriate for use for practical purposes. An ideal frame is described by concepts, but it is often not possible to use these concepts directly in practical situations. In contrast, a standard frame is intended to provide an accurate and unambiguous basis for the determination of positions. The positions in the catalogue are usually chosen so that the standard frame corresponds closely to the ideal frame, but they may be chosen to satisfy some other condition.

The ideal terrestrial coordinate frame may be considered to have its origin at the centre of mass of the Earth and be fixed in the Earth. The z-axis and the x,y-plane may be described as the axis of figure and the Greenwich meridian, but since the Earth is not a rigid body such a frame would not be fixed in the Earth. For example, tectonic motions could change the direction of the axis of figure as well as the position of Greenwich with respect to the interior of the Earth. Instead the standard frame used by IERS is based on the adoption of the coordinates of a set of stations where precise geodetic measurements are made regularly; these coordinates and their rates of change are chosen so that there is no apparent net rotation of the crust of the Earth with respect to the frame and so that the z-axis and x,y-plane are continuous with those previously in use. Thus the IERS pole corresponds to the CIO, and hence to the pole of the adopted mean coordinates of the stations of the original International Latitude Service; this pole now differs significantly from the mean position of the pole of rotation and hence from the current position of the axis of figure. The IERS prime meridian is not coincident with the meridian through the transit-circle at Greenwich and it might move further away from it. The reference timescale for the IERS terestrial frame is international atomic time (TAI).

The ideal celestial coordinate frame is usually considered to have its origin at the centre of mass of the Solar System (and so is described as "barycentric") and to be "inertial" or "non-rotating"; for many purposes such a frame may be regarded as being fixed in space. The plane and direction of the centre of the Galaxy may be used to specify such a frame, but in astrometry and geodesy the z-axis is usually taken to be the mean pole of rotation of the Earth at some arbitrary epoch, while the x-axis is in the direction of the mean equinox at this epoch. It is assumed that time may be measured in a "uniform" timescale. The IERS and IAU standard reference frames correspond to such an ideal frame. To the accuracy of measurement the directions of extragalactic radio sources should be constant with respect to such a frame, but the star catalogue defining the optical frame must include proper motions, parallaxes and radial velocities with the coordinates of the stars at the epoch.

The earth-rotation parameters that are determined by IERS include the coordinates of the true pole of rotation of the Earth with respect the terrestrial frame and the difference between UT and TAI. In effect, the parameters refer to an intermediate coordinate frame whose z-axis and x-axis are the true pole and true equinox of date. (The term "true pole" is here to be understood to be the "celestial ephemeris pole".) This intermediate frame has no diurnal rotation in space, but it is subject to the nutation and precession of the axis of rotation of the Earth. Sometimes the mean pole and the mean equinox of date are used in the specification of the intermediate frame since such a frame has no short-period motions in space.

4. The Connections between the Terrestrial and Celestial Frames

4.1 THE CONNECTIONS BETWEEN THE IDEAL COORDINATE FRAMES

The connections, or links, between the ideal celestial and terrestrial coordinate frames may be summarised as follows:

Celestial frame (barycentric, non-rotating)

      Orbital motion of the Earth around the barycentre

      Precession and nutation of the axis of rotation

Intermediate frame (geocentric, aligned to axis of rotation)

      Variations in the rate of rotation of the Earth

      Wobble of the axis of figure around the axis of rotation

Terrestrial frame (geocentric, aligned to axis of figure)

The links between the celestial frame and the intermediate frame can be calculated from numerical models with high accuracy over long periods, although further improvements in the theory of nutation are required. On the other hand, the links between the intermediate frame and the terrestrial frame must be monitored by regular observations; the results are analysed to provide information about, for example, the interactions between the atmosphere and the crust of the Earth and about the interior of the Earth.

4.2 THE CONNECTIONS BETWEEN THE STANDARD REFERENCE FRAMES

The connections between the standard celestial and terrestrial reference frames may be summarised from a different viewpoint as follows:

Standard celestial frame
    Reference-frame parameters for each technique

Technique-dependent celestial frames
    Observed objects   Technique-dependent
    Observing instruments   — earth-rotation
                                       parameters

Technique-dependent terrestrial frames
    Reference-frame parameters for each technique

Standard terrestrial frame

The reference timescale of the celestial frame is barycentric dynamical time (TDB), while the reference timescale of the standard terrestrial frame is TAI.

The analysis of the observations of each technique are used (a) to determine the earth-rotation parameters, (b) to derive improvements to the catalogue postions of both the observed objects and the observing instruments, and (c) to improve the models used in the analysis. The earth-rotation parameters and other results from the different techniques are combined and compared (a) to derive standard values of the earh-rotation parameters, (b) to determine the reference-frame parameters that specify the relative orientations of the standard and technique-dependent frames, and (c) to identify, if possible, any deficiencies in the reference systems for the techniques concerned.

5. Future Tasks on the Reference Systems for Astronomy and Geodesy

5.1 OPERATIONAL TASKS

The stucture and operational procedures of IERS have been developed in such a way as to encourage the steady improvement in the technique-dependent reference systems and in the standard reference frames. Some specific tasks that are in progress include:

(a) the densification of the IERS terestrial reference frame so as to make it more accessible for general use, although at a lower accuracy than for the primary observing sites;

(b) the densification of the IERS celestial reference frame and the improvement of the links between it and the stellar and dynamical reference frames; and

(c) the introduction of new IERS standards that will lead to reductions in the inconsistencies between the reference systems of different techniques.

5.2 THEORETICAL TASKS

The principal outstanding theoretical task appears to be to extend and clarify the application of relativistic concepts to astrometry and geodesy. It is first of all necessary to ensure that the theoretical models and the procedures for the reduction and analysis of the observations are consistent with an adopted relativistic theory. It is also important that the new approach be described and explained in ways that will be intelligible to those who will need to use the new concepts, models and procedures.

Some improvements in the theoretical bases of the links between the reference frames are required; in particular, the theory of the nutation of the axis of rotation of the Earth needs to be refined so as to match the sub-milliarcsecond precision of current VLBI observations. Further careful consideration of the theoretical models and of the corresponding numerical procedures will be required if optical-interferometric techniques can be utilised to make angular measurements to a precision that is of the order of microarcseconds.

5.3 ORGANISATIONAL TASKS

Project MERIT and the establishment of IERS have considerably strengthened international cooperation between the scientists and organisation whose primary interests have been in geodetic applications, such as the determination of the tectonic motions on the the Earth. It appears that the HIPPARCOS astrometric satellite mission is leading to a rejuvenation of international cooperation in astrometry. Special efforts must be made to bring together these two communities to discuss matters of common concern. Even now, many astrometrists appear to unaware of the advances that have been made in the development of a celestial reference frame based on extragalactic radio sources.

It is important that there should be full consultations about the use of the new IERS standards outside the context of the IERS. Agreement should be sought outside the IERS community so that the new standards will be suitable for general adoption by IAU and IUGG; otherwise there are risks that the standards will be ignored or modified by some groups.

Finally, efforts must be made to encourage the wider use of agreed standards. For example, even now confusion is caused and effort is wasted because many astronomers continue to use cgs units instead of SI units. The effects of using different values for astronomical and geodetic parameters are often very difficult to determine and can lead to unnecessary discrepancies between the results from different techniques of observation or analysis. Such discrepancies are minute in comparison with the differences in longitude and time that were common before the Washington conference of 1884, but I would like to think that the Leningrad conference of 1989 will lead to a similar unifying influence in astrometry and geodesy.

References

Danjon, A. (ed.) 1950. Colloque internationale sur les constantes fondamentales de l'astronomie, Observatoire de Paris, 1950. *Bull. Astron.* **15**, 163–292.

De Sitter, W., 1927. "On the most probable values of some astronomical constants", *Bull. Astron. Inst. Netherlands* **4**, 57–61.

Duncombe, R.L., Fricke, W., Seidelmann, P.K., and Wilkins, G.A., 1977. "Joint report of the working groups of IAU Commission 4 on precession, planetary ephemerides, units and time-scales", *Trans. IAU* **16B**, 56–67.

Fricke, W., Brouwer, D., Kovalevsky, J., Mikhailov, A.A., and Wilkins, G.A., 1966. "Report to the [IAU] Executive Committee of the Working Group on the System of Constants", *Trans. IAU* **12B**, 593–8.

Fricke, W., Schwan, H., and Lederle, T., 1988. Fifth fundamental catalogue (FK5): Part 1 The basic fundamental stars. *Veröff. Astron. Rechen-Institut*, Heidelberg, No. 32, 1-106. (Distributed 1989.)

Gondhalekar, P.M., (ed.) 1989. Astrometry into the 21st century. Proceedings of workshop held at Abingdon, Oxon, UK, on 1989 May 22-24. *Rutherford Appleton Laboratory Report* RAL-89-117.

Groten, E. 1989. "Geodesy and astrometry". In Gondhalekar 1989, 20–48.

Kovalevsky, J., (ed.) 1965. "XXIe Symposium de l'U.A.I. sur le systeme de constantes astronomique", *Bull. Astron.* **25**, 1–324.

Kovalevsky, J., Mueller, I.I., and Kolaczek, B., (eds) 1989. *Reference frames in geodesy and geophysics*, Kluwer Academic Publishers, Dordrecht.

McCarthy, D.D., (ed.) 1989. "IERS standards (1989)", *IERS Tech. Note* 3.

Melbourne, W., (ed.) 1983. "Project MERIT standards", *US Naval Observatory Circ.* No 167.

Mueller, I.I., 1988. "Reference coordinate systems: an update", *Dept. of Geodetic Science and Surveying, Ohio State University*, Report No. 394.

IAG 1971. "Geodetic reference system 1967", *Bull. Geodesique Publ. Spec.* No. 3.

Moritz, H. 1980. "Geodetic Reference System 1980", *Bull. Geodesique* **54**, 395–402.

NAO 1961. *Explanatory Supplement to the Astronomical Ephemeris*, H. M. Stationery Office, London.

NAO 1967. "The introduction of the IAU system of astronomical constants into the Astronomical Ephemeris", *Supplement to the A. E. 1968*.

NAO 1983. "The introduction of the improved IAU system of astronomical constants into the Astronomical Almanac", *Supplement to the Astronomical Almanac 1984*.

Wilkins, G.A., (ed.) 1980. *A review of the techniques to be used during Project MERIT to monitor the rotation of the Earth*. Royal Greenwich Observatory, England, & Institut für Angewandte Geodasie, Germany.

Wilkins, G.A., 1989a. "The connections between terrestrial and celestial reference frames", In Gondhalekar 1989, 8–13.

Wilkins, G.A., 1989b. "Standards for terrestrial and celestial reference frames", In Kovalevsky *et al.* 1989, 447–460.

Wilkins, G.A., and Mueller, I.I., 1986. "Joint summary report of the IAU/IUGG working groups on the rotation of the Earth and the terrestrial reference system", *Highlights of Astronomy* **7**, 771–788.

THE IAU WORKING GROUP ON REFERENCE SYSTEMS

James A. Hughes

U.S. Naval Observatory

ABSTRACT: The IAU Working Group on Reference Systems (WG) was founded by a Commission Resolution passed at the XXth General Assembly in Baltimore in 1988. Since most, if not all, participants in this meeting are familiar with that resolution, it will not be discussed here. Background material may be found, for example, in *Highlights of Astronomy*, XX General Assembly (1988, Ed. D. McNally). Particular attention is called to the reports by B. Morando, R. Duncombe and J.A. Hughes (pp. 482-500) which were given as part of the Joint Commission Meeting, *Towards Milliarcsecond Accuracy*, chaired by P.K. Seidelmann. This short presentation will provide a description of the current state of affairs of the WG, which is chaired by the speaker.

Introduction

The WG is charged with the consideration of four essential matters: Nutation of the Earth, Astronomical Constants, Time and Reference Frames/Origins. In order to address these diverse topics, several distinct areas of expertise are needed. Thus a single group would necessarily be quite large and would no doubt soon become unmanageable. For this reason, the Chairman opted to set up four subgroups, each with its own leader, to separately address the four matters listed above. After consultations with many colleagues, including Commission presidents, the Chairman drew up the membership lists which follow. The selection criteria are easy to describe. However, a perfect implementation based upon these criteria is impossible. The members of the subgroups were chosen primarily for demonstrated knowledge and interest regarding a specific area of concern to the WG. Next, consideration was given to achieving the greatest possible representation of all the Commissions involved, i.e., Commissions, 4,7,8,19,20,24,31,33 and 40; nine Commissions in all. Finally, in a completely similar way, every effort was made to achieve the best geographical distribution of the membership possible. The results of these efforts will now be presented. The membership of each subgroup will be displayed, each followed by brief remarks about the work of that particular subgroup.

J. H. Lieske and V. K. Abalakin (eds.), Inertial Coordinate System on the Sky, 47–50.
© 1990 *IAU. Printed in the Netherlands.*

Subgroup on Nutation

Leader: D. McCarthy; Members: V. Dehant, E. Groten, T. Herring, G. Kaplan, H. Kinoshita, M.G. Rochester, R. Vicente, J. Vondrak, J.M. Wahr, Ya.S. Yatskiv.

The work of this group is proceeding well. There appears to be a general consensus that for the immediate future the current IAU nutation be used. Those few users requiring the very highest accuracy should apply the IERS corrections to the IAU values. It is unlikely that any *ad hoc* adjustments to present coefficients will be recommended.

Souchay with Kinoshita as well as Groten have developed new theories for the celestial part of the nutation problem. I believe we will hear more on this topic later in this meeting. As far as a new Earth theory is concerned, we will assuredly hear more on this topic when Dr. Wahr speaks later this week. It may be remarked here that at present it appears that observation is driving geophysical theory in this area rather than the other way round.

Subgroup on Constants

Leader: T. Fukushima; Members: M. Bursa, J. Campbell, J. Chapront, W. Jin, G.A. Krasinsky, R. Reasenberg, P.K. Seidelmann, A.T. Sinclair, E.M. Standish, C. Veillet.

Although the work of all the subgroups is inter-related to some degree, there is an almost symbiotic relationship between Constants and Time which ought to be kept in mind. The current IAU time units are, in the end, proper units based upon International Standard (SI) values. The use of such "localized" units has led to inconsistencies in the definitions and usage of various constants. This has been clearly demonstrated by Fukushima *et al.* (1986). In addition, as has been pointed out by Murray (1983) and Fukushima *et al.* (*op. cit.*), the IAU has no clearly defined unit of length comparable to any time unit. These are the kinds of difficulties which must be addressed by the group. In short, some constants are inappropriately or inconsistently defined and a critical examination of the IAU76 System *vis a vis* General Relativity (GR) is necessary.

These brief comments cannot address all the issues. Examples of other questions are: What items should be included as constants? Which are the defining constants? How are values to be updated? How are the accuracies to be characterized?

Subgroup on Time

Leader: B. Guinot; Members: V.A. Brumberg, H. Deboer, M. Fujimoto, J. Lieske, P. Paquet, I. Shapiro, J. Taylor, G. Winkler, B. Xu, B. Yallop.

As already mentioned, the time scales currently in use are all ultimately based upon proper SI units. Since any realization of an observable time scale must be based upon some kind of a proper unit (event), the task is to define the interrelationships (transformations) between such scales and any functional arguments one might use in GR theory. Once any scale is adopted, provision must be made for the projection of that scale into the past. In addition, the complex process of the transfer of time must be addressed while considering the protocols which have already been established. Of course the "symbiotic" relationship with the Subgroup on Constants is a two-way street and this must be taken into account.

Subgroup on Frames/Origins

Leader: J. Kovalevsky; Members: V.K. Abalakin, S. Aoki, F. Arias, C. Boucher, N. Capitaine, Ch. de Vegt, K. Johnston, C. Ma, I. Mueller, C.A. Murray, H. Schwan, C.A. Smith, R. Wielen.

In my opinion the tasks of this group are perhaps the most difficult of all those facing the WG. Ironically, the idea of assigning positions and motions to celestial objects appears almost conceptually trivial and therefore many of our astronomical colleagues do not appreciate the intricacies involved, especially those difficulties encountered at the milli-arcsecond level of concern. For example, the very motions which are being characterized complicate matters tremendously.

The reconciliation of a milli-arcsecond radio coordinate frame with an optical frame known to be more than one order-of-magnitude less accurate is a central concern of this subgroup. Such a reconciliation may not be a simple matter since the needs of diverse groups must be satisfied. These groups may have completely different views. Indeed, within our own astrometric community, there is not even universal agreement on what constitutes a "reference frame", a "reference system", a "coordinate frame" and a "coordinate system"! Nevertheless, the essential issues are appreciated and the significance of various courses of action are becoming more widely known. The word "Origins" appears explicitly in the title of this subgroup since the choice of an origin, or origins, for the various possible coordinates is a distinct and to a great degree, a separable problem which can be addressed somewhat independently. For example, the origins of the coordinate frames used for celestial coordinates and the metering of the Earth's rotation need not, in principle, be identical, although there are clear advantages to such an identity. In this connection I call attention to the presentation to follow by Andrew Murray. He will undoubtedly illustrate the kinds of issues involved with his talk on the *Non-rotating Origin* (NRO), an origin originally proposed by Guinot (1979). (The concept of the NRO undertakes to remove the partial dependence of the origin of coordinates on the orbital motion of the Earth.)

Summary

I believe that the WG is off to a good start and that the people involved are capable of effectively crafting sensible proposals for the IAU based upon a thoughtful consensus. This is not to say that there are not many other individuals who could also make a good contribution. However, in order to have a WG of reasonable size, it was impossible to include all those who I would have otherwise chosen to participate. In any event, I do solicit the viewpoints of all those concerned with the work and its outcome. These may be addressed to me or to the leaders of the subgroups.

Finally, let me once again remind you of IAU Colloquium No. 127, *Reference Systems*, to be held at Virginia Beach, Va., 14-20 October 1990. The WG will formally meet at this Colloquium and those interested should plan on attending.

REFERENCES

Fukushima, T.,Fujimoto, M., Kinoshita H., Aoki, S. (1986), *Celestial Mechanics* **36**, pp. 215-30.

Murray, C.A. (1979), *Vectorial Astrometry*, Adam Hilger Ltd., Bristol.

Guinot, B. (1979), "Basic problems in the kinematics of the rotation of the Earth," in *Time and the Earth's Rotation*, IAU Symp. #82, D.D. McCarthy and J.D. Pilkington eds., D. Reidel Pub. Co., Dordrecht, pp. 7-18.

THE ORIGIN OF CELESTIAL COORDINATES

C. A. MURRAY
12 Derwent Road
Eastbourne
Sussex BN20 7PH
England, UK

ABSTRACT. In 1978, Guinot proposed that, for studies of Earth rotation, the zero point of the apparent "right ascension" coordinate on the true equator should be so chosen that the rate of change of its hour angle is exactly proportional to the inertial rate of rotation of the Earth. It has been subsequently suggested that this concept of the "non-rotating origin" supersede the equinox quite generally as the origin of celestial coordinates. Since this proposal was first put forward, there has been much discussion, and some criticism, from Aoki and his colleagues, both published and in private correspondence. Some of the arguments for and against Guinot's proposal are discussed, as a contribution to the wider debate on reference systems now being carried out under the auspices of the IAU.

1. Introduction

The celestial coordinate system which is currently used for recording observations and in published ephemerides is defined by the instantaneous directions to the celestial pole and the equinox. This coordinate system is, in principle at least, directly accessible to observation with minimum appeal to theory. The direction to the pole is obtained from successive meridian transits of stars at upper and lower culmination, and the direction to the equinox, relative to *a priori* assumed right ascensions of clock stars, is derived from observations of the Sun or planets relative to these stars.

In practice however, it is not quite as simple as this. Since this instantaneous coordinate system is continually in motion relative to inertial space, it is necessary to have a theory of this motion so that observations even over short periods of time can be referred to a common coordinate system. This requires theories of the Earth's orientation in space and of motions of the Sun and planets.

As the accuracy of observations has increased so has it become necessary to sharpen the geometrical definitions of the system. For example, Atkinson (1973) pointed out that even such a conceptually simple procedure as determining the direction to the instantaneous axis of rotation by observations at upper and lower culmination of the same stars is complicated by short period displacement of that axis within the Earth, and that this can be avoided if instead of the rotation axis one uses a pole which is based on the axis of figure. The pole is implicitly defined by the adopted nutation series; that which has been in use since 1984 (Seidelmann 1982) is based on Atkinson's proposal and is known as the Celestial Ephemeris Pole (CEP).

Another example is the origin of the instantaneous right ascension coordinate. Apparent sidereal time is obtained by comparing observed times of meridian transit with tabulated apparent right

51

J. H. Lieske and V. K. Abalakin (eds.), Inertial Coordinate System on the Sky, 51–59.
© 1990 *IAU. Printed in the Netherlands.*

ascensions measured from the equinox, which has a secular and oscillatory motion on the true equator due to precession and nutation in right ascension. This was of no practical consequence as long as clocks were not sufficiently precise to detect the shorter period terms. Atkinson and Sadler (1951) pointed out, however, that all the periodic terms would be avoided if, instead of apparent right ascension, the times of meridian transit were measured in terms of *mean* sidereal time; this is equivalent to adopting as the origin of "right ascension" a point on the equator which is displaced from the true equinox by exactly the amount of the nutation in right ascension.

Even following the proposal of Atkinson and Sadler, mean sidereal time, though sensibly uniform, measures the angular spin of the Earth relative to a moving origin on the equator. In 1978 Guinot (1979) proposed a logical extension of the definition of mean sidereal time to exclude from the origin not only the nutation but also the precession in right ascension. He thus defined the so-called "non-rotating origin" (NRO); the modified sidereal time, or "stellar angle", measured from this origin is a true measure of the Earth's spin.

Since Guinot first proposed the use of the NRO, there has been much discussion and some critcism of it, notably by Aoki and Kinoshita (1982). Although the latest published exposition of the proposal, by Capitaine, Guinot and Souchay (1986), appeared some years ago, the unpublished correspondence has continued at least until the time of the Baltimore IAU meeting in 1988. In this paper I shall, for brevity, refer to these two published papers as AK and CGS respectively, but will not give specific references to the unpublished correspondence.

The whole subject of astronomical reference systems is now under active examination by working groups of the IAU; it therefore seems appropriate that this particular topic should be widely discussed among the astrometric community before any final recommendation is made. The aim of the present paper, therefore, is to try to summarize the main points of the controversy as they now stand, rather than to make a judgement on the proposal. In view of the very extensive correspondence, much of it containing arguments of great mathematical complexity, I hope I may be forgiven if I inadvertently omit points whose significance I may not have fully appreciated.

2. Theory of the Non-Rotating Origin

In accordance with the notation used elsewhere (see Murray 1983, Appendix A), the prime (´) employed here denotes matrix transposition, so that if a and b are vectors then $a´ b$ is the same as the dot product $a \cdot b$. Since the scalar product of two vectors is essentially a transpose matrix multiplication, the prime replaces the more usual "dot" notation. Similarly, the scalar triple product $a \cdot (b \times c) = (a \times b) \cdot c$ is written here as $a \times b´ c$.

Let $\Sigma = [\sigma \ n \times \sigma \ n]$ be an orthogonal triad of unit vectors, where n is the direction to the celestial pole, and let Ω_Σ be the instantaneous angular velocity of Σ. If $\dot{n}$ denotes the inertial rate of change of the direction n, we can write

$$\dot{n} = \Omega_\Sigma \times n .$$ (1)

If $\dfrac{\partial}{\partial \tau}$ denotes rate of change relative to the triad Σ we have, quite generally, for any vector g,

$$\dot{g} = \Omega_\Sigma \times g + \frac{\partial g}{\partial \tau} .$$ (2)

In particular, if g is perpendicular to n, *i.e.* it lies in the equator, the component of $\dot{g}$ in the equatorial plane is

$$\mathbf{n} \times \mathbf{g}' \dot{\mathbf{g}} = \mathbf{n}' \Omega_\Sigma + \mathbf{n} \times \mathbf{g}' \frac{\partial \mathbf{g}}{\partial \tau} . \tag{3}$$

But since $\dot{\mathbf{n}}$ is perpendicular to $\mathbf{n}$, the component of Ω_Σ parallel to $\mathbf{n}$ can be chosen arbitrarily. Guinot (1979) proposed that this should be zero, or

$$\mathbf{n}' \Omega_\Sigma = 0 ; \tag{4}$$

in this case, we see from (3), that the angular velocity component about $\mathbf{n}$ relative to the moving frame Σ, of an equatorial vector, is equal to the same component of its inertial angular velocity. Multiplying (1) vectorially by $\mathbf{n}$, and using (4), we see that

$$\Omega_\Sigma = \mathbf{n} \times \dot{\mathbf{n}} . \tag{5}$$

If $\mathbf{g}$ is now identified with a direction which is fixed in the Earth, such as for example the equatorial point on the prime meridian, the angular velocity of $\mathbf{g}$ relative to the triad Σ is the same as the component about $\mathbf{n}$ of the inertial angular velocity of the Earth. The first axis, σ, of the triad Σ is directed toward the NRO.

CGS show that the direction to σ can be derived from a knowledge of the instantaneous direction to the pole $\mathbf{n}$ and its trajectory. Let Σ_0 be a fixed equatorial triad, at some initial epoch, and $\mathbf{n}_0$ be the direction to the corresponding pole. The direction cosines of $\mathbf{n}$ relative to Σ_0 can be expressed in the form

$$\mathbf{n}' \Sigma_0 = [\sin d \cos E \quad \sin d \sin E \quad \cos d]. \tag{6}$$

The direction to the node of the instantaneous equator on the equator of $\mathbf{n}_0$ is given by

$$\mathbf{m}_1 = | \mathbf{n}_0 \times \mathbf{n} |^{-1} \mathbf{n}_0 \times \mathbf{n} , \tag{7}$$

and its longitude along the equator, measured from the first axis of Σ_0, is $90° + E$. From (7) we have

$$\dot{\mathbf{m}}_1 = | \mathbf{n}_0 \times \mathbf{n} |^{-1} \{ \mathbf{m}_1 \times (\mathbf{n}_0 \times \dot{\mathbf{n}}) \} \times \mathbf{m}_1 = | \mathbf{n}_0 \times \mathbf{n} |^{-1} \mathbf{m}'_1 \dot{\mathbf{n}} \mathbf{n}_0 \times \mathbf{m}_1, \tag{8}$$

and the inertial rate of change of $\mathbf{m}_1$ along the fixed equator of $\mathbf{n}_0$, is then given by

$$\dot{E} = \mathbf{n}_0 \times \mathbf{m}'_1 \dot{\mathbf{m}}_1 = | \mathbf{n}_0 \times \mathbf{n} |^{-1} \mathbf{m}'_1 \dot{\mathbf{n}}. \tag{9}$$

The rate of change of $\mathbf{m}_1$ relative to the triad Σ is given by

$$\frac{\partial}{\partial \tau} \mathbf{m}_1 = \dot{\mathbf{m}}_1 - \Omega_\Sigma \times \mathbf{m}_1, \tag{10}$$

and hence the angular rate of change of $\mathbf{m}_1$ along the instantaneous equator, relative to the NRO, can be written

$$\dot{E} + \dot{s} = \mathbf{n} \times \mathbf{m}'_1 \frac{\partial}{\partial \tau} \mathbf{m}_1 \tag{11}$$

where

$$\dot{s} = -|1 + n' n_0|^{-1} n_0 \times n' \dot{n}.$$

(12)

Thus, at any time τ, the longitude of m_1 measured along the instantaneous equator from the NRO is given by

$$90° + E + s = 90° + E + \int_0^\tau \dot{s} \, d\tau.$$

(13)

The location of the NRO can therefore be determined in principle only from a knowledge of the direction of the pole n and its past history to time τ.

The hour angle, θ, of the NRO is the stellar angle and its time derivative is the inertial angular velocity component of the Earth about n. It can be expressed in the form

$$\theta = A + \lambda$$

(14)

where A is the instantaneous ascension of a star on the meridian through longitude λ.

3. Relationship between the Non-Rotating Origin and the Equinox

Let $N = [1 \ m \ n]$ be the instantaneous equatorial triad in which l is directed toward the true equinox and $m = n \times l$. If Ω_N denotes the instantaneous inertial angular velocity of N, due to precession and nutation, we have

$$\dot{n} = \Omega_N \times n,$$

(15)

hence, from (5),

$$\Omega_\Sigma = \Omega_N - n \, n' \, \Omega_N.$$

(16)

The inertial rate of change of the direction of the true equinox is

$$\dot{l} = \Omega_N \times l.$$

(17)

and its component in the equator is

$$m' \dot{l} = n' \Omega_N,$$

(18)

which represents the instantaneous rate of change of the equinox direction along the equator due to the combined effects of precession and nutation in right ascension.

But since l is an equatorial vector, its angular velocity relative to the moving frame Σ is equal to its inertial angular velocity component about n. Hence we also have

$$m' \frac{\partial l}{\partial \tau} = n' \Omega_N.$$

(19)

The right ascension of σ is then

$$\alpha(\sigma) = - \int_0^\tau \mathbf{n}' \Omega_N d\tau \tag{20}$$

and is the total displacement of the equinox due to the accumulated precession in right ascension from epoch $\tau = 0$, and the equation of the equinoxes.

The Greenwich Sidereal Time is

$$\varphi = \alpha + \lambda \tag{21}$$

where α is the right ascension of a star on the meridian through longitude λ and is related to the stellar angle by

$$\theta = \varphi - \alpha(\sigma). \tag{22}$$

Therefore, from (14), instantaneous ascension is related to right ascension by

$$A = \alpha - \alpha(\sigma). \tag{23}$$

4. Evaluation of $\alpha(\sigma)$

We can express the angular velocity vector Ω_N in the form

$$\Omega_N = \dot\chi \, \mathbf{n}_m - \dot\psi \, \mathbf{k} - \Delta\dot\psi \, \mathbf{k} + \dot\varepsilon_m \, \mathbf{l}_m - \dot\varepsilon \, \mathbf{l} \tag{24}$$

where $\dot\chi$ is the rate of planetary precession on the mean equator, $\dot\psi$ is the rate of precession in longitude on the ecliptic of date, (luni-solar *minus* geodesic precession), $\Delta\dot\psi$ denotes the rate of change of nutation in longitude, $\dot\varepsilon_m$, $\dot\varepsilon$ are the rates of change of mean and true obliquity respectively, and the unit vectors $\mathbf{k}$, $\mathbf{l}_m$ and $\mathbf{n}_m$ are directed toward the pole of the ecliptic, the mean equinox and celestial pole of date.

If $K_m = [\, \mathbf{l}_m \quad \mathbf{j}_m \quad \mathbf{k} \,]$ is the ecliptic triad of date we have

$$\mathbf{n}_m = \sin\varepsilon_m \, \mathbf{j}_m + \cos\varepsilon_m \, \mathbf{k} \tag{25}$$

and

$$\mathbf{n} = \sin\varepsilon \sin\Delta\psi \, \mathbf{l}_m + \sin\varepsilon \cos\Delta\psi \, \mathbf{j}_m + \cos\varepsilon \, \mathbf{k}, \tag{26}$$

and hence

$$\mathbf{n}' \Omega_N = - (\dot\psi + \Delta\dot\psi) \cos\varepsilon + \dot\chi \, (\sin\varepsilon_m \sin\varepsilon \, \cos\Delta\psi + \cos\varepsilon_m \, \cos\varepsilon) + \dot\varepsilon_m \sin\varepsilon \sin\Delta\psi. \tag{27}$$

This equation is quite rigorous. It is easy to show that it is equivalent to the equation (A2-32) of AK, when expanded to the second order in $\Delta\psi$ and $\Delta\varepsilon$, where

$$\Delta\epsilon = \epsilon - \epsilon_m. \tag{28}$$

Expanding (27) to the first order in $\Delta\psi$, $\Delta\epsilon$, and neglecting $\dot{\chi}$ and $\dot{\epsilon}_m$ which arise from the motion of the ecliptic, we obtain the approximation

$$\mathbf{n}'\,\Omega_N \approx -\dot{\psi}\,\cos\epsilon_m - \frac{d}{d\tau}(\Delta\psi\,\cos\epsilon_m) + (\dot{\psi} + \Delta\dot{\psi})\,\Delta\epsilon\,\sin\epsilon_m. \tag{29}$$

AK express the right ascension of σ in the form

$$\alpha(\sigma) = q + (\Delta q)_s + (\Delta q)_p \tag{30}$$

where, to the approximation in (29),

$$q = \int_0^\tau \cos\epsilon_m d\psi \quad, \tag{31}$$

$$(\Delta q)_p = \Delta\psi \cos\epsilon_m - \left[\int_0^\tau (\dot\psi + \Delta\dot\psi)\Delta\epsilon\,\sin\epsilon_m d\tau\right]_p \tag{32}$$

and

$$(\Delta q)_s = -\left[\int_0^\tau \Delta\dot\psi\,\Delta\epsilon\sin\epsilon_m d\tau\right]_s \quad ; \tag{33}$$

the subscripts p and s denoting periodic and secular terms respectively.

Equation (31) represents accumulated precession in right ascension. (It should be noted that in our notation ψ is the luni-solar precession on the *ecliptic of date*.)

Equation (32) is a modified equation of the equinoxes; the first term represents the first order expression, although the *Astronomical Ephemeris* gives $\Delta\psi$ cos ϵ.

AK make the point that the second order terms in (32) and (33) introduce both secular and periodic terms which have not hitherto been included in the calculation of the precession in RA and the equation of the equinoxes, although Woolard (1953) gave approximate values. Retaining only the largest terms it is easily shown that

$$(\Delta q)_p = \Delta\psi \cos\epsilon_m + 0\rlap{.}''00264 \sin\Omega + 0\rlap{.}''000063 \sin 2\Omega \tag{34}$$

where Ω is the longitude of the Moon's node, and

$$(\Delta q)_s = -0\rlap{.}''00385\,\tau \tag{35}$$

where τ is measured in centuries. AK gave $-0\rlap{.}''00388$ for this coefficient, but Aoki later revised it to $-0\rlap{.}''00386$; I find the value given here, which is derived from six periodic terms in the nutation series (Seidelmann 1982), and agrees with that given by CGS.

5. Discussion

The coordinate triad Σ has apparent theoretical advantages over the traditional celestial coordinate system, **N**, in that it depends formally on the direction of a single vector **n** and its instantaneous rate of change **ṅ**. It does not explicitly involve either the ecliptic or the equinox. We have seen in Section 2, above, how the direction to σ can be derived only from a knowledge of the direction of the pole relative to a fixed coordinate system based on the pole n_o at at some initial epoch. The direction to the pole at any instant relative to the fixed system must still be obtained from the theory of precession and nutation. Among the objections which have been raised against the proposal for the NRO by AK is that the motion of the pole is governed entirely by the action of the Moon and Sun on the Earth, and hence depends implicitly on the ecliptic as well as the equator; and therefore the supposed independence of σ from the location of the ecliptic is more apparent than real. In any case, the direction cosines of the pole **n** relative to the fixed frame Σ_o must be calculated from the usual precession and nutation quantities.

A practical difficulty with the current system is that the position of the ecliptic, and hence that of the equinox, is not well determined relative to the stars. However Aoki has pointed out that, in the foreseeable future, observations of pulsar positions by timing, combined with conventional interferometric positions, are likely to improve considerably the location of the ecliptic relative to the equatorial coordinate system.

We now discuss some of the specific points which have been raised in the controversy.

5.1 DEFINITION OF THE CELESTIAL POLE

One of the criticisms raised by AK, and repeated in subsequent unpublished correspondence, concerns the choice of pole for the reference system Σ. In the discussion following his original paper, Guinot said "I believe now that Atkinson's pole is the best"; in other words he supported the use of the CEP rather than the pole of rotation as the pole of the triad Σ. However CGS state "The pole of instantaneous rotation is the one which logically agrees with the following theoretical developments" and later "In principle, we should use here the coordinates of the pole of instantaneous rotation". It is clear from the development above in Section 2 that the choice of pole is irrelevant; at no point in the argument do we make any appeal to the angular velocity vector of the Earth. AK make the point, originally made by Atkinson that, by using the CEP, the "dynamical variation of latitude" is eliminated and furthermore that the spin about the CEP is theoretically constant, as I have also stressed elsewhere (Murray 1983), whereas the resultant total angular velocity of the Earth is not; it is thus exactly correct to use the CEP and not the rotation pole. This has now been accepted by Guinot and Capitaine.

5.2 MOTION OF THE NRO IN SPACE

Aoki makes the point that the name "non-rotating origin" gives a false impression because the NRO has a secular and oscillatory motion in space; he prefers the name "departure point", which is in accordance with the usual meaning of that term in celestial mechanics. It is easy to show that, while the right ascension of a fixed object increases by 2π during a complete precession period, the instantaneous ascension only increases by $2\pi \cos \varepsilon$. But there is also a more subtle point concerning the small secular term $(\Delta q)_s$ which comes from second order nutation. By considering the average over all right ascensions of fixed objects close to the equator, Aoki shows that the instantaneous ascension increases on the average by

$$\frac{1}{2}\int_0^\tau \left(\Delta\varepsilon\,\Delta\dot\psi - \Delta\psi\,\Delta\dot\varepsilon \right) \sin \varepsilon_n d\tau$$

which is exactly equal to $-(\Delta q)_s$; therefore this cancels in right ascension but not in instantaneous ascension.

This demonstrates that the reference frame based on the concept of the NRO is purely local, and has a net rotation relative to inertial space. This is accepted by Guinot and Capitaine, but it does not invalidate the principle that the stellar angle measured from the NRO is proportional to the orientation of the Earth about the CEP.

5.3 COMPUTATIONAL PROBLEMS

Among several computational difficulties, Aoki draws attention to the fact that mixed secular terms, namely periodic terms with secularly increasing coefficients, appear in s, and also in the nutation relative to the NRO, and that these are undesirable. Also, the node of the instantaneous equator on the fixed equator, m_1, which is used as an intermediate reference point in deriving σ, is numerically unstable because of the small inclination, d, between the two equators. Guinot and Capitaine however reject these points as giving no real computational problems.

5.4 CHANGE OF PRECESSION

One of the ideas behind Guinot's proposal is that the instantaneous ascension of an equatorial object will be unaffected by a change in the value of the precession, but Aoki points out that, away from the equator, an error in precession will introduce spurious proper motions in both coordinates.

6. The Space Reference System

Guinot's original paper contained recommendations for the definition and realization of the space reference system. The essential points may be summarized as follows:

The basic system should be defined by a catalogue of positions (and proper motions) of suitable objects; ideally these should be extra-galactic objects such as quasars whose proper motions can be assumed to be negligible.

The directions of the poles of the equator and ecliptic should be given explicitly as functions of time in this system. If subsequent observations show that the assumed coordinates of either pole relative to the basic system require correction, then the system should not be changed but corrections applied to the numerical expressions for the pole directions.

The NRO was apparently only considered by Guinot in connection with the instantaneous equatorial reference system and not specifically as a substitute for the equinox in any fixed epoch equatorial system. This is however envisaged by CGS who state that it "could *in a general manner* replace advantageously the equinox especially for observations which are not sensitive to the orientation of the ecliptic" (my italics).

In considering the possible implications for astronomy of redefining the right ascension coordinate, one should bear in mind what has already happened in the analogous case of the mean sidereal time proposal of Atkinson and Sadler. Nearly forty years later, published ephemerides still give apparent right ascensions referred to the true equinox! However, ephemeris computations for special purposes such as time services use mean sidereal time. It is after all merely a matter of computational convenience.

Guinot's suggestion that the directions to the poles be given explicitly in terms of the time is an excellent one which I also have advocated (Murray 1983). Nutation relative to the moving mean equatorial triad is already given in the form of harmonic series, but precession is concealed in elaborate trigonometric formulae and associated angular variables which disguise very effectively what is after all a simple physical phenomenon. This has been admirably demonstrated by Fabri (1980) who obtained numerical expressions for the direction cosines of the poles of the mean equator and of the ecliptic, by direct integration of the equations of motion of the poles in vector form. An example of the inadequacy of the traditional trigonometric representation is that the rate of luni-solar precession in longitude is given along the *fixed* ecliptic whereas the precessional motion of the mean celestial pole n is physically an instantaneous rotation about the pole of the ecliptic of date, k. This is not to imply that computations based on the traditional formulation are incorrect, but rather that they conceal unnecessarily the true physical situation.

In discussing revision of the astronomical reference system it is important to distinguish between fundamental changes, such as the proposal to define it by positions of extra-galactic objects rather than the kinematics of the Earth, and changes of procedure which are for computational convenience in particular applications, such as the form of presentation of the coordinate transformation for precession and nutation, or indeed the adoption of the NRO. Nevertheless, anything which makes fundamental astrometry more transparent to non-specialists is to be welcomed.

7. References

Aoki, S., and Kinoshita, H.: 1982, *Celes. Mech.*, **29**, 335.

Atkinson, R.d'E.: 1973, *Astron. J.*, **78**, 147.

Atkinson, R.d'E., and Sadler, D H.: 1951, *Mon. Not. Roy. Astron.Soc.* **111**, 619.

Capitaine, N., Guinot, B., and Souchay, J.: 1986, *Celes. Mech.*, **39**, 283.

Fabri, E.: 1980, *Astron. Astrophys.*, **82**, 123.

Guinot, B.: 1979, *Time and the Earth's Rotation*, 7, D.D. McCarthy, J.D. Pilkington (eds.) D Reidel Publishing Co., Dordrecht.

Murray, C.A.: 1983, *Vectorial Astrometry*, Adam Hilger, Bristol.

Seidelmann, P.K.: 1982, *Celes Mech.* **27**, 79.

Woolard, E.W.: 1953, *Astron. Pap. Amer. Ephem.* **15**, Part 1, 165.

Discussion

KOPEJKIN: How do you define the notion of inertial space? I know that general relativity gives some ambiguities in this notion.

MURRAY: I realize that in a relativistic context inertial space is difficult to define, but in this paper I have only been concerned with a purely classical problem.

STANDISH: The locations and motions of the equatorial plane and the ecliptic are now determined strictly from ranging observations of the Moon coupled with the lunar dynamics. The orientation of these planes in the ephemerides no longer depends upon the optical observations.

MURRAY: The problem is how to relate optical and radio observations of objects other than the Moon or planets to these well-defined planes.

Part 2
Pulkovo today

Plate IV. The original Pulkovo Observatory viewed from the north. Photograph provided by Pulkovo Observatory.

Plate V. The reconstructed Pulkovo Observatory viewed from the north. Photograph provided by Pulkovo Observatory.

THE PULKOVO PROGRAMME FOR THE STUDY OF VISUAL DOUBLE STARS

A.A. Kisselev, O.P. Bykov, O.A. Kalinichenko, O.V. Kiyaeva, L.G. Romanenko and N.A.Shakht
Central Astronomical Observatory, USSR Academy of Sciences
Pulkovo
196140 Leningrad
USSR

The observation of visual double stars at Pulkovo is a traditional work of Pulkovo astronomers started by W. Struve in 1830 at Dorpat (now Tartu). After the Second World War and restoration of the observatory in 1960, the observations of double stars have been carried out with help of 26-inch Zeiss refractor using the photographic technique. We observe the visual binaries which satisfy the conditions of highest accuracy of astrometric reduction: $d > 3"$, $m < 12.0$, $dm < 1.0$, $\delta > 20°$.

Annually we obtain about 400-500 plates with photographs of 100-200 binaries. Each plate contains 10-20 images of the binary. Each binary, being observed during 10-20 years, is provided by close set of accurate relative positions of the components. Then we compute the highly accurate apparent motion parameters (AMP), which form the basic data for further investigations.

The mean square errors of AMP values, as obtained by statistical adjustment are as follows: $\pm(0".003$ to $0".010)$ for the relative positions of components and $\pm(0".0003$ to $0".0020)$ for the relative annual motion of components.

The scientific goal of this programme is to find out kinematical and dynamical characteristics of visual double and multiple stars directly from observations. In the course of these investigations we determine the trigonometric and dynamic parallaxes, the orbit elements and the masses of some binaries within 50 parsecs from the Sun. Also we inspect a great number of known but forgotten (i.e., not observed for a long time) binaries, and among them we discover some new nearby stars and then we analyse the perturbations caused by dark companion of the multiple systems.

The Pulkovo AMP-method [1] serves as the theoretical foundation of the programme. This method provide the determination of orbit elements of a visual double star with long period on the basis of short arc observations. The AMP-method is effective, if the positional observation are aided by spectral ones in order to determine radial velocities of components. An appropriate programme of observations is carried out at the Special Astrophysical Observatory (Caucasus) with help of the 6-meter telescope.

63

J. H. Lieske and V. K. Abalakin (eds.), Inertial Coordinate System on the Sky, 63–64.
© 1990 *IAU. Printed in the Netherlands.*

The main results of the fulfillment of the Pulkovo double stars programme are presented by the catalogue of relative positions and relative motions of 200 visual double stars [2]. The catalogue contains also the new estimations of dynamical parallaxes for 50 binaries. Among them there are 11 binaries within 25 parsecs of the Sun, which are not included in Woolley's catalogue. The data of the Pulkovo catalogue were used for computing new orbits (by the AMP-method) for 9 visual double stars [1,3], three of them being determined for the first time.

The preliminary elements of these binaries are presented in Table 1.

Table 1.

ADS	Obser. time 1900+ ["]	a ["]	P [y]	e	i [°]	Ω [°]	ω [°]	T [y]	π ["]	M	dV [km/s]
2427	71–87	8.1	1250	0.42	62	358	21	2100	0.070	1.0	+3.3
8002	70–88	4.2	400	.64	42	351	283	1770	.077	1.0	−0.5
12169	61–83	12.9	3800	.44	124	65	323	2400	.042	2.0	−0.6

The data of the last columns represent the adopted values of the trigonometric parallaxes (π), the sum of the masses of components M (in the units of solar mass) and the relative radial velocities dV, which correspond with the orbit elements.

Also, the perturbations in apparent motion of the stars ADS 11632 and 5983 (δ Gem) were investigated. The detected waves made it possible to suppose the existence of a dark companion with mass 0.006 that of the Sun, belonging to the first binary mentioned, and the existence of a stellar-like component with mass about 0.2 that of the Sun, belonging to δ Gem [4].

Finally, the authors call attention of astronomers to the real problem of determination high accuracy radial velocities of components of double stars in the solar neighbourhood.

References

1. Kisselev, A.A., Kiyaeva, O.V.: 1980, *Astr. J. USSR*, **57**, 6, 1227–1241.
2. Kisselev, A.A., Kalinichenko, O.A., *et al.*: 1988, *Catalogue of relative positions and relative motion of 200 visual double stars*, Leningrad,"Nauka",1–40.
3. Kisselev, A.A., Kiyaeva, O.V.: 1988, *Astrophys. and Space Science*, **142**,181–183.
4. Shakht, N.A.: 1988, *Izv. GAO* (Pulkovo), #205, pp.5–14.

ON THE USE OF STATISTICAL PARALLAXES FOR THE PROPER MOTION REDUCTIONS

N.M. BRONNIKOVA, AND N.A. SHAKHT
Central Astronomical Observatory, USSR Academy of Sciences
Pulkovo
196140 Leningrad
USSR

The Pulkovo program of determination of proper motions with respect to galaxies approaches completion. It consists of 157 fields from +90° to –4° of declination. The proper motions and reductions with use of galaxies were derived for all areas. The errors of reductions are equal to ± 0.″006. In addition the statistical reductions R_x, R_y were computed for the control and for the zones of avoidance. The following formulae were used:

$$R_x = (\hbar/\rho)P + Q \,; \quad R_y = (\hbar/\rho)P' + Q' , \qquad (1)$$

where ($\hbar/\rho$) is the secular parallax according to Binnendijk [1]; P and P' are parallactic factors; Q and Q' are members dependent on the galactic rotation.

The nearby stars with large proper motions are able to give some systematic errors in reductions. The stars with proper motions greater than 0.05 arcsec per year were excluded in the Pulkovo catalogue treatment for the more uniform reference system.

At present the precise photographic magnitudes for all stars in 76 regions of the Pulkovo program were obtained. So it was possible to compute the values of secular parallaxes for reference stars with magnitudes from 12^m5 to 15^m5. The parallaxes were computed with the formulae (1), where in the left parts the reductions obtained by use of galaxies were taken. The parameters for galactic rotation Q, Q were adopted with standard values of galactic constants. The values of the parallaxes with their mean errors are given in Table 1.

Table 1.
The preliminary Pulkovo secular parallaxes (in 0.″0001).

m	12^m5	13^m5	14^m5	15^m5		12^m5	13^m5	14^m5	15^m5
$\hbar/\rho$	89	80	76	50		72	94	74	80
$|b| = 28°$	±26°	±26°	±19°	±20°	$|b| = 58°$	±25°	±20°	±20°	±16°
N	13	10	23	22		12	24	22	26

J. H. Lieske and V. K. Abalakin (eds.), Inertial Coordinate System on the Sky, 65–66.
© 1990 IAU. Printed in the Netherlands.

Here m, δ and N are the photographic magnitudes, average galactic latitude and numbers of areas. The total number of stars is about 9000.

The stars with large proper motions have been not excluded from the data on which the parallaxes by Binnendijk were based. But there are the values of secular parallaxes by Deits [2] and Klemola and Vasilevskis [3] which were obtained on the basis of proper motions with similar cut-off. The comparison of Pulkovo parallaxes with [1], [2] and [3] indicate that the influence of large proper motions is more considerable in bright stars and in high latitudes.

We have deduced some analytic formulae for paralaxes of [2] and [3] in the form:

$$\log p(m) = a + a_1 m + a_2 m^2.$$ (2)

where $p(m)$ is the mean parallax dependent on magnitude and calculated with velocity of Sun equal to 20 km/s. On the basis of formulae (2) some models of statistical reductions were made. Some differences between our galactic reductions and statistical reductions were considered.

In general these differences have reflected the distinction of parallaxes, but their behavior is complicated and the direct comparison of statistical reductions is difficult. A detailed discussion is planned in the future on the basis of all material. Nevertheless it seemed that the calculation of statistical reductions by means of the parallaxes from [2] and [3] based on the corresponding cut-off is more acceptable for our data.

References

[1] Binnendijk, L. (1943) "Mean parallaxes of faint stars, derived from a combination of the Pulkovo and Radcliffe Catalogue of proper motions", *Bull. of the Astr. Inst. of the Netherlands*, **10**, #362 , 9–18.
[2] Deits, A.N. (1947) "Secular parallaxes of faint stars as deduced from Pulkovo Catalogue data on proper motions in Kapteyn Areas", *Izv. GAO Pulkovo* **17**, #138, 2–59. (in Russian).
[3] Klemola, A.R., Vasilevskis, S. (1971) "A study of solar motion and galactic rotation", *Publ. Lick Obs.* **22**, Part 3, 1–13.

ABSOLUTE DETERMINATIONS OF STAR DECLINATIONS FROM POLE TO POLE

V.A. NAUMOV
Central Astronomical Observatory, USSR Academy of Sciences
Pulkovo
196140 Leningrad
USSR

The most widely spread method of determination of star declinations is Bessel's method. Alhough it gives good agreement between star declinations obtained from observations of stars in upper and lower culminations, it leads to great systematic errors if the system of the instrument has an error of the type

$$\Delta z = a \sin z \quad \text{or} \quad \Delta z = a \sin 2z,$$

such as flexure or errors of divided circle. Besides, in the determination of correction of latitude ($\Delta\varphi$) and refraction ($\Delta\mu$) considerable correlation between them exists.

It should be noted that the coefficients of correlation and the errors of unknowns increase while latitudes decrease. In particular, when the place of observation is changed from Pulkovo to Kislovodsk the error of $\Delta\varphi$ increases approximately two times, and the error of $\Delta\mu$ 1.5 times. When the place of observation changes from Pulkovo to Serro-Calan, the changes in $\Delta\varphi$ and $\Delta\mu$ are 5 and 3 times respectively.

To find a way out of this situation several modern methods of determination of absolute declinations of stars were proposed: from observations on equator or from observations in both hemispheres by vertical circles and zenith-telescopes.

The compound programme is carried out at Pulkovo Observatory: for southern hemisphere the observations by the photographic vertical circle at Serro-Calan in Chile are used. In the northern hemisphere the observations by the zenith-telescops on the island Spitzbergen, Pulkovo, Blagoveschensk, Kitab and the equator were proposed. The observations have already been completed: 30000 observations of stars were obtained. On the equator experimental observations were made only. The further work on equator remains undecided.

The coefficients of correlation between unknown corrections to 10° zones of star declinations and latitudes decrease by this method.

J. H. Lieske and V. K. Abalakin (eds.), Inertial Coordinate System on the Sky, 67.
© 1990 *IAU. Printed in the Netherlands.*

ON THE PHOTOGRAPHIC VERTICAL CATALOG

V.A.NAUMOV
Central Astronomical Observatory, USSR Academy of Sciences
Pulkovo
196140 Leningrad
USSR

The Photographic Vertical Circle (PVC) was constructed in Pulkovo shops in 1962 according to idea of M. Zverev for absolute determinations of declinations of stars. Observations were made at the observatory Serro-Calan in Chile by Chilean and Soviet astronomers in 1964–1966. The PVC is an experimental instrument: it is the first reflector and the first photographic instrument for absolute observations. It has two circles and each circle has four photographic microscopes.

The catalog contains approximately 700 stars from FK4, 700 stars from PFKSZ. Total number of observations is 12 762. The flexure was determined by horizontal collimators. Corrections of divisions of circles were determined by rotation of one circle relatively to the other one. Dependences of observing zenith distances from type of emulsion of photographic plates and colour indexes of stars were found and taken into account. The instrumental errors $O - C = 0\overset{\prime\prime}{.}65 \sin 2z$ and the jump of $O - C$ in September 1966 were found and taken into account, too.

There was a major problem in calculation of chromatic refraction, since we do not have coefficients of extinction of the Serro-Calan atmosphere. Investigation of corrections for chromatic refraction, using coefficients of extinction for 12 different stations and the theory of Pulkovo astronomers Zhilinsky *et al.* has shown that differences of this corrections submit to an empiric law , $\Delta z_{i,j} = C_{B-V}$ $\tan z$ if $2 \leq m \leq 9$, $z \leq 75°$, $h \leq 2\,000$m, where i,j are indexes of stations, $B-V$ are color indexes of stars, m is magnitude of stars, and h is altitudes of stations.

Therefore it is possible to use coefficients of extinction of some station (i) instead of coefficients of extinction of Serro-Calan atmosphere and to find C_{ij} from observations of lower and upper culminations in Serro-Calan by Bessel's method, if atmospheres of stations i and Serro-Calan submit to this empiric law. Five catalogs were formed by this method. It was found that differences of declinations of these catalogs are less than $0\overset{\prime\prime}{.}02$ at zenith distance of 45°. The declinations of final version of catalog PVC were formed as arithmetical mean of declinations of these catalogs. It was found that the systematic differences PVC – FK5 ($\Delta\delta_\delta$, $\Delta\delta_\alpha$, $\Delta\delta_m$, $\Delta\delta_{B-V}$) are less than $0\overset{\prime\prime}{.}1$.

Discussion

BASTIAN: You talked about colour-dependent refraction at quite some length. I understood all the details. But I did not get one point: did you discuss this because it is a completely new consideration in your work or because of the special problems at this very observatory?

NAUMOV: The latter. I discussed it because it could not be done in the usual way.

J. H. Lieske and V. K. Abalakin (eds.), Inertial Coordinate System on the Sky, 68.
© 1990 *IAU. Printed in the Netherlands.*

ON THE PROGRAMME OF GROUND-BASED OBSERVATIONS OF BRIGHT SELECTED MINOR PLANETS FOR 1991–2000

Yu.V. BATRAKOV, V.A. SHOR
Institute of Theoretical Astronomy USSR Academy of Sciences
10 Kutuzov Quay
191187 Leningrad
USSR

ABSTRACT. The programme of observations of 10 bright minor planets for 1991–2000 is proposed. One of the main aims of the programme is the determination of the orientation parameters and the systematic errors of star catalogues.

In 1976 during the IAU General Assembly in Grenoble, Commissions 8 and 20 approved the observational programme for 20 selected minor planets proposed by V.I. Orelskaya (1981). This programme was a significant broadening of the preceding ones proposed by D. Brouwer (1935) and by B.V. Numerov (1935) because it was aimed not only at determining the orientation parameters of the star catalogues, but at revealing the systematic errors of the latter as well. It was set up to 1990 and now it is near completion.

In the Institute of Theoretical Astronomy (ITA) there have been collected about 29000 observations of the 20 selected minor planets made at 35 observatories of the world within the framework of the programme mentioned and the earlier ones. Among these there are about 27000 observations of the 10 brightest minor planets covering the period 1949–1989. The error of unit weight of these observations when determining the orbital elements only is about 0".45. For the remaining 10 selected minor planets the number of observations is noticeably less, the O–C's are greater and the period of coverage, 1977–1989, is shorter.

The difference in accuracy owes partly to the fact that the minor planets of the second group are less bright, as a rule, as compared with those of the first group. To observe these minor planets one needs sufficiently long exposures. So their positions with respect to the reference stars are determined with rather large errors. At present the work is undertaken at the ITA for reducing the observations to the FK4 system and for determining the catalogue zero-point corrections and systematic errors which depend on the spherical coordinates.

When solving the problem two methods for processing observations are used. In the first one the usual normal places are formed within the areas of the celestial sphere. They are used to determine the orbital elements, the zero-point corrections and the systematic errors of the catalogue (Pierce 1971).

J. H. Lieske and V. K. Abalakin (eds.), Inertial Coordinate System on the Sky, 69–71.
© 1990 *IAU. Printed in the Netherlands.*

In the second one the generalized normal places are used (Batrakov et al 1987) to determine the same unknowns. Such a generalized normal place consists of the orbital parameters, the zero-point corrections, the systematic errors of the catalogue and their matrix of covariances determined from the restricted number of observations. To obtain these data the normal equations are subjected to regularization because of the bad conditioning. The method is quite correct as it gives the strict least squares solution based on all admissible observations and no information contained in them is lost. It is more laborious, of course, as compared with the first method, because for obtaining the generalized normal place the large dimension normal system must be solved. After completing the programme and collecting all the observations they will be checked and transferred to the Center of Stellar Data in Strassburg.

A question arises if the programme of observations of selected minor planets should be prolonged for the period after 1990. In our opinion, there are many reasons in favor of continuing it for 1991–2000, at least, though with some modifications. When using accurate observations the following problems can be solved:

- determining the minor planet masses,
- predicting the occultations of celestial bodies by minor planets,
- determining the star catalogue zero-point corrections and systematic errors,
- connecting the catalogues based on the ground-based observations with those obtained by astrometric satellites,
- connecting the coordinate frames based on ground-based optical observations with those based on VLBI observations of quasars, if radio-beacons are placed on some of the minor planets.

The limiting accuracy of $0''\!.1$ to $0''\!.2$ for ground-based observations can be reached when using telescopes of 5-10 m focal length (Duma 1985). With such instruments, however, it is difficult to have the required number of reference stars on the plate.

The smaller internal accuracy of $0''\!.3$ to $0''\!.5$ is attainable with telescopes of 3 m focal length which are more widely available. Many observatories have them and can take part in producing accurate positional observations. To obtain this accuracy short exposures must be used. So the minor planets observed must be as bright as possible.

Therefore, we propose to prolong the observation programme for the brightest minor planets only, namely numbers 1, 2, 3, 4, 6, 7, 11, 18, 39 and 40. The internal accuracy of individual observations in this programme must be $0''\!.3$ to $0''\!.5$. When observing the minor planets one must try to cover the widest possible arc of the planetary orbit, literally from quadrature to quadrature, in order to ensure good coverage of the star background and the separation of the unknowns. The most modern catalogues must be used. The observations and their dependences are to be published or delivered in a unified form.

The proposed programme for the 10 bright selected planets is a natural continuation of the preceding programmes mentioned above. The successful completion of it will give the proper base for solving many current scientific problems of importance.

References

Batrakov, Yu.V., Izvekov, V.A., Vashkevich, A.S.: 1987, in *Sovremennaya Astrometriya*, Pulkovo Observatory, Leningrad, pp. 284–296 (in Russian).

Brouwer, D.: 1935, *Astron. J.* **44**, 57–63.

Duma, D.P.: 1985, in *Problemy Astrometrii*, Moscow University, pp. 244–247 (in Russian).

Numerov, B.V.: 1935, *Astronom. Zh.* **12**, 339–345 (in Russian).

Orelskaya, V.I.: 1981, *Pis'ma v Astron. Zh.* **6**, 654–662 (in Russian).

Pierce, D.A.: 1971, *Astron. J.* **76**, 171–191.

Discussion

CORBIN: I know that L. Morrison will agree when I say that meridian circle observers support this program because of the increasing role that minor planets are playing in the determination of the equinox correction. This program will help improve the orbital elements, which in turn will give better results in our programs.

MORANDO: I support the programme of observations of minor planets. Such a programme of bright minor planets exists in France at Bordeaux and in Spain at La Palma.

YATSKIV: Are there any techniques which give improved accuracy for observation of minor planets? What about the possible role of CCDs?

SEIDELMANN: VLA observations of minor planets are accurate to about 0.01 arcsec. The problem with CCD observations is the need for good reference star catalog positions. The CCD field is very small.

IVASHKIN: When you constructed your proposed observation list of minor planets, did you take into consideration the possible candidates for proposed space exploration of minor planets? Also, did you consider employing the results of spacecraft-observations of minor planets in your ephemeris development work?

BATRAKOV: Data from spacecraft-based measurements of minor planets were not considered.

COMPLEX GEODYNAMICAL INVESTIGATIONS IN THE AREA OF THE USSR ACADEMY OF SCIENCES CENTRAL ASTRONOMICAL OBSERVATORY AT PULKOVO

V.K. Abalakin [1], V.I. Bogdanov [2], Yu.D. Boulanger [3], and V.A. Naumov [1]

[1] *Central Astronomical Observatory, USSR Academy of Sciences*
Pulkovo
196140 Leningrad, USSR

[2] *D.I. Mendeleyev Institute of Metrology*
Leningrad, USSR

[3] *O.J. Schmidt Earth Physics Institute*
Moscow, USSR

For astronomical, geodetical and geodynamical investigations as well as for practical applications the inertial coordinate system is widely used which is based on the Fundamental Star Catalogue FK5 together with local coordinate systems in observation stations on the Earth's surface which are intrinsically connected with the geometry of the gravitation field..

Instrumental observations forming the astronomical data bases are affected, however, by the influence of changes in local conditions which are generated by various endogenic, tectonic, exogenic, and/or anthropogenic (technogenic) factors on either of local, regional and global scales.

Construction of large industrial and social objects, growth of cities, intensive consumption of underground waters, oil, gas and other natural resources, ever growing economy engineering activities of man lead to changes in the conditions and characteristics of the environment, to distortion of natural physical fields, to local and regional deformations of the Earth's crust at the sites of fundamental astronomical, geodetic and geophysical precision measurements, locations of old observatories, stations and bench-marks, to distortion and disturbance in the homogeneity of long-term observational sets.

Pulkovo Observatory is situated near a large modern airport, a big industrial city, which has grown considerably for the last decades and in the area of the regional piezometric depression, formed in the Gdov horizon as a result of intensive consumption of underground waters.

A geodynamic polygon has been organized for investigations of anthropogenic processes, stability of the Earth's surface and the gravitational field in the region. The complex studies at the polygon are based on the Kronstadt depth gauge; on the levelling gauge and geodynamic complex at Shepelevo; on latitude, longitude and laser observations at Pulkovo; on metrological gravimetric and geodynamic studies at the underground observatory of D.I. Mendeleyev Institute of Metrology at Lomonosov. Geodetic, astronomical, gravimetric and hydrogeological studies are based on deep drill hole bench-marks, similar to bench-mark stations in Tallinn, which have proven to be rather efficient. In view of an increasing role of anthropogenic factors in the future it seems very appropriate that the IAU and IUGG discuss the problem of local condition change investigations on the sites of astronomical and geodynamical observatories at a particular symposium.

72

J. H. Lieske and V. K. Abalakin (eds.), Inertial Coordinate System on the Sky, 72.
© 1990 *IAU. Printed in the Netherlands.*

PULKOVO PROGRAMME FOR THE PHOTOGRAPHIC OBSERVATIONS OF SATELLITES OF PLANETS

T.P. Kisseleva, V.V. Bobylev, N.M. Bronnikova, A.A. Dementjeva,
O.A. Kalinichenko, A.A. Kisselev, H.I. Potter, S.V. Tolbin and N.A. Shakht
Central Astronomical Observatory, USSR Academy of Sciences
Pulkovo
196140 Leningrad
USSR

ABSTRACT. The satellites of Mars, the Galilean satellites of Jupiter and the first eight satellites of Saturn have been observed with the 26-inch refractor, the normal astrograph at Pulkovo and with the luni-planet telescope at the Ordubad station of the Pulkovo observatory since 1972.

Observations

Regular photographic observations of the planet satellites have been made at Pulkovo observatory since 1972. The observations were made with the 26-inch refractor (650/10000), the normal astrograph (300/3400), and the lunar-planet telescope (700/10000). The aim of the Pulkovo programme is to obtain highly precise positions of satellites relative to planets, which may be used for improving modern theories. The 26-inch refractor is the leading instrument in this programme. The most and the best observations have been made with this telescope (Kisseleva, 1986, 1987).

About ten exposures are taken on each plate nightly by the "scale-trail" technique. The exposure time is 30–60 sec for Galilean satellites and 1–3 minutes for Saturn's and Martian satellites.

For the Galilean satellites we used low sensitive ORWO WO-3 plates and highly sensitive NP-27 plates for Saturn's and Martian satellites combined with yellow filters. The evaporative chrome coated filters are used before the photographic plates in the filmholder for reducing the brightness of the planets by 5–6 magnitudes. The same filters are used for observations with the normal astrograph.

Astrometric Reduction

As mentioned above the "scale-trail" technique with two trails (before and after exposures) are used for observations by the long-foci telescopes: the 26-inch refractor and lunar-planetary telescope (Kisselev, 1989).

J. H. Lieske and V. K. Abalakin (eds.), Inertial Coordinate System on the Sky, 73–74.
© 1990 *IAU. Printed in the Netherlands.*

As regards the planets obtained with the help of the short-foci telescope the reference stars technique is used. The reference catalogues are the AGK-3 and FOCAT-S.

Two systems of reference stars are used for Saturn satellites. The plate measurements are carried out with the help of the semiautomatic machine "ASCO RECORD".

As the result of observations we have the planetocentric coordinates of satellites for every exposure on the plate. Then the average differences (Satellites – Planet) are calculated. The differences (Satellite – Satellite) are also determined.

In the "scale-trail" technique the effect of the variation of orientation due to the motion of planet system and other systematic errors are taken into account. The photographic phase effect of planets is taken into account by the methods proposed at Pulkovo (Kisseleva, 1986).

The Results and Their Comparison with Modern Theories

During 1972–1988 about 825 exposures of Martian satellites, 8000 exposures of Galilean satellites and 2000 exposures of Saturn's satellites were obtained.

Planetocentric coordinates of satellites were compared with modern theories of J.H. Lieske (E1, E2), J.E. Arlot (G5) for the Galilean satellites, V.A. Shor for the Martian satellites and G. Struve - T.K. Nikolskaya for the Saturn satellites. The ephemerides were calculated at the Institute of Theoretical Astronomy.

The O–C analyses permitted us to determine standard deviations of one observation which are $0''\!.3$ to $0''\!.4$ for Martian satellites, $0''\!.1$ - $0''\!.2$ for Galilean satellites and $0''\!.2$ - $0''\!.3$ for Saturn's satellites.

The observations of satellites of planets are planned to be continued.

References

Kisselev, A.A.: 1989, The theoretical foundations of photographic astrometry. Moscow, "Nauka".
Kisseleva, T.P.: 1986, Proc. IAU Symposium 141, *Relativity in Celestial Mechanics and Astrometry*, 129–134.
Kisseleva, T.P.: 1987, Isv. Gl. Astr. Obs. Pulkovo, **204**, 57–63.

Discussion

WILKINS: Have these observations been made available for analysis by other groups?
ABALAKIN: We can make them available.

CHEREPASHCHUK: Did you try to eliminate the scattered light from the central planet?
KISSELEVA: Yes, we have used filters for this purpose.

CONFOR: A NEW PROGRAM FOR DETERMINING THE CONNECTION BETWEEN RADIO AND OPTICAL REFERENCE FRAMES

V.S. GUBANOV [1] I.I. KUMKOVA [1], AND V.V. TEL'NYUK-ADAMCHUK [2]
[1] Institute of Applied Astronomy
8 Zhdanovskaya ul.
197042 Leningrad, USSR

[2] Astronomical Observatory of Kiev State University
Observatorna Str. 3
252035 Kiev, USSR

ABSTRACT. The program for establishment of a link between the fundamental system FK5 and the radioastronomical coordinate system is described. The program includes photographic and meridian observations of extragalactic radio/optical sources and intermediate reference stars. Observatories of the USSR, GDR and Yugoslavia are participating in the project.

One of the possible ways to establish the link between the radio and optical coordinate systems is determining extragalactic radio source positions with reference to fundamental catalogue stars by means of photographic astrometry. This method has been used in 1980–1985 when the program ROAS (Radio/Optical Astrometric Sources) was fulfilled. The observational part of the program was based on the list of extragalactic radio sources recommended by IAU Commission 24 working group and included the determination of positions of 80 objects brighter than 18th stellar magnitude which are available for observations from the territory of the USSR. The observations were carried out using the Zeiss-400 astrographs of the six observatories of the USSR. Final results of the program have shown the necessity of an expansion of the observations to more faint objects (fainter than 18^m) and also of providing a more reliable reference of intermediate stars to a fundamental catalogue.

Based on the results of ROAS [1] the new cooperative program was organized in 1988. It is called CONFOR – Connection of Frames in Optical and Radio regions. The program is supported by the Presidium of the USSR Academy of Sciences. This program is undertaken for the determination of relative orientation and errors of radioastronomical and optical coordinate systems and it consists of the following steps:
1. Determination of precise positions in the FK5 system of reference stars (up to 10^m) near radio sources;
2. Determination of positions of intermediate reference stars in the neighborhood of radio sources (12^m - 14^m, 16^m - 18^m) in selected areas;
3. Determination of the extragalactic radio source positions with respect to the fundamental coordinate system FK5;
4. Designing and constructing a photoelectric camera [2] for observation of very faint objects against the background of bright catalogue stars.

75

J. H. Lieske and V. K. Abalakin (eds.), Inertial Coordinate System on the Sky, 75–76.
© 1990 *IAU. Printed in the Netherlands.*

The observational part of the program CONFOR includes 200 central areas selected from the catalogue by Argue *et al.* [3]. The majority of objects on the list have brightness between 18 and 19 mag. The faintness of the objects is the main difficulty for obtaining their high-precision coordinates by means of photographic astrometry. That is why we use a two- or three-step method of reference. Constructing a new photoelectric camera which couples a CCD with a photographic plate will permit us to realize a direct connection between faint objects and bright catalogue stars.

Photographic observations of extragalactic radio sources are carried out by:
- 1-m R-C telescope of Sanglok Observatory, Astrophysical Institute of Academy of Sciences (AS) of Tajik SSR;
- 1-m R-C telescope of Astronomical Council AS of the USSR at Simeiz (since 1990);
- Schmidt telescope of Central Institute for Astrophysics AS of GDR;
- astrographs of Abastumani Observatory and at Maidanak;
- Shternberg State Astronomical Institute;
- 600-cm Zeiss telescope of the Main Astronomical Observatory of AS of the Ukrainian SSR.

The following telescopes are taking part in photographic observations of intermediate stars: astrographs of Astronomical Institute of AS of Uzbek SSR at Kitab, Astronomical Observatory of Kiev State University, of Abastumani observatory, of Shternberg State Astronomical Institute.

The meridian observations are being provided by astronomical observatories of Kiev, Odessa and Kazan' universities and Belgrad Observatory.

As result of the completion of the program, the following will be received:

- catalogues of intermediate reference stars for stars brighter than 10^m from meridian observations and for stars 12^m - 14^m, 16^m - 18^m from photographic observations;
- catalogue of extragalactic radiosources with respect to fundamental coordinate system FK5;
- the parameters of orientation and systematic errors of FK5-system reference to radioastronomical coordinate system;
- the model of the photoelectric camera and investigation of its astrometric possibilities.

References

1. Kumkova, I.I.: 1985, *Izv. GAO AN SSSR Pulkovo*, **203**, 28 (in Russian).
2. Gubanov, V.S., Kumkova, I.I., Malachov, E.I. and Shornikov, O.E.: 1988, Preprint *IPA AN SSSR*, **2** (in Russian).
3. Argue, A.N., de Vegt, C., *et al.*: 1984, *Astron. Astrophys.* **130**, 191.

Discussion

KOPEJKIN: How many reference objects have you in your program?
KUMKOVA: About twelve intermediate reference objects.

A PROPOSAL FOR ASTROMETRIC OBSERVATIONS FROM SPACE

M.S. CHUBEY, V.V. MAKAROV, V.N. YERSHOV, I.I. KANAYEV, V.A. FOMIN,
Yu.S. STRELETSKY AND A.V. SCHUMACHER
Central Astronomical Observatory USSR Academy of Sciences
Pulkovo
196140 Leningrad
USSR

ABSTRACT. Some aspects of the realisation of a celestial reference frame using a space astrometry facility are considered. An observational program is described, consisting of observing stars up to magnitude 14, radiostars and bright QSO's, planets, asteroids and of laser signals from the Earth. A scheme of an astrometric facility consisting of two telescopes on board the satellite is proposed. The overview strategy with *"inita"* is estimated, and the estimates of the accuracy of a single observation (0".02–0".05) and of the output catalogues (0".001–0".007) are made.

1. Introduction

The HIPPARCOS experiment has opened a way to the construction of a high-precision quasi-inertial reference frame by means of space facilities. New perspectives of astronomy in space are caused by a significant increase in the obtained data accuracy, by the number of the observed objects, and by a possibility to link directly the frames of reference (fundamental, dynamic, radiointerferometric, geodetic) which are employed .

We present here some goals for a proposed mission called AIST (from the Russian Астро-метрический Искусственный Спутник-Телескоп: Astrometric Artificial Satellite Tele-scope). The ideal program for the astrometric mission should include observations of the bright stars, QSO's, asteroids, planets, the Sun, and laser sources on the Earth. It is most convenient to use the astrometric satellite rotation (dynamic method) for realisation of the project which is designed for effective operation in orbit during the course of some years.

2. Method of Vlasov

One of the most appropriate in the dynamic method of angular distance measurements is the scheme of Vlasov (Transactions IAU 1970), depicted in Fig. 1. According to this approach six reference objects (*inita*) have been chosen on the celestial sphere so that the angular distances between the neighbouring inita be nearly 90 degrees. The coordinates, proper motions and parallaxes of the inita in the chosen frame of reference have to be measured several times more precisely than those of the objects in the output catalogue, that is not worse than 1 milliarcsec. Three arcs are measured from each object within one of the 8 inita's triangles to the vertices of this triangle. While knowing the coordinates of the inita one can compute those of the object at the moment of observation. Satellite

77

J. H. Lieske and V. K. Abalakin (eds.), Inertial Coordinate System on the Sky, 77–80.
© 1990 *IAU. Printed in the Netherlands.*

rotation causes the consecutive scans on the celestial sphere to intersect in the points near I₁ (Fig. 1) and then near I₂ and I₃ .

According to our calculations the method under consideration has secured high homogeneity of the determined precision of the astrometric parameters on the entire sphere. For instance the distribution of mean square position error within the *inita's* triangle is shown in Fig. 2 in the case

$$\text{var}(S) = 5.7s + 1,$$

where s is the true arc and S is the measured arc in term of radians.

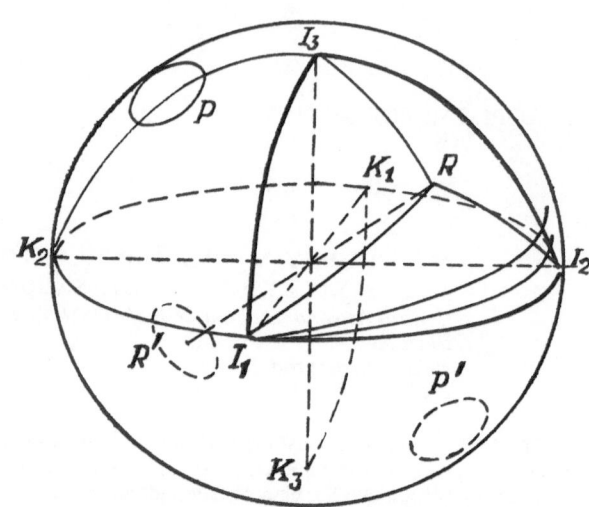

Fig. 1. Typical scans with AIST

3. Astrometry in space

A peculiarity of astrometry in space by the dynamic method requires that the satellite rotation has to be controlled with great accuracy by one's observations. The AIST frame (Chubey 1989; Chubey *et al.* 1989) Fig. 3, is available for ω (rotation vector) components reconstruction and measuring of the angular distances between the stars in each scan.

The frame consists of two telescopes: reflectors of the Ritchey-Chretien system with the working fields $2w = 60$ arcmin, foci 3 m, objective diameters 25 cm. These telescopes with the primary mirrors M_1 and M_2 and the secondary mirrors m_1 and m_2 are mounted rigidly on the general axis O_1-O_2 facing one another. A standard with the reference angles 180 and 190 degrees is placed in the middle of the apparatus between the objectives. This standard is comprised of two double-faced flat mirrors. Two-dimensional CCD detectors are mounted in the focal coordinate system O_1,X_1,Y_1 and O_2,X_2,Y_2. Reciprocal position of these systems is controlled through the using a light marks index sets L_1 and L_2 rigidly placed in the surfaces of detectors to gain advantage the collimation observation technique.

In such a manner possible displacements of the central angular standard block with circular and ring apertures are controlled. The optical system combines the images of four fields P, P', R, R' on the celestial sphere in the

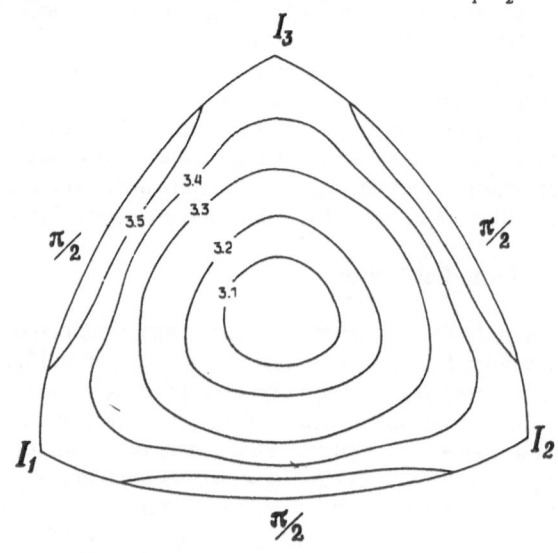

Fig. 2. Typical errors within an inita

focal plane. The satellite must be principally revolving around the axis W, as it is shown in Fig. 3, or around the axis V. In any case the vector's ω components may be determined if even only one object is observed at a given moment in any field from pairs P, P' and R, R'. The theoretically advantageous choice of principal moments of inertia $I_w = 2I_U = 2I_V$ or $I_V = 2I_U = 2I_w$ is assumed to be practically available in our design (Popov 1986).

In order to obtain the arcs between the *inita* and program objects it is necessary to know its local coordinates and rotation matrix, as it is done in HIPPARCOS "great circle reduction" procedure (van der Marel 1988). The Euler rotations around the U, V, W axes may be derived by integration of Euler's kinematic equations. It is possible to provide the acceptable conditions for separating of unknowns in the previously mentioned equations by dint of optimisation of satellite cover, form and mass distributions. Some systematic errors are reduced in the case where the direction of rotation is changed to the opposite; the scale error is essentially reduced due to the *inita* scheme of operation. The errors connected with discordance of optical elements are checked by special collimation observations (Chubey 1989).

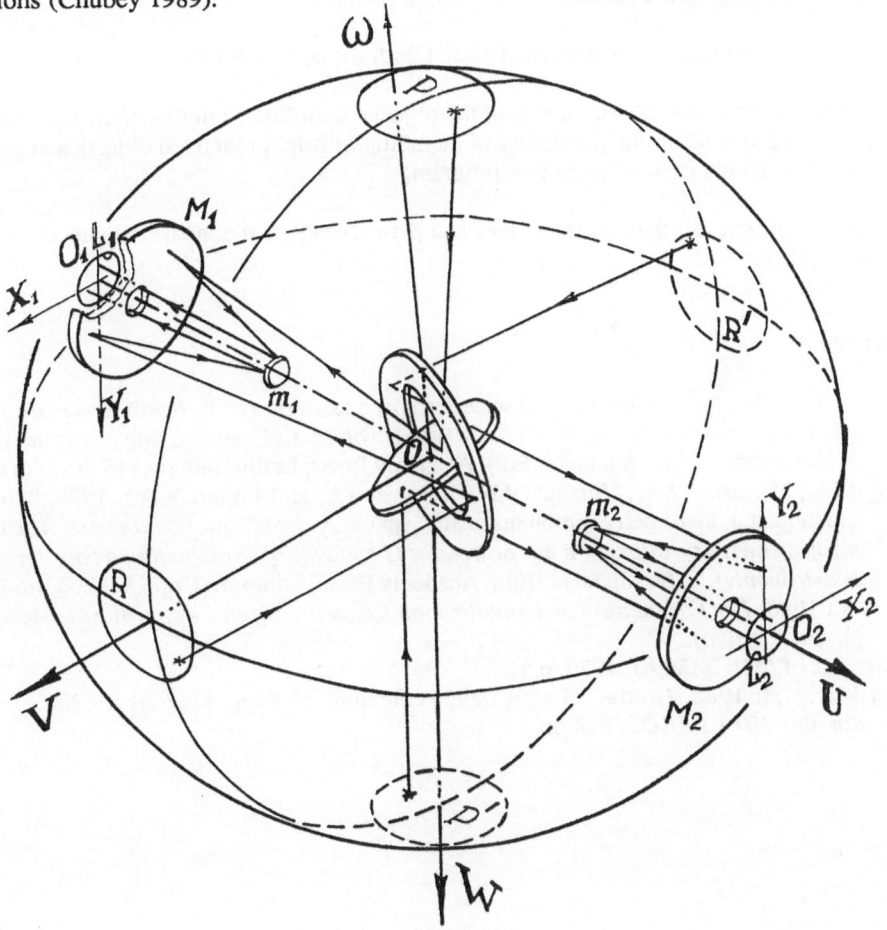

Fig. 3. AIST schematic

The optical limit of telescopes with declared parameters under rotational velocity being about 150 arcsec/sec will be 13–14 magnitudes, and the accuracy of single measurements will be about 0".15 under the condition that the integration time is 10 ms (Chubey *et al.* 1989). One star transit permits us to derive its image trail in the local reference frame with an accuracy of 0".01. The accidental errors of astrometric parameters in the output catalogue will be decreased by a factor of 7 during 4 years of AIST operation.

4. Summary

It is impossible to make comparison of the design in question with others within the limits of this concise report. So we shall emphasise only the peculiar features and advantages of the AIST design:

• high homogeneity of the accuracy of the output parameters ;

• controlling of all systematic errors connected with the optical scheme;

• Possibility of absolute measurements of the objects' coordinates in distances 180 and 90 degrees, and as a result the possibility of including of bright extended objects and ground-based light signals in the observation program;

• possibility of measuring of different arcs and permanency of the measurements in time and on the sphere.

References

Chubey M.S.: 1989, "A two-telescope device for space astrometry" in *Проблемы построения координатных систем в астрономии* (*Problems of constructing a coordinate system in astronomy*), V.K. Abalakin (Ed.), Academy Press, Leningrad, pp.245–252 (in Russian).

Chubey M.S., Makarov V.V., Yershov V.N., Kanayev I.I., and Fomin V.A.: 1989, "Solution of fundamental astrometry problems using space systems" in *Проблемы построения координатных систем в астрономии* (*Problems of constructing a coordinate system in astronomy*), V.K. Abalakin (Ed.), Academy Press, Leningrad, pp.252–265, (in Russian).

Popov V.I.: 1986, *The Orientation and Stabilisation Systems of Space Apparatuses*, Moscow, 183 pp. (in Russian).

Transactions of IAU, **XIV**(A), 1970, p.51.

van der Marel, H: 1988, On the "Great-Circle Reduction" in *Data Analysis for the Astrometric Satellite HIPPARCOS*, 228 pp.

ON THE CREATION OF A NEW CATALOGUE FOR USSR TIME SERVICE

V.L. GORSHKOV, D.D. POLOZHENTSEV (II), A.A. POPOV,
N.V. SHCHERBAKOVA
Central Astronomical Observatory, USSR Academy of Sciences
Pulkovo
196140 Leningrad
USSR
and
A.D. POLOZHENTSEV
Astronomical Observatory of Leningrad University
Leningrad
USSR

The R.A. system KSV [1] has been used in the USSR Time Service for observations with small transit instruments since 1970. Some methodological errors were made when creating the KSV [2] and the use of the proper motion system, distorted with the position errors of GC and KSV [3], make it necessary to improve the KSV system. Work on the creation of a new system KSV-2 was organized for this purpose. The catalogue included about 0.3 million computer-readable observations with small transit instruments of the USSR Time Service with mean standard deviations $\sigma = 11$ sec δ ms.

Sixteen individual catalogues have been constructed. In the process of this work the same methods were used for the recalculation of the original, checked and corrected material and for the adjustment of the observational results [4]. Besides the classical free chain adjustment of non-group observations there was used the total adjustment of all these series by the least squares method.

The following matrix was used:

$$(\Delta\alpha_i - \Delta\alpha_k) + \Delta a_j (A_k - A_i) = \Delta U_{ik,j} \mid P_{ik,j}$$

if $\Sigma \Delta\alpha_i = 0$. Here $\Delta U_{ik,j} = \Delta U_{ij} - \Delta U_{kj}$ for all the star pairs (i,k) on one night (j) and for all the nights of the series. Also, $P_{ik,j} = P_{ik} * P_j$, $P_{ik} = P_i * P_k /(P_i + P_k)$, $P_i = \sigma^2/ \sigma_i^2$ are weights of stars, σ^2 is the general dispersion of zenith stars of these series, $P_j = n_j$ are weights of the dates. The results obtained proved to be similar in the range of the free chain method applicability for both methods.

An analysis of the KSV errors $\Delta\alpha_\alpha$, derived on the basis of various catalogues shows the presence of a noticeable (unstable as a rule from year to year) seasonal influence on the observations with some transit instruments. A magnitude equation of the KSV (about 1 ms per 1 ph. mag.) and an insignificant dependence on the spectral class of the stars have been found.

81

J. H. Lieske and V. K. Abalakin (eds.), Inertial Coordinate System on the Sky, 81–82.
© 1990 *IAU. Printed in the Netherlands.*

82

A preliminary estimation of the R.A. system KSV is given in the Figure for the declination from 50° to 75°. The comparisons of the KSV with FK4 and FK5 for the same zone are given also.

It is easily seen that the KSV keeps better pace with modern observational data than the FK4 or even the FK5. This situation is unlikely to be changed in the near future, until a new reference frame is constructed by astrocosmic means. After the correcions for the above mentioned errors [2, 3] the KSV system practically coincides with the KSV-2.

Hence, we can make a conclusion that the small transit instrument observations can be effectively used for a construction of an independent coordinate system, especially after the organisation of star group observations, the inclusion of observations of stars at two culminations and the solar system bodies by all instruments, and coordination of observation programmes by different stations. The KSV-2 system is a good such independent coordinate system and would not require any improvement after the correction of proper motions.

The authors are grateful to all the astronomers of the USSR Time Service whose selfless labour helped make possible the present study.

References

1. Pavlov, N.N., Afanas'eva, P.M., Staritsin, G.V. (1971) "Compilation catalogue of USSR Time Service", *Trudy GAO* **78**, 59–98.
2. Afanas'eva, P.M., Gorshkov, V.L. (1973) "On a possibility to improve the R.A. catalogue of USSR Time Service", *Astronomicheskiy Circular* 1284, Oct. 21.
3. Vityasev, V.V. (1987) "On a systematic errors of proper motions of KSV stars", *Vestnic Leningrad. Gos. Univer.*, ser.1-2 (# 8), 86–92.
4. Gorshkov, V.L. (1984) "On R.A. catalogues Ph10 and Ph12 of Pulkovo Time Service", *Izvestiya GAO* **202**, 6–16.

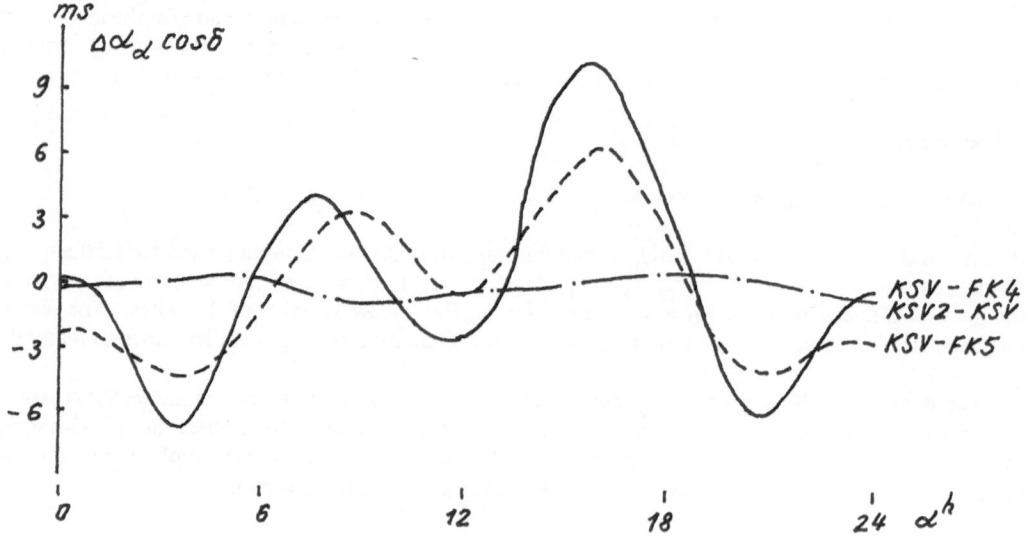

ON $\Delta\delta_\alpha$ SYSTEMATIC ERRORS OF THE FK5

G.S. KOSIN
Central Astronomical Observatory, USSR Academy of Sciences
Pulkovo
196140 Leningrad
USSR

An estimation of $\Delta\delta_\alpha$ systematic errors of the FK5 has been made. Systematic differences between the FK5 and FK4 and $\Delta\delta_\alpha$ corrections to the FK4 derived at Pulkovo have been used as given in the following series: $\Delta\delta_\alpha = A' \sin \alpha + B' \cos \alpha + C' \sin 2\alpha + D' \cos 2\alpha$.

The present paper is based on the use of the $\Delta\delta_\alpha$ correction to the FK4 (Kosin 1989). It was derived with the use of observations with the photographic zenith tube (PZT) and zenith telescope (ZT). The results of the observations with these instruments (derived on the basis of the chain adjustment of the PZT catalogue and half-sum of declinations of pair stars in the case of the ZT) are free from $\Delta\delta_\alpha$ errors. AGK3 and a differential (in the FK4 system) catalogue of star declinations on PZT and ZT programmes observed with the Vertical Circle at Pulkovo have been used as intermediate catalogues for our comparison. The FK4 $\Delta\delta_\alpha$ correction (for the epoch close to 1964) is

$$\Delta\delta_\alpha \, (\text{FK4}) = 0\rlap{.}''022 \sin \alpha -0\rlap{.}''040 \cos \alpha -0\rlap{.}''018 \sin 2\alpha \, .$$

A comparison of the FK5 and the FK4 has been made on the basis of the data and recommendations of the Astronomisches Rechen Institut (Heidelberg, FRG). The difference for the zone +20° to +70° and the above-mentioned epoch has proven to be FK5 − FK4 = $0\rlap{.}''002 \sin \alpha -0\rlap{.}''028 \cos \alpha -0\rlap{.}''002 \sin 2\alpha - 0\rlap{.}''009 \cos 2\alpha$.
The following two expressions have been compared for a determination of the FK5 correction:

$$\text{PZT(ZT)} - \text{FK4} = 0\rlap{.}''022 \sin \alpha - 0\rlap{.}''040 \cos \alpha - 0\rlap{.}''018 \sin 2\alpha$$
$$\text{FK5} - \text{FK4} = 0\rlap{.}''002 \sin \alpha - 0\rlap{.}''028 \cos \alpha -0\rlap{.}''002 \sin 2\alpha -0\rlap{.}''009 \cos 2\alpha \, .$$

Their difference is the correction in question:

$$\Delta\delta_\alpha \, (\text{FK5}) = 0\rlap{.}''020 \sin \alpha -0\rlap{.}''012 \cos \alpha -0\rlap{.}''016 \sin 2\alpha +0\rlap{.}''009 \cos 2\alpha \, .$$

References

Kosin, G.S.: 1989, "A determination of FK4 Systematic Errors $\Delta\delta_\alpha$", *Izvestia Gl. Astr. Obs. Pulkovo*, **206**, 3–5 (in Russian).

Translated from Russian by I.N. Voronina

J. H. Lieske and V. K. Abalakin (eds.), Inertial Coordinate System on the Sky, 83.
© 1990 *IAU. Printed in the Netherlands.*

ON THE ORGANIZATION OF ABSOLUTE COORDINATE DETERMINATION OF THE FK5 STARS

B.K. Bagildinsky, V.A. Fomin, I.S. Guseva, T.R. Kirian, A.A. Nemiro, G.M. Petrov, G.I. Pinigin, and V.D. Shkutov
Central Astronomical Observatory, USSR Academy of Sciences
Pulkovo
196140 Leningrad
USSR

The present-day realization of inertial coordinate system by means of traditional optical astrometry is the Fifth Fundamental Catalogue (FK5). The series of observations with new meridian instruments (CAMC, PMC-190, BAMC, HMC) show the significant correlated differences (up to $0''1$) of observational catalogues from the FK5. Moreover, FK5 mean epoch appears to be old (about 50 years ago) and FK5 proper motions would have essential errors owing to the fact that not many new original catalogues (only 25 in RA and 15 in DEC) were used when compiling FK5. It should be noted also that FK5 has a dissimilar accuracy of positions and proper motions of "old" and "new" stars.

The necessity of FK5 improvement is connected also with the problem of space catalogues orientation and with linking these catalogues to the classical ones. Abovementioned reasons require the re-observation of the entire FK5 and this work is currently performed at a set of observatories.

It seems to be necessary for USSR meridian instruments to take part in this work. The possible advantage of our original instruments (horizontal meridian circle, photographic vertical circle, axial meridian circle etc.) consists in their constructive differences from the classical meridian circles. Applying unusual instruments with minimal systematic errors (less then $0''05$) may allow one to improve the fundamental system.

It is important, in our opinion, to apply the new techniques of fundamental system creation. An optimal global distribution of instruments is assumed. In particular, it's of interest the experience of Pulkovo high-latitude observations on Spitsbergen. Moreover, it is necessary to get an agreement on a plan of observations at different instruments and to use a modern technique of observation reduction and catalogue compilation. Significant improvement may be achieved by choosing of observation place with a good astroclimate and by correct reduction for refraction.

The foregoing allows one to expect that the joined efforts on the FK5 re-observation will lead to improvement of fundamental system.

J. H. Lieske and V. K. Abalakin (eds.), Inertial Coordinate System on the Sky, 84.
© 1990 *IAU. Printed in the Netherlands.*

THE FK4 ORIENTATION PARAMETERS DERIVED FROM PHOTOGRAPHIC AND VLBI OBSERVATIONS OF RADIO / OPTICAL OBJECTS

V.S. GUBANOV, I.I. KUMKOVA AND N.I. SOLINA
Institute of Applied Astronomy
8 Zhdanovskaya ul.
197042 Leningrad
USSR

ABSTRACT. The photographic and VLBI positions of 59 quasars are used for determination of the FK4 orientation parameters with respect to the radioastrometrical system (RAS).

The difference $\Delta\alpha \cos \delta$ and $\Delta\delta$ between the equatorial coordinates of the quasars in RAS and FK4 in the sense (R – O) may be presented as functions of the mutual orientation parameters as follows:

$$\Delta\alpha \cos \delta = -a \sin \delta \cos \alpha - b \sin \delta \sin \alpha + c \cos \delta \qquad (1)$$
$$\Delta\delta = -b \cos \alpha + a \sin \alpha + d \qquad (2)$$

where

$$a = i \cos \omega, \, b = i \sin \omega, \, c = d \cot \varepsilon + \Omega - \omega, \qquad (3)$$

and where i is the mutual inclination of the FK4 and RAS equators, d is the constant correction to declinations in the FK4 system, ω and Ω are the longitudes of the point of intersection of the FK4 and RAS equators, and ε is the mutual inclination of the equator and ecliptic.

There have been selected 59 quasars with the coordinate differences (R – O) for RA and declination both less than the three-sigma value of the internal error of the optical positions, estimated as 0".20. These data are presented in Table 2. The results of separate solutions of the systems (1) and (2) obtained by the least-squares method and weighted values of twice-determined parameters are presented in Table 1.

Substitution of the weighted values for a, b, c and d in (3) will yield the orientation parameters of the FK4 coordinate system with respect to the RAS:
$$i = +0".09 \pm 0".04, \quad d = +0".08 \pm 0".03, \quad \Omega - \omega = -0".15 \pm 0".08 .$$
Then using these values we can obtain the correction of the RAS equinox with respect to the one of the FK4:
$$A = \Omega - \omega + i \sin \omega \cot \varepsilon = -0".26 \pm 0".10 .$$

Due to the non-uniform distribution of the considered quasars, it has not been possible to apply a more complicated expansion by spherical functions. We should notice that the external error of residuals (R – O), equal to 0".28, is larger than the internal error 0".20 — which suggests something

85

J. H. Lieske and V. K. Abalakin (eds.), Inertial Coordinate System on the Sky, 85–86.
© 1990 IAU. Printed in the Netherlands.

about the presence of local errors of the reference star coordinates.

Table 1. Parameters of Equations (1) and (2) in arcsec

Equations	a	b	c	d
(1)	+0.08 ± .11	−0.19 ± .11	+0.03 ± .03	
(2)	+0.07 ± .04	−0.03 ± .03		+0.08 ± .03
weighted	+0.05 ± .04	−0.05 ± .03	+0.03 ± .03	+0.08 ± .03

Table 2. Differences (R − O) of coordinates in units of $0.''01$

IAU number	Δα × cosδ	Δδ	Ref.	IAU number	Δα × cosδ	Δδ	Ref.	IAU number	Δα × cosδ	Δδ	Ref
0003−066	00	−20	1	0738+313	21	25	3	1328+307	50	−02	K
0106+013	36	−04	K	0823+033	00	20	1	1354−152	−40	20	1
0112−017	00	40	1	0851+202	00	35	3	1354+195	21	07	3
0119+041	−20	00	1	0859−140	20	−40	1	1510−089	−04	−48	2
0133+476	06	08	3	0906+015	40	20	1	1546+027	20	00	1
0135−247	40	10	1	0919−260	00	−40	1	1730−130	−45	39	2
0138−097	−20	50	1	0923+392	28	11	3	1741−038	51	−14	2
0153+744	−26	09	3	0941−080	−50	−20	1	1908−202	−30	10	1
0202−172	40	00	1	0952+179	−13	−29	K	1928+738	21	04	3
0319+121	47	−12	K	1015−314	−10	−59	1	1936−155	−40	10	1
0332−403	−34	−07	2	1104−445	−13	46	2	1958−179	00	−30	1
0420−015	06	−03	2	1145−071	−10	−50	1	2106−413	00	40	1
0438−436	−12	43	2	1148−001	15	−16	2	2203−188	00	−40	1
0440−004	−36	07	2	1219+285	02	18	3	2210−257	−20	−20	1
0457−024	10	−30	1	1226+023	37	04	3	2216−038	−20	10	1
0528−250	00	20	1	1237−101	50	−20	1	2245−328	30	−30	1
0552+398	42	−34	K	1243−072	−30	−50	1	2318+049	00	20	1
0642+449	51	05	K	1245−197	20	−10	1	2329−162	−10	50	1
0723−008	40	20	1	1302−102	20	00	1	2345−167	24	37	2
0736+178	53	00	K	1313−333	22	−27	2				

Notes: 1. The reference *K* marks unpublished results by I. Kumkova.
2. For the references *K*, and *3* the radio coordinates are taken into *[4]*.

References

1. Torres, C., Wrobiewski, H.: 1987, *Astron. Astrophys. Suppl Ser.* **69**, 23.
2. Walter, H., West, R.: 1986, *Astron. Astrophys.* **156**, 1.
3. Clements, E.: 1983, *Monthly Notices Roy. Astron. Soc.* **203**, 861.
4. Argue, A.N., *et al.*: 1984, *Astron. Astrophys.* **130**, 191.

EQUATOR AND ECLIPTIC FK5 POSITIONS ON THE BASIS OF SUN AND PLANETS OBSERVATIONS BY THE ERTEL-STRUVE VERTICAL CIRCLE AT THE KISLOVODSK MOUNT STATION

A.V. DEVYATKIN AND K.G. GNEVYSHEVA
Central Astronomical Observatory, USSR Academy of Sciences
Pulkovo
196140 Leningrad
USSR

The classical Ertel-Struve vertical circle of the Pulkovo Observatory was astablished at the Kislovodsk Mount Station. From 1984 to 1989 we have obtained 566 declinations of the Sun, 230 of Mercury, 413 of Venus, and 207 of Mars. Observations were visual, and the reductions were differential in the FK5 system. Only day-time observations of reference stars were used.

To eliminate the errors due to the phase of the planets, the new method was adopted which permits taking the distributions and the illumination over the image, the apparatus function and the means of the measuring into account.

The DE200 theory was used for the comparison. Assuming these data the corrections of FK5 zero-point $\Delta\delta$, inclination of the ecliptic $\Delta\varepsilon$ and Sun, and planets mean anomalies ΔM were obtained as the following values:

$$\Delta\delta \qquad = +0\rlap{.}''06 \pm 0\rlap{.}''01$$

$$\Delta\varepsilon \qquad = -0\rlap{.}''01 \pm 0\rlap{.}''01$$

$$\Delta M \text{ (Sun)} \qquad = -0\rlap{.}''02 \pm 0\rlap{.}''04$$

$$\Delta M \text{ (Mercury)} = +0\rlap{.}''10 \pm 0\rlap{.}''04$$

$$\Delta M \text{ (Venus)} = -0\rlap{.}''08 \pm 0\rlap{.}''08$$

$$\Delta M \text{ (Mars)} = -0\rlap{.}''03 \pm 0\rlap{.}''04$$

We are grateful to all observers of the vertical circle at the Kislovodsk Mount Station for obtaining the observations.

J. H. Lieske and V. K. Abalakin (eds.), Inertial Coordinate System on the Sky, 87.
© 1990 *IAU. Printed in the Netherlands.*

THE NIKOLAEV AXIAL MERIDIAN CIRCLE: THE PRESENT AND FUTURE STATUS

O.E. SHORNIKOV, A.V. SHULGA, N.S. LIADOVOI, S.L. KASHTALIAN, AND
P.V. MAIGUROV
Nikolaev Branch, Pulkovo Observatory
Observatory Str 1
327030 Nikolaev, USSR

ABSTRACT. Some information on the construction of the AMC was given at IAU Colloquium 100 in Belgrade. The present paper reports the first results of trial observations of star right ascensions. At present the instrument is being prepared for semiautomatic absolute and relative night-time and day-time observations. The scheme of the instrument, principal units and calculations of their parameters are estimated and an observational programme is proposed.

The visual observations of 136 FK5 stars were accomplished on the AMC at Nikolaev in 1987. Mean error of one observations is $\pm 0\overset{s}{.}0175 \sec \delta \sec z$. Now the second version of the instrument is erected for absolute observations. The construction of this instrument is represented in Fig. 1.

The AMC is equipped with eyepiece micrometeres installed on the telescope (T) and an autocollimator. The telescope (T) together with the pentag (P) can rotate around its optical axis for the possibility of star observation. The future systematic error of this AMC version will be less than $0\overset{''}{.}05$. This version of AMC will be mounted on the mountain station. The observing program for AMC consists of radio stars, some FK5 stars, and minor planets for the determination of planetary orbits and the fundamental reference system.

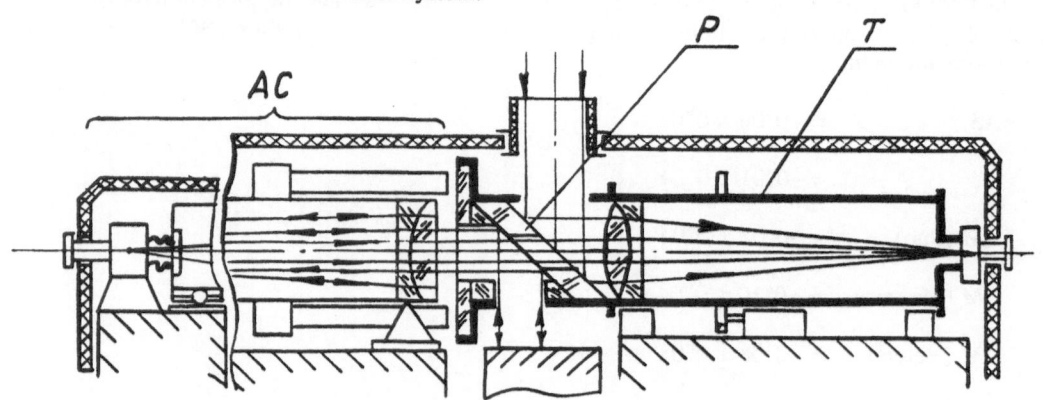

Figure 1. The Axial Meridian Circle.
AC: the immovable vacuum autocollimator (F = 12000 mm, D = 180 mm);
T: the telescope (F = 2500 mm, D = 180 mm); *P*: the titanium-glass pentag.

88

J. H. Lieske and V. K. Abalakin (eds.), Inertial Coordinate System on the Sky, 88.
© 1990 *IAU. Printed in the Netherlands.*

ON THE PROGRESS OF THE CONSTRUCTION OF THE SUKHAREV MERIDIAN AUTOMATIC HORIZONTAL INSTRUMENT (MAHIS)

R.I. GUMEROV [2], V.B. KAPKOV [2], T.R. KIRIAN [1], N.S. LIADOVOI [3], G.I. PINIGIN [3], A.A. POZHALOV [3], V.P. SIBILEV [3], A.W. SCHUMACHER [1], N.A. SHKUTOVA [1], AND O.E. SHORNIKOV [3]

[1] Central Astronomical Observatory (Pulkovo)
196140 Leningrad, USSR

[2] Engelhardt Astronomical Observatory
422526 Kazan, USSR

[3] Nikolaev Branch, Pulkovo Observatory
327030 Nikolaev, USSR

ABSTRACT. The MAHIS is the L.A. Sukharev meridian automatic horizontal instrument. It has all the advantages of the Pulkovo HMC and is devoid ,or nearly devoid ,of its disadvantages. A scheme of the MAHIS includes a flat mirror of D = 300 mm and two immovable horizontal tubes of D = 190 mm, focal length 8000 mm. There is a pavilion for the central unit and 2 roofed pavilions for eyepiece parts of the tubes. The MAHIS is provided with an automatic registering device, a system for meteorological data collection, and is computer controlled. The work on this telescope is being carried out at the Pulkovo, Kazan and Nikolaev observatories. The expected precision of the instrument is $\pm0\overset{''}{.}05$ in both coordinates with respect to systematic errors.

The Sukharev meridian automatic horizontal instrument (MAHIS) is intended for the determination of coordinates of celestial bodies: stars to 13th magnitude, planets and minor planets, star-like and disk-like objects. Photometry of celestial bodies to an accuracy of $0\overset{m}{.}05$, observations in different wavelength regions, and study of refraction can be performed with the use of MAHIS.

The design of the MAHIS is based on the Pulkovo HMC (Gumerov et al. 1986; Gumerov, Kapkov et al. 1987; Sukharev 1948). The essential part of the MAHIS is a monolithic flat mirror with an axis and the diameter of 300 mm. On both sides of the mirror graduated circles with a diameter of 412 mm and gears for setting the system are fixed. The automatic reading system of the divided circle consists of four photoelectric microscopes. Two additional microscopes are mounted for a study of division errors. The automatic setting system of the mirror along the zenith distance has two velocities of rotation: 10° /sec and 1 arcmin/sec, with a precision of setting of ±2". The mirror can be elevated with a special elevator — a counterpoise device meant to direct the horizontal tubes to each other. The position of the vertical line will be determined with the artificial horizons in nadir and zenith, with an accuracy of $0\overset{''}{.}01$ to $0\overset{''}{.}02$. MAHIS will be equipped with two vacuum immovable horizontal tubes (objective diameters of 190 mm, focal lengths 8000 mm). Star transit recording is made with the eyepiece photoelectric micrometer with an active analyser. The latter is a grid with a system of lambda-like slits. The time of registration can vary from 20 sec (for bright stars and light-marks) to 90 sec (for faint stars).

J. H. Lieske and V. K. Abalakin (eds.), Inertial Coordinate System on the Sky, 89–90.
© 1990 IAU. Printed in the Netherlands.

Computer control of the MAHIS makes it possible to realize two modes of star observations: *semiautomatic* control with the use of an observer for experimental observations and investigations, and *automatic* control for observations reduction and derivation of highly precise coordinates of a great number of stars.

There is a pavilion for the central unit and two roofed pavilions for eyepiece parts of the horizontal tubes. The observer and electronic facilities should be located in a separate room at a distance from the MAHIS.

The methods of investigations of the MAHIS are the same as for the automatic Puolkovo HMC (Gumerov *et al.* 1987a, 1987b; Pinigin 1976). It includes the determination and control of the variation of the instrumental parameters during star observation (*viz.*, the mirror azimuth relative to the horizontal tubes, the mirror tilt, the zero-point of the divided circle and the tube inclinations), collection of the meterological data and measurement of the temperature in the pavilion and instrument. The computer control allows one to include studies of the divided circle division errors, the pivot errors and the irregularities of the form of the relecting mirror's surface.

MAHIS enables not only the determination of differential coordinates of celestial bodies, but also the determination of absolute coordinates. For this purpose it is possible to determine a position of the divided circle relative to the vertical line with an artificial horizon, to receive a stable orientation of the two long-focus horizontal tubes as azimuth marks and to observe planets and minor planets (only at night). The expected precision of the MAHIS is $0''05$ in both coordinates with respect to systematic errors.

The observing program for MAHIS incorporates FK5 stars, IRS and other programs for the purpose of the derivation of a high accuracy star reference frame, of the determination of the interconnection between radio reference frame and optical reference frame and others. Simultaneous observations are planned to be made with two MAHIS' in the northern and southern hemispheres.

From the beginning of 1988 the design and model of some parts of the MAHIS and its electronic accessories have been elaborated.

References

Gumerov, R., Kapkov, V., and Pinigin, G.: 1986, "Automatic horizontal meridian circle at Pulkovo", in *Astrometric Techniques*, IAU Symposium 109, H. Eichhorn and R.J. Leacock (eds.), D. Reidel, Dordrecht, p. 206.

Gumerov, R., Kapkov, V., Kirian, T., and Pinigin, G.: 1987, "Potential of the computer controlled horizontal meridian circle at Pulkovo", in *Bull. Obs. Astron. Belgrade*, 137, 30.

Gumerov, R., Kirian, T., and Pinigin, G.: 1987a, Opredelenie popravok diametrov Pulkovskogo gorizontalnogo meridianogo kruga L.A. Sukhareva", in *Sovremenaja Astrometrija*, Akad. Nauk SSSR, Leningrad, p. 70 (in Russian).

Gumerov, R., Kirian, T., Korepanov, V., and Pinigin, G.: 1987b, "O sisteme skloneij gorizontalnogo meridianogo kruga Pulkovskoj observatorii", in *Sovremenaja Astrometrija*, Akad. Nauk SSSR, Leningrad, p. 256 (in Russian).

Pinigin, G.: 1976, "The results of determination of R.A. of 188 stars with −10° to +86° declination as obtained from observations with the Sukharev Horizontal Meridian Circle", in *Izv. GAO*, Nauka, Leningad, p. 105 (in Russian).

Sukharev, L.: 1948, "K voprosu o prinzipialnih preimushestvah i konstruktivnih osobenostjah horizontalnogo meridianogo instrumenta", in *Astron. Zhurnal* 25, No. 1, 59 (in Russian).

CONSTRUCTIONS OF ASTROMETRIC INSTRUMENTATION AT PULKOVO OBSERVATORY

V.N.YERSHOV [1], Yu.S.STRELETSKY [1] AND N.V.LEBEDEV [2]
[1] Central Astronomical Observatory USSR Academy of Sciences
Pulkovo
196140 Leningrad
USSR

[2] Television Institute
Leningrad, USSR

ABSTRACT. Past and present construction of astrometric instrumentation at Pulkovo is discussed.

From the very beginning of Pulkovo Observatory, special attention was paid by V.Ya. Struve to the construction of astrometric instrumentation. First, astrometric instruments ordered from Germany (such as the Vertical Circle and the Large Transit Instrument) were constructed under his close participation. The tradition of the participation of leading scientists in the production of astrometric instruments remains up to the present day.

Many fine instruments were created within the facilities of Pulkovo Observatory. For example, in 1904, the talented Pulkovo mechanic G.A.Freyberg fabricated a zenith-telescope ZTF–135 which has not lost its value as a scientific instrument even today. In the 1930s, N.N. Pavlov successfully used a method of photoelectric registration for transit instruments, thereby improving the time determination accuracy to 0.01 s.

In the 1950s and 1960s some new astrometric instruments were created, such as the thermally insulated transit instrument of N.N. Pavlov, A.A.Mikhailov's photographic vertical circle, and the large transit instrument for expeditions. Following the design of V.P. Linnik, the 6-meter baseline optical interferometer was created for the measurement of double stars. D.D. Maksutov, N.N. Mikhelson et al. designed at Pulkovo the fine two-meniscus wide field AZT–16 astrograph. During this period, Pulkovo astrometrists participated in the design process of the zenith-telescope ZTL–180 (V.I. Sakharov, I.F. Korbut) and of the photographic zenith tube (A.A. Mikhailov and V.A. Naumov).

The new instruments allowed the achievement of higher accuracy in stellar coodinates, latitude, longitude and time determination. Three instruments (AZT–16, the photographic vertical circle and the large transit instrument) were used in the Chile expedition.

In the 1970s and 1980s, the work of reconstruction and modernization of old instruments was carried out at Pulkovo. The design of the photographic vertical circle was essentially improved by B.K. Bagildinsky and V.D. Schkutov. The horizontal meridian circle in this period was fully automated. A scanning micrometer and a new photoelectric circle reading system were installed.

J. H. Lieske and V. K. Abalakin (eds.), Inertial Coordinate System on the Sky, 91–92.
© 1990 IAU. Printed in the Netherlands.

The modernization was carried out by G.I. Pinigin, T.R. Kiryan, R.I. Gumerov and other contributors of the Pulkovo and Engelhardt observatories.

The Toepfer meridian circle was used for a new semi- automatic instrument named MK–200. A scanning photoelectric micrometer was first installed on it and linear CCD's were used for the circle reading micrometers. The reconstruction of the Toepfer was done by Yu.S. Streletsky, Yu.G. Ostrensky, V.N. Yershov and others. Significant works on astrometric instrumentation were implemented at the Nikolayev department of Pulkovo Observatory. In the 1960s, a new circle reading device for the Repsold meridian circle was created and a circle division laboratory was formed. In 1980, the Repsold meridian circle was equipped with a photo-electric micrometer, giving an accuracy of 0".28. This was done by V.V. Konin, L.G. Karyakina, A.D. Pogony and others.

In this observatory, prototype of G.I. Pinigin's and O.E. Shornikov's axial meridian circle was designed and produced and the automatic measuring machine was made for astroplates (IFO–461). The main contributions to this work were made by A.V. Sergeev and O.E. Shornikov.

It is very urgent now to replace mechanical micrometers in meridian instruments by fixed micrometers based on the CCD. Accuracy of this micrometer increases at the cost of hard geometrical features and of obtaining more information about stellar images. With cooling and using a signal accumulation time of about 30 s, one can expand the magnitude limit of a standard meridian circle down to14–17 magnitude. This gives a possibility to solve new tasks by the means of the classical meridian method (*e.g.*, the connection of the optical reference frame to the inertial VLBI coordinate system). Additional efficiency can be obtained at the cost of increasing the variety of observing objects, such as close binary stars. New possibilities are opened when observing day-time objects (Venus, Mercury, and so on).

Some experiments were made in 1988 with CCD observations with the MK–200 meridian circle. This had a magnitude limit of about 8.8 . In 1989, some trial observations with a complex CCD micrometer were made in collaboration with the Leningrad Television Institute. The micrometer can automatically change its working mode and parameters. As a model of a meridian circle, a small telescope (D = 11 cm, f = 80 cm) was used. The estimation of accuracy made on the basis of $\alpha\, Oph$ observations gives a value 0".3 per 10 ms. During day-time observations of Polaris, the signal to noise ratio was about 50, which gives a day-time magnitude limit of about 4.5 .

Now some new astrometric instruments are being designed at Pulkovo Observatory. And following the tradition at Pulkovo, astrometrists participate directly in the creation of the instruments.

On the basis of experience with Sukharev's horizontal meridian circle investigations, a new automatic horizontal meridian circle (SMAHI) is being developed, which could give systematic errors less then 0".05. The organizer of the work is G.I. Pinigin.

Based on proposals of A.A. Nemiro and Yu.S. Streletsky, a reflecting automatic meridian circle is under construction. This instrument will give an improved limiting magnitude. Optical elements of the meridian circle will be checked by the special autocollimating systems and therefore the systematic errors will be minimized. A design of the photoelectric zenith tube is completed and is being implemented now. This instrument has an image dissector tube as a photodetector. As a prospect to the future, space astrometric telescopes are under consideration at Pulkovo Observatory.

POSITIONAL OBSERVATIONS OF COMPONENTS OF TRIPLE STARS FOR THE HIPPARCOS AND HST MISSIONS

J.P. ANOSOVA, V.V. ORLOV
Leningrad University Astronomical Observatory
Bibliotechnaya Pl. 2
198904 Leningrad, USSR

ABSTRACT. This work has been made in accordance with the plan of the International Working Group "Double Stars: HIPPARCOS Input Catalogue Consortium." The precise coordinates for the components of the triple stars from the programs of the Astronomical Observatory (AO) of Leningrad State University (LSU) and the Uccle zone (declination from +30° to +50°) are determined. The work has been made by two methods: (1) special astrometrical observations for about 70 triple stars; some colleagues from the observatories in Pulkovo, Nikolaev, and Goloseevo (USSR), as well as from Belgrade (Yugoslavia) have taken part in this work; (2) statistical treatment for about 200 triple stars, using the data of the catalogues ADS and WDS—all available positional observations of the relative coordinates of the components.

Also, the special observations of the triple stars in Pulkovo and Belgrade have been used, as well as the data from the catalogue by Ch. Worley from 1820 to 1988. The old observations by W. Struve in 1820-1830 play an important role. The data obtained are used for the compilation of the observational programs of the AO of LSU by the HIPPARCOS and HST.

PROBLEM OF USING LUNAR POSITIONAL OBSERVATIONS FOR DETERMINATION OF ZERO-POINTS OF FUNDAMENTAL STAR CATALOGUES

V.A. FOMIN
Central Astronomical Observatory, USSR Academy of Sciences
Pulkovo
196140 Leningrad, USSR

ABSTRACT. The long series of meridian observations of the Moon can be used for the precise determination of the equinox- and equator-corrections of a star catalogue. Systematic errors of different charts of the lunar marginal zone used for the reduction of the lunar limb observations have no influence on the determination of the secular variations of the zero-points of the fundamental coordinate system.

From meridian observations of the lunar limb made during the interval 1923–1977 in Washington, Greenwich, Cape and Tokyo the following estimate is found for the correction to the right ascension system of the FK4 catalogue:

$$\Delta\alpha(FK4) = -0\overset{s}{.}017 \pm 0\overset{s}{.}002 \ -0\overset{s}{.}019 \pm 0\overset{s}{.}018 \ (T - 19.46) \ \text{sec},$$

which is in disagreement with the values used for the compilation of the FK5 catalogue.

DENSITY FUNCTION OF THE FK4 SYSTEMATIC ERRORS

V.S. Gubanov, N.I. Solina
Institute of Applied Astronomy
8 Zhdanovskaya ul.
197042 Leningrad, USSR

ABSTRACT. A notion of density function of systematic errors of astrometric catalogues distributed on the unit sphere as a simple layer is introduced. Components of the catalogue errors in any direction are determined as partial derivatives of layer potential in the same direction. For example, the density function of FK4 errors is computed as an expansion of spherical harmonics.

ON THE TIE OF ASTROMETRIC RADIO SOURCES TO THE FK4 SYSTEM WITH ZENITH TELESCOPE ZTL-180 OBSERVATIONS

E.Ya. Prudnikova
Central Astronomical Observatory
Pulkovo
196140 Leningrad, USSR

ABSTRACT. A programme of Talcott pairs has been compiled at Pulkovo in order to tie bright radio sources to the FK4 system. The programme includes FK4 stars and radio stars to mag 9 (*Radio stars for astrometry*, Abh. Hamb. Sternw., 1982). About 150 observations of 18 pairs were made during 1985–1989. The correction

$$\Delta\delta = 0.5 \, (\Delta\delta_{FK4} + \Delta\delta_R)$$

to the declination of a radio pair has been found as a difference between the "reference" latitude and the latitude derived from the observation of the radio pair. The value of the "reference" latitude was determined from observations of ordinary Talcott pairs. The standard deviation does not exceed 0.09 arcsec. A radio star can be tied to the FK4 with an accuracy not worse than 0.1 arcsec.

CIRCLE-READING DEVICE OF THE MERIDIAN INSTRUMENT WITH SELF-INSTALLED ZERO-POINT

G.I. Pinigin
Nikolaev Branch, Pulkovo Observatory
327030 Nikolaev, USSR
and
Y.A. Bubnov, A.W. Shumacher
Central Astronomical Observatory, USSR Academy of Sciences
Pulkovo
196140 Leningrad, USSR

ABSTRACT. A circle-reading device of the meridian circle with a constant link to the vertical line and the method of using the device are proposed. The accuracy of a circle reading is expected to be 0.''02.

MODIFIED METHOD OF KREJNIN AND MURRI FOR THE DETERMINATION OF ABSOLUTE DECLINATIONS

S.A. TOLCHELNIKOVA-MURRI
Central Astronomical Observatory USSR Academy of Sciences
Pulkovo
196140 Leningrad, USSR

ABSTRACT. Krejnin and Murri's method (1973) enablesenables one to derive absolute declinations of stars in a narrow equatorial zone $|\delta| \leq 10'$ from observations near the Earth's equator $|\varphi| \leq 10'$. Some systematic effects, including the errors of the value of the micrometer screw for two equatorial instruments (or the scale error if one of the instruments is a PZT), might be determined if a global reduction is used for the original observations from the equator and from those of an astrolabe at latitude $|\varphi| \approx 20°$ to $23°$. Astrolabes—especially photoelectric ones (Hu 1988) are considered to be the most efficient for determination of absolute declinations of stars and absolute latitudes of the instruments in Tolchel'nikova-Murri (1985).

In *Izv. GAO* No. 206 the method will be published as well as the criterion for estimating the efficiency of different programs, which is required to improve planning in astrometry.

References

Hu Hui, Cai Xing, Wang Rui: 1988, *Acta Astron. Sci.* **4**, 333
Krejnin, E.I., Murri, S.A.: 1973, *Astron. zurn.*, **50**, 606
Tolchel'nikova-Murri, S.A.: 1985, "A Method for Determination of Variation of Mean Latitudes and the Secular Polar Motion," Dep. No. 150-185

ON THE DEFINITION OF AN "INERTIAL COORDINATE SYSTEM"

S.S. PERUANSKY
Central Astronomical Observatory
Pulkovo
196140 Leningrad, USSR

ABSTRACT. Astrometry is a branch of science which develops methods for the quantitative descriptions of places and time instants of astronomical events on the basis of observations of celestial bodies. For this purpose a theoretical coordinate system is introduced (*e.g.* equatorial α, δ). The aim of astrometry is to apply this system to the observed reference objects (stars, planets etc.) so that their coordinates $\alpha(t)$, $\delta(t)$ can be calculated according to the relations $\alpha(t) = f_1(P_k, t-t_o)$ and $\delta(t) = f_2(P_k, t-t_o)$ where P_k are parameters, t_o is the conventional time instant and t is the current time. In order to understand the term inertial coordinate system assume that the coordinates $\alpha(t_i)$, $\delta(t_i)$, $i=1,2,...n$ are used for plotting the coordinate origins. If these coincide then the system is conventional-fixed and therefore inertial. Thus, the inertial coordinate system in astrometry is a conventional-fixed reference frame reproduced with the use of celestial bodies whose law of motion is known with sufficient accuracy.

CONTRIBUTION OF SOVIET-BOLIVIAN ASTRONOMICAL OBSERVATORY IN THE CONSTRUCTION OF THE INERTIAL COORDINATE SYSTEM

D.D. POLOZHENTSEV, H.I. POTTER AND L.I. YAGUDIN
Central Astronomical Observatory
Pulkovo
196140 Leningrad, USSR
and
J.A. ZELAYA AND R.F. ZALLES
Observ. Astron. Nacional
Tarija, Bolivia

ABSTRACT. The work on photographic astrometry in Bolivia begun in 1983, when an expeditional astrograph was put there (D=23 cm, F=230 cm, working field of 4°x4°). The main purpose of the observations with this telescope is extending the inertial coordinate system to different classes of stars. For this purpose we have carried out and are currently carrying out the following observational astrometric programs:

1. The preliminary catalog of precise positions of 200 000 southern stars to mag 11 with declination in the range from 0° to –90° was created. Each star was observed four times with displacement of the photoplate by 2° in δ and by 8 secδ in α. The exposure time is from 4 to 12 minutes. The accuracy of the preliminary catalog is ±0.247 arcsec and the final one is ±0.15 arcsec.

2. We are not far from finishing the work on a catalog of 1900 bright stars (–90° < δ; 0°, m ≤ 6.05). Each star was observed not less than 2 times for different positions of a special filter which diminished the light of bright stars.

3. The work on astrometry of equatorial stars to mag 12 in the range of declination from –20° to +20° was begun. From the beginning of the observations on 1 May 1987, 1150 astroplates were obtained. On the whole it is necessary to obtain 3840 photoplates. All three programs use the reference catalog SRS, compiled by Pulkovo and Washington astronomers, in the FK5 system.

THE PROBLEM OF THE LEVELING OF THE RESERVED NET OF ARCS AT THE CELESTIAL SPHERE BY THE CONSTRUCTION OF THE REFERENCE COORDINATE SYSTEM

E.I. YAGUDINA
Institute of Applied Astronomy
8 Zhdanovskaya ul.
197042 Leningrad, USSR

ABSTRACT. The methods of the rational leveling of the results of the observations of the angular distances between quasars by VLBI method were suggested. The application of these methods on the model of reserved net on sphere shows a descrease of the standard error of the quasers' coordinates after leveling by a factor of three times. We employed 10 elementary configurations constructed on 20 chosen quasars in the Northern hemisphere.

Part 3
Concepts, Definitions, Models

Plate VI. The original Pulkovo Observatory viewed from the southwest. The large screen protects the prime vertical instrument from direct sunlight. Photograph provided by Pulkovo Observatory.

Plate VII. The reconstructed Pulkovo Observatory viewed from the southwest. Photograph provided by Pulkovo Observatory.

THE CELESTIAL REFERENCE SYSTEM IN RELATIVISTIC FRAMEWORK

HAN CHUN-HAO, HUANG TIAN-YI AND XU BANG-XIN
Department of Astronomy
Nanjing University
Nanjing 210008
People's Republic of China

ABSTRACT. The concept of reference system, reference frame, coordinate system and celestial sphere in a relativistic framework are given. The problems on the choice of celestial coordinate systems and the definition of the light deflection are discussed. Our suggestions are listed in Sec. 5.

1. Introduction

With the increasing of accuracy of observations, one should pay more and more attention to the relativistic effects in celestial reference systems. Some papers dealing on the definition and realization of reference system have been published (Moritz 1981; Brumberg 1981; Fujimoto *et al.* 1982; Eichhorn 1984; Fukushima *et al.* 1986; Brumberg *et al.* 1989). The monographs titled *Vectorial Astronomy* (Murray 1983), *Relativity in Astrometry, Celestial Mechanics and Geodesy* (Soffel 1989) and *Reference Frames in Astronomy and Geophysics* (Kovalevsky *et al.* 1989) provide the useful tools for construction of the modern astrometric theory in the framework of general relativity.

The concept of the work "reference frame" is used differently in physics and astronomy (Brumberg 1989) and there exist different opinions even among astronomers. According to the suggestion of Kovalevsky and Mueller (1981), "The purpose of a reference *frame* is to provide the means to materialize a reference *system* so that it can be used for the quantitative description of positions and motions on the Earth or of celestial bodies in space." For example, the catalogue of over 1500 star coordinates defines the FK4 frame, materializing the FK4 system. An alternative definition is defined by a set of (physical) points which constitutes a well-defined spatial configuration. To specify the location of a point, which is not an element of the defining set within this frame of reference, one needs a construction that sets numbers to specify completely and uniquely the location of the point with respect to the specific frame of reference. Such a construction is called a coordinate system (Eichhorn 1984). According to this definition the FK4 frame is defined by the barycentre and the FK4 stars themselves rather than their coordinates. It seems that attention must be paid to the distinction between these definitions.

One would also question the conceptual aspects of the barycentric celestial sphere as well as the global equatorial coordinate system, because the equator and the ecliptic could only be defined locally at the Earth when the spacetime is curved.

99

J. H. Lieske and V. K. Abalakin (eds.), Inertial Coordinate System on the Sky, 99–110.
© 1990 *IAU. Printed in the Netherlands.*

Another questionable concept is the light deflection, the most important relativistic effect for optical observations. Because of the space curvature, the two tangent vectors of a light path, at the observer and at the remote source respectively, belong to different tangent spaces, and there is no exact definition to determine the angle made by these two vectors.

In the present paper, we try to make one step to solve these problems. For simplicity our discussions are mainly limited to optical astrometry. The reference system and celestial sphere will be discussed in Section 2, the choice of coordinate systems and the light deflection will be treated in section 3 and 4 respectively. Section 5 lists our conclusions.

2. The Reference System and the Celestial Sphere

In some parts of this section we are much influenced by the textbook of Sachs and Wu (1977), but we use our own words to explain the concepts in order to be understood by astronomers more easily. If there is some misunderstanding we should take the full responsibility.

2.1 OBSERVER

The worldline of any pointlike object is an *observer* (Sachs and Wu 1977, p. 41). Here the word "observer" has a more general meaning than that in astronomy. It could be a real observer or an observational object.

Every observation is made by an instantaneous observer, which could be defined as $(\mathbf{E}, \mathbf{U})$, where $\mathbf{E}$ represents the event of the observation and the $\mathbf{U}$ is the instantaneous 4-velocity of the observer that passes through $\mathbf{E}$ (Sachs and Wu 1977 p. 43).

2.2 REFERENCE FRAME

Reference frame is a group of selected observers, to which all the observations or motions are referred. The only restriction to these observers is that they can not cross each other, otherwise at one event there could exist two different instantaneous observers that move with respect to each other. The observers, of which the reference frame is comprised, are called stationary observers with respect to the very reference frame.

For astronomical application a reference frame must be realizable and be carefully chosen (Kovalevsky 1989). The stationary observers could be real celestial bodies such as remote stars or fictitious bodies connected with the real objects such as the barycentre in the solar system.

2.3 COORDINATE SYSTEM

A *coordinate system* draws a 4-dimensional network on the spacetime to make the quantitative measurement of the events and the other physical quantities possible.

After a reference frame has been selected, there still remain infinite choices of possible coordinate systems. On the contrary any coordinate system implies one reference frame, which is sometimes not physically realizable. A family of coordinate systems belong to the same reference frame if the transformation between them is as follows:

$$t = \varphi(t', x_i') ,$$
$$x_i = f_i(x_i') \tag{1}$$

where i and i' run from 1 to 3, φ and f are scalar functions (Møller 1973). Eq. (1) tells us that $\partial / \partial t$ and $\partial / \partial t'$ are along the same direction.

A coordinate system is essentially only a mathematical tool for convenience. In traditional astrometry coordinate systems have been usually defined physically. In the language of this paper we could say that the coordinate system defined by the equinox and the equator belongs to the reference frame specified by them. They could be considered as connected with some celestial bodies.

A local tetrad is a local coordinate system, whose reference frame is represented by the instantaneous observer at its centre and some observers in its infinitesimal neighbourhood.

2.4 REFERENCE SYSTEM

A *reference system* includes three parts:

1)   a reference frame:
2)   a coordinate system;
3)   a recommended set of constants, theories and procedures of data processing.

Parts 1) and 2) construct an idealized reference system, in which the reference frame represents the physical and fundamental part and the coordinate system is its mathematical aspect. Part 3) makes the reference system real. Astronomers make great efforts to improve that in part 3) from time to time and carefully keep the same reference frame and the same coordinate system. It is possible to happen that astronomers want to change the reference frame itself, for example, from the stellar frame to the radio source frame.

2.5 OBSERVER'S CELESTIAL SPHERE

Let (E, U) be an instantaneous observer, then the *local rest space* of (E, U) is defined as the 3-dimensional subspace of the tangent space at E, in which every vector is orthogonal with U (Sachs and Wu 1977, p. 45). The light direction observed by (E, U) is the projection vector of the 4-dimensional tangent vector of the light path at E into its local rest space.

It is natural to define the *local celestial sphere* of (E, U) to be all the unit vectors as a whole in its local rest space. We use the word "unit" here in order to remove one dimension and keep the direction only in the concept of the celestial sphere. Actually the observation is always taken as the event E when the photon arrives and only the velocity of the photon, which is different from U, is meaningful.

One can also talk about the *celestial sphere of an observer*. It can be defined in the same way as above. It is moving together with the observer. Then one should define the correspondence between the vectors of the local celestial spheres at one instant and another. Different correspondences could represent diferent local reference frames.

There is not much sense to talk about the stellar reference system in the local celestial sphere, in which the star light is changing from time to time. This is not only due to the aberration or parallax but also to the curvature of the spacetime. The invariant direction of the stars can only make sense in the barycentric celestial sphere described in the next subsection.

2.6 BARYCENTRIC CELESTIAL SPHERE

The *barycentric celestial sphere* can be defined as the reference frame that takes the remote sources and the barycentre as its stationary observers. As in the previous subsection one dimension representing the distance from the barycentre has been suppressed.

To include the barycentre in this definition is due to the fact that in relativity there would be different space and time separation for different Lorentz reference frames even though in a flat spacetime.

In defining the barycentric celestial sphere, there is no restriction on the barycentric reference frame inside the solar system except the barycentre, but it does put some limitation on the barycentric reference frame to assure that the remote sources are stationary. Here and after we will consider the barycentric reference frame that meets this restriction only.

3. Choice of the Coordinate System

3.1 THE PROBLEM

In solar system dynamics, the PPN metric is popularly adopted (Will 1981). The general form of a metric is

$$ds^2 = g_{\mu\nu} \, dx^\mu \, dx^\nu \tag{2}$$

and the coefficients of the PPN metric are estimated to be

$$g_{00} = -1 + O(2), \qquad g_{0i} = O(3)$$

$$g_{ii} = 1 + O(2), \qquad g_{ij} = O(4), \ i \neq j \,. \tag{3}$$

Here the Greek letters run from 0 to 3 and the Latin letters from 1 to 3. The PPN metric is quasi-isotropic and its coordinate system is quasi-Cartesian.

The coordinates appear in $g_{\mu\nu}$ only in their differences so that a rigid rotation of the space axes will not change $g_{\mu\nu}$. Traditionally one should choose the direction of the mean equinox of J2000 to be x^1 axis and the mean equator of J2000 to be $x^1 x^2$ plane. The curvature of the spacetime brings in problems to realize this tradition. Actually the equatorial and the ecliptic polices can only be defined locally at the Earth as two vectors and they do not have global definition. To solve this problem we should 1) define a definite connection between the global coordinate system and the local tetrad centered at the Earth, 2) adjust the directions of the space base-vectors of the local tetrad according to the locally defined equinox and the equator, and then 3) make the corresponding adjustment of the space axes of the global coordinate system.

3.2 THE CONNECTION BETWEEN THE GLOBAL AND THE LOCAL SYSTEM

The connection constructed in the following holds not only for the local tetrad at the Earth, but also for that at any observer. It could be considered as the definition of the local tetrad when a global coordinate system is given.

At every event x^α there is an instantaneous observer whose 4-velocity $\mathbf{e}_{(0)}$ is tangential to the t-axis of the global coordinate system and can be expressed as

$$\mathbf{e}_{(0)} = \frac{1}{\sqrt{-g_{00}}} \frac{\partial}{\partial t} \tag{4}$$

The local celestial sphere, *i.e.* the local rest space, of $(x^\alpha, \mathbf{e}_{(0)})$ can be the space expanded by three mutually orthogonal unit base vectors, $\mathbf{e}_{(i)}$. They are also orthogonal to $\mathbf{e}_{(0)}$ according to Section 2. We name this kind of tetrad, $\mathbf{e}_{(\alpha)}$, natural tetrad (NT) or natural frame after Murray (1983). Obviously there are infinitely possible choices of $\mathbf{e}_{(i)}$, which are connected by rigid rotations.

Soffel (1989, p. 77) recommended choosing

$$e_{(m)i} = \left(\hat{g}^{1/2} \right)_{mi} \tag{5}$$

where the 3x3 matrix

$$\hat{g}^{-1} = \left(g^{ij} \right). \tag{6}$$

Soffel pointed out that "if the metric $g_{\mu\nu}$ is diagonal the induced tetrad is simply given by

$$e_{(\alpha)} = |g_{\alpha\alpha}|^{-1/2} \frac{\partial}{\partial x^\alpha} \qquad ." \tag{7}$$

In the general case it is impossible to construct the relation between $\mathbf{e}_{(m)}$ and $\partial/\partial x^\alpha$ in a closed form based on (5) and (6). Also $\mathbf{e}_{(1)}$ usually does not agree with $\partial/\partial x^1$ and the plane generated by $\mathbf{e}_{(1)}$ and $\mathbf{e}_{(2)}$ does not coincide with that generated by $\partial/\partial x^1$ and $\partial/\partial x^2$.

We suggest an alternative choice of NT. By Gram-Schmidt orthonomalization (Horn and Johnson 1985) we can construct a mutually orthogonal and normalized tetrad $\mathbf{e}$:

$$\begin{aligned}
\mathbf{e}_{(0)} &= \left(\frac{\partial}{\partial x^0} , \frac{\partial}{\partial x^0} \right)^{-1/2} \frac{\partial}{\partial x^0} = (-g_{00})^{-1/2} \frac{\partial}{\partial t} \\
\mathbf{e}_{(\alpha+1)} &= \left(\mathbf{e}'_{(\alpha+1)} , \mathbf{e}'_{(\alpha+1)} \right) \mathbf{e}'_{(\alpha+1)} \\
\mathbf{e}'_{(\alpha+1)} &= \frac{\partial}{\partial x^{\alpha+1}} + \sum_{\beta=0}^{\alpha} a_{\alpha\beta} \, \mathbf{e}_{(\beta)} \\
a_{\alpha\beta} &= - \left(\frac{\partial}{\partial x^{\alpha+1}} , \mathbf{e}_{(\beta)} \right) \left(\mathbf{e}_{(\beta)} , \mathbf{e}_{(\beta)} \right)^{-1}
\end{aligned} \tag{8}$$

Keeping only up to $0(3)$ terms, they are

$$\begin{aligned}
\mathbf{e}_{(0)} &= (-g_{00})^{-1/2} \frac{\partial}{\partial t} \\
\mathbf{e}_{(1)} &= g_{01} \frac{\partial}{\partial t} + g_{11}^{-1/2} \frac{\partial}{\partial x^1}
\end{aligned} \tag{9}$$

$$\mathbf{e}_{(2)} = g_{02}\frac{\partial}{\partial t} - g_{12}\frac{\partial}{\partial x^1} + g_{22}^{-1/2}\frac{\partial}{\partial x^2}$$

$$\mathbf{e}_{(3)} = g_{03}\frac{\partial}{\partial t} - g_{13}\frac{\partial}{\partial x^1} - g_{23}\frac{\partial}{\partial x^2} + g_{33}^{-1/2}\frac{\partial}{\partial x^3}$$

(9, cont.)

For time-orthogonal coordinate system g_{oi} are equal to zero, $\mathbf{e}_{(1)}$ is along the direction $\partial /\partial x^1$ and $\mathbf{e}_{(2)}$ is on the plane generated by $\partial /\partial x^1$ and $\partial /\partial x^2$. Though it is quite complicated, Eq. (8) could be written in a closed form.

Here and after the word "observer" will refer to a real observer. It may be in motion with respect to the background coordinate system. Let τ be its proper time, then its 4-velocity is

$$\tilde{\mathbf{e}}_{(0)} = \frac{\partial}{\partial \tau} = \frac{dx^\alpha}{d\tau}\frac{\partial}{\partial x^\alpha}$$

(10)

One would construct another tetrad, $\tilde{\mathbf{e}}_{(\alpha)}$, called proper tetrad (PT) or proper frame (Murray, 1983), which is centered at the instantaneous observer, $(x^\alpha, \tilde{\mathbf{e}}_{(\alpha)})$, and is related to the above NT by a Lorentz transformation (Soffel, 1989):

$$\tilde{\mathbf{e}}_{(\mu)} = \Lambda_{\tilde{\mu}}^\nu \mathbf{e}_{(\nu)}$$

(11)

where the Lorentz matrix is

$$\Lambda^0_{\tilde{0}} = \tilde{\gamma}$$
$$\Lambda^0_i = \Lambda_{\tilde{0}}^i = \tilde{\gamma}\tilde{\upsilon}^i$$
$$\Lambda^i_{\tilde{j}} = \delta_{ij} + \frac{(\tilde{\gamma}-1)}{\tilde{\upsilon}^n \tilde{\upsilon}^n}$$

(12)

and

$$\tilde{\upsilon}^i = \frac{\tilde{e}_0^\mu e_{(i)\mu}}{\tilde{e}_0^\mu e_{(0)\mu}}$$

$$\tilde{\gamma} = \left(1 - \frac{1}{2}\tilde{\upsilon}^n \tilde{\upsilon}^n\right)^{-1/2}$$

(13)

PT is the comoving tetrad with the observer. It should be considered as the coordinate system in which the astrometrists measure their optical data. Similarly to NT, there are infinitely possible PT. Astronomers have to choose one as conventional when processing their data. Formulae (9) – (13) are our recommendation.

3.3 THE ADJUSTMENT OF THE SPACE AXES

The classical precession and nutation can be considered as the motion of the celestial pole with respect to a tetrad called a Fermi tetrad (FT) or Fermi frame at the Earth. FT is a tetrad that is centered at an observer $x^{(0)}$. Its time axis is $\tilde{e}_{(0)}$ and space axes could be realized by three gyroscopes that are mutually orthogonal. After correction of the relativistic precession, of which the main term is usually called geodetic precession, one can define the celestial equatorial pole in PT of the Earth.

To define the celestial ecliptic pole is more difficult. During the construction of DE ephemerides as described by Standish (1980), a vector $r \times dr/dt$ is defined as the direction of the normal of the instantaneous orbit plane of the Earth, where the components of r and dr/dt are x^i and dx^i/dt of the Earth respectively in a simplified PPN coordinate system. In the relativistic framework dx^i/dt could be replaced by the 4-velocity $dx^\alpha/d\tau$ that is coordinate system independent. But r or x^i is coordinate system dependent. This fact shows the difficulty to define the ecliptic in a coordinate system independent way. We have not solved this problem.

Now let us assume that the celestial equatorial and the ecliptic pole of epoch have been defined in the NT at $(x^\alpha, e_{(0)})$. One could determine the dynamic equinox and the equator of epoch in the very NT, then adjust $e_{(1)}$ along the direction of the equinox of epoch and $e_{(2)}$ on the equator of epoch. This adjustment is a rotation in the local rest space (the local celestial sphere) of $(x^\alpha, e_{(0)})$. Let P be the orthogonal matrix representing the rotation for the adjustment of NT. In order to keep the relation between the local NT and the background coordinate system as described by Eq.(9), one has to rotate the background space axes. Let M represent the rotation matrix of the background system. It is necessary to find the relation between M and P. Within post-newtonian accuracy we found that

$$M = P - PH + H'P, \tag{14}$$

and

$$P = M - H'M + MH \tag{15}$$

where

$$H + H^T = G - I,$$

$$H' + H'^T = G' - I,$$

$$G = (g_{ij}) \qquad G' = (g'_{ij}) = M G M^T \tag{16}$$

$$I = (\delta_{ij}),$$

and H^T is the transposed matrix of H. In other words

$$H = \begin{vmatrix} \frac{1}{2}h_{11} & 0 & 0 \\ h_{12} & \frac{1}{2}h_{22} & 0 \\ h_{13} & h_{23} & \frac{1}{2}h_{33} \end{vmatrix} \tag{17}$$

$$h_{ij} = g_{ij} - \delta_{ij}$$

H' is defined by a similar formula.

The elements of H and H' are all in the magnitude of $0(2)$. Since the adjustment from the catalog equinox and equator to the dynamic equinox and equator is a very small quantity, which is usually less than 0".1, one can neglect the difference between M and P.

Eqs. (14) and (15) can also be used when the transformation of the coordinate system between different epochs is necessary. One important nature of Eqs. (14) and (15) is that the relation between M and P is location dependent. Factually, M – P depends on the metric coefficients, $g_{\mu\nu}$, which are functions of the location of the observer. Here we would like to point out that M is equal to P exactly if $g_{oi} = 0$ and $g_{ij} = 0$ $(i \neq j)$, which implies that the background coordinate system is isotropic. The real metric in the solar system does not meet this demand but could satisfy it within a certain accuracy. We call the metric that nearly meets this condition quasi-isotropic metric and it is preferable.

4. The Light Deflection

4.1 THE PROBLEM

To complete the construction of the celestial reference system a conventional procedure for data processing has to be founded. For optical observations the most important relativistic effect is the light deflection. Here we limit ourself to the discussion of this effect.

Many works on the light deflection have been published. The following lists some of the recent publications. Murray (1981) derived his formulae within the post-newtonian accuracy. Epstein and Shapiro (1980), Fischbach and Freeman (1980) worked up to post-post-newtonian terms and later Xu et al. (1984) extended their results to the case that treats a source and an observer that are both inside the solar system. All of them dealt with a spherically symmetric metric only. Richter and Matzner (1982) demonstrated that knowledge about light propagation in the solar system to any given order requires knowledge of every term in the metric to that same order. They extended the PPN metric to a metric called parametrized post-lineal (PPL) metric by including the second-order relativistic contributions from the Sun into g_{ij}. But they calculated the gravitational deflection of a photon in the equatorial plane of the Sun only.

Being different from physicists, astronomers are not so interested in the total deflection of a light from a remote source to a remote observer. The most accurate observations are made on or near the Earth. This fact poses a problem. The two tangent vectors of the light path that are at the source and at the observer belong to different tangent spaces at different locations. There is no known definition of the angle between the two tangent vectors. Atkinson (1963) has pointed out that the relativistic effect in a stationary gravitational field with spherical symmetry can be calculated by pure Euclidean geometry. The real metric in the solar system is much more complicated. It is neither stationary nor spherically symmetric. When astronomers have to pursue higher precisions in their data processing, it is necessary to give a definition of the light deflection that can be followed in calculation for any case.

4.2 THE DEFINITION OF THE LIGHT DEFLECTION

Let P_s and P_o be the two 4-velocities of the light path from a source to an observer and let them be in the corresponding tangent spaces at two events, E_s and E_o, respectively. Here E_s is the event that the source launched the photon; E_o is the event that the observer received the same photon. It is evident that P_s and P_o are independent of coordinate systems and also of reference frames.

Assume $\mathcal{F}$ to be a family of barycentric coordinate systems that imply the same barycentric celestial sphere as introduced in Section 2. Also, without loss of generality, we assume that they are all quasi-Cartesian coordinate so that the corresponding metric at infinity is Minkowsky metric and the base vectors, $\partial /\partial t$ and $\partial /\partial x^i$, are all the same at infinity among this family. The coordinate resolution of $\mathbf{P}_s$ and $\mathbf{P}_o$ in one member of $\mathcal{F}$ are

$$\mathbf{P}_s = P_s^o \frac{\partial}{\partial t} + P_s^i \frac{\partial}{\partial x^i} \tag{18}$$
$$\mathbf{P}_o = P_o^o \frac{\partial}{\partial t} + P_o^i \frac{\partial}{\partial x^i}$$

The 3-dimensional vector $P_s^i \partial /\partial x^i$ can be defined as the position of the source in the barycentric celestial sphere and it is coordinate system independent among $\mathcal{F}$. And $P_o^i \partial /\partial x^i$ represents the observed direction by an instantaneous observer $(\mathbf{E}_o, \mathbf{U}'_o)$ where $\mathbf{U}'_o$ is along the direction of $\partial /\partial t$. It is evident that $P_o^i /\partial x^i$ is generally coordinate system dependent among $\mathcal{F}$, but it is coordinate system independent among a subset of $\mathcal{F}$, between which the transformation takes the form of Eq. (1).

Several authors (Soffel, 1989; Brumberg, 1989) calculated $P_s^i - P_o^i$. Obviously it is coordinate system dependent among $\mathcal{F}$ and even among its subset mentioned above because the spatial base, $\partial /\partial x^i$, at $\mathbf{E}_o$ is generally different from one coordinate system to another.

Astronomers possibly would not like to define $P_s^i - P_o^i$. as the light deflection. The problem is that the spacial base, $\partial /\partial x^i$, is neither orthogonal nor normal and $P_o^i P_o^i$. is generally not equal to 1. On this kind of base one can not introduce the spherical equatorial coordinates, the right ascension and the declination. A more natural and practical definition we suggest here is the following. Firstly, construct NT at every event over the spacetime, which is related to the local base, $\partial /\partial x^\alpha$, by a set of defined formulae. Here and after we use $\mathbf{e}_{(\alpha)}(\mathbf{E})$ to denote the NT at event $\mathbf{E}$. Resolving the vectors $\mathbf{P}_s$ and $\mathbf{P}_o$ with respect to $\mathbf{e}_{(\alpha)}(\mathbf{E}_s)$ and $\mathbf{e}_{(\alpha)}(\mathbf{E}_o)$ respectively, we have

$$\mathbf{P}_s = n_s^o \mathbf{e}_{(0)}(\mathbf{E}_s) - n_s^i \mathbf{e}_{(i)}(\mathbf{E}_s) \tag{19}$$
$$\mathbf{P}_o = n_o^o \mathbf{e}_{(0)}(\mathbf{E}_o) - n_o^i \mathbf{e}_{(i)}(\mathbf{E}_o)$$

and $n_s^i n_s^i = n_o^i n_o^i = 1$ because the measured light speed in a local inertial frame is equal to 1. Then we define the light deflection as $\mathbf{n}_s^i - \mathbf{n}_o^i$ and the angle between the two directions as

$$\Delta\theta = | \mathbf{n}_s \times \mathbf{n}_s | \tag{20}$$

$$\mathbf{n}_s = [n_s^1, n_s^2, n_s^3]^T, \qquad \mathbf{n}_o = [n_o^1, n_o^2, n_o^3]^T.$$

This is a pure Euclidean definition when the two NT are fictitiously coincided with each other. Actually this definition is a new version of that described by Murray (1981, 1986) but in a more general case and in a more clear and definite way.

It is evident that $n_s^i - n_o^i$ and $\Delta\theta$ both are coordinate system dependent. They depend on the choice of NT and the choice of the barycentric coordinate system. We should choose NT to assure that $\mathbf{e}_{(\alpha)}(\mathbf{E}_s)$ coincides with the base vectors at $\mathbf{E}_s$: $\partial /\partial x^\alpha$. With this limitation the natural tetrads are restricted in a family called $\mathcal{N}$ here and after. The natural tetrads recommended by Soffel (1989) or by this paper in Eq. (9) both meet this demand. To be consistent with Section 3, we recommend

108

that defined in Eq. (9).

It has to be mentioned that the angle between two light directions from two sources is coordinate system independent whether at infinity or at the observer. Therefore the correction of the light deflection of this angle is also coordinate system independent. In astronomical practice one needs to do the correction of the observational data one by one so that the discussion in this section is necessary.

The above definition has been applied to the Schwarzschild metric. We found that the formulae for the light deflection in isotropic coordinates is much simpler than that in the standard coordinates.

In Murray's paper (1981) the formulae for the light deflection in these two coordinates are the same. This is due to the fact that his definition of the natural frame brings in the same tetrads for these two coordinates and is different from ours, but it would not be the same in the general case. We do not think that one definition is superior to another solely so far as the light deflection is concerned. The most important thing is that the position of the source in the barycentric celestial sphere after the correction of the light deflection and other system effects is unique and invariant. To this end any selection of the barycentric coordinate system among the family $\mathcal{F}$ and any definition of the natural tetrad among the family $\mathcal{N}$ is allowed.

5. Conclusions

Our main conclusions are

1) Astronomers should have an agreement on the implication of "reference system", "reference frame" and "coordinate system". We suggest that the terminology of astronomers should be close to that of physicists if possible. It is proper to distinguish the physical aspect "reference frame" from the mathematical aspect "coordinate system". We suggest that a "reference system" include these two components and a recommended procedure of data processing, a set of constants and theories that are involved.

2) Astronomers should have an agreement on the relation between the barycentric coordinate system and the local natural tetrad in order to keep a consistent procedure of data processing. We recommend that defined in Eqs. (8) and (9).

3) The barycentric quasi-isotropic coordinate is preferable because the adjustment rotation of the space axes in the global and the local system will be almost the same just as in the classical case and the calculation of the light deflection is simpler as well.

4) We suggest that the light deflection be defined as follows. When a barycentric coordinate system is chosen, the 4-velocity of a photon at an event can be projected into the local space in the local NT to get a 3-vector. The light deflection is the difference between the direction cosines of the two 3-vectors, at the source and at the observer, in their corresponding NT. It also could be described as that the angle between them when the two corresponding NT are fictitiously moved to coincide with each other.

ACKNOWLEDGEMENT

This research is supported by the Foundation of Natural Science of China.

References

Atkinson, R.d'E. (1963), "General relativity in Euclidean terms", *Proc. R. Soc. Lond. A*, **272**, 60.

Brumberg, V.A. (1981), "Relativistic reduction of astronomical measurements and reference frames", in E.M. Gaposchkin and B. Kolaczek (eds.) *Reference Coordinate Systems for Earth Dynamics*, D. Reidel Publishing Company, Dordrecht, pp. 283–294.

Brumberg, V.A. and Kopejkin, S.M. (1989), "Relativistic theory of celestial reference frames", in J. Kovalevsky, I.I. Mueller and B. Kolaczek (eds.) *Reference Frames in Astronomy and Geophysics*, Kluwer Academic Publishers, Dordrecht.

Eichhorn, H. (1984), "Inertial systems — definitions and realizations", *Celest. Mech.* **34**, 11–18.

Epstein, R. and Shapiro, I.I. (1980) "Post-Post-Newtonian deflection of light by the Sun", *Phys. Rev.* **22D**, 2947.

Fischbach, E. and Freeman, B.S. (1980) "Second-order contribution to the gravitational deflection of light", *Phys. Rev.* **22D**, 2950.

Fujimoto, M.K., Aoki, S., Nakajima, K., Fukushima, T. and Matzuzaka, S. (1982) "General relativistic framework for the study of astronomical/geodestic reference coordinates", *Proc. Symp. No. 5 of IAG*, pp 26–35.

Fukushima, T., Fujimoto, M.K., Kinoshita, H. and Aoki, S. (1986) "Coordinate systems in the general relativistic framework", in J. Kovalevsky and V.A. Brumberg (eds.) *Relativity in Celestial Mechanics and Astrometry*, D. Reidel Publishing Company, Dordrecht, pp. 145–168.

Horn, R.A. and Johnson, C.R. (1985) *Matrix Analysis*, Cambridge University Press.

Kaplan, G.H. (ed.) (1981) "The IAU Resolutions on Astronomical Constants, Time Scales and the Fundamental Reference Frame", *U.S. Naval Observatory Circular* No. 163.

Kovalevsky, J. and Mueller, I.I. (1981) "Comments on conventional terrestrial and quasi-inertial reference systems", in E.M. Gaposchkin and B. Kolaczek (eds.) *Reference Coordinate System for Earth Dynamics*, D. Reidel Publishing Company, Dordrecht, pp 375–384.

Kovalevsky, J. (1989) "Stellar reference frames", in J. Kovalevsky, I.I. Mueller and B. Kolaczek (eds.) *Reference Frames in Astronomy and Geophysics*, Kluwer Academic Publishers, Dordrecht.

Kovalevsky, J., Mueller, I.I. and Kolaczek (eds.) (1989) *Reference Frames in Astronomy and Geophysics*, Kluwer Academic Publishers, Dordrecht.

Møller, C. (1972) *The Theory of Relativity*, 2nd ed., Clarendon Press, Oxford.

Moritz, H. (1981) "Relativistic effects in reference frames" in E.M. Gaposchkin and B. Kolaczek (eds.) *Reference Coordinate System for Earth Dynamics*, D. Reidel Publishing Company, Dordrecht, pp. 43–58.

Murray, C.A. (1981) "Relativistic Astrometry", *Monthly Not.ices Roy. Astron. Soc.*, **195**, 639–648.

Murray, C.A. (1983) *Vectorial Astrometry*, Adam Hilger Ltd. Bristol.

Murray, C.A. (1986) "Relativity in astrometry" in J. Kovalevsky and V.A. Brumberg (eds.) *Relativity in Celestial Mechanics and Astrometry*, D. Reidel Publishing Company, Dordrecht, pp. 169–175.

Richter, G. W. and Matzner, R.A. (1982) "Second-order contributions to gravitational deflection of light in the parameterized post-Newtonian formalism", *Phys. Rev.* **26D**, 1219.

Sachs R.K., Wu, H. (1977) *General Relativity for Mathematicians*, Springer-Verlag, Heidelberg.

Soffel, M.H. (1989) *Relativity in Astronomy, Celestial Mechanics and Geodesy*, Springer-Verlag, Heidelberg.

Standish Jr., E.M. (1982) "Conversion of positions and proper motions from B1950.0 to the IAU system at J2000.0", *Astron. Astrophys.* **115**, 20–22.

Will, C.M. (1981) *Theory and Experiment in Gravitational Physics*, Cambridge University Press.

Xu, C.-M., Xu, J.-J., Yang, L.-T. and Huang, Z.-H. (1984) "The PPN light deflection in any region", *Fudan Journal (Natural Science)* **23**, 228. (in Chinese).

Discussion

TURYSHEV: The author tells us only about his concept of the description. I would like to ask him about the calculation in a real situation, such as for the solar system.

XU: Please read our full paper which will be published in the proceedings.

KOPEJKIN: It is very difficult to use the local tetrad approach for a description of the gravitational field of the solar system. For example, in the construction of the geocentric coordinate system, we must take into account the gravitational field of the Earth. Therefore, in my opinion, it is better to use coordinate systems which are not local in the tetrad sense.

XU: For description of the motion of the celestial bodies we cannot use the tetrad approach and you are correct. But for the description of the observational data, which are the events at the station, the local (tetrad) approach is adequate and enough.

GENERAL RELATIVISTIC DESCRIPTION OF CELESTIAL REFERENCE FRAMES

A.V.Voinov
Institute for Applied Astronomy, Leningrad

ABSTRACT. The astonomical consequences of recently developed theoretical methods of relativistic astrometry are discussed. The set of practically important reference systems is described. These reference systems generalize the locally inertial frames of general relativistic test observer, the hierarchy of Jacoby coordinates for dynamical problems and the dynamically inertial reference systems of fundamental astrometry. In practical application of this formalism much attention is paid to relativistic transformation functions relating the ecliptical coordinates corresponding to the barycenters of the Solar system, the Earth-Moon subsystem and the Earth. Solutions to several kinds of relativistic precession are also presented.

The ultimate aim of astrometry is to set up an inertial reference frame. Traditionally, the problem is to introduce a coordinate system which does not move and rotate with respect to very remote light emitters. Within the framework of classical mechanics such a kinematical construction immediately provides the nesessary dynamical properties of an inertial system - namely the absence of translatory, centripetal and Coriolise inertial forces.

General Relativity prohibits the classical inertiality. Only in the case of weak gravitation one may construct a system which retains some particular properties of an inertial one. Thus, if a system moves, it cannot be sumutaneously dynamically and kinematically inertial.

If we consider the Solar system as a whole, we can use its *Barycentric Reference System* (*BRS*). It may be regarded as completely inertial at the sufficient level of accuracy.

On the other hand, most of astronomical techniques and applications are concerned with the Earth, its close vicinity and the Earth-Moon subsystem. Therefore it is reasonable to consider a set of quasi-inertial reference frames which are related to these bodies. Evidently, most important of them would be the *Geocentric Reference System* (*GRS*) and *Terrestrial-lunar Reference System* (*TRS*). The latter

111

J. H. Lieske and V. K. Abalakin (eds.), Inertial Coordinate System on the Sky, 111–114.
© 1990 *IAU. Printed in the Netherlands.*

is related to the Earth-Moon barycenter.

Let us consider in more detail the *dynamically inertial* terrestrial-lunar reference system *(TRS)* $(x^{\tilde{0}}, x^{\tilde{i}})$ in its relation to the BRS (x^0, x^1) of the Solar system.

To define a coordinate system in General Relativity is to maintain the corresponding metric tensor.

Since the Earth-Moon subsystem is compact with respect to its distance to the Sun, we may take an advantage to treat separately the gravitation of the internal bodies (the Earth and the Moon) and of those external (the Sun and planets). Then the required metric in *TRS* may be found as a post-Newtonian solution to the Einstein equations in harmonic coordinates where the boundary conditions are used to account for the dynamical inertiality of the spatial axes of *TRS*. Detailed description of this techniques may be found in [1,2,3,4].

As a result we obtain some general form of metric tensor both in BRS and in *TRS*:

$$g_{00} = 1 - 2\varphi + 2\varphi^2 - 2\chi_{,00} - 2\psi, \quad g_{01} = 4\varphi_1 \quad , \quad g_{1j} = -\delta_{1j}(1 + 2\varphi)$$

where all the "potentials" φ, φ_1, χ and ψ are represented as sums of its internal and background parts. Internal components define the gravitational field of the Earth-Moon subsystem in the post-Newtonian limit. The background field is produced by the Sun and the planets. The background potentials are "direct" in BRS and "tidal" in *TRS*. Thus, the background solar potential takes the form:

$$w_{\tilde{k}} x^{\tilde{k}} + \frac{1}{2} w_{\tilde{k}\tilde{l}} x^{\tilde{k}} x^{\tilde{l}} + \frac{1}{6} w_{\tilde{k}\tilde{l}\tilde{m}} x^{\tilde{k}} x^{\tilde{l}} x^{\tilde{m}} + \dots \quad ,$$

where

$w_{\tilde{i}}$ is the covariant acceleration of the *TRS* origin (point T),

$$w_{\tilde{i}\tilde{j}} = E_{\tilde{i}\tilde{j}} + 3 w_{\tilde{i}} w_{\tilde{j}} - w_{\tilde{m}} w_{\tilde{m}} \delta_{1j} \quad , \quad w_{\tilde{i}\tilde{j}\tilde{k}} = E_{\tilde{i}\tilde{j}\tilde{k}} - \frac{10}{3} \delta_{(1j} w_{\tilde{k})\tilde{n}} w_{\tilde{n}} + 4 w_{(\tilde{i}} w_{\tilde{j}\tilde{k})}$$

$E_{\tilde{i}\tilde{j}}$ is the "eleertic" part of background curvature, which leading terms are:

$$R_{0\tilde{1}0\tilde{j}} x^{\tilde{i}} x^{\tilde{j}} = E_{\tilde{i}\tilde{j}} x^{\tilde{i}} x^{\tilde{j}} = \frac{m_S}{R_T^3} P_2 [\cos(\mathbf{R_T} \cdot \tilde{\mathbf{r}})] + \dots$$

The structure of this equation is similar to that of traditional expansion on powers of parallax.

As a side product of these techniques we immediately obtain the transformation functions which relate *TRS* to BRS:

$$x^\alpha = x^{\tilde{\alpha}} + L^{\tilde{\alpha}}(x^{\tilde{v}}) + P^{\tilde{\alpha}}(x^{\tilde{v}}) + T^{\tilde{\alpha}}(x^{\tilde{v}})$$

It contains the Lorentz bust (L), the relativistic precession (P) and the terms (T), which are nesessary to reduce the background potentials to the tidal ones.

Relativistic precession (especially - geodetic) determines the difference between the kinematically and dynamically inertial orientations of the moving reference systems. This precession may be expressed in terms of two angular quantities (γ and ε), which define correspondingly the relativistic precession in longitude and inclination. The laws of Fermi-Walker transport in the background metric provide two equations for these angles:

$$\frac{d\gamma}{dx^{\tilde{0}}} = \omega_G^3 + \omega_T^3 \; , \frac{d\varepsilon}{dx^{\tilde{0}}} = (\omega_G^1 + \omega_T^1)\cos\gamma + (\omega_G^2 + \omega_T^2)\sin\gamma \; .$$

where

$$\omega_G = \frac{3}{2}V_T \times \overline{\nabla b} + \overline{2rotb} \; , \quad \text{(geodetic + Lense-Thirring)}$$

$$\omega_T = \frac{1}{2}V_T \times w \; . \quad\quad\quad \text{(Thomas)}$$

$$V_T = dr_T/dx^0, \quad w = dV_T/dx^0$$

In virtue of a perturbation method we find the solutions for these equations:

$$\gamma = \gamma_p + \gamma_N \; ,$$

$$\gamma_p = \left[\frac{3}{2}\frac{n'^3 a'^2}{1-e'^2} + \sum_{b=1}^{N} \nu_b \left(-\frac{3}{2}n_b^2 n' a'^2 - 2n_b^3 a_b - \frac{27}{32} n_b^2 n' \frac{a'^4}{a_b^2} - \right.\right.$$
$$\left.\left. - \frac{3}{2}\nu_b n_b^3 a_b^2 \right) + \dots \right] x^{\tilde{0}} = (19\overset{s}{.}192996/1000\text{years}) \; x^{\tilde{0}} \; ,$$

$$\gamma_N = N_0 \sin(E-\pi') + 0.00192\sin(2E-2\pi') -- 0.00106\sin(E-J) + \dots \; ,$$

$$\varepsilon = \varepsilon_0 + 0.00011\cos(E+J) + 0.00010\cos(E-J) + \dots \; ,$$

$$N_{\underline{\;}} = -e'^2 a'^2 e' \left(\frac{9}{2} + \frac{45}{16}e'^2 + \dots + \frac{39}{8}\sum_{b=1}^{N} \nu_b \frac{n_b^2}{n'^2} + \dots \right) = 0.15321 \; ,$$

$$\nu_b = \frac{m_b}{m_s} \; ,$$

Here the "primed" quantities describe the heliocentric motion of the Earth-Moon barycenter. Besides, n_b is the mean motion of the planet b, a_b is the semimajor axis of the planet b ($a_b > a'$), E is the mean longitude of T, J is the mean longitude of Jupiter, π' is the longitude of perihelion of T. ε_0 is the inclination constant. For example,

$$\varepsilon_0 = 0 \; , \quad\quad \text{means ecliptical orientation,}$$

$$\varepsilon_0 = 23^0 \; 27' + \dots \; \textit{may mean the equatorial one.}$$

Hereafter all the angular coefficients are expressed in milli arc seconds.

We can construct analogous analytical or semianalytical expansions for transformation between *BRS* and *TRS*. It seems convenient to express them in terms of spherical coordinates.

Let us adopt for the sake of simplicity the ecliptical orientation of both *BRS* and *TRS* and introduce the following:

$$r - r_{\mathfrak{r}} = r \ (cos\lambda \ cos\beta, \ sin\lambda \ cos\beta, \ sin\beta),$$

$$\tilde{r} = \tilde{r} \ (cos\tilde{\lambda} \ cos\tilde{\beta}, \ sin\tilde{\lambda} \ cos\tilde{\beta}, \ sin\tilde{\beta}),$$

Then the relativistic transformation mentioned above is reduced to (note that $\varepsilon_0 = 0$):

$$\lambda = \tilde{\lambda} - \gamma + \delta\lambda, \qquad \beta = \tilde{\beta} + \delta\beta, \qquad r = \tilde{r} + \delta r,$$

where

$$cos\tilde{\beta} \ \delta\lambda = L_1 cos\tilde{\beta} + L_2 \frac{\tilde{r}}{a},$$

$$\delta\beta = B_1 cos\tilde{\beta} \ sin\tilde{\beta} + B_2 \frac{\tilde{r}}{a} cos\tilde{\beta} + B_3 \frac{\tilde{r}}{a} sin\tilde{\beta},$$

$$\delta r = \tilde{r}(R_1 + R_2 cos^2\tilde{\beta} + R_3 \frac{\tilde{r}}{a} cos\tilde{\beta} + R_4 \frac{\tilde{r}}{a} sin\tilde{\beta}),$$

$$L_1 = 0.50884 sin(2\tilde{\lambda}-2E) + 0.01701 sin(2\tilde{\lambda}-3E+\pi') +$$

$$+ \ 0.00046 sin(2\tilde{\lambda}-4E+2\pi') - 0.00041 sin(2\tilde{\lambda}-E-J) + \ldots,$$

Analogous expansions may be written for the other coefficients in above formulae.

Since this transformation is relativistic, it must contain the appropriate time component (see e.g. [ ]).

Expression for λ contains explicitly the relativistic precession and nutation in longitude. That for the inclination occurs completely negligible. Therefore we can easily obtain a more practical reference system, which is related to *BRS* with only the periodic terms of the above transformation:

$$\lambda = \lambda - \gamma_N + \delta\lambda, \qquad \beta = \tilde{\beta} + \delta\beta, \qquad r = \tilde{r} + \delta r,$$

This system is seen to be kinematically inertial in average. It also meets the modern IAU standards which combine the secular part of geodetic precession with that Newtonian.

Nevertheless, it should be noted, that initial definition of *TRS* is more theoretically consistent from the point of view of General Relativity.

References

1. Thorne K.S., Hartle J.B. (1985). Laws of Motion and Precession for Black Holes and Other Bodies. Phys.Rev., D31, 1815-1837.

2. Brumberg V.A., Kopejkin S.M. (1989). Relativistic Theory of Celestial Reference Frames In Kovalevsky J., Mueller I.I., Kolaczek B. (eds) Reference Frames. Kluwer Academic Publishers.

3. Voinov A.V. (1988). Motion and Rotation of Celestial Bodies in the post-Newtonian Approximation. Celestial Mechanics, 42, 293-307.

4. Voinov A.V. (1989). Relativistic Equations of Motion of an Earth Satellite. Manuscripta Geodactica (in press).

RELATIVISTIC REFERENCE FRAMES OF LOCAL OBSERVER AND SPACE RADIOINTERFEROMETER

A.N. ALEXANDROV, S.L. PARNOVSKY, V.I. ZHDANOV
Astronomical Observatory of Kiev University
Observatorna St. 3
252053 Kiev, USSR

In a considerable number of works on relativistic astrometry (see, *e.g.* Kovalevsky and Brumberg 1986) the reference frames (RFs) are introduced either by means of coordinate representation of a space-time metric, such as using harmonicity conditions (Brumberg and Kopejkin 1989), or on the basis of invariant constructions like Fermi coordinates (Synge 1960; Ashby and Bertotti 1986; Boucher 1986). Both approaches must, probably, be combined in applications. We consider the local observer RFs (LORFs) based on the Fermi coordinates and on the optical ones (Synge 1960), which are rigorously defined for a general metric and are directly related to observable quantities. In particular, the optical RF operates with the observed direction of the light source, whereas the Fermi RF seems to be a natural generalization of the classical Cartesian RF.

The specificity of the radiointerferometry in space leads to the following questions dealt with in this report.:

(i) In view of a large baseline (10^5 to 10^6 km) of the space radiointerferometer (SRI) a more accurate account of relativistic effects in calculation of the delay time τ is needed in comparison with VLBI on the Earth. As in Finkelstein *et al.* (1983) and in Zeller *et al.* (1986), we are guided by accuracy of 1 ps; however, for the above baseline size this requires the relative accuracy of 10^{-12}, that is two orders higher than that discussed by Finkelstein *et al.* and by Zeller *et al.*

(ii) In construction of LORFs in the nearby space taking into account of the Earth-Moon gravitational field is desirable. This is not allowed by the methods of Ashby and Bertotti (1985) and Boucher (1986) where the Fermi RF in a fictitious background metric has been treated by using the Taylor expansion in spatial coordinates. We used the other way to construct the transformation to the Fermi and optical RFs, which embraces the case when the distance from gravitating bodies is comparable with the SRI base. The transformations are obtained according to the definitions of Synge (1960) by means of direct solution of the geodesic equations in the first weak-field approximation. If one rules out the Earth-Moon field, then in the case of geocentrical Fermi RF the transformations can be reduced to the results of Ashby and Bertotti (1985) and Boucher (1986).

J. H. Lieske and V. K. Abalakin (eds.), Inertial Coordinate System on the Sky, 115–117.
© 1990 *IAU. Printed in the Netherlands.*

On the basis of these results we obtain the relation for τ in the LORF, in which at the limit of the accuracy there is the input of the geodesic precession of the LORF for the measurement time of 10^3 sec, and of the homogeneous part of the solar gravitational field. For VLBI on the Earth (relative accuracy 10^{-10}) the relation for τ in the Fermi RF agrees with the result of Zhu and Groten (1988) and differs from that of Finkelstein *et al.* (1983) and Zeller *et al.* (1986) by the scale factor $(1 - \varphi_o)$ where φ_o is the solar gravitational potential. The reason for the discrepancy is that another normalization of the distance was employed in Finkelstein *et al.* (1983) and Zeller *et al.* (1986).

For SRI with a large baseline the calculation of the gravitational input τ_{grav} in τ is somewhat different from that of Finkelstein *et al.* (1983). Apart from the Sun and the Earth, the gravitational contributions of Jupiter and Saturn are essential, the inputs of the other planets are essential only for the received signals passing in the close neighborhood of the planets. In analogous cases, (a) the post-post-Newtonian corrections to the solar gravitational field and (b) angular momentum of the Sun give also the input of some ps. The input (b) linear in J is derived in harmonic coordinates, the PPN contribution (a) is given by the results of Brumberg (1987).

With the aim of rigorous formulation of some LORF problems, we have considered the methods of construction of the Fermi and optical RFs using the exponential mapping technique (EMT) in curved space-times. It gives a consistent description of extended relativistic systems in the closed form, including the cases when their sizes are comparable with the characteristic scale of the gravitational field. EMT allows one to specify an even P by means of a four-vector y at the position of the observer, the components of y being Riemannian normal coordinates of P. The algorithms to find all physical characteristics in this coordinate system through observables are known (Alexandrov 1981). To apply these to the above LORFs one must take into account the variation of the mapping $P \rightarrow y$ under the shift along the observer's trajectory and perform $(3+1)$-splitting of the space-time. This is described by Jacobi fields which satisfy the geodesic deviation equation. By means of EMT we have obtained the transformation of observables to the LORFs and the exact equations of the test body motion in the LORFs. The resulting expressions are presented in the covariant form and may be computed either by means of covariant Taylor expansion or in the weak-field approximation.

References

Kovalevsky, J. and Brumberg, V.A.: 1986, *Relativity in Celestial Mechanics and Astrometry*, Proc. IAU Symp. No. 114, D. Reidel, Dordrecht.

Brumberg, V.A. and Kopejkin, S.M.: 1989, "Relativistic theory of celestial reference frames", in J. Kovalevsky, I.I. Mueller, B. Kolaczek (eds.) *Reference Frames in Astronomy and Geophysics* , Kluwer Academic Publishers, Dordrecht, pp. 115–141.

Synge, J.L.: 1960, *Relativity: The General Theory*, North-Holland, Amsterdam

Ashby, N., and Bertotti, B.: 1986, *Phys. Rev.* D **34**, 2246.

Boucher, C.: 1986, "Relativistic effects in geodynamics", in J. Kovalevsky and V.A. Brumberg (eds.) *Relativity in Celestial Mechanics and Astrometry* , D. Reidel, Dordrecht, pp. 241–253.

Finkelstein, A.M., Kreinovich. V.Ya., and Pandey, S.N.: 1983, *Ap. Sp. Sci.* **94**, 233.

Zeller, G., Soffel, M. Ruder, H., and Schneider, M.: 1986, *Veröff. d. Bayer Komm. Int. Erdmes.* Munich 48, 218.

Zhu, S.Y., and Groten, E.: 1988, *Manuscr. Geod.* **13**, 33.

Brumberg, V.A.: 1987, Кинемат. и физ. небес. тел **3**, 8.

Alexandrov, A.N.: 1981, *Acta Phys. Polon.* **B12**, 523.

Discussion

KOPEJKIN: I want to point out that the method of construction of local coordinate systems in the vicinity of a massive body was developed by Fukushima *et al.* as well.

ZHDANOV: The difference of our consideration from other works is that we avoid expansion in spatial coordinates and therefore we do not need to separate the gravitational field of the Earth in construction of the LORFs.

KLIONER: I think that to derive the relativistic correction due to the rotation of the bodies and in particular due to the rotation of the Sun, you had to obtain the trajectory of the light considering the bodies' rotation. Did you really obtain this trajectory and, if yes, what kind of coordinates (*i.e.* cartesian or spherical) did you use?

ZHDANOV: We actually have obtained this trajectory in the cartesian coordinates.

THE OPTIMUM CONVENTIONAL TERRESTRIAL SYSTEM DETERMINED BY VLBI AND SLR STATIONS *

W. KOSEK, B. KOLACZEK
Space Research Centre, Polish Academy of Sciences
Bartycka 18
00-716 Warsaw
Poland

The optimum Conventional Terrestrial System (CTS) can be defined by accurate coordinates of some number of stations distributed homogeneously all over the world. There are scores of laser and VLBI stations whose coordinates are known with high accuracy of the order of 1–2 cm. There are many CTS defined by the sets of station coordinates determined in the process of determination of the Earth rotation paramaters and the Earth gravity field. Presently existing stations are not distributed homogeneously on the Earth. They are located mostly in Europe and in North America. In this situation, the errors of orientation of axes and origin positions are not equal. Some of them, based on a small number of not homogeneously distributed stations, are not well-defined (stable).

In the complete paper various sets of different number of stations from 5 to 50, defining the optimum CTS, have been chosen from the IRIS nets of stations and GSFC sets of VLBI and SLR stations, GSFC88L01, GSFC88R01 respectively (BIH 1987; Carter *et al.* 1988; Ma *et al.* 1988).

Accuracies of CTS considered here were investigated through errors of 7 parameters of transformation of a CTS defined by adopted coordinates of considered stations. Results show that 15–20 well-distributed stations can define an optimum CTS with accuracy of the order of 1 cm or better. Further increasing the number of stations does not improve the accuracies very much.

Discussion

HUGHES: I neglected to mention yesterday that the WGRS is also concerned with terrestrial systems in the sense that cooperation is maintained with the IAG/IUGG.

YATSKIV: I was surprised to find in the optimal solution that the stations Onsala and Wettzell are so closely situated to one another.

KOLACZEK: It may be due to the high accuracy of the stations' relative coordinate determinations.

---

* Full paper published in the *Proceedings of IAG Symposium No. 105: "Earth Rotation and Terrestrial Reference Frames"* of the IAG General Assembly in Edinburgh, 2–12 August 1989.

J. H. Lieske and V. K. Abalakin (eds.), Inertial Coordinate System on the Sky, 118.
© 1990 *IAU. Printed in the Netherlands.*

THE CELESTIAL SYSTEM OF THE INTERNATIONAL EARTH ROTATION SERVICE

E.F. ARIAS
Central Bureau of IERS, Paris, France
and La Plata Observatory/CONICET, Argentina

M. FEISSEL
Central Bureau of IERS - Observatoire de Paris/URA 1125 du CNRS
Observatoire de Paris - 61, avenue de l'Observatoire, F-75014 Paris, France

ABSTRACT. The celestial system of the International Earth Rotation Service (IERS) is materialized by the J2000.0 positions of more than 250 extragalactic compact radio sources observed by VLBI. The source coordinates are evaluated from the combination of individual celestial frames obtained by the Goddard Space Flight Center, the Jet Propulsion Laboratory and the U.S. National Geodetic Survey.
    The combination model and the maintenance algorithm are described. To free the IERS celestial frame from inconsistencies due to the inaccuracy of the IAU conventional models for precession and nutation, it is implemented on individual frames which have been obtained in parallel to the adjustment of corrections to the direction of the celestial pole.
    The IERS celestial reference frame is consistent with FK5 at a few milliarcsecond level. To be made denser and more accessible for astronomical uses, it will be related to the HIPPARCOS stellar frame.

Introduction

Between 1978 and 1980, several multibaseline VLBI observation programs were initiated for space navigation, geophysics and astronomy; the Deep Space Network (DSN) is operated by the Jet Propulsion Laboratory (JPL), the Crustal Dynamics Project (CDP) is operated by NASA, and the International Radio-Interferometric Surveying (IRIS) program is operated by agencies in the United States and other countries. A common requirement for the scientific use of these observations is the establishment of an accurate celestial reference system, materialized by a reference frame, *i.e.*, a catalog of positions of extragalactic radio sources and a set of astronomical and geophysical models (Ma, 1989).
    The current practice in the analysis of VLBI observations consists of a global adjustment of the terrestrial coordinates of the observing stations, the celestial coordinates of the radio sources and the series of the Earth Orientation Parameters (EOP) which relate the terrestrial to the celestial frame as a function of time. These adjustments are performed over the complete period of observations.
    The International Earth Rotation Service (IERS) was established in 1987 by the IAU and the IUGG and it started operation on 1988 January 1st (IAG, 1988). It replaces the International Polar Motion Service (IPMS) and the earth-rotation section of the Bureau International de l'Heure (BIH). IERS is a member of the Federation of Astronomical and Geophysical Data Analysis Services (FAGS). It

119

J. H. Lieske and V. K. Abalakin (eds.), Inertial Coordinate System on the Sky, 119–128.
© 1990 *IAU. Printed in the Netherlands.*

is responsible for defining and maintaining a conventional terrestrial reference system based on observing stations that use the high-precision techniques in space geodesy; defining and maintaining a conventional celestial reference system based on extragalactic radio sources, and relating it to other celestial reference systems; and determining the Earth orientation parameters connecting these systems, *i.e.*, the terrestrial and celestial coordinates of the pole and universal time.

Several laboratories contribute to the work of IERS by analysing the above mentioned VLBI observations on the basis of common models and constants. The Central Bureau of IERS performs a global combination of these results with similar ones obtained by laser ranging to the Moon (LLR) and to artificial satellites (SLR). By the nature of the observing techniques, the celestial frame realized in this global analysis is based solely on VLBI, while the terrestrial frame and the EOP include a large contribution of satellite geodesy (IERS, 1989).

This paper first describes the IERS System in some detail. The implementation by the Central Bureau of the system maintenance process is then outlined, with application to the most recent realization of the celestial frame. Finally the link to FK5 and the future extension to the HIPPAR-COS stellar frame are discussed.

The IERS Reference System

The IERS Reference System is composed of several parts: the IERS standards, the IERS reference frames, and the corresponding series of the Earth Orientation Parameters, which are consistent with one another at a few milliarcsecond level.

IERS STANDARDS

The IERS standards used in 1988 are the MERIT Standards (Melbourne et *al.*, 1983) including *Update No 1*; the IERS Standards (1989) are under preparation (McCarthy, 1989). They consist of a set of constants and models used by the IERS Analysis Centres for VLBI, LLR and SLR, and by the Central Bureau in the combination of results. The values of the constants are adopted from recent analyses; in some cases they differ from the current IAU and IAG conventional ones. The models represent, in general, the state of the art in the field concerned.

IERS REFERENCE FRAMES

The IERS reference frames consist of the IERS Terrestrial Reference Frame (ITRF) and IERS Celestial Reference Frame (ICRF); both frames are realized through lists of coordinates of fiducial points: terrestrial sites or compact extragalactic radio sources.

Terrestrial frame. The origin of the ITRF is located at the center of mass of the Earth, with an uncertainty of 10cm. The length unit is the metre (SI). The IERS Reference Pole and Reference Meridian are consistent with the corresponding directions in the BIH Terrestrial System (BTS) within 0.''003. The BIH reference pole was adjusted to the Conventional International Origin (CIO) in 1967; it was then kept stable independently until 1987. The uncertainty of the tie of the BIH reference pole with the CIO was 0.''03 (Boucher and Feissel, 1984).

Celestial frame. Guinot (1979) proposed that the definition of the celestial reference frame be attached to the directions of extragalactic sources. The maintenance of the frame would then be based on the stability of the set of the selected objects, no attempt being made to re-adjust the origins of the frame. The recommendations of the IAU/IUGG Working group MERIT (Wilkins and Mueller, 1986), endorsed by the IAU in 1985, follow the same lines. The analyses of the Central Bureau of IERS implement these recommendations.

The origin of the ICRF is at the barycentre of the solar system. The direction of the polar axis is the one given for epoch J2000.0 by the IAU 1976 Precession and the IAU 1980 Theory of Nutation. The origin of right ascensions is such that the right ascension of 3C 273 is $12^h 29^m 6.6997$. The initial definition was based on 23 radio sources available in three VLBI reference frames derived by the Goddard Space Flight Center (GSFC), JPL and the U.S. National Geodesy Survey (NGS) from observations from 1978 through 1987; it is described by Arias *et al.* (1988).

THE EARTH ORIENTATION PARAMETERS

The IERS Earth Orientation Parameters (EOP) are the parameters which describe the rotation of the ITRF to the ICRF as a function of time, in conjunction with the conventional precession and nutation models. They model the unpredictable part of the Earth's motion. They consist of five parameters: the coordinates of the terrestrial pole (x, y) ; universal time (UT1), related to the Greenwich mean sidereal time (GMST) by a conventional relationship (Aoki *et al.*, 1982); the offsets in longitude and in obliquity ($d\psi$, $d\varepsilon$) of the celestial pole with respect to its position defined by the conventional IAU precession/nutation models.

Shortly after the adoption of the IAU 1980 Theory of Nutation, (Seidelmann, 1982), the VLBI observations started to show evidence of inaccuracies in some of its components, reaching 0.01 in the principal term, 0.005 in the annual term and 0.001 in several other terms in longitude; some of the obliquity components also have inaccuracies of $0.001 - 0.002$; in addition, the precession constant might require a correction of about 0.003 per year (Herring *et al*; 1986). In the analysis of observations, these model deficiencies would propagate unevenly in the source positions, giving rise to local systematic errors and to misorientation of the frame, both at the level of several milliarcseconds. To avoid these effects, the VLBI analyses in the framework of IERS include the estimation of $d\psi$, $d\varepsilon$, for each session. Such solutions entail fixing *a priori* the values of $d\psi$ and $d\varepsilon$ for one date; the numerical values chosen reflect themselves in the orientation of the derived celestial frame (an alternative procedure is the direct adjustment of corrections to some of the terms of the models).

Maintenance of the IERS Celestial Reference Frame

The initial realization of the IERS Celestial Reference Frame, RSC (IERS) 88 C 01, (Arias *et al.*, 1988) included a total of 228 extragalactic radio sources.

In the normal process of IERS global analyses, the ICRF can be re-evaluated whenever new information is available. This is a maintenance process, in which strict rules are applied, with the aim of insuring the consistency of the direction of axes with the initial definition, while improving the accuracy of the existing sources and extending the frame to new sources. The practical implementation which led to RSC (IERS) 89 C 01, published in the IERS Annual Report for 1988, is described below.

The adjustment of the celestial reference frame is one part in a process which also embodies the terrestrial frame and the series of EOP over several years. This approach makes it possible to compare the corresponding elements in the various solutions, thus providing an external check of the consistency of the models used. In the case of the celestial frames, the model for comparisons is a three-angle rotation (see below). The consistency checks are based on a comparison of rotation angles between frames with systematic differences between time series of universal time and of the celestial pole offsets (Altamimi *et al.* 1989). The results of the comparisons characterize the link of the individual systems with the IERS System.

CLASSIFICATION OF CONTRIBUTING CATALOGS

The individual frames which are to contribute to the *maintenance* of axes are selected according to the consistency of their link to the IERS System, and to their density and precision. Another class of frames, still dense and precise but less consistent with the IERS System, is used to *densify* the combined frame when it contains additional sources. Finally, the less dense and consistent frames are only *compared* to the combined one. The individual frames available for the 1989 analysis are described in Table 1.

CLASSIFICATION OF SOURCES

The individual catalogues include well observed pointlike sources, with small uncertainties in their coordinates, and other sources, less observed or having sizeable structure, with coordinates of greater uncertainty. It is therefore necessary to select carefully the *primary sources* to be used in the maintenance of the frame. The 23 primary sources in the initial definition of the IERS frame were

Table 1 - Individual reference frames: number of sources (n), rms formal uncertainty (σ) and declination interval (δ). Unit for σ: $0\rlap{.}''001$. (*1*)

Frame		Total n	Primary n	σ	Secondary n	σ	Complem. n	σ	declination limits (°)
Maintenance									
RSC (GSFC)		89 R 01	64	20	0.2	36	0.6	8	0.8	−30,+81
RSC (JPL)		89 R 02	189	16	2.5	173	2.7	-	-	−45,+85
RSC (NGS)		89 R 01	50	20	0.2	30	0.7	-	-	−30,+79
Densification									
RSC (JPL)		89 R 03	195	-	-	173	4.2	4	15.0	−45,+85
Comparison									
RSC (NAOMZ)	(*2*)	89 R 01	20	10	-	10	-	-	-	−4,+79
RSC (SHA)	(*3*)	89 R 01	14	6	-	8	-	-	-	0,+79

(*1*) The classification of sources is explained in the text.
(*2*) NAOMZ : National Astronomical Observatory, Mizuzawa (Japan)
(*3*) SHA : Shanghai Observatory (China).

selected on the basis of structure, sky coverage, time stability and number of observations (Carter, 1987).

Radio sources common to at least two of the frames used for the maintenance or densification, but not in the primary list, are defined as *secondary sources*, while those belonging to only one individual frame are the *complementary sources*. Neither secondary nor complementary sources are used to define the directions of axes of the combined frame; only their positions relative to it are evaluated.

For the implementation of RSC(IERS) 89 C 01, the list of primary sources was re-examined. For this purpose, a preliminary combined frame was calculated with axes aligned on the initial definition by use of the 23 initial primary sources; then the differences in coordinates in the two frames 88 C 01 and 89 C 01 for all common sources were compared to their standard errors derived from the combination. Former primary sources were kept in this category only if their change in coordinates was smaller than twice their standard error; former secondary sources with a change in coordinates less than their standard error were upgraded to the primary status.

Following these criteria, four radio sources were deleted from the primary list: 0212+735, 0454–234, 1034–293 and 1803+784, and a new one was included: 1749+096. The sources 0212+735 and 1803+784 showed right ascension differences of 2.9σ and 3.5σ respectively, confirmed by structure in right ascension observed in their maps (Charlot, 1989).

THE COMBINATION MODEL

The differences between catalogs are expressed in a direct trirectangle coordinate system with the x-axis and z-axis in the respective directions to the origin of right ascensions and pole of the IERS celestial frame. The model currently used considers a three-angle rotation matrix, assuming no systematic local deformation; the validity of the latter assumption is a part of the examination of the results.

Let A_1 (i), A_2 (i), A_3 (i) be the rotation angles between the combined reference frame and the individual frame i ; α_{ij}, δ_{ij} are the coordinates of the source j in the frame i, and α_{cj}, δ_{cj} its coordinates in the combined celestial reference frame. The respective equations of observation in right ascension and declination are :

$$+ A_1 (i)\ \tan \delta_{ij} \cos \alpha_{ij} + A_2 (i)\ \tan \delta_{ij} \sin \alpha_{ij} - A_3(i) + \alpha_{cj} = \alpha_{ij}$$

$$- A_1 (i) \sin \alpha_{ij} + A_2 (i) \cos \alpha_{ij} + \delta_{cj} = \delta_{ij}$$

(1)

The combination is performed in three steps.

1-   Using only the three frames selected for the maintenance of the system, the coordinates of the primary sources in the combined frame and the relative orientations $A_1(i)$, $A_2(i)$, $A_3(i)$ between it and the three individual frames are evaluated in a weighted least squares fit. The directions of axes of the combined frame are constrained to be aligned with the axes of RSC (IERS) 88 C 01.

2-   The orientation of the densification frame relative to the combined frame is evaluated on the basis of the primary radio sources available in it.

3-   The coordinates of secondary and complementary sources in the combined frame are obtained by applying to the individual frames the rotation angles derived in steps (1) and (2).

124

In the three steps, the equations of observation are weighted according to the formal uncertainties of the individual coordinates; the formal uncertainties in the individual source positions below 0".00025 have been set to this value.

RSC (IERS) 89 C 01

The combination described above led to the positions of 209 compact extragalactic radio sources of which 20 are primary, 177 secondary and 12 complementary. The formal uncertainties of the primary sources are in the range 0".0002 to 0".0004, which is consistent with the formal uncertainties in the individual frames. They agree well with the respective values for RSC (IERS) 88 C 01 given in the BIH Annual Report for 1987. For the secondary sources belonging only to JPL frames, we have adopted as formal uncertainty the largest value in the individual frames; for the other secondary sources, the consistency between coordinates in the individual frames ranges from 0".0003 to 0".02.

57 addditional extragalactic sources had coordinates in RSC (IERS) 88 C 01, whereas they were not present in the frames of Table 1; they can be considered as a complement of the new realization. The 266 sources of RSC (IERS) 89 C 01, including this complementary list, are plotted in Figure 1. Its extension south of its present limit, −45°, on the basis of recent observations (Ma, 1989), is planned.

The rotation angles of the frames of Table 1 to RSC (IERS) 89 C 01 are given in Table 2. The major part in these angles reflects the fact that the *a priori* fixing of the values of $d\psi$ and $d\varepsilon$ for one of the observing sessions is made independently in the various analyses. After taking out the corresponding biases, the disagreement of the maintenance frames axes with the IERS one is under 0".0005, except for still unexplained discrepancies in the celestial pole offsets in longitude angles at the level of 0".005.

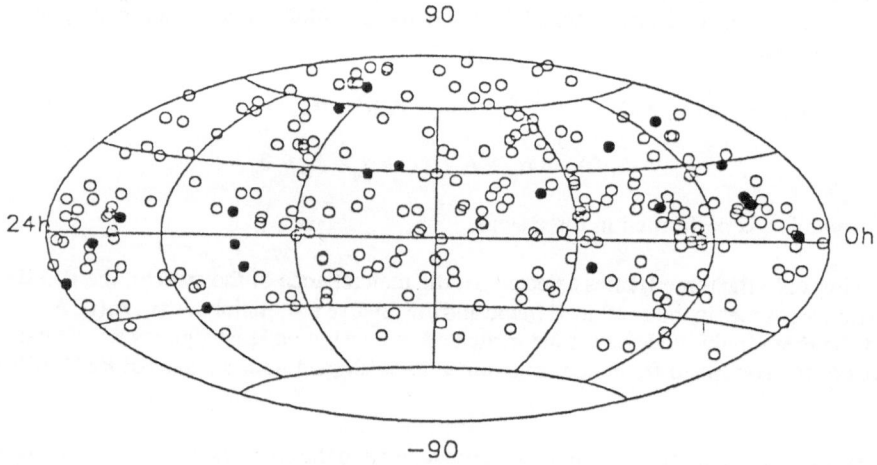

Figure 1. Sky distribution of the 266 radio sources of RSC (IERS) 89 C 01.
Dark circles represent the primary sources.

Comparisons between the individual frames, and between them and the combined one, show that local systematic deformations of the IERS celestial frames can be expected to be lower than 0″002 (Arias and Lestrade, 1989).

Table 2 - Relative orientation between the individual frames and RSC (IERS) 89 C 01.
The rotation angles A_1, A_2, A_3 transform the coordinates from the individual VLBI frames to the combined frame. N is the number of common radio sources.
Unit : 0″001.

Individual frame		N	A_1	A_2	A_3
Maintenance					
RSC (GSFC)	89 R 01	64	+1.28 ± 0.09	+2.46 ± 0.10	−0.15 ± 0.07
RSC (JPL)	89 R 02	162	−0.04 ± 0.06	−1.15 ± 0.07	+0.07 ± 0.04
RSC (NGS)	89 R 01	50	+0.31 ± 0.10	−0.07 ± 0.10	−0.24 ± 0.08
Densification					
RSC (JPL)	89 R 03	176	+0.76 ± 0.08	−3.03 ± 0.08	+0.72 ± 0.24
Comparison					
RSC (NAOMZ)	89 R 01	20	−1.30 ± 0.37	−2.08 ± 0.37	−0.19 ± 4.06
RSC (SHA)	89 R 01	14	−1.85 ± 0.41	−5.19 ± 0.38	−1.63 ± 4.33

Extension of the IERS Celestial Reference Frame to galactic frames

The densest VLBI reference frames presently available contain a few hundreds of sources. The extragalactic frame is considered the best realization of a quasi-inertial reference frame because of its properties: it is kinematically stable (nuclei of galaxies and quasars being at cosmological distances, their apparent proper motions can be considered nonexistent), and its precision is at the milliarcsecond level. Nevertheless, it has two disadvantages relative to the stellar frames :

a) it does not supply a large number of fiducial points (200-300 sources), and
b) extragalactic radio sources are very faint objects (m > 13) which are not optically observable, thus it is accessible only with the VLBI technique.

The observations of the astrometric satellite HIPPARCOS will contribute to the densification and accessibility of the extragalactic frame, if the satellite can accomplish its *nominal* mission. On the other hand, the astrometric qualities of the HIPPARCOS frame will be exploited only if it is linked to a quasi-inertial frame of similar precision. The HIPPARCOS reconstructed sphere will be affected by the galactic rotation at a rate of about 0.007"/year (Lestrade *et al.* 1985). Being derived from a space telescope, its reference system will not be linked to the terrestrial equator and the ecliptic. For stars brighter than m = 9, the expected mean precision is 0''.002 in the positions and 0.002"/year in the proper motions. It will observe about 120 000 stars brighter than m = 13.

Several possibilities have been envisaged to make the HIPPARCOS reference frame inertial (Mignard, 1989) :

- To link it to the VLBI frame by radio stars. Radio stars are members of our galaxy. Most are quite bright optically. The satellite HIPPARCOS has over one thousand of them in its observational program. They provide a direct link between the radio and optical frames.

- The Hubble Space Telescope (HST) will determine relative positions between extragalactic objects and nearby HIPPARCOS stars. There are 92 extragalactic objects for the HST and 173 HIPPARCOS "Super High priority" stars at small angular distances to link them. 52 radio sources in RSC (IERS) 89 C 01 had already been selected for the HIPPARCOS / ST proposal, giving prospect for a good link by this route (Argue, 1989).

Until the HIPPARCOS stellar frame becomes available, the IERS celestial frame can be made accessible by considering the rotation angles between it and the FK5 stellar frame (Fricke *et al.* 1988). The uncertainty of individual positions is evaluated to ±0''.05 for the present epoch by Ma *et al.* (1989), and Morrison *et al.* (1989) have shown that regional systematic errors are present in the FK5 at the level of 0''.05-0''.10. A preliminary evaluation, based on a comparison of 28 quasar positions brought to the FK5 frame through the AGK3RN catalog (Ma *et al.*, 1989) with the IERS ones, indicates that the rotation angle of the IERS celestial frame to the FK5 frame in the equator is –0''.005 and that the polar axes of the two frames are consistent within 0''.04, with an uncertainty of about 0''.02 due to the noise in the optical positions of quasars.

Summary

The current realization of the IERS Celestial Reference Frame contains 266 extragalactic sources of which 76 have position uncertainties smaller than 0''.001 and 117 sources are between 0''.001 and 0''.003. The algorithm by which it is maintained insures the consistency of the successive versions at the best level achievable with the existing data (0.1 mas in 1989). It allows the continuous improvement of the accuracy and distribution of the frame as new data become available.

Through the global IERS analysis, the IERS Celestial Reference Frame is indirectly related to the dynamical celestial reference frames used in satellite geodesy and Lunar Laser Ranging, and it is related directly to the IERS Terrestrial Reference Frame at the milliarcsecond level.

The IERS Celestial Reference Frame is the realization of a quasi-inertial frame at the level of 0''.002 in the individual source positions. The FK5 is another realization at the level of 0''.05-0''.10. The axes of the two frames are in agreement within 0''.04. Future connection with the HIPPARCOS stellar frame through VLBI observations of radio stars and/or Hubble Space Telescope observations of quasars should extend the accessibility of the IERS celestial frame to optical observation of stars at the level of 0''.002, if the nominal HIPPARCOS mission is achieved.

Acknowledgment. We are grateful to Chopo Ma and J.G. Williams for enlightening discussion of our work. This work was carried out during a stay of the first author at Paris Observatory and Bureau des Longitudes.

References

ALTAMIMI, Z., ARIAS, E.F., BOUCHER, C., FEISSEL, M., 1989 : Earth Orientation determinations : some tests of consistency. *IAG Symposium 105*. Wilkins (ed.).

AOKI, S., GUINOT, B., KAPLAN, G.H., KINOSHITA, H., McCARTHY, D.D., SEIDELMANN, P.K.:1982 : The new definition of Universal Time. *Astron.and Astrophys*. **105,** 359.

ARIAS, E.F., FEISSEL, M., LESTRADE, J.-F.: 1988. An extragalactic celestial reference frame consistent with the BIH Terrestrial System (1987). *BIH Annual Report for 1987*, p.D-113. Observatoire de Paris.

ARIAS, E.F., and LESTRADE, J.-F., 1989 : Extended astrometric comparisons between two dense VLBI extragalactic reference frames elaborated at the GSFC and the JPL. Submitted to *Astron. and Astrophys.*

ARGUE, N., 1989 : The strategy adopted for linking HIPPARCOS to an extragalactic reference frame. *This volume.*

BOUCHER, C., and FEISSEL, M., 1984 : Realization of the BIH Terrestrial System. *Proc. Internat. Symp. on Space Techniques for Geodynamics*, Somogyi and Reigber (eds). Res. Inst. of the Hungarian Acad. of Sci. (Sopron, Hungary), Vol. **1,** 235.

CARTER, W.E., 1987 : IERS memorandum, 1 December 1987.

CHARLOT, P., 1989 : 14 extragalactic radio sources mapped at 2.3 and 8.4 GHz with a 24 hour Crustal Dynamics Program VLBI experiment. *Astron.and Astrophys*. (in press).

FRICKE, W., SCHWAN, H. and LEDERLE, T., 1988 : *Fifth Fundamental Catalogue (FK5)*, Part 1, Veröffentlichungen ARI, Heidelberg.

GUINOT, B., 1979 : Basic problems in the kinematics of the rotation of the Earth, *IAU Symposium 82*, McCarthy and Pilkington (eds.), Reidel.

HERRING, T.A., GWINN, C.R., and SHAPIRO, I.I., 1986 : Geodesy by radio interferometry : study of the forced nutation of the Earth, 1, Data analysis. *J. Geophys Res*. **91,** 4745.

IAG, 1988 : *Geodesist's Handbook*, Bull. Géod., **62,** 337.

IERS, 1989 : *Annual Report for 1988*, Observatoire de Paris.

LESTRADE, J.-F., PRESTON, R.A., MUTEL, R.L., NIELL, A.E. and PHILLIPS, R.B., 1988 : Linking the HIPPARCOS Catalog to the VLBI Inertial Reference System, high angular resolution structures and VLBI positions of 10 radio stars, *The Second F.A.S.T. Thinkshop*, J. Kovalevsky (ed).

128

MA, C., SHAFFER, D.B., DE VEGT, C., JOHNSTON, K.J., and RUSSELL, J., 1989 : Precise radio source positions determined by Mark III VLBI : observations from 1979 to 1988 and a tie to the FK5, *Astron. J.* (in press).

MA, C., 1989 : The realization of an inertial reference frame from VLBI. *This volume.*

McCARTHY (ed.), 1989 : IERS Standards. *IERS Technical Note No 3*, Observatoire de Paris.

MELBOURNE,W. (ed.), 1983 : Project MERIT Standards. *USNO Circular 167.*

MIGNARD, F., 1989 : HIPPARCOS and reference frames. *This volume.*

MORRISON, L.V., GIBBS, P., HELMER, L., FABRICIUS, C., REQUIEME, Y., RAPAPORT, M., 1989: *Proc. IAU Colloquium 100*, Reidel, *in press.*

SEIDELMANN, P.K.,1982 : IAU 1980 Theory of nutation: the final report of IAU working group on nutation. *Celest. Mech.* **27**, 79.

WILKINS, G.A., and MUELLER, I.I., 1986 : Joint summary report of the IAU/IUGG Working Groups MERIT and COTES, *Bull. Geod.* **60**, 85.

Discussion

HUGHES: For how many sources, especially primary sources, have you maps ?

ARIAS: In addition to the two sources mentioned, the Charlot (1989) paper includes maps of 0229+131, 0234+285, 0528+134, 0552+398, 0851+202, 1404+286, all of them being primary sources.

ARGUE: I have examined the IERS Catalogue. I do not have the actual figures with me, but my recollection is that there are some 50-60 sources on the Hubble Space Telescope selected for linking to HIPPARCOS.

ARIAS: There are 52 radio sources in the IERS Celestial Reference Frame which have been selected for the HIPPARCOS/Hubble Space Telescope link.

WESTERHOUT: Are you planning to introduce a new version of the celestial reference frame each year?

ARIAS: The analysis will be done every year, but a new version will be computed only when justified by the expected improvement in quality or entension of the frame.

WALTER: You process several independent observation catalogues. To which extent have these catalogues been reduced in compliance with the current IAU conventions of astronomical constants ?

ARIAS: The most questionable part of IAU conventions in this work is the precession and nutation models. While these models are used in the individual data analyses to refer the frames to the J2000.0 System, the selected individual frames had been obtained by adjusting for each session the celestial pole offsets, which frees the frames from related systematic errors.

A NEW CONCEPT OF CONSTRUCTING AN ACCURATE COORDINATE SYSTEM FROM GROUND-BASED OPTICAL OBSERVATIONS

MAO WEI, HU XIAOCHUN, GUO XINJIAN AND FAN YU
Yunnan Observatory
P.O. Box 110
Kunming, Yunnan Province
China, P.R.

ABSTRACT. Based on the expected precision and characteristics of the Low Latitude Meridian Circle (LLMC), and the development of CCD astrometry at Yunnan Observatory, an internally consistent and non-rotating optical celestial coordinate system can be set up through observations with the LLMC and CCDs. To obtain this goal, the main work we plan to do are (1) to establish a fundamental stellar reference system of several thousand stars based on the absolute obsrvations with the LLMC; (2) to provide the accurate zero-point corrections for the system from observations of minor planets with the LLMC and CCDs; (3) to determine the precessional rotation of the system with respect to an extragalactic reference system with the LLMC and CCDs, thus transforming the system into a quasi-inertial coordinate system; and (4) to obtain the atmospheric refraciton corrections from the observations with the LLMC.

1. Introduction

It is necessary that a milliarcsecond quasi-inertial celestial coordinate system be realized for the development of astronomy and relevant subjects. However, it is difficult to satisfy this demand only by means of traditional ground-based optical astrometric measurements. Therefore, many astrometrists are investigating space astrometry and VLBI. Undoubtedly, the realization of the new ideas and techniques will bring great changes for fundamental astrometry. Under these circumstances, it is one of the tasks in astrometry whether one could make use of new methods of ground-based optical fundamental astrometry and develop new instruments by new ideas and techniques, and then improve the celestial coordinate system, which can be not only compared with the systems of space astrometry and VLBI, but also be more readily applicable. According to our work during the past decade, we think that this idea can be realized with the LLMC and CCDs.

2. Realization of an internally consistent stellar reference system

The method of meridian absolute determination of the LLMC (Mao *et al.* 1986a) makes it possible to observe absolutely the stars of thirteenth magnitude in low latitude areas (including the equator area). This is very important to link the observations between the northern and the southern hemispheres, and ensures the consistency of the observations. Of course, the method can also be used in higher latitude areas. With the method, we can obtain the absolute azimuth and instantaneous latitude of the instrument through the alternate observation of a star in the prime vertical and meridian directions of the LLMC. Especially, we can determine several absolute latitude values during a night, and then obtain a curve of instantaneous latitude variation after a period of observation. Thus, the

129

J. H. Lieske and V. K. Abalakin (eds.), Inertial Coordinate System on the Sky, 129–130.
© 1990 *IAU. Printed in the Netherlands.*

positions of celestial bodies can be determined absolutely. The method can completely avoid the influence of various errors in the culmination observations of circumpolar stars.

Some instrumental errors, for example, the tube flexure, the circle division errors, etc. of the LLMC can be determined accurately and eliminated based on the practical measurement rather than the mathematical model in various zenith distances. From the charactaeristics of the LLMC, we can determine the corrections of atmospheric refraction in real time and also obtain better corrections even close to 80° zenith distance (Mao *et al.* 1986b). This allows us to obtain the local refraction corrections.

We adopt the full-star-adjustment method in constructing a catalogue (Fan *et al.* 1988). The method overcomes the difficulty due to non-continuous observations. It does not require observation of regular sets of stars, so it raises the efficiency of compiling a catalogue and can eliminate the influence of the variations of clock errors and seasonal variations, which are the main sources of systematic errors in right ascension. Thus, an internally consistent stellar reference system will be obtained.

3. The zero-point corrections and the elimination of precessional rotation of the stellar system

According to the application of CCDs in astrometry, we apply an overlapping exposure method for the observation of a CCD (Mao *et al.* 1989). By means of the method, we can determine the relative positions of fainter extragalactic objects and minor planets with respect to two reference stars (the stars are about tenth magnitude and about one degree apart) provided directly by the LLMC. In the measurement, the influence of atmospheric agitation in low altitude can be eliminated. The accuracy of observations in a night will be able to reach about 0.″003. With the LLMC in combination with the CCD, we are going to observe minor planets in the same instrument system in order to determine the zero-point corrections to the stellar system (Xu *et al.* 1989; Hu *et al.* 1989). We also plan to observe extragalactic objects for obtaining the corrections of lunisolar, planetary precession and nutation constants with the LLMC and the CCD in order to determine the rotation of the system. The position errors of reference stars can be eliminated in the determination of the objects (Guo 1989).

References

Fan, Y., Du, M., Mao, W., Zhou, Y., Zhang, Y.: 1988: "Full-star-adjustment method for calculating meridian and astrolabe catalogues", *Acta Astronomica Sinica*, **29**, 318–325.

Guo, X.: 1989, "Various methods for determining corrections for the precessional constant: I and II, *Publ. Yunnan Observ.*, No. 2, 102–121.

Hu, X., Mao, W.: 1989, "A method for determining the corrections for the equinox and equator by separating the disturbing forces", *Publ. Yunnan Observ.*, No. 2, 83–95.

Mao, W., Li, Z., Fan, Y., Hu, X., Du, M., Li, H.: 1986a, "A new method of determining absolute azimuth and latitude and suggestions for a new type of meridian circle", *IAU Symposium 109*, 551–555.

Mao, W., Li, Z., Fan, Y., Hu, X. Du, M.: 1986b, "The quasi-absolute determination of the refraction correction by means of the Low Latitude Meridian Circle", *Publ. de l' Observ. astronom. de Belgrade*, No. 35, 315–317.

Mao, W., Guo, X., Xu, S., Wu, G., Lu, R.: 1989: "Construction of an inertial coordinate system using a CCD", *Astron. Astrophys.* **215**, 190–194.

Xu, S., Hu, X., Zhang, J.: 1989, "A new method for observing asteroids", *Publ. Yunnan Observ.*, No. 2, 77–82.

TRANSFORMATION OF THE MEAN PLACE FROM FK4 TO FK5

Mitsuru SÔMA and Shinko AOKI
National Astronomical Observatory
Mitaka, Tokyo 181, Japan

ABSTRACT. Transformation procedures from the FK4 reference system of B1950.0 to the FK5 reference system of J2000.0 have been developed by Standish (1982) and by Aoki *et al.* (1983). We review here these procedures and discuss the differences between them. Especially we note that among researchers of this field a misunderstanding still exists in the problem at which stage the equinox correction should be applied. We show that the equinox correction should be applied in the precessing frame as Aoki et al. did. We also show that the epoch of the transfer from the FK4 to the FK5 in the transformation procedure is related to the systematic and individual corrections to the FK4.

1   Introduction

The matrix formulation of the transformation of the mean places and proper motions from the system of the FK4 at B1950.0 to that of the FK5 at J2000.0 has been developed by Standish (1982) and by Aoki *et al.* (1983). The main differences between these transformations are (1) in the application of the equinox correction and (2) in the epoch of the transfer from the FK4 system to the FK5 system.

Murray (1989a, b) claims that the transformation by Standish is correct in both points mentioned above.

In this paper in Sect. 2 we discuss the matter of the equinox correction and show that the transformation by Aoki *et al.* is correct as for the application of the equinox correction. In Sect. 3 we deal with the matter of the epoch of the transfer and show that it is related to the systematic and individual corrections to the FK4.

Details about these matters will be published in a volume of *Astronomy and Astrophysics*.

2   Application of the equinox correction

In the Standish's or Murray's transformation the equinox correction is applied in the fixed frame whereas in the Aoki *et al.*'s transformation it is applied in the frame rotating by the precession. These transformation procedures are expressed as follows at the epoch of the transfer from the FK4 to the FK5:

131

J. H. Lieske and V. K. Abalakin (eds.), Inertial Coordinate System on the Sky, 131–136.
© 1990 *IAU. Printed in the Netherlands.*

Standish or Murray (the epoch is B1950.0):

$$\alpha^N = \alpha^O + E, \tag{1a}$$
$$\delta^N = \delta^O, \tag{1b}$$
$$\mu_\alpha^N + m^N + n^N \sin(\alpha^O + E) \tan \delta^O = \mu_\alpha^O + m^O + n^O \sin(\alpha^O + E) \tan \delta^O + \dot{E}, \tag{1c}$$
$$\mu_\delta^N + n^N \cos(\alpha^O + E) = \mu_\delta^O + n^O \cos(\alpha^O + E), \tag{1d}$$

Aoki *et al.* (the epoch is 1984 January 1.0):

$$\alpha^N = \alpha^O + E, \tag{2a}$$
$$\delta^N = \delta^O, \tag{2b}$$
$$\mu_\alpha^N + m^N + n^N \sin(\alpha^O + E) \tan \delta^O = \mu_\alpha^O + m^O + n^O \sin \alpha^O \tan \delta^O + \dot{E}, \tag{2c}$$
$$\mu_\delta^N + n^N \cos(\alpha^O + E) = \mu_\delta^O + n^O \cos \alpha^O, \tag{2d}$$

where m and n are the rates of the general precession in right ascension and declination based on the Newcomb's precession (superscript O) and the IAU 1976 precession (superscript N). Here the time unit is supposed to be the same in both sides of the equations for proper motions. The difference between Eqs. (1) and (2) is in the terms including n^O in the equations for proper motions.

Murray (1989a, b) considers the special case in which there is no change in the precession constant $(m^N = m^O, n^N = n^O)$, and the equinox correction is independent of time ($\dot{E} = 0$). He supposes that in this case both coordinate systems are inertial, in other words proper motions are the same in the two systems. The Eqs. (1c) and (1d) support his supposition while the Eqs. (2c) and (2d) don't, and he concludes that Standish's procedure is correct. Smith *et al.* (1989, in Note added in proof) and Yallop (1989) support this Murray's inference. But his inference is erroneous, because the precession is dependent on the location of the equinox (mean pole moves always toward the equinox at each instant) and therefore the equinox correction E affects the obtained proper motions even if the equinox motion $\dot{E}$ is zero. Detailed considerations are given in the followings.

Proper motions of stars (μ_α, μ_δ) are not directly determined from observations but are determined from the observed variations $(d\alpha/dt, d\delta/dt)$ of the mean places (α, δ) from the following expressions:

$$\frac{d\alpha}{dt} = m + n \sin \alpha \tan \delta + \mu_\alpha,$$
$$\frac{d\delta}{dt} = n \cos \alpha + \mu_\delta.$$

The right ascension α appearing in the right sides of the above equations is, of cource, the observed right ascension in the catalog (FK4 or FK5) system and not in the dynamical system. Therefore if the catalog includes an error E in the right ascension, it also affects the obtained proper motions. From this consideration the equations (2c) and (2d) are derived.

The correctness of the equations (2c) and (2d) can also be confirmed using the formulae of spherical triangles.

Thus we can conclude that as for the application of the equinox correction the procedure by Aoki *et al.* is correct.

Murray insists that the equinox corrections determined from analyses of observations of the Sun, Moon and planets are corrections to the right ascension system of FK4 *in the B1950.0 frame*, but this statemant is not correct. In fact, the equinox corrections so determined are corrections to the right ascension system of FK4 at the epoch of observation *in the frame of date*. For example, in the Eq. (5) given by Fricke (1985) and applied to observations, for example, by Hughes and Scott (1982):

$$\Delta\alpha = -E + \cos\epsilon\sec^2\delta\Delta L - \cos\alpha\tan\delta\Delta\epsilon$$
$$+ 2\sin\alpha\sec\delta\Delta h - 2\cos\alpha\sec\delta\cos\epsilon\Delta k,$$

the quantity E is the equinox correction at the epoch of observation in the frame of date. In the above equation $\Delta\alpha$ is the $O - C$ in the apparent right ascension of the Sun, ϵ is the obliquity of the ecliptic and ΔL is the correction to the mean longitude of the Sun.

It is obvious that Fricke (1982) assumed that the equinox correction $E(t)$ at any epoch *in the frame of date* can be expressed as

$$E(t) = E_{1950} + \dot{E}t,$$

and obtained the values of E_{1950} and $\dot{E}$ from analyses of observations of the Sun, Moon and planets. Because $\dot{E}$ in the above equation is theoretically equal to the value obtained from an analysis of FK4 proper motions in the frame of 1950.0, Fricke (1982) compared these values and obtained the final value of $\dot{E}$.

3 Epoch of the transfer from the FK4 system to the FK5 system

In the transformation by Aoki *et al.* the transfer from the FK4 system to the FK5 system is performed at the epoch of 1984 January 1 when the FK5 came into effect. Motivation of their transformation is that observed values such as stars' positions, UT1, etc. are not changed when the system of the catalog is changed, except stars' right ascensions and proper motions in right ascension, which are intentionally changed by the values obtained by Fricke (1982). Especially we note that their transformation is consistent with the new expression for the relationship between UT1 and GMST (Aoki *et al.*, 1982). On the other hand Murray insists that the transfer should be performed at the epoch of B1950.0 as Standish did. The two 6×6 transformation matrices are given in Tables 1 and 2. In calculating these matrices the application of the equinox correction is changed from Murray for the reasons mentioned in Sect. 2 and for the values of the equinox correction to the FK4 we adopt the values $E = 0\overset{s}{.}035$ at J1950.0 and $\dot{E} = 0\overset{s}{.}085$ / Julian century as Aoki *et al.* did in order to be consistent with the new expression for the relationship between UT1 and GMST. For the precession formulae based on the Newcomb's precession constant we use those given by Kinoshita (1975). The formulae for obtaining the position and velocity vectors from the mean place, proper motion, radial velocity and parallax, and vice versa, are given by Aoki *et al.*

The role of fundamental catalogs such as FK4 or FK5 is to define the equator and the equinox among stars on the celestial sphere *at any epoch* (see e.g. Woolard and

Table 1. Transformation Matrix (Transfer at 1984 January 1)

$$
\begin{pmatrix}
+0.9999256782 & -0.0111820610 & -0.0048579477 & +0.00000024239502 & -0.0000000271066 & -0.000000000117766 \\
+0.0111820609 & +0.9999374784 & -0.0000271765 & +0.0000000271066 & +0.0000024239788 & -0.000000000000659 \\
+0.0048579479 & -0.0000271474 & +0.9999881997 & +0.0000000117766 & -0.0000000000658 & +0.0000024241017 \\
-0.00055 & -0.23854 & +0.43574 & +0.99994704 & -0.01118251 & -0.00485767 \\
+0.23849 & -0.00267 & -0.00854 & +0.01118251 & +0.9995883 & -0.00002718 \\
-0.43562 & +0.01225 & +0.00212 & +0.00485767 & -0.00002714 & +1.00000956
\end{pmatrix}
$$

Table 2. Transformation Matrix (Transfer at B1950.0)

$$
\begin{pmatrix}
+0.9999256782 & -0.0111820602 & -0.0048579481 & +0.00000024239502 & -0.0000000271054 & -0.000000000117789 \\
+0.0111820602 & +0.9999374784 & -0.0000271632 & +0.0000000271054 & +0.0000024239788 & -0.000000000000658 \\
+0.0048579481 & -0.0000271607 & +0.9999881997 & +0.0000000117789 & -0.0000000000659 & +0.0000024241017 \\
-0.00055 & -0.23803 & +0.43551 & +0.99994704 & -0.01118172 & -0.00485911 \\
+0.23802 & -0.00266 & -0.00023 & +0.01118172 & +0.9995884 & -0.00002716 \\
-0.43549 & +0.00395 & +0.00212 & +0.00485911 & -0.00002717 & +1.00000955
\end{pmatrix}
$$

Clemence, 1966, pp. 376–377; or Fricke and Kopff, 1963, p.1). Whether the equator or the equinox defined by a fundamental catalog is coincident with the real equator or equinox is another question (the differences are called the systematic error of the catalog); absolute or fundamental observations will answer that question. In fact because we now know that the adopted precession constant in the FK4 has an error, if the equator defined by the FK4 is coincident with the real equator at *some* epoch, the equator defined by the FK4 at any other epoch is not coincident with the real equator. This fact suggests that the systematic error of the fundamental catalog depends on time. (The error of the first power of time can be eliminated by adjustment of proper motions, but the error of the second and higher power of time cannot be eliminated.) Besides systematic error there is an individual error for each star. This depends also on time. Murray implicitly assumes that the systematic and individual errors in the FK4 are zero at the epoch of B1950.0, but this cannot be justified, because positions and proper motions in the FK4 were not determined from observations only at the epoch of B1950.0.

If one uses the transformation matrix given in Table 1, one must apply the systematic correction to the FK4 system at the epoch of 1984 January 1, and if one uses the transformation matrix given in Table 2, one must apply the systematic correction to the FK4 system at the epoch of B1950.0. We can show that the matrix given in Table 2 can be derived from the matrix given in Table 1 by applying the systematic and individual corrections to the FK4 system.

If one ignores the systematic and individual corrections to the FK4 in the transformation to the FK5 system, the matrix given in Table 1 (Aoki *et al.*'s procedure) is recommended, because it is consistent with the definition of UT1. Also in the transformation of precise positions obtained by VLBI, the matrix given in Table 1 is recommended, because most of those observations are performed in the 1980's.

References

Aoki, S., Guinot, B., Kaplan, G. H., Kinoshita, H., McCarthy, D. D., Seidelmann, P. K., (1982), 'The new definition of Universal Time', *Astron. Astrophys.* **105**, 359–361.

Aoki, S., Sôma, M., Kinoshita, H., Inoue, K. (1983) 'Conversion matrix of epoch B1950.0 FK4-based positions of stars to epoch J2000.0 positions in accordance with the new IAU resolutions', *Astron. Astrophys.* **128**, 263–267.

Fricke, W. (1982) 'Determination of the equinox and equator of the FK5', *Astron. Astrophys.* **107**, L13–L16.

Fricke, W. (1985) 'Fundamental Catalogues, past, present and future', *Celes. Mech.* **36**, 207–239; or *Veröff. Astron. Rechen-Inst., Heidelberg* No. 31, Verlag G. Braun, Karlsruhe.

Fricke, W., Kopff, A. (1963), 'Fourth fundamental catalogue (FK4)', *Veröff. Astron. Rechen-Inst., Heidelberg* No. 10, Verlag G. Braun, Karlsruhe.

Hughes, J. A., Scott, D. K. (1982) 'Results of observations made with the six-inch transit circle 1963–1971', *Publ. U. S. Naval Obs.* Sec. Series Vol. XXIII, Part III.

Kinoshita, H. (1975) 'Formulas for Precession', *SAO Special Report* No. 364, Smithsonian Institution Astrophysical Observatory, Cambridge, Massachusetts.

136

Murray, C. A. (1989a) 'The transformation of coordinates between the systems of B1950.0 and J2000.0, and the principal galactic axes referred to J2000.0', *Astron. Astrophys.* **218**, 325–329.

Murray, C. A. (1989b) 'The transformation between the coordinate systems of FK4 at B1950.0 and FK5 at J2000.0', *Highlights of Astronomy* Vol. 8, pp. 482–483, Kluwer Academic Publishers, Dordrecht.

Smith, C. A., Kaplan, G. H., Hughes, J. A., Seidelmann, P. K., Yallop, B. D., Hohenkerk, C. Y. (1989) 'Mean and apparent place computations in the new IAU system. I. The transformation of astrometric catalog systems to the equinox J2000.0', *Astron. J.* **97**, 265–273.

Standish, E. M. Jr. (1982) 'Conversion of positions and proper motions from B1950.0 to the IAU system at J2000.0', *Astron. Astrophys.* **115**, 20–22.

Woolard, E. W., Clemence, G. M. (1966), *Spherical Astronomy*, Academic Press Inc., New York.

Yallop, B. D. (1989) 'Apparent place reduction', *Highlights of Astronomy* vol. 8, pp. 481–482, Kluwer Academic Publishers, Dordrecht.

Discussion

SEIDELMANN: The 1984 date is only an arbitrary date introducing the change. It is involved in the UT1 equation to maintain continuity across the change. It is not a part of the transformation of the star catalog mean position. It was not part of the determination of the equinox correction in the definition of the J2000 system of the FK5.

SÔMA: The catalog positions have a close relation to the sidereal time and UT1. Therefore, it is important that the catalog positions should be determined in a consistent way with the equation for UT1 and sidereal time. But, as we have shown in this paper, the positions in the FK5 system do not depend on the epoch of transfer, if one applies the systematic correction to the FK4 properly.

MURRAY: In the first part of your paper you gave a fair description of the difference between us, but I still maintain that Fricke's motion of the equinox must be interpreted as being in the *fixed* frame B1950.

My motivation in entering into this subject was to determine the transformation to galactic coordinates for J2000. Using the transformations by Aoki *et al.*, the galactic coordinates of an object calculated from its equatorial coordinates in the J2000 frame differ from those calculated from the B1950 frame, which is absurd!

SÔMA: As shown in our paper, the equinox corrections must be applied in the moving frame. Aoki *et al.*'s transformation gives no inconsistency even in the galactic coordinates.

THE COMPUTATION OF MEAN POSITIONS AND PROPER MOTIONS

Carl S. Cole
U. S. Naval Observatory
Washington, DC

ABSTRACT: This paper investigates the estimation of mean positions and proper motions given independent solar-system barycentric positions observed at different epochs, reduced to the same equator and equinox and freed of systematic errors. Past practices are reviewed and the relative quality of the data is studied to determine the appropriate form of the model.

1. INTRODUCTION

A program is in progress at the U.S. Naval Observatory which will result in a new fundamental catalog of approximately 40,000 stars. This catalog will establish a dynamical reference frame of stellar positions and proper motions based on conventional values of astronomical constants and a new analysis of observations of solar system objects incorporating corrections to current planetary theories. The catalog will also provide a more direct connection between the optical and radio reference frames by combining the current fundamental stars (FK5) and fainter reference stars into a single fundamental system. As part of this work, an investigation into the method of estimating mean positions and proper motions was conducted.

2. PAST PRACTICES

Traditional practice has been summarized by Newcomb (1906). One uses initial estimates of the position and proper motions (α_0, δ_0, μ and μ') and computes corrections to these quantities ($\Delta\alpha_0$, $\Delta\delta_0$, $\Delta\mu$ and $\Delta\mu'$). For each independent observation, observed minus computed residuals ($\Delta\alpha_\iota$ and $\Delta\delta_\iota$) are calculated. Condition equations are formed:

$$\Delta\alpha_\iota = \frac{d\alpha}{d\alpha_0}\Delta\alpha_0 + \frac{d\alpha}{d\delta_0}\Delta\delta_0 + \frac{d\alpha}{d\mu}\Delta\mu + \frac{d\alpha}{d\mu'}\Delta\mu' ,$$

$$\tag{1}$$

$$\Delta\delta_\iota = \frac{d\delta}{d\alpha_0}\Delta\alpha_0 + \frac{d\delta}{d\delta_0}\Delta\delta_0 + \frac{d\delta}{d\mu}\Delta\mu + \frac{d\delta}{d\mu'}\Delta\mu'$$

and solved for by weighted least squares. The partial derivatives being evaluated at the epoch of observation.

J. H. Lieske and V. K. Abalakin (eds.), Inertial Coordinate System on the Sky, 137–140.
© 1990 IAU. Printed in the Netherlands.

Newcomb used this method for 4 northern polar stars in his "Catalogue of Fundamental Stars" (Newcomb 1898). Unfortunately, this is the only catalog compilation solving for both coordinates simultaneously, subsequent compilers either being unaware of or ignoring Newcomb's work.

Several simplifying approximations in the application of this method are still in general use:

$$\frac{d\delta}{d\alpha_0} = \frac{d\alpha}{d\delta_0} = \frac{d\delta}{d\mu} = \frac{d\alpha}{d\mu'} = 0,$$

$$\frac{d\alpha}{d\alpha_0} = \frac{d\delta}{d\delta_0} = 1 \text{ and} \qquad\qquad (2)$$

$$\frac{d\alpha}{d\mu} = \frac{d\delta}{d\mu'} = \tau_\iota$$

where τ_ι is the epoch difference between the observation and initial estimate. Thus, the two coordinates are decoupled and solved for separately. Although the use of normal points to combine data is no longer common, many authors still use catalog weights, rather than residuals, to determine error estimates (cf. Cole in press).

With recent advances in computational ability, the assumptions of equations (2) are no longer necessary. Once the relationships between the observed coordinates and the star parameters are stated, they can be linearized similar to equations (1) and the star parameters, with the associated covariance matrix, can be estimated using least squares. However, the use of subjective judgment is still required to form these relationships.

3. THE FORM OF THE MODEL

The question arises how best to model the relationship between the coordinates (α and δ) at any arbitrary epoch and the relevant star parameters (α_0, δ_0, μ and μ'). (It is assumed that the measured epoch of observation is exact.) Several options are uniform motion in each coordinate, uniform motion on a great circle, uniform rectilinear motion and circular motion about the center of the galaxy.

The rotation of the galaxy and the finite speed of light (cf. Stumpff 1985) were ignored and uniform rectilinear motion was assumed. The ignored effects were deemed insignificant given the time span and accuracy of the observations. The problem with the model of uniform rectilinear motion is that six parameters are needed to describe the motion, and parallax and radial velocity data are scarce. One could estimate six parameters from the positional data given at least six angular measurements, but this would reduce the error degrees of freedom by two, and would give very poor estimates of the radial parameters.

Schlesinger (1917) suggested solving for radial velocities given accurate parallaxes and Eichhorn (1981, 1982) suggested solving for parallaxes given accurate radial velocities. If radial data (parallaxes and radial velocities) were available, they could be combined with the angular measurements into a single adjustment. This approach was rejected for use in the current program. The angular measurements are often determined to one part in 10^7, whereas the radial data are sometimes known only to one part in ten or one part in a hundred. Thus the weights can easily vary by ten orders of magnitude. In combining data of unequal weight, a factor of two error in the relative weights can lead to useless results. The weights can be iteratively determined, but then one must form a data base of all parallax and radial velocity measurements made over the last century. One must also become intimately familiar with the various systematic errors of these data.

The situation remains that the best methods of computing parallaxes and radial velocities are the conventional ones. In computing the mean position and proper motions, therefore, one should use modern, accurate values for the parallax and radial velocity and assume them to be exact. In the event that either value is unavailable, one should set it equal to zero.

Rigorous formulae for expressing the two spherical coordinates in terms of the initial position, proper motions, parallax and radial velocity under the assumption of uniform rectilinear motion are given by Eichhorn and Rust (1970) and will not be repeated here.

With the longer time span and greater accuracy of future observations, one can conceive of solving for positions, proper motions, parallaxes, radial velocities, galactic rotation, the speed of light, etc. in a single adjustment. The size and complexity of data adjustments will continue to grow, but this growth will require the matching of computational ability with theoretical knowledge and, as always, careful analysis of the data.

4. PRACTICAL CONSIDERATIONS

There are several interesting consequences of solving for both coordinates in a single adjustment. The error sums of squares and degrees of freedom are pooled, resulting in more degrees of freedom when making tests of hypotheses such as the determination of outliers. Also, the variance of the mean squared error is reduced, which gives rise to fewer cases of estimates of zero error. The possibility of estimates of zero error is the leading argument for estimating errors from catalog weights rather than from residuals.

The combined adjustment also results in a single mean (or central) epoch. Thus all four star parameters are correlated with each other. But this only tells us in a quantitative way what we already know: one coordinate is not independent of the proper motion in the other coordinate. But, as we also know, these correlations are very small except in special situations.

Since the sum of the residuals of both coordinates is being minimized, the error estimates of the coordinates are very nearly equal given similar numbers of observations and weights. One could argue that agreement in one coordinate would be adversely affected by scatter in the other. But it is generally agreed that very small error estimates are artificial. Also, a better job of outlier rejection is done with the combined error sum of squares and degrees of freedom.

Several caveats are now in order. The right ascension residuals must be scaled to the declination residuals by multiplication by the cosine of the declination. The declination used to scale the right ascension should be a fixed, reference declination and not allowed to vary in the adjustment. Care must be taken with units. Angular data must be expressed in radians and parallax and radial velocity must be appropriately scaled. Along with having to estimate the relative quality of the various input catalogs, the investigator must now also be concerned with the relative quality of the two angular measures within a catalog.

Two methods of adjustment have been tried, both using commercially available software. One method (IMSL 1987) uses a nonlinear regression routine which approximates the partial derivatives using finite differences and a second method (REDUCE 1987) uses an algebraic manipulator to give an exact representation of the partial derivatives. In comparing these methods with conventional ones, all results agree within the estimated errors.

5. SUMMARY

In solving for mean positions and proper motions, the model used is uniform, linear motion in 3-dimensional space. Positions and proper motions are solved for using fixed values for the parallaxes and radial velocities. This uniform rectilinear motion model leads to non-linear condition equations which require linearization and initial estimates of the star parameters. Both coordinates are solved for in a single adjustment, resulting in a single mean epoch and a pooling of the error sums of squares and degrees of freedom.

REFERENCES

Cole, C.S. (in press), in *Erreurs, Biais et Incertitudes en Astronomie*, C. Jaschek and F. Murtagh, eds.
Eichhorn, H. (1981), *Astron. J.* **86**, 915-917
Eichhorn, H. (1982), *Sitzungsber., Österr. Akad. Wiss., Math.-Naturwiss. Kl. Abt. II*, 191, 429-459
Eichhorn, H. and Rust, A. (1970), *Astron. Nachr.* 292, 37-38
IMSL (1987), *User's Manual*, IMSL Inc., Houston
Newcomb, S. (1898), *Astron. Pap.* **VIII**, pt. 2
Newcomb, S. (1906), *A Compendium of Spherical Astronomy*, MacMillian, New York
REDUCE (1987), *User's Manual*, The Rand Corp., Santa Monica
Schlesinger, F. (1917), *Astron. J.* **30**, 137-138
Stumpff, P. (1985), *Astron. Astrophys.* 144, 232-240

CAN WE DEFINE AN INERTIAL REFERENCE SYSTEM?

LI LINGHUAI , TONG FU
Purple Mountain Observatory
Academia Sinica
Nanjing, China

The topic (Seidelmann, 1986) is one of the unresolved questions in the field of celestial mechanics because the definitions of an inertial system are all controversial (Mach, 1893; Eichhorn, 1984). Eichhorn's definition enlarges its connotation and is not acceptable either. Connotation of a concept is endowed through defining it. One of the reasons why the existing definitions of an inertial system are all controversial is that the connotations that are endowed by them are not unique, in other words, they do not draw up the same area of ideas. Taking the interplay of all of them for its connotation is an acceptable solution.

According to general relativity, as far as a reference system is concerned, we always divide the Universe into two parts: the background and the test bodies we are interested in. According to the metric theories of gravity, the action of matter in the Universe to the test particles is not direct but through thespace-time which is determined by it. Thus reference systems are space-time in nature and should be classified in terms of its intrinsic nature, the scalar curvature. As a result, an inertial system can be defined as: Flat background space-time is called an inertial reference system.

In order to make the definition not so abstract in appearence, let's associate it with something familiar to us through discussing its coordinate representation. To specify a reference system is to give a scalar curvature field $R(p)$, for example, $R(p)$ equals zero for the inertial system according to the definition above. According to Riemannian geometry, the curvature tensor R_{ij} can be constructed through making use of R and g_{ij}: $R_{ij} = g_{ij}R/4$.

On the other hand, R_{ij} is defined as functions of the metric and its first and second derivatives, $R_{ij} = R_{ij}(g_{\mu}, Dg_{\mu}, D^2g_{\mu})$. Thus we have following equations: $R_{ij}(g_{\mu}, Dg_{\mu}, D^2g_{\mu}) = g_{ij}R/4$, which are independent of any coordinate system. In order to specify a coordinate system, a coordinate condition consisting of four additional equations must be given. Consequently, it can be inferred that different coordinate conditions in GTR are related to different measurement conventions and the coordinate transformations are related to transformations between measurement conventions.

References

Seidelmann, P.K.: 1986, *Celes. Mech.*,39, 141.
Mach, E.: 1893, *The Science of Mechanics* , trans. by T.J. McCormack (2nd ed., Open Court Publishing Co.).
Eichhorn, H.: 1984, *Celes. Mech.*, 34, 11.

J. H. Lieske and V. K. Abalakin (eds.), Inertial Coordinate System on the Sky, 141.
© 1990 *IAU. Printed in the Netherlands.*

THE ROLE OF OBSERVATION WITH HORIZONTAL MERIDIAN CIRCLE IN CHINA FOR ESTABLISHING AN INERTIAL FRAME

Li Zhi-gang and Qi Guan-rong
Shaanxi Astronomical Observatory
Lintong, Xian, China, P.R.

ABSTRACT. While HIPPARCOS is expected to measure positions and proper motions with more accuracy than those obtained by ground-based instruments, what can we do in the future for ground-based instruments? The observations with them still are important for establishing an inertial frame because of the long history of observations with them and improvements in the instruments. Moreover, it is necessary to have data of observations from them for research on problems related to the Earth. The horizontal meridian circle in China (DCMT) is expected to have advantage over the classical meridian circles. The DCMT will be assembled and tested this year. It should work in the following fields: (1) observing radio stars, (2) observation of minor planets, (3) absolute determinations of IRS.

ON THE PROBLEM OF THE DEVELOPMENT OF ASTRONOMICAL REFRACTION METHODOLOGY ON THE BASIS OF NEW TECHNOLOGY

V.I. Sergienko
NPO Etalon
Borodina 57
664018 Irkutsk, USSR

ABSTRACT. The work on construction and operation of automated systems for collection and processing astrometric and meteorological information and also of automated television systems with remote control have placed before us a number of fundamental difficulties in that hundreds of correct solutions to these problems require a systematic approach.

The most suitable cybernetics-mathematical composite for solving the problems of astronomical refraction may become the so-called hybrid-expert systems — a union of traditional expert systems with calculation-logical ones. In these systems logical-linguistic models are used together with mathematical ones.

The basic problem of ARA-C consists in obtaining a model of the environment in which light propagates and also to give a notion of interaction between electrons (environment) and photons (light).

It is proposed to use in this system a model of light interaction with environment built up on special formalisms of artificial intellect (logical-linguistic model). The basic principal in hybrid-expert system ARA-C operation must be mathematical modelling and calculation experimentation. In the proposed hybrid-expert system knowledge is presented on three levels — object, mathematical and programmed. Functioning of the lower levels of ARA-C is connected with formulation of the calculation problems. A calculation problem presented to a technical model determines a partial order on a twopart column of mathematical correlations

DETERMINATION OF REFRACTION AND PROGRESS OF FUNDAMENTAL ASTROMETRY

A.Yu. YATSENKO
Engelhardt Astronomical Observatory
422526 Kazan
USSR

1. Introduction

The potentional possibility of high precise instruments of fundamental astrometry cannot be realized because we have a bad refraction determination: the instrumental accuracy of these telescopes is about 0".01 (Yoshizawa 1987) and the real accuracy of determination of declinations is 0".20. The theoretical possibility of determining real refraction with a precision of 0".01 to 0".05 is shown. According to Kolchinskij (1987) astronomical refraction can be divided into the following components:

Table 1. Refraction fluctuations

N	Types	Quasi-period, sec	Notes
1	High frequency (image motion)	0.001–1	r.m.s. deviation computed by special formulas
2	Medium frequency (accidental refraction)	1–1000	Gravitational waves. Not computed by the theory of refraction.
3	Low frequency diurnal variation, annual	1000–10000000	Computed by the theory of refraction or tables (normal refraction)
4	Low frequency (anomalies)	- " -	Not computed by the theories of refraction

The first type of fluctuations can be reduced with the impersonal photoelectric micrometer if it has about 100 registrations of star positions during an observation. Computations by Yatsenko (1989) show that this refraction type can be reduced to about 0"03 during one observation with probability 0.90. A part of the second type is reduced in this manner also. The remainder of the second type and the fourth type of the fluctuations are computed by using aerological and meteorological information in order to do its correct calculation.

2. Determination of accidental refraction and anomalies of refraction

Let's divide the atmosphere into two refraction parts:
1. *Surface Layer* is up to the altitude at which the influences of buildings and relief peculiarities to the horizontal temperature gradients disappear . Depending on locality it is from

J. H. Lieske and V. K. Abalakin (eds.), Inertial Coordinate System on the Sky, 143–144.
© 1990 *IAU. Printed in the Netherlands.*

0.02 to 0.1 km.

2. *Free Atmosphere* is from 0.1 to 82 km.

2.1. THE INFLUENCE OF THE FREE ATMOSPHERE

For the accurate determination of this influence one should obtain the answers to the following questions: (1) What is the tolerable absolute error of aerological probe sensors? (2) Does it correspond to the real error of these devices? (3) Is the number of levels in the current aerological practice enough? (4) What is the greatest tolerable distance of an aerological station from an observation place?

For obtaining answers to these questions Yatsenko (1987, 1988, 1989) has fulfilled particular computations by the using spatial (three-dimensional) theory of refraction. The results have shown the following: (1) The error of 0.8 °K leads to the error of a refraction computation equal or less than 0."004. (2) It corresponds to the real error of aerological devices. (3) The number of levels in the current aerological practice is enough for evaluating the refraction in the free atmosphere with accuracy better than 0."01. (4) A distance to 400 km between the aerological station and observing site allows one to evaluate the refraction with error less than 0."01.

For observations at zenith distances 0–60° a spatial area which propagates a light ray, is a cylinder with radius 140 km and altitude 82 km. But 98% of the refraction effect appears in a smaller area: a cylinder with radius 46 km and altitude 27 km. Thus aerological probes actually provide us with information about the whole second refraction part of the atmosphere and with sufficient accuracy.

2.2. THE INFLUENCE OF THE SURFACE LAYER

An altitude of a surface layer for every place on the Earth is particular. Evidently, anomalies of the refraction in surface layer is the main cause of the whole refraction anomalies. For its determination one needs a spatial set of meteorological sensors. A system for the collection of meteorological data should consist of 6 masts or fasted balloons with height 0.02–0.1 km equipped mainly with temperature sensors. The masts should be disposed in N-S direction by pairs: the one in the plane of the first vertical and two others at a distance 170 m from an instrument.

The accuracy of three sensors above the system should be 0.02 °K (for evaluating the main part of astronomical refraction), the others from 0.8 °K (at the top) to 0.02 °K in a vicinity of an instrument (for obtaining of the tilts of the equal refractive index surfaces which can be vertical (Yatsenko 1985).

The processing of the whole information should carry out with the "cube"-method (Yatsenko 1985) by using spatial theory of refraction.

References

1. Kolchinskij, I.G. (1987) "On terminology in the theory of atmospheric refraction", *Publ. Observ. Astron. Beograd* **35**, 332–337.
2. Yatsenko, A.Yu. (1985) "On the room refraction", *Bull. Observ. Astron. Beograd*, **135**, 16–20.
3. Yatsenko, A.Yu. (1988) "Refraktsiya w meridional'noj modely atmosfery Teoreticheskiye osnowy", *Kinematika i fizika nebesnikh tel* **4**, 2, 59–66. (In Russian)
4. Yatsenko, A.Yu. (1989a) "Refraktsiya w meridional'noj modely atmosfery. Algoritm wychislenij", *Kinematika i fizika nebesnikh tel* **5**, 1, 68–74. (In Russian)
5. Yatsenko, A.Yu. (1989b) "Refraktsionnoye obespecheniye wysokotochnikh meridiannykh nablyudenij", *Kinematika i fizika nebesnikh tel* **5**, 3, 84–89. (In Russian)
6. Yoshizawa, M., Suzuki, S., Fukaya, R. (1987) "Tokyo PMC catalog 85", *Ann. Observ. Astron. Tokyo* **21**, 393–421.

PRECISION OF STUDIES FROM ASTROMETRIC OBSERVATIONS: RECENT PROGRESS AND FUTURE PROSPECTS

Z. X. LI
Shanghai Observatory
Academia Sinica
Shanghai   200030
P. R. of China

ABSTRACT.   Recent progress of precision in astrometric studies is summarized and the weights of systematic errors in astrometric observation are analysed.

Precision improvement of certain studies based on a re-determination of 1962-1982 Earth Rotation Parameter(ERP) is described in Table 1.

TABLE 1. Improvement of precision based on improved ERP of the BIH

No.	Item	Improvement of precision	Reference
1	ERP: X,Y,UT1-UTC	24%,26%,29%	(Feissel,1985)
2	Auxiliary parameters: Z,W	about 30%	(Li,1988)
3	Primary nutation constants	over 50%	(Capitaine,1988)
4	Love number of the Earth	about 50%	(Nei,1988)
5	Plate motion	about 30%	(Li,1989)
6	Motion of the mean pole	good accordance with the IRIS's and LAGEOS's	(Markowitz,1988)

In the re-determination, average rms for the 24 most important instruments, after correcting their observational series by Z,W and Group Unknowns G,Ġ which have been evaluated in a "global reduction" of almost 500,000 observations during the 20 years, diminish 22%,11% for a time, latitude group-observation respectively. It may be used to explain partly the above improvement.

In order to discuss the further possible improvement in studying from astrometric observations, the observational series in 1976-1977 of the photoelectric astrolabe at Shanghai Observatory is used to estimate the weights of different sources of error in astrometry. The ERP of the 20 years is used as a reference and rms of a group-observation is estimated after the adding of different kinds of correction. "Perfect correction" means only the residuals of a same group in a year are used and then estimate the rms after a fitting line is used.

J. H. Lieske and V. K. Abalakin (eds.), Inertial Coordinate System on the Sky, 145–146.
© 1990 IAU. Printed in the Netherlands.

TABLE 2. rms of a group observation in the case of
the photoelectric astrolabe at Shanghai

Correction	Time	Latitude
No correction	$0^s.0078$ (100%)	$0''.082$ (100%)
FK5-FK4	0.0074 (94%)	0.075 (91%)
G, $\dot{G}$	0.0070 (90%)	0.071 (87%)
G,$\dot{G}$ and W or Z	0.0069 (89%)	0.069 (84%)
Perfect correction	0.0064 (82%)	0.060 (73%)

From the above figures, the following points for the instrument are presented:

-- After the using of FK5 Catalog, rms of a group observation will
decrease 6-9%;
-- Even the using of G,$\dot{G}$ alone can give a greater decrement to the
rms (10-13%) than that of FK5 Catalog, but it should be mentioned here that G,$\dot{G}$ contain the catalog correction as well as
some of the systematic local error existing in the observations;
-- The combine using of G,$\dot{G}$ and W,Z corrections will diminish the
rms furthermore (11-16%), while Z does more (3%) than W (less
than 1%) to the decrement of rms;
-- The difference of rms decrement between the using of G,$\dot{G}$,W,Z
corrections and that of the "perfect correction" (7-11%) reflects the existence of unmodeled local errors. It is expected
to be able to model them better in the future;
-- Future prospects of precision improvement in astrometry depends
on the studies of local error, astronomical constants, and star
catalog.

REFERENCE

Capitaine,N. et al.(1988) 'Determination of the principal term
of nutation from improved astrometric data',Astron.Astrophys.
202, 306-308.
Feissel,M. et al.(1985) 'Precision and accuracy of Earth rotation
determination at BIH from optical astrometry', Proceeding of
the International Conference of Earth Rotation and the Terrestrial Reference Frame, 3-14.
Li,Z.X.(1988)'A homogeneous z-term series of astrometric latitude', Astron. Astrophys. Suppl. Ser.75,151-156.
Li,Z.X.(1989)'Current plate motionfrom astrometry and Doppler
satellite observations',ACTA GEOPHYSICA SINICA,Vol.32,No.6,
660-676.
Markowitz,W.(1988)'Motion of the mean pole from ILS,BIH,IRIS,and
Lageos', Report to IAU Commission 19.
Nei,S.Z.(1988)'Analysis of short period terms in the BIH new UT1
series', Annuals of Shanghai Observatory, No.9,78-82.

NUTATION THEORY FOR A NON-RIGID EARTH: PRESENT STATUS AND FUTURE PROSPECTS

J. WAHR
Department of Physics
Univ. of Colorado
Boulder, Colorado 80309 USA

ABSTRACT. The 1980 IAU nutation series includes effects of the Earth's non-rigidity. Certain simplifying assumptions were made in the model, including that the Earth is elastic and hydrostatically pre-stressed. The rationale was that geophysicists didn't understand the Earth well enough to justify a relaxation of those assumptions, and that the nutation observations were not accurate enough to detect the difference, anyway. Geophysicists are still not able to construct a more accurate model that is independent of the nutation observations, themselves. But, the observations have improved enormously. Recent results obtained from VLBI data show discrepancies with the IAU model of close to 2 milliseconds of arc: many times the observational uncertainty. Thus, the nutation observations are beginning to tell us things about the Earth that cannot presently be inferred as accurately from other techniques. I will discuss some of these possible geophysical applications. Among them are the shape and internal structure of the core, and the Earth's anelasticity.

Discussion

KAPLAN: Are the theoretical problems likely to be resolved within the next few years? Have we reached the point where the observational accuracy will continue to exceed the theoretical capabilities? Specifically, for astrometric use, perhaps should the IERS publish corrections to nutation angles based upon observations?

WAHR: I'm not prepared to recommend what the IERS should do. But my guess is that it will be several years before geophysicists can independently determine the important parameters well enough to give significant improvement in the nutation theory.

TREUHAFT: Is it possible that moving the 460-day period to a 430-day period is an over-correction, causing the six-month term to pop up? Or do all the observational data sufficiently constrain the shift from 460 days to 430 days, causing you to look elsewhere for the 6-month discrepancy?

WAHR: The shift to 430 days could, conceivably, be somewhat of an over-estimate. For example, if the effects of mantle anelasticity are added prior to fitting the period, then the new result could be a little larger than 430 days. But, you are still stuck with the problem of finding an effect that can perturb the retrograde annual term much more strongly than any other term. And, I see no reasonable alternative for solving this problem, than to change the period to something close to 430 days.

J. H. Lieske and V. K. Abalakin (eds.), Inertial Coordinate System on the Sky, 147.
© 1990 *IAU. Printed in the Netherlands.*

PROPOSALS FOR AN IMPROVED NUTATION FORMULA

E. Groten, S.Y. Zhu *
Institute Physical Geodesy
Petersenstrasse 13
D-6100 Darmstadt
Germany, Federal Republic

* on leave from Shanghai Observatory, Shanghai, China

There are a variety of reasons why in geodesy an improved formula for nutation is needed; related topics of interest are the determination of time-dependences in low degree zonals of the earth gravity field, ocean tide modeling, determination of odd harmonics of gravity field etc. in satellite geodesy. A combined model of deterministic and stochastic components is used in order to evaluate two new nutation series where, in an adjustment, mainly VLBI data (IRIS, GSFC, IERS) have been applied. Contrary to earlier revisions of the present nutation formula, not only the five significantly affected waves (annual, semi-annual, FCN etc.) are corrected but rather all constituents are revised in such a way that white noise residuals result from the adjusted observations, based on the new formula. Still remaining problems (such as the separation of long-period terms from precession etc.) are outlined.

Alternatives to the approach applied in (Zhu et al., 1989) have been considered; one possibility would be to dispose otherwise of the FCN-period, differently from the aforementioned paper. However, due to the still inherent uncertainties in polar motion etc. at this time, a substantial change in comparison to the aforementioned paper does not seem to be available now. Consequently, also a solution for the free core nutation period (FCN) does not yet appear appropriate with the existing VLBI-data at hand.

References
Zhu, S.Y., E. Groten and Ch. Reigber, "Various aspects of numerical determination of nutation constants part II", to appear in *Astron. J.*, 1989.

J. H. Lieske and V. K. Abalakin (eds.), Inertial Coordinate System on the Sky, 148.
© 1990 *IAU. Printed in the Netherlands.*

THE RESULTS OF THE RECONSTRUCTION OF THE COEFFICIENTS OF THE NUTATION FOR THE RIGID EARTH MODEL AND THEIR COMPARISON WITH NUMERICAL INTEGRATION

J. SOUCHAY and H. KINOSHITA
Tokyo National Astronomical Observatory, Mitaka Shi
Tokyo 181, JAPAN

ABSTRACT. In view of the present accuracy of the astrometric observations and of the development of the theory of the nutation, it became necessary to make a complete revision of this theory for a rigid Earth model. We present the results of our recent one (Kinoshita and Souchay, 1989), which includes planetary effects and second order effects no considered in the previous tables (Kinoshita, 1977). We analyze the difference between these tables and the new ones providing from the revision above and the comparison between the theory and numerical integrations recently performed (Kubo and Fukushima, 1987; Shastok et al., 1987; Shastok et al., 1989). The results of this comparison are much better after revision than before.

Introduction

A complete reconstruction of the theory of the nutation for the rigid Earth model was recently performed by Kinoshita and Souchay (1989). At first they calculated the terms due to the first-order potential of the Moon and of the Sun with a level of truncation of 0.01 milliarcsecond instead of 0.1 milliarcsecond (Kinoshita, 1977), and by using up-to-date semi-analytical theories VSOP82 (Bretagnon, 1982) and ELP2000 (Chapront-Touzé and Chapront, 1983). Moreover, they accounted for scond-order terms due to smaller components of the gravitational field of the Earth, and to a coupling effect between the orbital motion of the Moon and the rotational motion of the Earth. At last, they included planetary influences which can be selected into three categories: their direct torques on the Earth, their perturbation on the orbital motion of the Moon, and their perturbation on the orbital motion of the Earth. In the following we summarize the consequences of all these contributions.

Comparison between Old and New Tables

The Hamiltonian associated with the motion of rotation of the Earth can be divided in two parts according to their nature. The first-order one is the part (not combined) directly coming from the main component of the disturbing potential, that is saying the component containing the zonal harmonic J_2 of the Earth. Thus, rigorously speaking, we can include in this part the direct and the indirect effects of the planets. In his old tables Kinoshita (1977) included only the terms due to the lunisolar potential. He found 106 coefficients

J. H. Lieske and V. K. Abalakin (eds.), Inertial Coordinate System on the Sky, 149–152.
© 1990 *IAU. Printed in the Netherlands.*

in longitude, and 54 in obliquity, up to 0.1 mas (notice that because of the truncation, Kinoshita kept in fact any term bigger than 0.05 mas). Within the same range, we find two supplementary terms in longitude (and the two corresponding terms in obliquity), which are written, in mas:

$$\Delta\psi = 0.13\sin(2l - 2F + 2D - \Omega) + 0.12\sin(-l + l' + D + \Omega) \tag{1}$$

As far as the lunisolar effect only is concerned, 148 terms in longitude and 95 terms in obliquity must be added only by accounting for all the coefficients up to 0.01 mas instead of 0.1 mas. Notice also that the total planetary effect is notable, with 180 terms at all, and some of them bigger than 0.1 mas.

Now, if we consider the revision of the theory at the second order in Kinoshita and Souchay (1989), we observe that it produces important changes in the low frequencies, when inserting a contribution due to the dynamical interaction between the figure of the Earth and the orbital motion of the Moon (Kubo, 1982; Kinoshita, 1988). Besides this main effect, other subsequent terms are providing from the influence of tesseral (C_{22} and S_{22}) and zonal (J_3) harmonic coefficients of the gravitational field of the Earth on the expression of the lunisolar potential.

In table 1, we indicate the number of new coefficients involved in the theory both at the first order and at the second order, according to their origin. The most significant corrections after reconstruction concern the terms of argument Ω and 2Ω, principally because of the dynamical interaction described above. They are written:

$$\delta(\Delta\psi) = -0\!''\!00026\sin\Omega + 0\!''\!000117\sin 2\Omega$$

$$\delta(\Delta\varepsilon) = 0\!''\!00095\cos\Omega - 0\!''\!00022\cos 2\Omega \tag{2}$$

Comparison with Numerical Integration

To compute the nutation by means of numerical integration is a very useful task in order to check the results given by the theory. This has already been done by Kubo and Fukushima (1987) and Schastok et al. (1987) before the revision of the theory (Kinoshita and Souchay, 1989), and again by Schastok et al. (1989) after this revision. All the authors used Woolard's theory as a basis. The two former works look very much in agreement one to each other, showing relatively big discrepancies with the values given by Kinoshita's tables (1977). Besides, Kubo and Fukushima give explicitly the following differences for the terms of argument Ω and 2Ω :

$$\delta(\Delta\psi) = 0\!''\!0006\sin(\Omega - 26° + 0\!''\!0013\sin(2\Omega - 2°)$$

$$\delta(\Delta\varepsilon) = 0\!''\!0008\cos(\Omega + 26°) - 0\!''\!0003\cos(2\Omega + 37°) \tag{3}$$

which are very similar to the corresponding analytical corrections (2). Furthermore, in the curves given by Schastok et al. (1987) as well as in the power spectra given by Kubo and Fukushima (1987) semi-annual discrepancies appear clearly whose the amplitudes (about 0.1 mas in longitude, 0.05 mas in obliquity) correspond exactly to the amplitudes of the quasi semi-annual corrections between Kinoshita (1977) and Kinoshita and Souchay (1989). At last, when comparing their results of numerical integration with the new series by Kinoshita and Souchay (1989), Schastok et al. (1989) show that the discrepancy is no more than 0.3 mas in longitude and 0.2 mas in obliquity, after fitting some of the biggest coefficients. The same comparisons with the old series of Kinoshita (1977) gives discrepancies of the order of 1.5 mas and 1. mas respectively. This big improvement is partly due to the introduction of the planetary terms in the new series. It is also worthy to remark that the big discrepancy for the term of argument 2Ω noticed by Kubo and Fukushima (equations (3)) disappears completely after the analytical corrections (2), both in longitude and in obliquity.

Notice that we will soon achieve ourselves a complete determination of the nutation for the rigid Earth model by numerical integration, starting from Kinoshita's basic equations (Kinoshita, 1977). Preliminary comparisons with the tables given by Kinoshita and Souchay (1989) within a few years and without any fit, show very small dicrepancies, with a mean square error lower than 0.05 mas for $\Delta\psi$ and $\Delta\varepsilon$.

Part	Longitude		Obliquity	
	Number of new terms ≥ 0.01 mas	Sum (in phase)	Number of new terms ≥ 0.01 mas	Sum (in phase)
First order				
Moon (main problem)	145	2.5 mas	93	1.6 mas
Sun (keplerian motion)	3	0.1 mas	2	0.1 mas
Planets (indirect effect)	93	1.6 mas	37	0.4 mas
Planets (direct effect)	36	0.8 mas	14	0.2 mas
Second order				
Triaxiality	7	0.1 mas	3	0.05 mas
J_3	6	0.2 mas	3	0.1 mas
Coupling effect	7	1.9 mas	7	0.5 mas
Total	297	7.2 mas	159	3.0 mas

Table 1. Number of new terms included in the new tables of the nutation (Kinoshita and Souchay, 1989) and their sum, in phase.

152

References

Bretagnon P. (1982), *Astron. Astrophys.* **114**, 278-288

Chapront-Touzé M., Chapront J. (1983), *Astron. Astrophys.*, **124**, 50-62

Kinoshita H. (1977), *Celest. Mech.* **15**, 277-326.

Kinoshita H. (1988), *BIH : Annual Report for Year 1987*, D-103.

Kinoshita H., and Souchay J. (1989), "Theory of the Nutation at the Second Order",*Celest. Mech.*, in press.

Kubo Y. (1982), *Celest. Mech.* **26**, 96-112.

Kubo Y., and Fukushima T. (1987), in: G. Wilkins and A. Babcock (Eds), *The Earth's Rotation and Reference Frames for Geodesy and Geodynamics*, Proc. of IAU Symposium. No. 182, Reidel.

Schastok J., Soffel M., Ruder H. (1987), "Variations in Earth Rotation", *Proc. IUGG Symposium*, U4

Schastok J., Soffel M., Ruder H. (1989), "Numerical Derivation of Forced Nutation Terms for a Rigid Earth," Comm., in press.

Discussion

HUGHES: Can you state which of the three new effects—J_3, indirect perturbations, and direct perturbations—has contributed most to the improvement, or are they more-or-less equal?

SOUCHAY: The effect of J_3 is small compared with the two others (≈0.1 mas). The indirect and direct effects are of the same order, but the indirect one contributes the most.

BARKIN: The unperturbed rotational motion of the Earth in your paper is the Eulerian motion of a rigid body. How does the difference between the Euler period and the Chandler period in the pole motion influence the perturbed rotation of the Earth, which is constructed in your paper in analytical form?

SOUCHAY: Since the change of period between the Euler period and the Chandler period characterizes the nature of the model of the Earth which is chosen, it would be nonsense to study the difference mentioned because we only deal with the rigid-Earth model. Anyway, it seems that this difference should not modify the equations for the perturbed motion.

YATSKIV: (1) Zhu and Groten have pointed out some differences between your calculations of nutation and their work. What are the causes? (2) Could you explain the differences between the new analytical nutation and numerical integrations?

SOUCHAY: (1) Zhu and Groten did not take into account some terms. (2) It is due to the truncation in the series, or to the combination of terms of very small amplitude (≈0.01 mas) in the second-order theory.

THE DETERMINATION OF THE PRINCIPAL NUTATION TERMS FROM THE OBSERVATIONS WITH ZTF–135 IN 1948.7–1989.0

L.A. GLEBOVA, L.D. KOSTINA, Z.M. MALKIN AND N.R. PERSIYANINOVA
Central Astronomical Observatory, USSR Academy of Sciences
Pulkovo
196140 Leningrad
USSR

ABSTRACT. A re-reduction of observations made with ZTF–135 in Pulkovo during 1948.7-1989.0 (all together about 87 000 observations) has been carried out with the aim of determing the amplitude of the principal terms of nutation in longitude and obliquity. The derived coefficients of nutation M=6''.8445±0''.0033 and N=9''.2028±0''.0025 are in good agreement with similar astrometric determinations.

The two observational programmes were used for ZTF–135 in 1948.7– 1989.0. During the first programme (Gordon, 1948) from 1948.7 to 1968.1 about 48 000 observations were made and during the second one (Kostina, 1970) from 1968.1 to 1989.0 about 39 000 observations were made. All individual latitudes were recalculated using the MERIT Standard algorithm and were corrected for polar motion. As the observations are not yet reduced to a uniform system, we estimated nutation coefficients for the two programmes independently. For further culculations we have formed the normal points for each Talcott pair as the mean value from observations made in one season, *i.e.* one normal point in a year for each pair.

Two methods were used in the present work. According to the first one, previously used for ZTF–135 observations (Kulikov, 1949; Romanskaya, 1961; Kostina et al., 1988), we solved by least squares the equation:

$$\varphi = \varphi_0 - \Delta\delta - \Delta\mu(t - t_0) - dN(\sin\alpha\cos\Omega - m\cos\alpha\sin\Omega)$$

where dN is the correction to the convential coeffficient of nutation in obliquity, $m = 0.7434$ is the relation between the axes of nutation ellipse and the other designations are obvious.

Thus we found for the two programmes
    N = 9''.2028 ± 0''.0022 (epoch 1958.4),
    N = 9''.2147 ± 0''.0049 (epoch 1978.5)
and the weighted mean
    N = 9''.2049 ± 0''.0020 (epoch 1968.5).

However, as analysis of the algorithm shows, the coefficient of nutation obtained using such a method significantly depends on the conventional value m. For control we applied this method to

153

J. H. Lieske and V. K. Abalakin (eds.), Inertial Coordinate System on the Sky, 153–154.
© 1990 *IAU. Printed in the Netherlands.*

the latitudes reduced using the Woolard nutation coefficients. In this case we found N=9".1955. It is clear that the method discussed is not satisfactory.

Therefore we estimated the coefficients of nutation solving by least square the well known equations:

$$\varphi = \varphi_0 - \Delta\delta - \Delta\mu(t - t_0) - dN \sin\alpha \cos\Omega + dM \cos\alpha \sin\Omega$$

where dM and dN are the corrections to the convential coefficients of nutation in longitude and obliquity respectively.

Thus we found for the same epochs, given earlier, the values

$M_1 = 6".8414 \pm 0".0035$, $N_1 = 9".2009 \pm 0".0028$,
$M_2 = 6".8645 \pm 0".0094$, $N_2 = 9".2101 \pm 0".0056$

and the weighted mean values

$M = 6".8445 \pm 0".0033$, $N = 9".2028 \pm 0".0025$,

which are in good agreement with other astrometric determinations (Capitaine *et al.*,1988).

The authors wish to express their thanks to A.R. Valsky for his help in calculations.

REFERENCES

Capitaine, N., Li, Z.X., and Nie, S.Z.: 1988, *Astron. Astrophys.*, **202**, 306–308.
Gordon, Ya.E.: 1948, *Izvestiya GAO AN SSSR*, **141**, 130–139 (in Russian).
Kostina, L.D.: 1970, *Izvestiya GAO AN SSSR*, **185**, 49–69 (in Russian).
Kostina, L.D., Malkin Z.M., and Persiyaninova N.R.: 1988, In: *The study of the Earth as a planet by methods of geophysics, geodesy and astronomy*, Kiev, 8-10 (in Russian).
Kulikov, K.A.: 1949, *Astron. Zhurnal*, **26**, 165–174.
Romanskaya, C.V.: 1961, In: *Preliminary results of the investigations of the latitude oscillations and Polar motion*, Moscow, 81–85 (in Russian).

CORRECTIONS TO THE IAU 1980 NUTATION SERIES FROM LAGEOS OBSERVATIONS

V.K. TARADY
Main Astronomical Observatory
Ukrainian Academy of Sciences
252127 Kiev
USSR

The attempt was made to determine the corrections to the coefficients of the IAU 1980 nutation series together with other geodesical and geodynamical parameters by the use of the Kiev-Geodynamics-3 program complex for LAGEOS laser data analysis. This complex is based on the MERIT standards and on numerical integration techniques for satellite orbit computation. An integration is carried out in cartesian coordinates by the Adams method of variable step and order.

At the first stage of the data analysis the 3-minute normal points of ranges to LAGEOS (laser observations from 1986–1987) were used for deriving the orbital elements and station coordinates.

Then at the second stage the Earth rotation parameters and the orbital elements were determined on the basis of the 5-day arcs.

Finally, using the arcs of about 14 days the following parameters were adjusted: Earth rotation parameters; fortnightly nutation term; Love numbers h,k,l; geocentric constant of gravitation, etc.

The final r.m.s. residuals of the orbital fits were about 8 cm. We have found the following corrections to the 13.7-day nutation term:

in longitude: $-0\overset{''}{.}00047 \pm 0\overset{''}{.}00013$

in obliquity $+0\overset{''}{.}00009 \pm 0\overset{''}{.}00002$.

Discussion

He: (1) What kind of approach is used for the solution of the LAGEOS orbit from 1986–1987—a single long arc or multiple short arcs?
(2) Did you include the changes of station coordinates due to the motion of tectonic plates?

Tarady: (1) A multiple short arc pproach was used for the solution of the LAGEOS orbit from 1986 to 1987. In the procedure of ERP determinations we have used 5-day arcs, but for the determination of short-period nutation we have taken 13.7-day arcs.
(2) The changes of station coordinates due to the motion of tectonic plates were taken into account in accordance with the MERIT standards using the AMO2 model.

155

J. H. Lieske and V. K. Abalakin (eds.), Inertial Coordinate System on the Sky, 155.
© 1990 *IAU. Printed in the Netherlands.*

OCEANIC INFLUENCES ON THE ANGULAR VELOCITY OF THE EARTH

P. Brosche
Observatorium Hoher List
Univ. Sternwarte Bonn
D-5568 Daun
Germany, Fed. Rep.

ABSTRACT. Hydrodynamical computations of the major partial tides in the oceans have been evaluated for the changes both in moment of inertia and relative angular momentum due to ocean currents. If the system solid Earth plus oceans is seen as an isolated system for these time scales, the oceanic variations lead to mirror-like changes in the rotation of the solid Earth. Amplitudes are of the order of 0.1 ms in Universal time. In contrast to the effects of solid Earth tide, phases are away from equilibrium phases.

THE PROGRAM COMPLEX "KIEV — GEODYNAMIC R1"

M.M. Medvedsky
Main Astronomical Observatory
Ukrainian Academy of Sciences
252127 Kiev, USSR

ABSTRACT. The software for the reduction of the VLBI observations of the Main Astronomical Observatory of the Ukrainian Academy of Science is described.

Our program complex has three parts:
1) calculation of geometrical delay τ,
2) calculation of required corrections to values of τ,
3) data analysis procedures.

This software has been tested using the NGS VLBI data for the obtaining of corrections to the ERP. The parameters of the clock were derived. Our results were compared with similar results, which were published by Shanghai observatory (China) and IRIS bulletin.

The differences between our results and the data given in the IRIS bulletin are within 5 mas, and for UT are within 0.1 ms.

CURRENT STATE OF THE THEORY OF NUTATION
S.M. Molodenskij
O.Yu. Shmidt Institute for Earth Physics
Bol. Gruzinskaya 10
123810 Moscow D-242, USSR

Paper not available.

THE STUDY OF THE STRUCTURE OF 142-YEAR SERIES OF POLE COORDINATES

L.V. RYKHLOVA AND G.S. KURBASOVA
Astronomical Council USSR Academy of Sciences
Pyatnitskaya ul 48
109017 Moscow
USSR

ABSTRACT. The 142-year series of pole coordinates in the system of mean pole of epoch of observation are used for the investigation of the Chandler period. The method of the best approximation of the initial data with the step of 1.5 years by means of quasi-polynomials is used.

1. Guidelines

Searching for the correlation between polar motion and different types of geophysical phenomena, we must take into account the presence of systematical errors and homogeneity of the data lists which are used. From this point of view, the 142-year series of pole coordinates in the system of the mean pole of epoch of observation, which was published by the Main Astronomical Observatory of the Ukrainian Academy of Sciences [1], is very suitable for the study of the Chandler period of polar motion, since it is very homogeneous and it does not have long-period components.

In [2] we have proposed a numerical method of the prediction of polar motion and UT1–UTC values using a quasi-polynomial set. The number of variable parameters and their values was found as a result of the best approximation to the initial data series. The predictions were made using published Rapid Service IERS data. A comparison with USNO predictions showed good qualities of the quasi-polynomial model for approximation and prediction.

We used the method for searching for the Chandler period in that 142-year series.

2. Mathematical model

The quasi-polynomial model is a set of polynomials, exponential, and trigonometric functions:

$$Z = \sum_{i=1}^{n} A\, t^{\eta} e^{\xi t} \left(\cos \frac{2\pi}{P_i} t + \sin \frac{2\pi}{P_i} t \right)$$

where A, ξ, η and P_i are variable parameters which are determined during the solution of the problem of obtaining the best approximation to the coordinates x, y on a moving 1.5-year interval.

J. H. Lieske and V. K. Abalakin (eds.), Inertial Coordinate System on the Sky, 157–159.
© 1990 IAU. Printed in the Netherlands.

Existing uncertainties in the periodic components, in quasi-periodic and aperiodic components are taken into account during the choice of the exponential functions' variable parameters.

3. Results

For the investigation of the pole coordinate series x, y $N = 81$ sets of quasi-polynomial parameters were obtained. In each of these sets of parameters, annual and Chandler components were also obtained. The annual period has inessential changes near the mean value of 365.24 days. The estimations of the Chandler period can be seen in Fig. 1.

The correlation and inter-correlation functions $Cp_x p_y(\tau)$, $Cp_y p_y(\tau)$ and $Cp_x p_x(\tau)$ are shown in Fig. 2. From that we can conclude that the p_x and p_y sets do not contain well-determined components other than non-zero means. The values of these means are

$$\overline{P}_x = 419.4 \text{ d} \pm 18.6 \text{ d} \quad \text{and} \quad \overline{P}_y = 419.6 \text{ d} \pm 18.7 \text{ d}.$$

In Table 1 the statistical characteristics of the deviations ΔP_x and deviations ΔP_y are given,

Table 1.

	A	E	χ^2_E	$\chi^2_T (L = 99\%)$
ΔP_x	+0.3 ±0.46	−0.48±0.63	9.18	11.30
ΔP_y	−0.2±0.46	−0.62±0.46	9.27	11.30

where A is the asymmetry; E is the excess; χ^2_E is the experimental value of the χ^2 criterion; and χ^2_T is the theoretical value of the χ^2 criterion for significance level $L = 99\%$. There are not any serious deviations of the ΔP_x, ΔP_y deviations of the experimental distribution from the theoretical one.

4. Conclusions

Using the 142-year series of coordinates, a statistically steady estimation of the Chandler period is 419.4 d ± 18.6 was obtained. The period was confirmed and specified via the independent method of spectral analysis of zero intersections. The analysis of the Chandler period has not given any basis for speaking of two or more close frequencies.

References

Korsun, A.A., Rykhlova, L.V., Cholij, V.J.: 1989, "Earth pole coordinates in observation epoch of mean pole system (A.J. Orlov pole) during the period 1846–1988", *Kinematics and physics of space bodies*, **5**, No. 3, 14.

Rykhlova, L.V., Kurbasova, G.S., Tajdakova, T.A.: 1990, "Earth rotation parameters forecasting", *AG* **67**, 151–159.

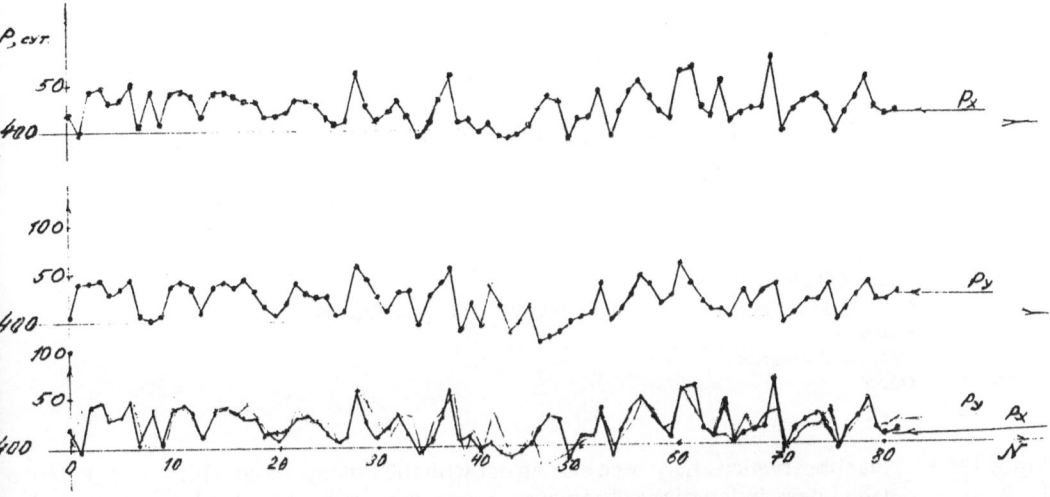

Figure 1. Estimates of the Chandler Period in the interval 1846–1987.

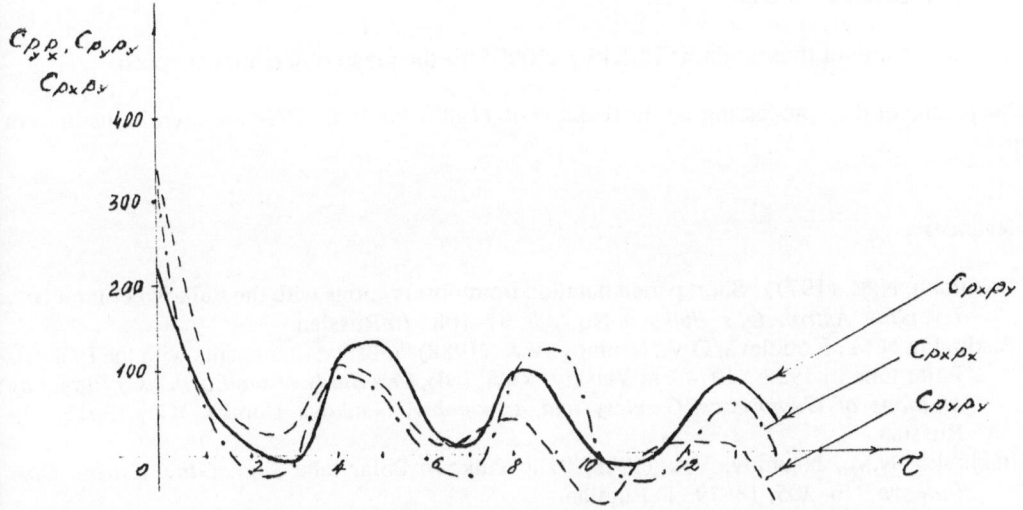

Figure 2. Correlation functions Cp_{xy}, Cp_{yy} and the mutual correlation function Cp_xp_y.

DETERMINATION OF NUTATION BASED ON OBSERVATIONS WITH THE PULKOVO POLAR TUBE

O.V. KOTRELEVA
Central Astronomical Observatory, USSR Academy of Sciences
Pulkovo
196140 Leningrad
USSR

Since 1953 regular observations have been carried out with the Polar photographic tube at Pulkovo [1, 2, 3]. A detailed description of the observational procedure, the data reduction technique, and error sources were given by Potter [4].

During 1953–1987 848 observing nights were obtained and 9911 polar distances of stars were measured (m.e. of $\pm 0''.05$). The analysis of the material yelds the correction to the Woolard principal nutation term

$$\Delta N = -0''.0160 \pm 0''.0054 \, ,$$

the nutation constant thus attaining $9''.1945 \pm 0''.0054$ for the mean observational epoch.

The results of data processing by Bakhrakh *et al.* (1988) for 1953–1974 are used in the present paper.

References

1. Bakhrakh, N.M. (1971) "Short period nutation from observations with the Pulkovo Polar tube", *Izv. Glav. Astron. Obs. Pulkovo*, No. 187, 97–108. In Russian.
2. Bakhrakh, N.M., Kotreleva, O.V., Naumov, V.A. (1988) "Observation results with the Pulkovo Polar tube in 1953 - 1974", in Yatskiv, Ya.S. (ed), *The Study of the Earth as a Planet by Methods of Geophysics, Geodesy and Astronomy*, Naukova Dumka, Kiev,13–15. In Russian.
3. Bakhrakh, N.M., Naumov, V.A. (1988) "The Pulkovo Polar tube", *Izv. Glav. Astron. Obs. Pulkovo*, No. 205, 18–19. In Russian.
4. Potter, H.I. (1956) "A method for reducing photographs obtained with a Polar tube", *Izv. Glav. Astron. Obs. Pulkovo*, No. 157, 101–119. In Russian.

J. H. Lieske and V. K. Abalakin (eds.), Inertial Coordinate System on the Sky, 160.
© 1990 *IAU. Printed in the Netherlands.*

CAN THE CORE-MANTLE BOUNDARY TOPOGRAPHY INFLUENCE THE EARTH'S NUTATION?

V.V.Bykova
Institute of Earth Physics
Bolshaya Gruzinskaya 10, 123810 Moscow D-242, USSR

Abstract. The nutation of the Earth with slightly nonelliptical liquid core is investigated by the perturbation theory method. It is shown that first-order terms affect the core ellipticity and its triaxiality. The most sensitive nutation terms in the second approximation were found to be retrograde 18.6-year term and retrograde annual term. The observed nutation amplitude values can be satisfied by special core-mantle boundary form.

The estimates of the forced nutation amplitudes display deviation between calculated and observed values. This deviation was supposed to be explained, for example, by assuming the higher value of the core ellipticity than that of the hydrostatic theory [1], the effects of the core viscosity, or the internal structure of the core [2]. The influence of the core-mantle boundary topography on the nutation is investigated in the present paper.

Model and method. The Earth is considered to consist of the rigid shell and liquid core of the shape S (the liquid is homogeneous and incompressible). The problem is solved by the perturbation theory with small parmater equal to the ratio of the inclination angle of the core-mantle boundary surface element to that of the ellipsoid. The Poincare solution is used as an initial approximation. If the velocity vector is represented as a sum $\vec{v} = \vec{v}_0 + \vec{v}_1 + \vec{v}_2 + \dots$, and normal vector is $\vec{n} = \vec{n}_0 + \vec{n}_1$, where $\vec{n}_0$ is normal vector to ellipsoid S_0 : $r = r_0 (1 + e \sin^2\theta)$, $\vec{n}_1$ is one to the real boundary S :

$$r = r_0 (1 + e \sin^2\theta) + \sum_k \sum_1 a_{k1} \ Y_{k1} (\theta,\varphi) \qquad (1)$$

r_0 , e are the mean radius and the ellipticity of the core, Y_{k1} are spherical harmonics, normalised as in [3], then considering the terms of the same order of smallness,

$$(\vec{v}_0, \vec{n}_0)_{s_0} = 0 \qquad (2a)$$

$$(\vec{v}_1, \vec{n}_0)_{s_0} + (\vec{v}_0, \vec{n}_1) + (\delta\vec{v}_0, \vec{n}_0) = 0 \qquad (2b)$$

$$(\vec{v}_2, \vec{n}_0)_{s_0} + (\vec{v}_1, \vec{n}_1) + (\delta\vec{v}_1, \vec{n}_0) = 0 \qquad (2c) \quad \text{etc.,} \quad \delta\vec{v}_1 = \vec{v}_i|_s - \vec{v}_i|_{s_0}$$

$\vec{v}_1|_s$ is defined by equation (2b) with known $\vec{v}_0$, $\vec{n}_1$, $\vec{v}_0$, $\vec{v}_2|_s$ is defined by (2c) ,etc. So every $\vec{v}_i$ can be found as an expansion in inertial modes V_{nm} in rotating liquid ellipsoid [4]. $V_{nm}|_s \sim Y_{nm}(\theta,\varphi)$ and are the full orthogonal function system. Each inertial mode V_{nm} is induced by the spherical harmonic Y_{nm} with the corresponding number.

Results. The calculations were carried out for the eight main nutation terms. In the first approximation the nutation is influenced by the

J. H. Lieske and V. K. Abalakin (eds.), Inertial Coordinate System on the Sky, 161–162.
© 1990 *IAU. Printed in the Netherlands.*

modes V_{20}, V_{40}, affecting the seaming increasing of e and V_{22} resulting in the ellipticity of the angular velocity vector moment in rotating coordinate system: $(A_x-A_y)/2A_0 \sim 10^{-2}$, where A_x, A_y and A_0 are the nutation amplitudes for the Earth with liquid core in x and y directions and for the absolutely rigid Earth, respectively.

Second-order terms consideration shows that in this case the significant role is played by boundary harmonics, inducing the inertial modes with eigen frequencies closest to the frequencies of forced nutation. The calculations reveal that these are the harmonics with l=0 and k=164,78,250,... for different nutation terms. Then to satisfy the observed nutation amplitude values it must be accepted that the boundary S_0 is superimposed by the wave Y_{k0} with amplitudes dr given in the table.

nutation terms	corrections to nutation amplitudes(mas)		k	dr (km)
	I	II		
+6800	-0.03	-0.02 ± 0.0004		
-6800	0.23	0.27 ± 0.05		
-365.3	0.17	0.61 ± 0.15	164	1.5
+182.6	-0.04	-0.02 ± 0.01	250	13.8
-182.6	0.01	0.02 ± 0.01	78	2.9

I-1 approximation, the results obtained for mode Y_{40}, a_{40}=4.4 ± 2.4 km according to [3]. II- 2 approximation, the core-mantle boundary coefficients are assumed to be distributed randomly with mean value equal to zero and dispersion $=(h/n)^2$, where h=10 km is boundary relief [3], n is the quantity of coefficients, n=200.

Discussion. Therefore the results obtained demostrate, that the observed nutation amplitude deviations can't be explained by the core-mantle boundary topography influence in the first approximation. The choice of the special core-mantle boundary form can yield the observed results in the second approximation without contradiction to modern seismic data [3]. However it is very doubtful. If the core-mantle boundary expansion coefficients are assumed to be distributed randomly the probabilistic estimates can be made. The probabilities of the dr from the table are 0.00001. So such a situation seems to be physically unrealizible.

Acknowledgements. I am very grateful to Prof. Molodenskii S.M. for scientific guidance and helpful comments.

References.
1.Gwinn C., Herring T., and Shapiro I. (1986) 'Geodesy by radio interferometry : Studies of the forced nutations of the Earth, 2,Interpretation , J. Geoph. Res. 91, 4755-4765.
2.Wahr J. and de Vries D.,'The possibility of lateral structure inside the core and its implications for nutation and Earth tide observation'.Geoph. J. R. Astr. Soc, in press.
3.Morelli A. and Dziewonski A. (1987) 'Topography of the core-mantle boundary', Nature 325, 678-683.
4.Гринспен X.(1975) 'Теория вращающихся жидкостей'.Л.Гидрометеоиздат.

ON THE CALCULATION OF LOW FREQUENCY OSCILLATIONS OF THE EARTH'S CORE

S.V. DYAKONOV
O.Yu. Shmidt Institute of Earth Physics
Bolshaya Gruzinskaya 10
123810 Moscow D-242
USSR

ABSTRACT. The problem of calculation of low frequency oscillations of an ideal rotating compressible fluid is investigated. An original method of solving such a problem, based on using characteristic functions of the Poincaré operator is proposed. An efficient scheme of calculating the characteristic numbers and functions of the Poincaré operator is worked out. The high speed of convergence of the method is shown. An essential influence of compressibility on the theoretical nutation amplitude is found.

While calculating low frequency oscillations of the Earth's fluid core (*e.g.*, for calculation of the forced nutation amplitude), spherical harmonic representation of the deformation field is usually used [1–3]:

$$u = \sum_l \sum_m [S_l^m + T_l^m] \tag{1}$$

Substitution of Eq. (1) into the equations of motion gives an infinite system of differential equations for the scalar functions S_l^m and T_l^m. Approximate solutions of such a system are obtained by truncating the system of equations. Probably one cannot prove the convergence of the set of such approximate solutions. The author has proposed an alternative method for calculation of low frequency oscillations of a fluid Earth core.

Let us calculate the forced nutation amplitude for the Earth model consisting of an absolutely solid mantle and a compressible liquid core with the simple fluid density distritution

$$\rho(R) = \rho_o \left(1 - \delta (R/R_n)^2 \right) \tag{2}$$

where R_n is the core radius. Geometry of the equal-value density surfaces may be obtained from Clairaut's equation. Let the Brunt-Väisälä frequency equal zero. Then the fluid oscillations are described by the equations

$$\Delta \psi - \frac{4\omega^2}{\sigma^2} \frac{\partial^2 \psi}{\partial z^2} = -(\sigma^2 - 4\omega^2)\frac{P_1}{\rho \alpha^2} - \frac{\sigma^2 - 4\omega^2}{i\sigma} v \frac{\nabla \rho}{\rho} \tag{3}$$

$$\Delta V_1 = 4\pi G \alpha^{-2} (\psi - V_1 + V_t) \qquad \psi = P_1/\rho + V_1 + V_t \tag{4}$$

where ω is the angular velocity, σ the frequency, α the speed of sound, and v, P_1, V_1 are perturbations of velocity, pressure and gravitational potential, respectively; V_t is the tide-generating potential.

J. H. Lieske and V. K. Abalakin (eds.), Inertial Coordinate System on the Sky, 163–164.
© 1990 *IAU. Printed in the Netherlands.*

The first term in the right hand of Eq. (3) is negligible. Thus, only Eq. (3) needs to be solved. Let the solution be represented by expansion in characteristic functions of the Poincaré operator:

$$\psi = \left(a_0 (xz - iyz) + \sum_l \sum_k \sum_m a^m{}_{lk} \, \Psi^m{}_{lk} \right) \exp(i\sigma t) \tag{5}$$

$$\left(\Delta - \frac{\partial^2}{\partial z^2} \right) \Psi^m{}_{lk} = \lambda^m{}_{lk} \, \Psi^m{}_{lk}$$

Substitution of Eq. (5) into Eq. (3) gives a system of equations for $a^m{}_{lk}$. The non-diagonal elements of the matrix of this system are small in comparison with the diagonal ones. This makes it possible to truncate the expansion Eq. (5) in order to obtain an approximate solution. The first 36 characteristic numbers of the Poincaré operator for one of the frequencies are given in Table 1.

Table 1. Characteristic numbers of Poincaré operator ($m = 1$)

$l \backslash k$	1	2	3	4	5	6
1	−16.1103	0.1355	76.6711	235.8353	489.7864	843.0952
2	−58.0358	−46.4465	8.3694	83.8804	197.9489	366.1997
3	−133.1034	−126.2940	−60.1717	15.9224	116.7159	243.3884
4	−245.1963	−242.0729	−155.8808	−80.9797	23.6751	152.8642
5	−396.6794	−395.5154	−277.3416	−208.0382	−101.7706	31.6236
6	−588.3208	−587.9475	−425.0769	−364.3752	−258.0745	−122.5184

Nutation amplitudes for different maximum numbers of characteristic functions in Eq. (5) are given in table 2. For comparison, there are shown nutation amplitudes for the absolutely solid and the Poincaré models.

Table 2. Nutation amplitudes [$N = \max \{l, k\}$]

	Solid model	Poincaré model	Compressible core model $l = 1, k = 2$	N = 2	N = 4	N = 6
−6800	8051.05	7999.60	7992.19	7992.32	7992.42	7992.43
−365.3	24.91	−38.633	−28.757	−28.905	−29.028	−29.042
−182.5	22.60	28.197	29.242	29.223	29.208	29.205

A good convergence of the proposed method makes it possible to use it while calculating the Earth nutation amplitudes and free nuclear oscillations. The method may be extended for the case of non-zero Brunt-Välsälä frequency. Extension of the method for the case of a solid inner core will be the subject of future investigations. *Acknowledgements*: The author is grateful to Prof. S.M. Molodenskii and to E.V. Andrjievskii, I.V. Kurushkin, V.R. Leschuk, and V.V. Zuikin.

References

1. Smith, M.L. 1974, "The scalar equations of infinitesimal elastic-gravitational motion for a rotating slightly-elliptical Earth", *Geophys. J. Royal Astron. Soc.* **37**, 491–526.
2.. Crossly, D.J., Rochester, M.G. 1980, "Simple core undertones", *Geophys. J. Royal Astron. Soc.* **60**, 129–161.
3. Wahr, J.M. 1981, The forced nutation of an elliptical, rotating elastic and oceanless Earth, *Geophys. J. Royal Astron. Soc.* **64**, 705–727.

THE MEAN POLE OF THE MOON'S ROTATIONAL AXIS AND GENERAL SELENOCENTRIC COORDINATE SYSTEM

Yu.V. Barkin
Moscow State Technical University
2nd Bauman Street 5
Moscow
USSR

One of the fundamental problems of lunar astronomy is the reduction of the coordinates of the Moon's surface, found by astronomical methods, to its mean pole. The instantaneous poles of the rotation axis and the instantaneous equator move in the Moon's body. The unstable position of this equator does not allow one to use in selenodesy the instantaneous spherical coordinates which have not been preliminarily transformed into some unified system of coordinates. Such a reduction can be made to the system of coordinates connected with the mean pole—to be definite, we shall speak about the Moon's North pole.

According to theoretical studies [1], it can be asserted that owing to the resonant character of the Moon's rotational motion, the influence of various perturbing factors (in the first case, the influence of the third and higher harmonics of the Moon's force function), the Moon's mean pole is displaced with respect to the poles of the polar axis of inertia mainly in the plane of the zero meridian by the angle 74".72 away from the Earth and by the small angle 0".025 eastward. However, it should be noted that the displacements are to a considerable degree dependent upon the choice of a model of the Moon's gravitational field.

The Earth and Moon theories of rotation for nonprincipal axes of inertia can be applied to the solution of numerous direct and inverse problems of the Earth-Moon system: geodynamics, geophysics, and physics; in particular forr the determination of the main geocentric and selenocentric systems of coordinates, *e.g.* for the determination of the principal central axes of inertia of the mean poles of the Earth and the Moon. If precice enough observational data are available, the theories allow also to solve problemes of determination of the parameters of the Moon's gravitational field and of various parameters of its rotation.

References

[1] Barkin, Yu.V.: 1989, "Dynamics of a system of non-spherical celestial bodies and the theory of the Moon's rotation", PhD Thesis Sternberg Astronomical Institute, Moscow State University, 412 pp.

J. H. Lieske and V. K. Abalakin (eds.), Inertial Coordinate System on the Sky, 165.
© 1990 *IAU. Printed in the Netherlands.*

ON THE ASTROMETRICAL METHODS

V. SHKODROV AND V. IVANOVA
Department of Astronomy, Bulgarian Academy of Sciences
Blvd. Lenin 72
1784 Sofia
Bulgaria

ABSTRACT. In this paper some problems are discussed concerning the connection between the highly accurate optical and radio observations in the case when the waves propagate in an anisotropic medium.

There are several particular circumstances that justify the discussion given in this paper. These are the HIPARCOS satellite project, the results from interferometry in radio and optical wavelengths and the discussion of ESA Workshop on Interferometry in Space held in Granada (June 16–18, 1987), [1], [2], [3]. However, the general circumstance is the significant increase in the accuracy of astrometrical observation (<0''.001). As some other authors [4], remark, this will be relevant to the discussion of some traditional definitions in astronomy. In the paper we discuss some problems of the connection between the high-accuracy optical and radio observations in the case when the waves propagate in an anisotropic medium.

The question of "What is an exact location?" has been put by E. Preuss [4]. If we start from the electromagnetic nature of the light, the direction $\mathbf{\hat{r}^0} = \mathbf{r}/r$ to the heavenly body S will be determined from $\mathbf{\hat{r}^0} = -\mathbf{\hat{g}}$ where $\mathbf{\hat{g}} = \mathbf{G}/G$, with $\mathbf{G}$ being the well known Poynting vector. The direction of $\mathbf{G}$ gives the direction of the energy flux F and its magnitude is the amount of the energy F crossing an unit area normal to the direction of propagation of the flux in one second.

In optical astronomy we consider electromagnetic waves with different frequencies which are time averaged. Until now, optical astronomy has used idealized conditions for the determination of $\mathbf{\hat{r}^0}$. In this sense, astrometry takes into account only the refraction in the lower layers of the Earth's atmosphere and the phenomena associated with the light transmission through the optics of the telescope. This has been acceptable because of the low accuracy with which $\mathbf{\hat{r}^0}$ is determined.

The new accuracy of $\mathbf{\hat{r}^0}$ achieved in observational astrometry needs a new method for analysis. Probably it is essential to derive a general theory to connect $\mathbf{\hat{r}^0}$ with the electromagnetic nature of the light more definitely than it is now.

References

1. Kovalevsky J., 1978, IAU Colloquium No. 48, p. 573.
2. ESA Workshop on Optical Interferometry in Space, Granada, Spain, June 16–18, 1987.
3. Kovalevsky J., 1984, *Space Science Rewiews*, **39**, 1.
4. Preuss E., 1987, ESA Workshop on Optical Interferometry in Space, Granada, Spain, June 16–18, 1987.

J. H. Lieske and V. K. Abalakin (eds.), Inertial Coordinate System on the Sky, 166.
© 1990 *IAU. Printed in the Netherlands.*

REFERENCE SURFACES AND GLOBAL FIGURE PARAMETERS OF THE TERRESTRIAL PLANETARY BODIES

K.K. KAMENSKY
SKTB of Institute for Applied Problems of Mechanics and Mathematics
Ul. Lermontova 15
290005 Lvov
USSR

For connecting the inertial coordinate system with the non-inertial planetocentric one we need to know values of the fundamental astronomical constants as well as global figure parameters of the planet. We distinguish among planetary figures a dynamical, a level, and hypsometric figures.

Parameters of the last two figures can be determined by using the least squares method under the condition:

$$\int_0^{2\pi} \int_{-\frac{\pi}{2}}^{+\frac{\pi}{2}} p\,(\varphi, \lambda)[\, r_e\,(\varphi, \lambda, a, b, c, \lambda_a, \varphi_c, \lambda_c) - r\,(\varphi, \lambda)]^2\, d\varphi\, d\lambda = min,$$

where p is weighting function averaging for the trapezium with the centre coordinates φ, λ; r and r_e are the corresponding mean values of the radii of the surface being approximated, and of the approximating free-oriented triaxial ellipsoid with semiaxes $a > b > c$, and three orienting angles, for instance, λ_a, φ_c, λ_c.

The results of computations (see the table) show that orientation of the level figures is very close to the orientation of the dynamical ones, determined by other authors using different methods. Flattenings of these figures are also near to the values recommended by the IAU as the reference surfaces parameters [1]. The smallest axes are close to the rotation axes for the planets with fast axial rotation and pronounced biaxiality (*e.g.*, the Earth and Mars). The angle between the axes of figure and of rotation for Venus reaches the value of 3.6. Using other data we have obtained an even larger value of 11° [2].

In an inertial frame of reference a level surface becomes an equipotential surface of attractive potential because there is no inertial force and other fictitious forces in this frame. Numerical investigations of this phenomenon were conducted in [3]. Analysis of the results obtained shows that in the case of the Earth and Mars the flattening of the attractive potential figure is about two times greater than for the gravity potential figure. The axis of the former figure is closer to the rotation axis than for the latter one.

J. H. Lieske and V. K. Abalakin (eds.), Inertial Coordinate System on the Sky, 167–168.
© 1990 *IAU. Printed in the Netherlands.*

The smallest axes of hypsometric figure (see numerical quantities in [3]) diverge considerably from the rotational axes and the inertia axes of the planets . At the same time, flattenings of hypsometric surfaces are much closer to flattenings of attractive potential figures than of gravity ones.

The Earth's level figures investigation shows different orientation of the smallest axis for different geopotential models. The last circumstance can be explained by the complex motion of the terrestrial coordinate system relative to the inertial one and by the discrepancy of geopotential models.

Table. Global gravity figure parameters of the terrestrial planets and the Moon

Parameter	Dimension	Venus	Earth	Moon	Mars
GM	10^9 m³/s²	324 858.15	398 600.448	4 902.799	42 828.44
ω	10^5 rad/s	0.029 924	7.292 115	0.266 170	7.087 919
a	m	6 051 473.	6 370 831.	1 737 861.	3 391 212.
b	m	6 051 448.	6 370 762.	1 737 630.	3 389 929.
c	m	6 051 422.	6 349 458.	1 737 212.	3 372 848.
λ_a	deg	353.5 E	345.0576 E	0.02585 E	104.905 W
φ_c	deg	86.4 N	90.0000 N	89.98000 N	89.999 N
λ_c	deg	75.4 E	69.2912 E	0.88594 E	195.634 W

References

1. Davies, M., Abalakin, V.K., Burša, M. *et al.* (1986) "Report of the IAU/IAG/COSPAR working group on cartographic coordinates and rotational elements of the planets and satellites, 1985", *Celestial Mechanics* **39**,103–113.
2. Kamensky, K.K. (1988) "The research of global pecularities of Venus equipotential and hypsometric surfaces", in Ya.S. Yatskiv (ed.), *The study of the Earth as a planet by methods of geophysics,geodesy and astronomy*, Naukova Dumka, Kiev, pp. 80–83, (in Russian).
3. Kamensky, K.K., Kislyuk, V.S., Yatskiv, Ya.S. (1988) "Geometrical and dynamical characteristics of the Earth, the Moon and the terrestrial planets: 1. Topographic surfaces and gravitational fields", Preprint *Int. Theoretical Physics* ИТФ-88-86P, Kiev, (in Russian).

POSSIBLE CONTRIBUTIONS OF CLASSICAL ASTROMETRIC INSTRUMENT TO THE STUDY OF GEOSCIENCE AND REQUIREMENTS ON STAR CATALOGUES

HAN YANBEN and LI ZHISEN
Beijing Astronomical Observatory
Chinese Academy of Sciences
Beijing 100080
CHINA

ABSTRACT. In this paper possible contributions of classical instrument of time and latitude determinations in the study of geoscience is discussed. Some research results indicate that the observations of the classical instruments contain rich information on geoscience study. They can play a certain role in research of crust structure, plate motion, vertical variation, topography of core-mantle boundary, and in study of seismicity and major earthquake forecasting. It is important that a stable reference system is established and maintained and a relevant observed programme is formulated upon a star catalogue with good quality for geoscience and astro-geodynamics. Some instruments with high precision, especially located in seismic regions, would be continuously operated after the end of their task in ERP determination.

1. Introduction

For a long time, large contributions have been made to the time and latitude services and the study of catalogues by the classical astrometric observations and researches. The precision of the classical data was progressively improved along with improvements of the instruments and data processing. Now the mission of the classical instruments in determination of ERP will be replaced by new techniques with higher precision, such as VLBI, LLR and SLR, etc. The relating of the space reference system and terrestrial reference frame will still need the classical observations. Futhermore, can the classical instruments make more contributions?

Even if the more early classical observations were not concerned, the history of their systematic and extensive observation is near one hundrend years, and a wealth of data were accumulated. A result of the ILS in a homogeneous system (1899.9–1979.0) was given by Yumi[1], and astronomers have done a lot of work with the data. Li Zhengxin re-sorted out the data of BIH from 1962 to 1982 and provided a new evaluation[2]. Li's result showed that the development of the classical astrometry went through a period of great prosperity from 1960's to 1970's and there were so much data. The new techniques can provide timely ERP with high precision, but their series are short. As it is often in the case of astrometry and geophysics, many scientists are very interested in the long series data in their researches. So the classical observations of different systems (*e.g.* ILS, BIH and IPMS, etc.) are seemingly

169

J. H. Lieske and V. K. Abalakin (eds.), Inertial Coordinate System on the Sky, 169–172.
© 1990 *IAU. Printed in the Netherlands.*

still valuable. On the other side, the datum line of the classical instrument is the local vertical, the observations were inevitably affected by some geophysical factors. Although these factors have caused many troubles in the improvement of ERP determination, they set up the close ties between classical astrometry and geoscience. The study of some geoscientific subjects with the classical observations become feasible.

2. Possible Contributions of Classical Instruments to Geoscience Research

The continuous variations of local vertical reflect the fine structure of geoid surface. Developments of modern topography and gravimetry set a still higher demand on the determination of the vertical variation. A concept of the vertical deviation was continued to use in geodesy, it is different between astronomical coordinates and geodetic ones and reflects the deviation between the geoid and reference ellipsoid. The classical instruments measure the variation of angle between the local vertical and the Earth's axis, and the axis can be measured exactly by other methods (e.g. new techniques), therefore, the determination of vertical variation with the classical instruments is a quasi-absolute at least. Mironov has pointed out the possibility that the vertical variation was determinated by classical instruments[3]. Melchior has introduced the related works[4].

When the horizontal components of the variations in the gravitational field are about 0.2 mgal, the estimate of related variations of residuals RTg in time and RFg in latitude will be about $0\overset{s}{.}003$ and $0\overset{''}{.}04$ respectively. According to the present precision of classical instruments, their observations may reflect the long term vertical variation.

Classical data may also play a role in research and forecasting of major earthquakes. On 28 July 1976, a major earthquke ($M = 7.8$) occurred in Tangshan region of China. An anomaly appearing in the observations of Danjon Astrolabe installed at Shahe Station of the Beijing Observatory about $160km$ away from the epicentre was found by Zhang [6], and Li et al. [7]. Through the analysis of many major earthquakes and data of the RT(or $O - A$) and RF, They pointed out that the interrelationship between the short-term anomalies of the RT and RF and the major earthquakes which occurred in regions around the classical instruments was confirmed. The relationships between RF and earthquakes have been investigated also by others[8,9]. Over thirty major earthquakes which occurred in the regions around ten instruments have been studied, the observations of other instruments situated in places without major earthquakes were used for contrast. The results may be summarized as the following:

(1) The anomalies usually become apparent simultaneosly or separately in the RT and RF of the classical instrument before the occurrence of a major earthquake of $M \geq 7.0$ within the distant of $300km$, or $M \geq 6.0$ within the distant of $100km$; (2) The double peaks in different direction usually appear in RT and only single peak in RF; (3) The beginning time of the anomalies were several weeks or months before the occurrence of the earthquake; (4) The vertical variation due to the motion of groundwater probably before the earthquake was considered as the main cause of the anomalies[6,10]; (5) The fluctuations of residuals of some instruments installed in the regions where no major earthquakes were relatively smaller.

Han et al. have studied the possibility that the information of earthquake forecasting was provided by classical instruments, and concluded that the instrument with high precision and observations can make the contribution to the major earthquake forecasting[11]. The result has been tested and verified by the photoelectrical astrolabe (PA) Mark-2 of Yunnan Observatory of China before a major earthquake[12].

Another area in which classical observations may play a role is the detection of plate motion. The motion of the Earth's plate would result in the long drift of observatory site, the information of which may be contained in the observation residuals of ERP. In 1960's, the theory of plate tectonics was developed. After then, attempts were made to monitor the plate motion with classical instruments, but satisfactory results haven't been obtained, because of the limitation of measuring precision and the processing method of the observations[13].

The recent studies of some authors show broad prospects in this field. Li Zhengxin [14] gave one new method of total solution for reduction of the observations of 136 classical instruments in the whole world for 1962–1982.

New techniques have more precision, but with shorter history. It is important and helpful for the study of plate motion to use jointly the observations in different approaches.

3. Requirements to Star Catalogue and Observational Program

Due to the positive role of the classical instruments in the work of star catalogue, they will be retained to observe continously. As mentioned above, the classical instruments can make contributions to geoscience, and the new techniques cannot replace them completely. Therefore, the instruments can play double task of observing star catalogue and providing data for geoscience study simultaneously.

Since the geophysical effects usually are smaller except the anomalous refraction and vertical variation sometimes are larger, only a lot of observations with high precision are more significant for the geoscience study. The following points should be paid attention to: (1) The chain observing of several groups will be achieved in every clear night as far as possible; (2) In order to ensure internal precision of a group, there should be enough stars in each group; (3) In order to reduce affection of change of meteorological condition and as far as possible to observe complete group or near complete one at short clear night, time interval owned by a group should be shorter; (4) In order to reduce effects of change of instrument constant and refraction, it is necessary that stars situated at different azimuth should be observed alternately for the astrolabe; (5) The program should be used continuously in longer period. This is important for ensured long-term stability of observations. Effects of precession and proper motion must be considered carefully.

Therefore, we need a catalogue with high quality in which precision is higher (position precision is better than ±0″.05), so that the catalogue errors are reduced and we can compile a suitable program which can play double task in longer period. At present, the position precisions of bright stars of $FK5$ may be satisfactory, but the stars are not so many.

4. Concluding Remarks

The positive role of classical instruments in geoscience study has initially been manifested, and their importance will be embodied by more detailed works. Acording to the comparison and analysing of the data, astrolabe is more suitable for geoscience study among the classical instruments. Some astrolabes have been used for geoscience study in USA, USSR,

DDR, and Japan, and a lot of interesting results have been obtained. Chinese Mark-2 PA have been modified in automation and photo-counting detector will be adopted. The precision and magnitude of star (up to 9.5mag.)will be raised. Mark-3 PA will observe fainter stars (11mag.). It is necessary that classical instruments joint intimately with new techniques and geoscience instruments, and astronomers cooperate closely with geophysicists. Then more contributions of the classical instruments will be seen in the offing.

References

[1] Yumi, S. and Koyoyama, K. (1980), *Results of the international latitude series in a homogeneous system 1899.9-1979.0*, P167–172, Mizusawa.

[2] Li Zhengxin (1986), *New determination of the Earth rotation parameters from optical astrometry observations, 1962.0-1982.0*, Shanghai Technique and Science Press, Shanghai, China, (in Chinese).

[3] Mironov, N. T. *et al.* (1974), 'On the relative displacements of zenith of astronomical observatories', *Proc. Geod. and Phys. Earth* (2nd Inter. Symp. Geod. and Phys. Earth, May, 1973), pp. 173–181.

[4] Melchior, P. (1978), *The tides of the planet Earth*, Pergamon Press.

[5] Ma Zongjin *et al.* (1982), *Nine major Earthquakes of China (1966-1976)*, Seismic Press, Beijing, (in Chinese).

[6] Zhang Guodong (1981), 'Deviation of the vertical caused by change of ground water level before a strong earthquake', *Acta Seismologica Sinica*, **3**, No. 2, (in Chinese).

[7] Li Zhisen, Zhang Guodong *et al.* (1978), 'Correlation between the short anomalies of residuals astronomical time and latitude and the major earthquakes around the observatories', *Acta Geophysica Sinica*, **21**, No. 4, pp. 278–290, (in Chinese).

[8] Mavliakov, G. A. *et al.* (1980), 'Anomalous variation of latitude and warning sign of the major earthquake of Alays mountain', *J. Uzbekistan Geology*, No. 2, pp. 66–70.

[9] Han Yanben *et al.* (1987), 'Occurrence of short-period anomaly of residuals of astronomical time-latitude at Yunnan Observatory preceding the Luquan Earthquake ($M_L = 6.3$)', *Kexue Tongbao* (Science Bulletin), **32**, No. 17, (in Chinese).

[10] Li Zhisen and Han Yanben, (1988), 'A possible geophysical interpretation of the anomalies in the residuals of astronomical time and latitude determination', *Vistas in Astronomy*, **31**, pp. 671–675.

[11] Han Yanben *et al.* (1987), 'Possible warning sign of major earthquake in the observations of astronomical time and latitude', *Publ. Beijing Astron. Observ.*, No. 10.

[12] Li Zhisen, (1988), 'Anomalous phenomena of time-latitude residual before earthquakes directly verified at Yunnan Astronomical Observatory', *Recent Developments in World Seismology*, No. 11, pp. 38–39, (in Chinese).

[13] Proverbio, E. F. *et al.* (1975), 'Astronomical evidence of change in the rate of the Earth's rotation and continental motion', In *Growth Rhythms and History of the Earth's Rotation*, (ed. Runcorn), pp. 385–395.

[14] Li Zhengxin (1989), 'Current plate motions from astrometry and Doppler Satellite observations', *Acta Geophysica Sinica*, **32**, No. 5, pp. 567–573, (in Chinese).

DYNAMICAL REFERENCE FRAMES IN THE PLANETARY AND EARTH-MOON SYSTEMS

E. M. STANDISH AND J. G. WILLIAMS
Jet Propulsion Laboratory
California Institute of Technology
4800 Oak Grove Drive
Pasadena, CA 91109 USA

ABSTRACT. We summarize our previous estimates of the accuracies of the ephemerides. Such accuracies determine how well one can establish the dynamical reference frame of the ephemerides. Ranging observations are the dominant data for the inner four planets and the Moon: radar-ranging for Mercury and Venus; Mariner 9 and Viking spacecraft-ranging for the Earth and Mars; lunar laser-ranging for the Moon. Optical data are significant for only the five outermost planets. Inertial mean motions for the Earth and Mars are determined to the level of 0″003/cty during the time of the Viking mission; for Mars, this will deteriorate to 0″01/cty or more after a decade or so; similarly, the inclination of the martian orbit upon the ecliptic was determined by Viking to the level of 0″001. Corresponding uncertainties for Mercury and Venus are nearly two orders of magnitude larger. For the lunar mean motion with respect to inertial space, the present uncertainty is about 0″04/cty; at times away from the present, the uncertainty of 1″/cty² in the acceleration of longitude dominates. The mutual orientations of the equator, ecliptic and lunar orbit are known to 0″002. The inner four planets and the Moon can now be aligned with respect to the dynamical equinox at a level of about 0″005.

1. Introduction

By estimating the accuracies of the planetary and lunar ephemerides themselves, the authors have estimated the accuracy of the inherent dynamical reference frame (Williams and Standish, 1989, hereafter referred to as Paper I). The present paper gives a summarized version of the former estimations, briefly listing the observational data in order to emphasize the variety of data types, presenting a table of the estimated accuracies and noting some recent ephemeris improvements. An attempt is again made to illustrate why the mean motions of the inner planets and the Moon are determined almost exclusively by ranging data, not by optical observations. Finally, the paper discusses a long-known discrepancy: mean motions determined by fitting to only optical observations differ by 1″/century from those determined by fitting to ranging data.

The observational data are listed in Section 2; Section 3 notes recent improvements to the ephemerides of the outer planets; the determination of inertial mean motions using ranging data is discussed in Section 4; Section 5 presents a summary of the error estimates derived from Paper I and Section 6 discusses the 1″/cty discrepancy.

J. H. Lieske and V. K. Abalakin (eds.), Inertial Coordinate System on the Sky, 173–182.
© 1990 *IAU. Printed in the Netherlands.*

2. The Observational Data

The most important ingredient in the process of creating modern-day ephemerides is the collection of the observational data: accuracy, variety and coverage. With modern computers, there is no longer any problem correctly integrating the equations of motion: no worry about non-converging expansions, neglected terms, truncated series, etc. Numerical integration programs have been tested; they provide sufficient accuracy. Also, it seems safe to assume that our equations of motion properly describe the physical laws of gravitation and that all of the significant forces which affect the motions of the planets and Moon are known.

Table 1 lists the observational data now being fit to generate ephemerides. During the present decade, in preparation for the Voyager encounters, newer and more accurate observations of the five outermost planets have produced significant improvements to those ephemerides. These are discussed in the next Section.

The acquisition and utilization of observational data continues to be the most vital part of the ephemeris creation process.

Table 1. The sources of the observational data.

		Sun	Mer	Ven	Mar	Jup	Sat	Ura	Nep	Plu	Moon
Optical transits	1911-	S	M	V	M	J	S	U	N		
Photoelectric transits	1982-				M	J	S	U	N	P	
Astrolabe	1969-				M	J	S	U			
Radar ranging	1964-		M	V	M						
Mariner 9 Ranges	1971-72				M						
Mariner 10 Ranges	1974-75		M								
Viking Lander Ranges	1976-82				M						
Radio Astrometry	1983-					J	S	U	N		
Ring Occultations	1977-							U			
Disk Occultations	1968-								N		
Pioneer Tracking Data	1973-80					J	S				
Voyager Tracking Data	1979-89					J	S	U	N		
Pluto Astrometry	1914-									P	
Lunar Laser Ranging	1969-										M

3. Outer Planet Ephemerides

It is apparent in Table 1 that there are now newer and more accurate types of observational data than what were available just a decade ago. This is especially true for the outer planets where a concentrated effort was made for the Voyager encounters with the Jovian planets. Indeed, there are significant differences between the ephemerides DE202, JPL's latest, and its predecessor at JPL, DE200, the standard planetary ephemeris of the Astronomical Almanac. Since DE202 is believed to be much more accurate than DE200 for the present epoch, most (80+ percent?) of the difference may be attributed to errors in DE200.

The differences, DE202–DE200, are plotted in a recent paper by Standish (1989). For Jupiter, the right ascension varies over the planet's 12-yr period between –0".1 and –0".2 throughout the century. For Saturn, the right ascension also varies over the 30-yr period, but in addition, drifts down to –0".25 at present. The difference for Uranus is small through the first half of the century, but reaches nearly –0".5 by the year 2000. For Neptune, the error is +0".6 for 1900 and is near –1".0 by 2000. The error for Pluto exceeds +2".0 by the end of the century and is rapidly increasing — not surprising since this is the first time that JPL has fit the orbit of Pluto to observations. The declination errors are generally periodic and smaller than those of right ascension.

For the four Jovian planets, consistency of all of the new observations taken during the present decade indicate that over that time-span, DE202 has position errors of 0".05 or less. For Pluto, observed only photographically to date, the ephemeris is expected to have an uncertainty of about 0".5 during the present era, due mainly to catalogue errors. This uncertainty may decrease in future ephemerides, now that photoelectric transit observations of Pluto are being obtained at La Palma (Morrison *et al.*, 1990).

4. Inertial Mean Motions

There are a number of explanations designed to show why inertial mean motions are determined almost exclusively from the ranging data; why the mean motions of the four inner planets and the Moon are almost entirely independent of the optical observations (Standish 1982, 1985, 1988; Paper I). We give one more here.

Consider the Newtonian two-body problem: True, the distances are invariant under a rotation of the ellipse. However, a rotating ellipse is not a possible Newtonian two-body motion.

The actual case of the solar system is similar; the planetary orbits are nearly non-rotating ellipses. The departures from Keplerian motion, for the most part, are small and accurately computed. It is the accuracy of computing these departures which determines the major uncertainties in the mean motions for the moon and planets; these were discussed in Paper I.

In the Newtonian case, the lunar range is given by

$$r = a - ae \cos E \approx a - ae \cos n(t - T) ;$$

i.e., there is a constant term, an amplitude, a phase and a period (inertial mean motion). Given a reasonable span of observations, each of these parameters can be determined to an accuracy comparable to or better than the 3 centimeter accuracy of the present data. For the actual case, the non-Keplerian features of the lunar motion include solar and planetary perturbations, general relativity, the tidal deceleration in longitude and the precession of the perigee and node. As shown in Paper I, the greatest uncertainty in the lunar motion is in the perigee's precession, 0".04/cty, for times near the center of the data span; for remote times, the tidal deceleration uncertainty of 1"/cty² dominates.

The case for planetary ranging is analogous, though here the ranging occurs between bodies moving on two ellipses instead of one. Nevertheless, the same principle applies, as described in Paper I. The observational data for Earth and Mars are dominated by the six years of ranging to the Viking Landers at 10 meter accuracy. The major uncertainty in the motion with respect to an inertial frame is due to uncertainties in Mars' perihelion rate due to mass uncertainties of the planets. When the motion is extrapolated decades to centuries beyond the span of the ranging data, the dominant error source becomes the long-period perturbations from many asteroids whose masses are not known better than a factor of about two.

5. Ephemeris Error Estimates

In Paper I, an analysis of the ephemeris parameter uncertainties was made by considering the accuracies of the relevant observational data and by considering how sensitive such observations are to changes in each of the parameters. Those results are used here in producing Table 2, where the estimates are intended to be realistic uncertainties. We also include our estimates of the orbits of the outer planets, realizing that the uncertainty of extrapolating into the future is largest for the outermost planets.

Table 2. Estimated ephemeris errors.

	Moon	Merc & Ven	Mars	Jup...Nep	Plu
longitude					
wrt earth in 1980	[0."001]	0."002	0."00002	0."05	0."5
wrt earth in 1990	[0."001]	0."02	0."001	0."05	0."5
wrt 1980 dyn eq	0."005	0."005	0."005	0."05	0."5
wrt 2000 dyn eq	0."02	0."02	0."02	0."05	0."5
latitude	0."002	0."02	0."0005	0."05	0."5
mean motion	0."04/cty $1''/cty^2$	0."2/cty	0."01/cty	0."5/cty	2"/cty

6. Optical Problem

Something is wrong with the optical transit observations and/or with the reference systems with which they are reduced.

The optical residuals for the Sun and inner planets show secular drifts in right ascension of approximately one arcsecond per century. An example is given in Figure 1a showing the 20th century solar residuals of observations taken with the USNO 6" and 9" transit circles. Without the presence of ranging data, the drift would be removed by the least squares adjustment of the orbits, primarily by changes in the semi-major axes. This is essentially what happens with the outer five planets, for which there are no ranging measurements. However, in the presence of the dominating range data, the orbits for the inner four planets are adjusted primarily to the ranges, not to the optical observations.

This problem has been recognized for many years (Stumpff and Lieske, 1984; Seidelmann *et al.*, 1985; Seidelmann, 1986). The source, however, remains unknown. Two things are certain: the error does not come from the newly adopted value of precession (Fricke, 1971); the error does not come from the mean motions of the inner planets in the modern ephemerides. The first assertion is valid since both lunar laser ranging and VLBI have now verified that the error in Fricke's value of precession is no more than about 0''.3/cty. For the second assertion, we here perform an experiment.

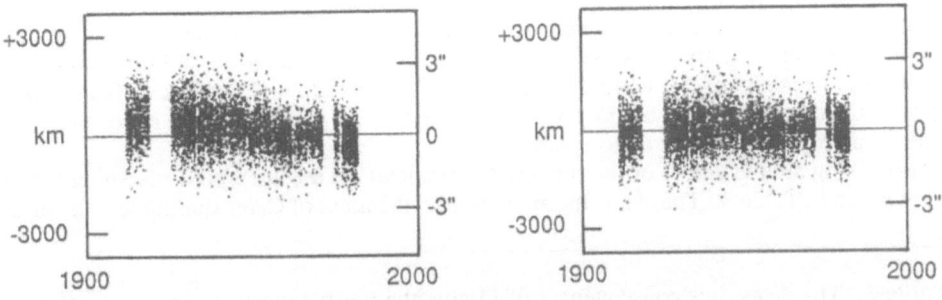

Figures 1a and 1b. The right ascension residuals of the Sun from the USNO 6-inch and 9-inch meridian circle observations. A downward slope of about 1"/cty is noticeable in Figure 1a, where the residuals are plotted with respect to a present day ephemeris. The slope is gone in Figure 1b where the Earth's mean motion has been forced to change by 1"/century.

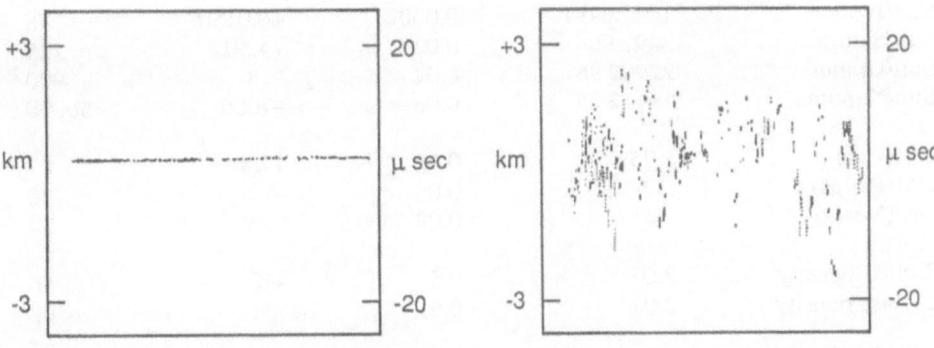

Figures 2a and 2b. The range residuals from the Viking Lander on Mars. Figure 2a shows the residuals from a present-day ephemeris. Figure 2b shows the residuals from an ephemeris which has been forced to fit a change of 1"/century in the Earth's mean motion.

6.1 FORCING THE EARTH'S MEAN MOTION TO FIT THE OPTICAL DATA

In order to illustrate how sensitive the mean motions are to ranging data, we artificially force a change of 1"/cty into the mean motion of the Earth and then perform a full least squares adjustment of all of the other ephemeris parameters.

Figures 1a and 1b show the right ascension residuals of the Sun before and after forcing the Earth's mean motion. One can detect the left-to-right downward slope of about 1"/cty in Figure 1a as it appears in present-day ephemerides; in Figure 1b, the slope is gone since the Earth's mean motion has been adjusted artificially. The scale shows that the differences in question correspond to the level of a few hundred kilometers; the differences are not blatantly obvious in these residuals.

In contrast, Figures 2a and 2b show the corresponding residuals of the Viking Lander ranging where the change is gigantic. Previous residuals of 10 meters can no longer be fit better than a full kilometer; further, they would obviously become even worse if extrapolated beyond the six-year interval. This fit is completely unacceptable.

Just as striking are the values of the ephemeris parameters in the attempted adjustment. Some of these are given in Table 3. These results are nonsense: the mass of Ceres quintupled, the masses of

---

Table 3. The disastrous consequences of forcing the Earth's mean motion to change by 1"/century. Shown are the resultant values for the solar system scale parameter, the Sun/planet-system mass ratios, the masses of the three major asteroids [in $au^3/day^2 \times 10^{-13}$], the densities of the S-type and C-type asteroids and the locations of the Deep Space Network radio antennas.

Parameter	Present Value	Std. Dev.	Forced Change	Ratio
scale (m/au)	149597870660	20	−1340	67
Sun/Venus	408523.5	0.5	+8	16
Sun/Jupiter	1047.3491	0.0002	+0.0150	75
Sun/Saturn	3497.90	0.02	+4.50	225
Sun/Uranus	22902.96	0.02	−18	900
Sun/Neptune	19412.25	0.06	−7000	350000
GM(Ceres)	1.75	0.09	+7.8	87
GM(Pallas)	.32	0.05	−1.0	20
GM(Vesta)	.41	0.08	−1.9	24
S-class density	2.0	0.5	−28	56
C-class density	2.0	0.5	−31	62
DSN longitudes [cm]	—	30	100000	3300
spin radii [cm]	—	30	100000	3300
z-heights [cm]	—	50	200000	4000

Pallas and Vesta became negative, etc. Most of the changes to the parameters are on the order of 100 times their presently accepted standard deviations.

The result is clear: there is no way that the mean motion of the Earth in modern ephemerides can be in error by 1"/cty.

An error of 1"/cty is certainly greater than one hopes for the optical system; however, it is completely intolerable for modern planetary ranging.

7. Conclusions

The ephemerides continue to improve as more and newer data measurements are acquired. In particular, over the past decade, substantial improvements have been made to the outer planet ephemerides. Corresponding refinements to the ephemerides of the inner planets and the Moon are expected within the next year.

A number of different methods have been considered for estimating the errors in the modern-day ephemerides. They seem to be consistent with each other; their results were presented here.

Not all of the observational data sets are consistent, however. In particular, there is still the puzzling drift in the optical residuals of about 1"/century for the inner bodies of the solar system.

The forcing of a 1"/century change in the mean motion of the Earth required the other ephemeris parameters to change by about 100 times their standard deviations in a futile attempt to re-fit the data. This tends to indicate that the mean motion of the Earth has been established to a level of about 0".01/century.

One must conclude that there is something wrong with the optical data and/or the reduction of it: not precession certainly, but possibly the equinox drift; perhaps not the values themselves, but possibly where and when the parameters are applied in the reduction processes. It is too early to speculate further; it is intended that this will be investigated soon.

ACKNOWLEDGMENT. The work described in this paper was carried out by the Jet Propulsion Laboratory, California Institute of Technology, under contract with the National Aeronautics and Space Administration.

8. References

Fricke, W.: 1971, "A Rediscussion of Newcomb's Determination of Precession", *Astron. Astrophys.* **13**, 298–308.

Fricke, W.: 1982, "Determination of the Equinox and Equator of the FK5", *Astron. Astrophys.* **107**, L13–L16.

Morrison, L.V., Helmer, L., Fabricius, C., Einicke, O., Quijano, L., Muinos, J.L., Argyle, R.W.: 1990, "Optical Reference Frame Defined by Carlsberg Meridian Catalogues for the Years 1984–1987", these proceedings.

Seidelmann, P.K.: 1986, "Unsolved Problems of Celestial Mechanics — The Solar System", *Cel., Mech.*, **39**, 141–146.

Seidelmann, P.K., Santoro, E.J. and Pulkkinen, K.F.: 1985, Systematic Differences between Planetary Observations and Ephemerides", in *Dynamical Astronomy*, V. Szebehely and B. Balazs, Eds., University of Texas Press, Austin, 55–65.

Standish, E.M.: 1982, "The JPL Planetary Ephemerides", *Cel., Mech.* **26**, 181–186.

Standish, E.M.: 1985, "On the Orientation of Ephemeris Reference Frames", *Cel., Mech.* **37**, 239–242.

Standish, E.M.: 1988, "Celestial Reference Frames: Definitions and Accuracies", in *The Impact of VLBI on Astrophysics and Geophysics* (M.J. Reid and J.M. Moran, Eds.), D. Reidel, publ., 309–315.

Standish, E.M.: 1989, "An approximation to the outer planet ephemeris errors in JPL's DE200", *Astron. Astrophys*, in press.

Stumpff, P. and Lieske, J.H.: 1984, "The Motion of the Earth-Moon System in Modern Tabular Ephemerides", *Astron. Astrophys.*, **130**, 211–226.

Williams, J.G. and Standish, E.M.: 1989, "Dynamical Reference Frames in the Planetary and Earth-Moon Systems", in *Reference Frames* (J. Kovalevsky, I.I. Mueller and B. Kolaczek, eds.) Kluwer Academic Publishers, Dordrecht, 67–90.

Discussion

KRASINSKY: We have obtained similar results on the inconsistency of planetary mean motions determined from radar and optical observations. In our opinion the discrepancy may be explained by secular errors in the ephemeris time scale before 1959. There are indications that these errors do exist (from analyses of Mercury and Venus solar transits). These errors influence the mean motions in the same way as an error in the precessional constant.

STANDISH: In order to produce an error of 1"/century in the solar residuals, the ET-UT tables would have to be in error by 24 time seconds/century. That is too large for most people to accept.

HUGHES: As is well known, observing the Sun is difficult. I would comment that the new observing system on our instrument in New Zealand has increased our ability to observe day-stars. We commonly observe up to 50 stars in a day including azimuth stars. This certainly will help with the perennial problem of relating day and night observations.

STANDISH: Those observations certainly will be instructive in a number of ways. However, I do not believe that the 1"/century is a problem of data observation; rather, one of data reduction.

KAPLAN: Could you comment on the possibility—or impossibility—of significant amounts of unobserved mass in the solar system? In particular, a tenth planet?

STANDISH: Yes, I often make such comments. Certainly, there is nothing to say that there could not be significant amounts of mass outside of Pluto, even a concentration large enough to be a planet. However, certain regions seem to be excluded by the different searches (Tombaugh, Kowal, IRAS, etc.).

Further, there is nothing in the observational planetary data which needs Planet X to explain it. With a 20th century error as gigantic as 1"/century, it seems silly to invoke a Planet X to explain only a couple of tenths of an arcsecond from the 19th century data. What's even worse, the newer data are greatly improved in comparison to the older data which are fraught with known systematic errors.

SEIDELMANN: I predict that in ten years the observations of Neptune and Pluto will deviate from the ephemerides as they have each time in the past.

STANDISH: I agree; accurate extrapolation can not be expected. We have less than a period's worth of observations for these planets, and the observations that we do have contain systematic errors.

There is no question, though; we continue to need the optical observations, especially those of the outer planets.

KLIONER: What is the present accuracy of the barycentric position of the Sun in kilometers?

STANDISH: The former (0.5%) error in the mass of Neptune used to be the dominant error in the position of the barycenter, amounting to about 1200 km. Now, after the Voyager determination, the uncertainties due to the uncertainties of the masses of Jupiter through Pluto are 0.1, 2.4, 0.1, 0.7 and 10.4 km, respectively.

THE MAIN STAGES OF THE CONSTRUCTION OF AE89 — THE NUMERICAL EPHEMERIS
OF THE PLANETS AND THE MOON

M.L. Belikov, V.N. Boyko, N.I. Glebova, G.I. Eroshkin, L.I. Rumyantseva, M.L.
Sveshnikov, E.S. Sveshnikova, R.I. Smekhacheva, A.A. Trubitsyna, M.A. Fursenko,
L.I. Chunaeva, and A.A. Shiryaev
Institute for Theoretical Astronomy, USSR Academy of Sciences
10 Kutuzov Quay
191187 Leningrad
USSR

The realization of theoretical and applied researches in the domain of ephemeris astronomy,
connected with analysis of precision of existing planetary and lunar theories, the construction of an
inertial coordinate system and investigation of physical properties of space-time, necessitated the
elaboration in ITA of the numerical theory of the motion of heavenly bodies suitable for calculation
of high-precision ephemerides at large time-spans, and fit also for the maintenance of space
experiments.

In the course of construction of this theory, named AE89, the tasks to be solved are as follows:

a) the construction of the dynamically consistent model of the orbital motion of the planets and
the orbital-rotational motion of the Moon;
b) the elaboration of an effective numerical integration method fit to the development of the
ephemerides;
c) the working out of a base of astrometric solar, lunar and planetary data;
d) the determination of the dynamical parameters of the Solar system from the observations.

1. Mathematical model

The adequate mathematical model of the motion of the Solar system bodies constitutes the
framework of the numerical ephemeris. This model is accomplished in the form of a system of
differential equations describing the orbital motion of 9 major planets, 5 of the most massive
asteroids, and the orbital-rotational motion of the Moon, in the barycentric ecliptic reference system
defined by the ecliptic and equinox of the epoch J2000.0.

The orbital motions of the heavenly bodies considered as point masses, are presented by Einstein-
Infeld-Hofmann equations. In the equations concerning the Earth and the Moon, the perturbations
caused by their gravitational interaction as nonspherical rotating rigid bodies were added; the
terrestrial zonal harmonics through 4-th degree and all the lunar harmonics through 4-th degree were
also included. The rotation of the Moon about its centre of masses is described in terms of Rodrigues-

183

J. H. Lieske and V. K. Abalakin (eds.), Inertial Coordinate System on the Sky, 183–185.
© 1990 IAU. Printed in the Netherlands.

Hamilton's parameters, defining the position of the lunar principal axes of inertia with respect to the ecliptical reference frame defined earlier.

The position of the Earth's true equator is defined by the precession parameters of Lieske and the theory of nutation of IAU (1980). The independent variable of the equation is barycentric dynamical time TDB. The system of constants adopted in this model (planetary and Moon's masses, geo- and selenodynamical parameters etc.) coincide with that of the DE200/LE200 numerical ephemeris. The initial conditions of the orbital motion of the planets and the Moon are taken from the same ephemeris; those of the asteroids are calculated by means of programme used in preparation of the *Ephemerides of minor planets*. The initial values of Rodriques-Hamilton parameters are defined by the series which represent the components of the lunar physical libration for the values of the DE200/LE200 selenodynamical constants [1].

2. The numerical integrator

For the integration of the equations the integrator RA15 [2] was used, as well as the integrator INCH7 [3] elaborated in ITA especially for this work. The integration and the comparison of the results with DE200/LE200 and with the series [1] was carried out at the time-span 1960–1990.

The deviations of the coordinates of AE89 from those of DE200/LE200 for the 1980–1990 time-span are shown in the table. For the planets in most cases they don't exceed tens of meters and 1.5 meters for the Moon (single precision integration: 12 digits). For double precision integration the deviations diminish considerably.

In view of the difference between the models adopted in AE89 and in DE200/LE200, one may consider these results as satisfactory. The comparison also made obvious the advantages of the INCH7 over RA15 with respect to the precision and speed of itegration. Moreover, in INCH7 the original method is employed for the construction of the coefficients of Chebyshev's polynomials, representing the coordinates of the heavenly bodies in the ephemeris by means of the coefficients of the interpolation polynomials used in the process of numerical integration.

3. Treatment of observations

The treatment of the observations for the improvement of the Solar system parameters will be performed by means of data base DVA (Disk Version of Archives) and Planetary Data Management System (PDMS). At present the construction of this system is completed, and so are three specialized archives of the observations, *viz.*:

a) the archive of optical angular observations of the Sun and major planets, which contains 66000 observations made at various observatories from 1774 to 1988.
b) the archive of planetary radar observations (7500 measurements) made in USSR and USA from 1964.
c) the archive of lunar laser observations (5500 measurements) made in USA, France and USSR from 1969 to 1985.

Table. The comparison with DE200 / LE200 and semianalytic theory of the physical libration of the Moon [1] at 10-years time-span from JD 2444 400.5
(Maximum and minimum deviations for the single precision integration are indicated)

	$\Delta X(m)$	$\Delta Y(m)$	$\Delta Z(m)$	$\Delta R(m)$	$\Delta\lambda(1''\times 10^6)$	$\Delta\beta(1''\times 10^6)$
Mercury	+650.86	+562.66	+60.84	+124.60	–	+201
	–431.93	–624.51	–86.80	–121.50	–3040	–338
Venus	+28.78	+32.45	+1.87	+1.75	+1	+4
	–32.64	–30.11	–1.86	–0.46	–68	–4
Earth	+14.85	+24.25	+0.53	+3.89	+5	+1
	–28.80	–22.00	–0.56	–2.38	–40	–1
Mars	+431.51	+346.57	+4.74	+73.23	+19	+6
	–398.35	–550.70	–8.57	–52.85	–561	–8
Jupiter	+172.43	+54.10	+1.61	+48.53	–	+1
	–68.29	–171.45	–2.80	–23.76	–53	–1
Saturn	–	+3.24	+6.54	+70.78	–	+1
	–71.42	–107.38	–	–	–15	–
Uranus	–	+3.31	+5.26	+32.54	–	–
	–41.18	–34.56	–	–0.79	–3	–
Neptune	–	+3.33	+5.66	+21.77	–	–
	–41.56	–29.13	–	–2.68	–2	–
Pluto	–	+3.33	+7.12	+48.73	–	–
	–45.24	–28.76	–	–	–1	–
Moon	+1.30	+1.35	+0.25	+0.12	+303	+140
	–1.59	–1.31	–0.25	–0.14	–914	–139

The physical libration of the Moon

$\Delta\tau$	$\Delta\rho$	ΔI_σ
+0''231	+0''050	+0''266
–0''153	–0''476	–0''160

References

1. Fursenko, M.A.: 1989, "The physical libration of the Moon. M.Moons' 1985 theory for the Stokes constants DE200/LE200", In press.
2. Everhart, E.: 1985, "An efficient integrator that uses Gauss-Radau spacings", In: *Dynamics of comets: Their origin and evolution* (Ed. A.Carusi and G.B.Velsecchi), D. Reidel Publ. Co., Dordrecht, pp. 185–202.
3. Belikov, M.V.: 1989, "An efficient one-step integrator with Chebyshev's approximation (INCH7)", In press.

EFFECT OF THE NEW EQUINOX DEFINITION ON THE ZERO-POINT OF LONGITUDE OF THE INDIAN CALENDAR

A.K. BHATNAGAR
Positional Astronomy Centre
P 546, Block N, 1st Floor
New Alipore, 700053 Calcutta, India

ABSTRACT. Indian calendars follow a sidereal system of astronomy taking a fixed initial point on the ecliptic as the origin from which the longitudes are measured. Its position for the official Indian Calendar has been defined by the Calendar Reform Committee (1955) as the point on the ecliptic whose true tropical longitude was 23°15'00" as on 21 March 1956, 0^h UT. Its position was determined upto the year 1984 in accordance with Newcomb's value for general precession using the relation

$$A = 22°27'37''65 + 5025''75 \, T + 1''11 \, T^2$$

where T is in centuries of 36525 ephemeris days from 1900 January 0.5 ET. Recent changes in the location and the motion of the equinox with reference to the epoch J2000.0 have necessitated corresponding changes to be included in the determination of the mean and true positions of the above initial point. The new algorithm worked out is

$$A = 23°51'25''532 + 5029''0966 \, T + 1''11161 \, T^2$$

where T is in Julian centuries of 36525 days from J2000.0.

THE CONSTRUCTION OF A FRAME OF HOMOGENEOUS ACCURACY BASED ON SPACE OBSERVATIONS

B.I. VLASOV
VNIIFTRI
Institute for Physical, Technical and Radio Measurements
141570 Moscow, USSR

ABSTRACT. A definition of the initium-method of spherical arcs is proposed. Theoretical and operational aspects of the method are discussed. Geometrical homogeneity is achieved by measuring arc distances between vertices of the octant (a right-angle spherical triangle) and the object in question. The measurements are made with the use of the standard angle γ on board a satellite with rotation rate reconstructions with splines. The chosen standard angle is discussed. The accuracy estimations are given.

HIGH ACCURACY ALGORITHMS OF NUMERICAL PREDICTION OF THE MOTION OF SOLAR SYSTEM BODIES

T.V. BORDOVITSYNA AND V.A. SHEFER
Research Institute of Applied Mathematics and Mechanics
Tomsk University
634050 Tomsk
USSR

ABSTRACT. A brief summary of results obtained by the authors for investigations of the efficiency of numerical prediction algorithms for natural and artificial minor bodies using regularizing and stabilizing transformations is given.

The problem of high accuracy interpretation of observations of solar system bodies is closely connected with the problem of the accurate prediction of the motion of these objects.

This paper presents a brief summary of results obtained by the authors (Bordovitsyna 1984; Shefer 1989; Bordovitsyna *et al.* 1989) for the development and investigation of effective algorithms for the numerical prediction of the motion of minor bodies of the solar system. Research on the efficiency of the set of regularizing and stabilizing methods in the problem of numerical prediction of the motion of celestial bodies was carried out by the authors. It was shown that the introduction of integrals of motion, KS-transformation and time transformation $dt = r^n ds$ with $n = 1$ in the equations of motion is the most effective way of stabilization and regularization of these equations. These transformations do not destroy the numerical stability of the calculation algorithm and at the same time they stabilize and regularize the equations of motion.

The research on numerical methods has shown that Everhart's implicit single sequence methods and the modified algorithm of rational extrapolation are the most effective ones.

As a result of these investigations, high accuracy algorithms and software were developed. There are two application packages for prediction of the motion of natural minor bodies of the solar system and of artificial satellites.

The research on the efficiency of the application package "The dynamics of minor bodies of the solar system" was carried out on the examples of the motion of the unusual minor planets Icarus and Geographos as well as comets Halley, Honda-Mrkos-Pajdusakova and Gehrels 3.

An analogous research on the application package "The numerical model of artificial satellite motion" was made in solving the problem of numerical prediction of the motion of artificial satellites

J. H. Lieske and V. K. Abalakin (eds.), Inertial Coordinate System on the Sky, 187–188.
© 1990 *IAU. Printed in the Netherlands.*

with different kinds of orbital parameters as well as representing observations obtained by the MERIT program.

The accuracy of representation of observations of minor MERIT program objects equals ±50 cm.

References

Bordovitsyna, T.V.: 1984, *Modern Numerical Methods in Celestial Mechanics*, Nauka , Moscow (in Russian).
Bordovitsyna, T.V., Bykova, L.E., Boronenko, T.S., Tamarov, V.A., Sharkovsky, N.A., and Shmidt, Yu.B.: 1989, *Numerical and Semi-analytical Algorithms for Prediction of the Motion of Artificial Satellites*, Tomsk State University, Tomsk (in Russian; in press).
Shefer, V.A.: 1989, *Celest. Mech.* (in press).

ON THE ERRORS OF THE EPHEMERIDES DERIVED FROM OPTICAL OBSERVATIONS OF PLANETS

A.S. KHARIN
Main Astronomical Observatory
Ukrainian Academy of Sciences
252127 Kiev
USSR

Yu.B. KOLESNIK
Astronomical Council
USSR Academy of Sciences
109017 Moscow
USSR

ABSTRACT. On the basis of about 40 000 optical observations of the Sun and major planets obtained with 33 meridian and photographic instruments during last 3 decades a comparative consistency analysis of old and new ephemerides with these observations has been made.

For the inner planets significant improvement in RA is confirmed while in DEC it is less apparent. For outer planets the improvement is strongly marked in DEC for Jupiter and Saturn and espesially for Neptune in both coordinates. Significant systematic differences between meridian and photographic observations are detected.

During the last 3 decades a large amount of a newer planetary optical data have been obtained with meridian and photographic methods.

Only a small part of them has been used for creating of a new fundamental ephemeris DE200 [1]. So we have analysed the agreement of all available modern optical observations in the time span 1960–1987 obtained with 19 meridian instruments and 14 astrographs with the old standard ephemerides (Newcomb's theory with Ross corrections for the four inner planets and the Eckert-Brouwer-Clemence numerical ephemeris for the five outer planets) and with the new one — DE200, adopted as standard by IAU since 1984.

Two homogeneous sets of observation data have been formed by reducing the published values (O-C) to the corresponding ephemeris and fundamental catalogue. For this purpose the differences between apparent places based on the old and new ephemerides and constants of the IAU were added or subtracted, according to the case, to/from (O-C) for all objects. Fricke's equinox correction [2] was applied in the same way for RA. Systematic differences between reference star catalogues are ignored at this stage of analysis. For each object, method of observation and ephemeris, the curves of normal points in 30 degree zones of orbital longtitude have been obtained. Dispersion of each curve may be interpreted as the combined effect of systematic errors of the ephemeris and observations. The ratio of dispersions of the ephemerides compared should reflect the degree of their relative proximity to the observational data. Cross-correlation of curves will

J. H. Lieske and V. K. Abalakin (eds.), Inertial Coordinate System on the Sky, 189–190.
© 1990 *IAU. Printed in the Netherlands.*

display the same systematic errors of both ephemerides. After all computations are made, the results are represented in Table 1. Values having the significance not exceeding the 5% level of F-criteria are marked with an asterisk.

Table 1. Ratios of dispersions of normal point curves and cross-correlation coefficients for old and new ephemerides

| | Ratios of dispersions | | | | Cross-correlation coefficient | | | |
| | Meridian | | Photographic | | Meridian | | Photographic | |
Objects:	RA	DEC	RA	DEC	RA	DEC	RA	DEC
Sun	11.2*	3.0*	-	-	0.28	0.82*	-	-
Mercury	2.1	1.0	-	-	0.65*	0.58*	-	-
Venus	2.1	1.0	8.8*	2.2	0.74*	0.87*	−0.49	0.40
Mars	3.0*	1.8	7.6*	4.3*	0.34	0.75*	0.39	0.72*
Jupiter	1.6	13.5*	0.9	1.7	0.52	−0.58	0.93*	−0.24
Saturn	0.5	17.4*	1.3	5.3*	0.43	−0.04	0.98*	0.31
Uranus	2.2	1.2	12.1*	2.0	−0.39	0.78*	−0.94*	0.93*
Neptune	70.9*	0.3	123.8*	1.2	0.81	−0.52	1.00*	1.00*

The examination of this table leads us to the following conclusions:

1. Significant improvement of the new ephemerides for the four inner planets in RA is confirmed by the two methods. An apparent correlation of DEC curves for all these planets is remarkably noticeable while both ephemerides were deviating systematically from observations by approximately 0".5.

2. As to the 4 outer planets the analysis reveals rather considerable improvement in DEC for Jupiter and Saturn. The same is noticeable for Uranus RA according to photographic data. The very large (about 10" in RA and 1" in DEC) discrepancies of the old Neptune ephemeris are eliminated by using the new one.

3. Some discordance of cross-correlation coefficients—and ratios of dispersions between meridian and photographic measurements—reveal their significant relative systematic differences.

References

1. Standish, E.M. (1986), in *Proc.IAU Symp.114, Relativity in Celest. Mech. and Astrometry*, J. Kovalevsky and V.A. Brumberg (eds.), D. Reidel, Dordrecht, 71–83.
2. Fricke, W. (1982), *Astron. Astrophys.* **107**, L13–L16.

DIFFERENCE BETWEEN DYNAMIC AND STELLAR REFERENCE FRAMES: RESULTS BASED ON ASTROMETRIC OBSERVATIONS OF MARTIAN SATELLITES

N.V. EMELYANOV [1], Yu.F. KOLYUKA [2], S.M. KUDRYAVTSEV [2], K.V. KUIMOV [1], V.F. TIKHONOV [2] AND V.V. CHAZOV [1]

[1] *Sternberg State Astronomical Institute*
Universitskii Prosp. 13
119899 Moscow, USSR

[2] *Mission Control Centre*
Kaliningrad
Moscow Region, USSR

ABSTRACT. The possibility of using precise astrometric observations of Martian satellites in 1988 for improving the mutual orientation of the dynamic and stellar reference frames is discussed. The results obtained lead to the conclusion that the stellar reference frame is rotated about the X axis of the dynamic one. However, the value of this rotation (~2".5) proves to be unrealistically large and repetitions of precise observations of Martian satellites are needed in the future for verifying this result.

During the period from July up to November 1988 Sternberg Astronomical Institute (SAI) astronomers made a large series of photographic measurements of Martian satellites coordinates (rms error is 0".12) at the Mt. Maidanak obsevatory.

These data [1] are presented not as the satellite coordinates relative to Mars or another satellite but as the absolute measurements of the right ascension and declination in the SRS catalog reference frame which is an expansion of the FK4. Attempts to use this kind of measurements of Phobos and Deimos for improving its orbital parameters display systematic differences (up to 0".4) in the measurements with their calculated values. These differences cannot be removed by correction of satellite orbital parameters without contradiction with another observations. But in the case of "Phobos – Deimos" relative coordinates, the measurements are in full correspondence with the theory [2] of Martian satellites motion worked out in the Mission Control Centre (MCC).

These are the possible explanations of the phenomenon:

1. *Systematic errors of observation.* The satellite coordinates were first determined relative to distant stars fixed on plates. Star positions were then connected to the SRS catalog reference frame. As a result the systematic observation error is evaluated to be less then 0".1" to 0".15.

J. H. Lieske and V. K. Abalakin (eds.), Inertial Coordinate System on the Sky, 191–193.
© 1990 *IAU. Printed in the Netherlands.*

2. *Errors in Earth and Mars ephemerides.* In our calculations we used the American DE118 and/ or DE200 ephemerides and/or Soviet ephemerides [3]. These ephemerides are mainly based on precise radar measurements of Mercury, Venus and Mars — so relative distances between those planets are well determined. The errors in the position of Mars relative to Earth are evaluated to be not more than 10-15 km. Equal angular errors are less than 0".05.

3. *The difference between the dynamic and stellar coordinate systems as a whole.* Connection of those systems was based on optical planetary observations which have 0".5 to 1".0 errors. Only as a result of the 1988 observations did it become possible to use high-precision Martian satellites measurements for this purpose.

The authors have investigated a conclusion that one should be able to give if one accepts the 3rd explanation of the phenomenon. The obtained results are based on two separately derived theories of satellite motion and various planet ephemerides. In MCC an analytical theory of the 2nd order of Martian satellites motion [2] based on all known ground-based and space observation of satellites in 1877–1989 and high-precision radio-tracking of the spacecraft "Phobos-2" in the vicinity of Phobos was used.

The calculation precision of the satellite ephemerides on the theory base is evaluated as (1-σ) 2-3 km for Phobos and 6-9 km for Deimos. An analytical theory of Phobos motion [4] was developed in SAI. The features of this theory are utilization of solving the general problem of two static centres for an intermediate satellite orbit and the determination of all perturbations from gravity factors with required precision. DE118 and/or soviet ephemerides in MCC and DE200 in SAI were used to represent the planet coordinates.

The table shows values of rotation angles of the stellar reference frame relative to the axes of DE118 and/or DE200 one obtains on the basis of different measurements files. The results obtained with use of soviet ephemerides are very close to those which were obtained in the case of using DE118.

Table 1. Rotations of the stellar reference frame relative to axes of the dynamical one

Rotation (right)	Observations involved, method, planet ephemerides			
	Phobos, MCC, DE118	Phobos, SAI, DE200	Deimos, MCC, DE118	Phobos+Deimos MCC, DE118
Wx	−2".40 ± 0".33	−2".14 ± 0".27	−2".70 ± 0".33	−2".56 ± 0".24
Wy	+0".04 ± 0".05	+0".04 ± 0".06	+0".01 ± 0".05	+0".02 ± 0".05
Wz	−0".10 ± 0".03	+0".56 ± 0".06	−0".09 ± 0".03	−0".10 ± 0".03

The Wz values in the MCC and SAI alternatives differ significantly because of utilization of different planet ephemerides. The DE118 reference frame differs from the DE200 one mainly by precession and additional rotation around the Z-axis at 0".53 angle.

As soon the results that are given in the table contradict with optical observations of Mars, the existence of systematic errors in photographic stellar catalogs may be assumed.

The orientation of the stellar reference frame relative to the dynamic one may be well determinated on the basis of precice Mars satellite observations in the future.

References

1. Bugaenko, O.I., Evstigneeva, N.M., Kudryavtsev, S.M. *et al.* (1990) *Astron. Asrophys.* (submitted for publication).
2. Kudryavtsev, S.M., Kolyuka, Yu.F., Tikhonov, V.F. (1989) SAI preprint **2**, 19–25.
3. Akim, E.L., Brumberg, V.A., Kislik, M.D. *et al.* (1986) in J. Kovalevsky and V.A. Brumberg (eds.), *Relativity in Celestial Mechanics and Astrometry*, (proceedings of 114th symposium of IAU), D.Reidel, Dordrecht, 63–68.
4. Emelyanov, N.V., Nasonova, L.P. (1989) *Sov. Astron. J.*, **66**, 850–858.

Discussion

STANDISH: There is no way for the plane of the Martian orbit to be mis-aligned by 2.5 arcsec in the ephemerides. This is two orders of magnitude too much. You have tilted the orbit of Mars by that large amount in order to fit a small time-span of residuals that are only about 0.4 arcsec. From the type of astrometry that you have done, 0.4 arcsec is not an unreasonable error to expect. Additional observations, especially ones that are in other parts of the orbit, would be most informative.

KUDRYAVTSEV: We agree with you that the obtained value of ~2.5 arcsec for the discrepancy between a stellar reference frame and a dynamical one is too large. Nevertheless, the results which we discussed give a reason for the conclusion that a small rotation of a stellar reference frame about the X-axis of a dynamic one really exists. We consider, as mentioned in our report, that the repetition of photographic observations of Martian satellites in the near future would be very desirable. The task of connecting different celestial coordinate systems is important for space missions.

KISSELEV: How large was the working field of the telescope? I think that there may be a mistake caused by a two-step reduction.

KUDRYAVTSEV: There were two telescopes: one was an astrograph AFR (5°x 5°) and the other was Zeiss-1000 (40' x 40'). As for the second question, we agree with you that a mistake in reduction is possible.

BRONNIKOVA: Why have you used the FK4 system?

KUDRYAVTSEV: The observations were reduced in that system.

THE POSSIBILITY TO LINK RADIO AND OPTICAL REFERENCE FRAMES WITH ARTIFICIAL SATELLITES

TONG FU
Purple Mountain Observatory,
Academia Sinica,
Nanjing, China.

Based on extragalactic radio sources, a new high precision extragalactic radio reference frame can be established from radio interferometric measurements. To link the optical fundamental reference frame presently represented by the FK4/5 to the extragalactic radio frame, the optical counterparts of extragalactic radio sources (quasars, BL Lac objects *etc.*) and radio stars are the most important classes of objects. Besides these two classes of objects, are there any other objects which can be used to link the optical and radio frames? A posible answer is that artificial satellites could be a candidate class of objects contributing to this subject.

Because of the motion of an artificial satellite on the sky, long-term observation will cover a large area of the sky. From radio observation, the orbit of the satellite can be derived relative to the extragalactic radio frame accurately. The differences between observed optical positions in the optical fundamental frame and radio positions calculated from the theory of orbital motion of the satellite derived from radio observation, $\Delta\alpha(t)$ and $\Delta\delta(t)$, consist of local errors of the optical reference system, the effect of the rotation of the optical frame as well as the errors of the theory of orbital motion of the satellite. If the errors of the theory of orbital motion are small enough, neglecting the local errors of the optical reference system, only the rotation of the optical frame is taken into account, $\Delta\alpha(t)$ and $\Delta\delta(t)$ can be expressed as follows:

$$\Delta\alpha(t) = C_\alpha + R_\alpha\, t$$
$$\Delta\delta(t) = C_\delta + R_\delta\, t$$

where R_α and R_δ are the rotation of the optical frame relative to the radio frame, C_α and C_δ are the differences of origins of both frames. With long period and large amount of observation, C_α, C_δ, R_α and R_δ can be determined with high precision, thus the link between both frames can be established.

There are many advantages to link radio and optical frames with satellites. In the case of extragalactic radio sources and radio stars, the structure of the sources and the inconsistency of radio and optical positions are problems to be solved. The faint optical brightness of quasars and the weak radio emission of radio stars make observation difficult. T hese problems do not occur in the case of satellites. They are point objects. Their emission in both the radio and optical domains can be strong enough. The observation is relatively convenient. If the satellite is bright enough—for example, brighter than 10th magnitude—it can be observed with transit type instruments, such as meridian circle and/or astrolabe.

On the other hand, for some satellites with large orbital obliquity, such as GPS, $i > 60°$, the observation of a satellite covers quite a large area of sky. Most of the sky can be covered only with a few satellites, unlike the use of radio sources and radio stars, which need a large number of objects homogeneously distributed on the whole sky.

J. H. Lieske and V. K. Abalakin (eds.), Inertial Coordinate System on the Sky, 194.
© 1990 *IAU. Printed in the Netherlands.*

ANCIENT CHINESE ASTRONOMICAL OBSERVATIONS RELATED TO THE STELLAR BACKGROUND ON THE SKY

WU SHOUXIAN AND LIU CIYUAN
Shaanxi Observatory
Academia Sinica
Lintong, Shaanxi
China, P.R.

1. Stellar reference system in ancient China

Ancient Chinese astronomers had special interest to measure positions of celestial bodies. "Shishi Xingjing" including at least 115 stars was produced in the 4th century BC. By the 11th century, the measure of stellar coordinates was in full swing and 5 detailed measures were carried on in only 100 years. So several ancient stellar catalogs have been retained up to now. As an ancient reference system, many astronomical phenomena have been recorded on it. This is meaningful for modern astronomers to do some modern research.

An equatorial coordinate system was adopted in ancient China. The whole sky is divided into about 300 asterisms. There is a determinative star in each asterism to determine its position. The twenty-eight lunar mansions are 28 asterisms along the equator, which are the basic frame of traditional Chinese coordinate system. Differences of right ascension between the determinative stars of adjoining lunar mansions are called "xiudu" (width of a mansion) which were measured repeatedly in the history. That the winter solstice is in how many degrees of a lunar mansion became the indicator of zero of the system. A statement "an object is in certain degrees of certain mansion (ruxiudu)" means their difference of right ascension. Declination is expressed by the distance from the north pole (qujidu).

Two sets of units were adopted. One set is "degree" (*du*). A whole circle is divided into 365.25 "degrees". The other set is a set of length units (1 *zhang* = 10 *chi* = 100 *cun*).

2. Positions of moving celestial bodies

The position measurement for moving bodies is usually related to the stellar background. "Ruxiudu" and "qujidu" were rarely used in these records.

Many eclipses were recorded, but they usually were very simple. "Ruxiudu" were roughly recorded only in a few cases. Lunar and planetary occultations and "fan" (close approaches) of planets or stars occupy an important proportion in ancient astronomical records. Some of them recorded the Moon or a planet covered a certain star or planet; others reported a star or a planet was "fan" by the Moon or another planet, sometimes even with direction and distance measured with "chi, cun".

J. H. Lieske and V. K. Abalakin (eds.), Inertial Coordinate System on the Sky, 195–196.
© 1990 *IAU. Printed in the Netherlands.*

Comets are changing their position and shape, therefore their records are detailed. Its date, position, length of tail, and shape are often recorded. Positons usually are expressed by "ruxiudu" or some "chi" to a certain star while length of tail by "zhang, chi, cun". Novas were usually called guest stars. Their positions were reported relative to stars. In meteor records "zhang, chi, cun" often were used to express the positions and lengths.

3. Research on some terms

We have seen that the measurement of positions of astronomical phenomena in ancient China was mainly dependent upon the stellar background and expressed by a set of angular units, 1 *zhang* = 10 *chi* = 100 *cun*, and some special terms.

Zhang, chi and *cun* are a set of length units, but they were used as angular units in astronomy for a long time. Their definition has not been found in early books and they are rarely used in the measurement of stars. A book of *Qing* dynasty (17th century) says one *chi* is equal to one degree, but modern studies have got different conclusions in which one *chi* is from $1.^{\circ}24$ to $1.^{\circ}5$. We consider them unreliable because few and unsuited historical records were used.

It is effective to employ planetary records in this research. We have found 156 planetary records with epochs dated from Han to Yuan dynasty (147 BC to 1364 AD). From these records we get
$$y = 0.14 + 0.93x,$$
that is 1 *chi* = $0.^{\circ}93 \pm 0.^{\circ}04$. The constant 0.14 can be explained by an error of computed position and visual error of observers. Variation dependent on dynasties has not been found. The statistics suggest that as an angular unit one *chi* is just equal to one degree.

Of terms in ancient records, "fan" was used the most frequently. Meng kang (3rd century) said: *fan* means within 7 *cun*, rays touching each other. This definition was mentioned repeatedly in later books. But one *chi* was repeatedly recorded as the limit of *fan* in practical historical records. On the other hand, what we need to know is not only the definition of *fan*, but also the practical situation in ancient observations. For this reason, we have made a statistical analysis of two groups of material: 128 *fan* of stars by Saturn in "Songshi" (972–1266 AD) and 767 *fan* of stars by the Moon in "Yuanshi" (1250–1350 AD).

The two bar-graphs (frequency *vs* distance) gotten from the two groups of records are quite different in their shape, but it can be well explained by the moving velocities of the Moon and Saturn. Both graphs show that most events of *fan* fall within one degree. From definitions in ancient books and our research on *chi*, as well as our statistical evaluation of historical records, we can see that the definition of *fan* which appears frequently in ancient Chinese astronomical records should be one degree. Some error exists in practical observations, but for 98% of the records it is within $1.^{\circ}25$.

Historical astronomical records have offered valuable information for modern astronomy and our work is aimed for this purpose.

SIDEREAL YEARS — CATALOGUE USES IN ARCHAEOASTRONOMY
How astrometry can help in solving archaeoastronomical problems

K. BARLAI
Konkoly Observatory
P.O.B. 67
H-1525 Budapest XII, Hungary
and
I. ECSEDY
Orientalist Research Centre
Budapest, Hungary

In ancient societies the Sun was the natural time marker in daytime and base for the calendar. At night, however, the stars, their movements, risings and settings, disappearence and reappearence served quite obviously as time-keepers, and people were not always conscious that they measured time using two slightly but fundamentally different clockworks.

A sidereal year, the period in which the Earth completes a revolution relative to a fixed point in the ecliptic, is longer than a complete revolution from equinox to equinox: the solar tropical year. In astronomical practice of ancient cultures instead of this unobservable point on the ecliptic a visible "fixed" star was observed: a complete revolution of the Earth relative to the same fixed star. On a historic time scale the length of the sidereal year varies and its variation is determined by the proper motion and the precession of the star chosen as time marker.

In one period of the Egyptian history the heliacal rising of the Sirius occurred at the same time as the summer solstice and by coincidence at the same time of the Nile inundation. From this time on Sirius — a star of considerable proper motion and far outside the ecliptic — became extremely important to Egyptians. Its heliacal rising has been watched and the time interval between two heliacal rises measured. The calendar was calibrated and the year began with the heliacal rising of Sirius, although heliacal rising and summer solstice have slowly got out of step.

Let us see some data concerning the length of the tropic solar year in the past. It has slightly shortening character [1].

Gregorian year	length of the tropical year
− 3000	365.24249964 d
− 1500	365.24240755 d
+ 1	365.24231545 d
+1500	365.24222335 d

The tropical solar year should have been 365.24244 days in −2000 B.C., and the sidereal year determined by successive heliacal rises of the Sirius about 365.25059 days at 30 degrees northern latitude (it is the ancient Memphis where observations took place in Egypt), thus by about 11 minutes longer. In the Gregorian reform of the calendar the length of the Julian year was diminished by this

197

J. H. Lieske and V. K. Abalakin (eds.), Inertial Coordinate System on the Sky, 197–198.
© 1990 *IAU. Printed in the Netherlands.*

amount in order to "correct" the "error" of the ancient observers and to obtain the true length of the tropical solar year [2].

It is unbelievable, however, that such a big failure could have been made by skilled ancient observers and that by coincidence their error equals to the difference between the "Sirius year" and the tropical solar year. It is much more likely that Julius Cesar in performing his calendar reform introduced the Sirius year of 365.25 d length. He consulted an Egyptian astronomer, Sosigenes from Alexandria, familiar with Egyptian time reckoning. Julius Cesar, being a politician, did not care of the nature of the year recommended to him. Presumably had no idea of the precession of the equinoxes. My suggestion is that the Julian year has been the sidereal year of Sirius.

If it is so, Christian churches which got out of the authority of Rome, among them e.g. the Russian Orthodox Church, are not using an "incorrectly determined" solar year but a correct Sirius year. The variation of the Sirius year in time should be calculated by means of accurate star catalogue to settle the problem finally.

Going farther from Europe, but in the same belt of geographic latitude, exciting problems of Chinese history face us. From ancient Chinese documents one can guess the existence of a very early peasant calendar based on heliacal rising of the "Fire Star" (very probably Antares) at the spring equinox [3]. These archaic records of Chinese chronicles, however, claim that certain spring festivities called "Pure Brightness" (ch'ing ming) were held about two weeks later in the country (at the beginning of April) than the beginning of spring in the official (solar) calendar. The records are from historical past, from about the middle of the dynasty Chou (11th to 3rd century B.C.). This "being late" of the popular festivities preserves an earlier situation, when heliacal rise of Antares and spring equinox coincided. When did the heliacal rise of Antares mark the spring equinox and when did the chronicles depicting this delay between the "official" (solar) and the traditional peasant (sidereal) calendar originate? Calculating the length of the Antares year and its variation in time this question can be settled.

Among the oldest variants of Chinese characters — from the last centuries of the second millennium B.C — the constellation Scorpion (Ch'en) and Orion (Shen) are represented "pictorially". The central three stars of the Scorpion and three stars of Orion (probably its belt) are related to different Chou time principalities defending the people but being hostile to one another, i.e. a symbol of a fratricidal war. The visibility of these constellations, their regular appearence or disappearence recorded in mythic agricultural tradition may establish the probable extent in geographic latitude of ancient Chinese civilization.

These few examples may outline what kind of problems are to be solved in archaeoastronomy. Going back to far historic (or prehistoric) times and using only precessed coordinates we can't obtain correct results in determining the exact date of a certain heavenly phenomenon. Astrometry delivering exact coordinates and proper motions has a key role in archaeoastronomical calculations. Futhermore, astrometry is in the position to answer the question, how long one may venture to go back into the remote past, when dealing with archaeoastronomical problems while still getting reliable results.

References

1. W.M. O'Neil: 1978,*Time and Calendars*, Sidney University Press.
2. G.V. Coyne S.J., M.A. Hoskin and O. Pedersen (eds.): 1983, "Gregorian Reform of the Calendar", *Pontifica Academia Scientiarum*, Specola Vaticana.
3. I. Ecsedy, K. Barlai, R. Dvorak, R. Schult: 1988, "Antares Year in Ancient China", in: *World Archaeoastronomy*, A.F.Aveni (ed.), Cambridge University Press, p. 183.

IMPROVED METHOD FOR ORBITAL ELEMENTS DIFFERENTIAL CORRECTION.

A. López García[1], J. A. López Ortí[1)2)], R. López Machí[1)2)].

[1] Observatorio Astronómico de la Universidad de Valencia.

[1] Departamento de Matemática Aplicada y Astronomía.

[2] Colegio Universitario de Castellón.
   Universidad de Valencia.

ABSTRACT
As was pointed out by IAU in 1976, an important problem in fundamental astronomy is the improvement of the vernal equinox and equator positions. To this aim, it is necessary to know the accurate values of minor planets orbital elements. The classical methods of orbital elements differential corrections are based on linking the observations at different epochs considering equal derivatives respect to the initial and osculating elements.

In this paper we present an improved method in which the least squares matrix coefficients is calculated from the integration of the Lagrange planetary equations and its derivatives.

1. INTRODUCTION
One of the fundamental problems in positional astronomy is the determination of the equator and the vernal equinox. This problem has been faced traditionally by meridian observations of the Sun and the inner planets. Actually, other alternative methods are used to this aim, as the ocultations of stars by the Moon and the programs of astrometry of asteroids (coordinated by the ITA of Leningrad).

In this way, as a natural continuation of the method developed to the improvement of minor planets orbital elements (López García *et al.*; 1989), we have studied its inclusion in the method of vernal equinox correction (Calaf, J.; Catalá, Mª.A.; 1983).

As it is known the Minor Planets elements change in time due to planetary perturbations and it is no possible, strictly speaking, to apply differential corrections taking these elements as constants, mainly when the observations are far in time.

The main limitation which arises in the mean least-squares classical methods to the differential elements correction is the difficulty of to replace the partial derivatives of the O-C residual, respect to the epoch elements by its derivatives respect to the initial epoch elements, to be able to obtain the correction of these.

In the present work we have completed our orbital elements correction algorithm, including the corrections to the catalog equinox. This algorithm allows to use ephemeris far in time, linking them by means of the integration of the planetary equations of Lagrange and its derivatives. We use the planetary theory VSOP87 (Bretagnon, P.; Francou, G., 1988), and we finally get a linear system to improve the initial elements for the epoch t_0 and the vernal equinox of the reference used catalog.

J. H. Lieske and V. K. Abalakin (eds.), Inertial Coordinate System on the Sky, 199–200.
© 1990 *IAU. Printed in the Netherlands.*

2. FORMATION OF NORMAL EQUATIONS.

The starting point is to consider a set of N asteroids and to have n_r $(r=1,..., N)$ observations of each one. Taking $\vec{\sigma_r^0}$ the orbital elements of a minor planet r for the epoch t_0, (α_i^r, δ_i^r) the observed positions for other t_i^r epoch, $\alpha(\vec{\sigma_r^0}, t_i^r)$, $\delta(\vec{\sigma_r^0}, t_i^r)$ the positions for t_i^r, integrated from the elements $\vec{\sigma_r^0}$, and $\Delta\xi$, $\Delta\eta$ the errors in RA and Dec, respectively, for the vernal equinox in the used catalog, for the epoch t_0 the true elements will be $\vec{\sigma_r^0} + \Delta\vec{\sigma_r^0}$, which will be determinated by the condition of minimum of

$$R(\Delta\vec{\sigma_1^0}, ...,\Delta\vec{\sigma_N^0}, \Delta\xi, \Delta\eta) = \sum_{r=1}^{N} \sum_{i=1}^{n_r} [(\Delta\alpha_i^r)^2 \cdot \cos^2\delta_i^r + (\Delta\delta_i^r)^2] \tag{1}$$

where

$$\Delta\alpha_i^r = \alpha(\vec{\sigma_r^0} + \Delta\vec{\sigma_r^0}, t_i^r) - \alpha_i^r + \Delta\xi, \quad \Delta\delta_i^r = \delta(\vec{\sigma_r^0} + \Delta\vec{\sigma_r^0}, t_i^r) - \delta_i^r + \Delta\eta$$

The condition of minimum in Eq. 1 can be obtained in a linear form for $\Delta\vec{\sigma_r^0}$, $\Delta\xi$, and $\Delta\eta$, by means of the integration of Lagrange's planetary equations and its derivatives. For this purpose we may use the technique shown in (Simon, J.L.; 1987).

With this we finally get a 6N+2 linear system equations.

The solution of this linear system provides the initial elements correction $\Delta\vec{\sigma_r^0}$, so as the corrections $\Delta\xi$ and $\Delta\eta$. The integration of the Lagrange's equations from t_0 to t taking $\vec{\sigma_r^0} + \Delta\vec{\sigma_r^0}$ as initial values gives the corrected elements $\vec{\sigma_r}$ for the desired osculating epoch.

3. CONCLUSIONS

In the present paper we develop an algorithm for the correction of minor planet elements varying with time, that allows to link all the observations available for each minor planet in an only adjust that also includes the vernal equinox corrections. The main advantage of this algorithm is that allows to calculate the derivatives of the residuals in α and δ respect to the elements of the initial epoch, instead of respect to the osculating elements.

4. REFERENCES

Bretagnon, P.; Francou, G. *Astron. Astrophys.*, **202**, 309-315. (1988).

Brower, D.; Clemmence, G. M. METHODS OF CELESTIAL MECHANICS. Academic Press. (1961).

Calaf, J.; Catalá, Mª.A., Actas IV Asamblea Nacional de Astronomía y Astrofísica. Vol. II, 941-954. Santiago, (1983).

López García *et al.* Asteroids, Comets, Meteors III. Com. nº 160. Uppsala, (1989).

Simon, J. L. *Astron. Astrophys.*, **175**, 303-308. (1987).

ON THE LUNAR SECULAR ACCELERATION: A POSSIBLE APPROACH *

V. Protitch–Benishek, M. B. Protitch
Astronomical Observatory
Volgina 7
Yu–11050, Belgrade
Yugoslavia

ABSTRACT. The secular quadratic term in the expression of the Moon's longitude has been introduced empirically after the conclusion that its mean motion is not constant (Halley, 1695).

But, the explanation of this term and also of its numerical evaluation presented and still presents in our time great difficulties. All efforts, namely, to obtain an exact agreement between observed and theoretical value of Moon's secular acceleration were unsuccessful: the first of these two values exceeds always the second one by a very large amount. This discordance and unexplained residuals $(O - C)$ in the mean longitude of the Moon gave rise finally to the statement that these are due to a retardation and irregularity in the Earth's rotation. But, after hardly a fifty years, this hypothesis revealed even more new difficulties and questions concerning also the problem of stability of the Earth–Moon system. It seems that there is a true reason for which this problem occurs as one of the unsolved problems of Celestial Mechanics (Brumberg and Kovalevsky, 1986; Seidelmann, 1986).

However, such an inconvenience can be, nevertheless, avoided if the conditioned periodical functions are applied as theoretical basis to find the solution of the perturbed motion of celestial bodies relatively to their central mass. Without entering here into details, only recalling that in this case the analytical expression for the longitude is defined by:

$$L = L_o + nt + \sum A_i \sin(\alpha_i t + \beta_i) , \qquad (i = 1, 2, \ldots, n)$$

i. e. without any other t–term except the linear one.

The analogous relations are deduced for the elements $\bar{\omega}, \Omega$, longitude of pericenter and node, respectively. The orbital elements a, e, i, semi–major axis, eccentricity and inclination, appear as cosine functions, while t–terms do not exist in an explicit form. Generally, the unknowns L_o, n, A, α, β, are to be found from the observations.

If we assume these solutions as correct and very well adapted to describe also the motion of the Moon, the unexplained residuals in its longitude can be expressed as:

$$\delta L = \delta L_o + \delta n\, t + \sum C_i \sin(\alpha_i t + \beta_i) ,$$

---

* This paper in extenso will be published in the *Bull. Obs. Astron. Belgrade*, **142**, 1990.

201

J. H. Lieske and V. K. Abalakin (eds.), Inertial Coordinate System on the Sky, 201–202.
© 1990 *IAU. Printed in the Netherlands.*

where the sum of periodical terms represents the inequalities not involved in actual theory of lunar motion (Main problem, omitting secular acceleration as a whole).

In view of the above comments, we shall now summarise the main result found on the basis of residuals $(O - C)$ relatively to Brown's tables of the Moon. The reduction of the residuals $(O - C)_{Brown}$ to the quasi–keplerian values δL was performed applying the corrections for a secular quadratic term, the great empirical term and the great Venus term (A), introduced by Hansen to remove the large excesses in Airy's series of Greenwich meridian observations of the Moon during 1750–1830. Then, we find:

$$\delta L = (O - C)_{Br} + 7.14'' T^2 + 10.71'' \sin(140.0°T + 240.7°) + A ,$$

(T counted in Julian centuries from 1900.0), or:

$$\delta L = B + G + A ,$$

if the values of the fluctuation B and corresponding corrective term :

$$G = +4.65'' + 12.96'' T + 12.36'' T^2$$

are introduced instead of forgoing numerical part in expression for δL.

The data we have utilised were: (1) the values B, adopted from Brouwer's paper (1952), for a period 1685–1950, and (2) the same values, but derived on the basis of the corrections $\Delta T = ET - UT$ for the following thirty five years. After removal of the long-periodical inequality, which satisfies entirely all Ptolemaic, Arabian and medieval lunar eclipses, the reminder part of residuals δL was analysed. The mentioned inequality of a period of 64686 lunations or, 5230 years, comprises also the corrections δL_o and δn. Therefore:

$$\Delta(\delta L) = \delta L - \delta L_1 .$$

Determining the parameters of periodical terms, the sixteen smoothed and equally displaced values of residuals are chosen. Kühnen's method and the least square method were applied in two steps: firstly, with a four periods and then, with three additionals, derived from the remain residuals. It is, however, important to notice that the computed periods are very close to the times needed for restoring the identical conjunction (or opposition) of the lunar ascending node and also of the perigee with the Sun at the same point of the ecliptic. During the determination of the amplitudes C these last intervals of time were accepted as true.

Reversing the process, the fluctuations B_C, corresponding to a five–years intervals and also to all of the utilised data were calculated. The computed values are in a good agreement with the observed ones. This fact enables us to expect that here presented approach can contribute not only to solve the problem of discordances between observations and theory, but also to find the true answer to many important questions of Celestial Mechanics.

REFERENCES

Halley,E. : 1695, *Phyl. Trans. Roy. Soc. London*, **218**.
Brumberg,V. A. ,Kovalevsky,J. : 1986, *Celest. Mech.* , **39**, 133–140.
Seidelmann,P. K. : 1986, *Celest. Mech.* , **39**, 141–146.

FLUCTUATIONS OF THE EARTH'S ROTATION BY STELLAR OCCULTATIONS

C. JORDI, G. ROSSELLO
Dept. Física de l'Atmosfera, Astronomia i Astrofísica
Universitat de Barcelona
Avda. Diagonal, 647, E-08028 Barcelona, Spain.

ABSTRACT. Observations of lunar occultations made between 1800 and 1955 are analyzed in order to determine the fluctuations of the Earth's rotation before the International Atomic Time was established. The reduction is made on the basis of the DE200/LE200 ephemeris and the FK5 system. The analysis of the fluctuations shows a main period of 11.7 lunations caused by the error in the stars places.

1. INTRODUCTION

Fluctuations of the Earth's rotation before the International Atomic Time scale was established, were studied by several authors (Brouwer, 1952; Martin, 1969; Morrison, 1979; Morrison & Stephenson, 1981). Main limitations on these works were the accuracies of both the Moon's positions given by the ephemeris and the reference system.

The introduction in 1984 of the FK5 reference system and the new ephemeris DE200/LE200 build at JPL by numeric integration, allowed us to study these fluctuations with more detail. About 50000 observations of occultations of stars by the Moon made between 1800 and 1955.5 have been analyzed with this purpose. The stars positions and proper motions were taken from SAOC.

2. REDUCTION AND ANALYSIS

Reduction method is described in detail in Jordi and Rossello (1987). If the systematic errors in the FK5 reference system and in Watts' datum (Jordi & Rossello, 1987; Rossello et al, 1989), are subtracted from the residuals ($\Delta\sigma$), the new residuals ($\Delta\sigma'$) can be interpreted as due only to the error in the time-scale, i.e. due to the fluctuations of the Earth's rotation, $\Delta\sigma'=(\partial\sigma/\partial T)\Delta T$.

Our first purpose was to analyze the new residuals each lunation. However, the number of observations decreases backwards in time, so, some compromise has to be taken between the length and the number of observations within the period to be analyzed. Regarding the distribution of the observations along the time, we took annual

J. H. Lieske and V. K. Abalakin (eds.), Inertial Coordinate System on the Sky, 203–204.
© 1990 *IAU. Printed in the Netherlands.*

solutions between 1800 and 1830, four lunations solutions between 1830 and 1890 and two lunations solutions between 1890 and 1955.5. We computed the average of ΔT each period rejecting those observations showing a difference with the average ΔT greater than 2.5 the standard deviation. The coefficient $\partial\sigma/\partial T$ is clearly correlated with the angle of the occultation measured from the North Pole of the Moon, and it decreases at the Poles. On the other hand, the Watts charts have poor limb at these positions. So, this coefficient was used as a weight when the means were computed.

A polynomial of second degree was fitted to seven points, and the curve ΔT(t) and its first derivative, that represents the change of the length of the day (l.o.d.), were obtained. The amplitude of our values for the change of l.o.d. is about 3 ms/day while the BIH values for the years after 1955.5, have an amplitude of 1.5 ms/day. This means that some systematic errors are still present in the residuals. When applying a Fourier analysis, a main period of 11.7 lunations appears.

Within a lunation the majority of the observations are made around the first quarter. The Moon moves along its orbit and that means a progressive change in right ascension and declination from a lunation to the next one. As the stars catalogues have systematic errors in $\Delta\alpha \cos\delta$ depending on declination, and an error as such is fully related with ΔT, the application of Fourier analysis to the mean declination of each period, reproduces the period of 11.7 lunations. When performing the reduction taking the positions of the stars from other catalogues (AGK3, CAMC, FK5), the behavior of ΔT(t) and its first derivative changes from one catalogue to another. This is a clear confirmation of the effects of the errors in the stars places when computing the fluctuations of the Earth's rotation.

We know the differences between the time-scales after 1955.5. Then the observations afterwards are only affected by the systematic errors in the stars places. We are working now on these observations in order to determine them and in this way, to improve the values of ΔT before 1955.5.

Acknowledgements

We wish to thank L.V. Morrison for his valuable comments on this work. This work has been supported by the Comisión Interministerial de Ciencia y Tecnología under contract ESP88-0731.

References

Brouwer, D., 1952: Astron. J., 57, 125.
Jordi, C. Rosselló, G., 1987: M.N.R.A.S, 225, 723.
Martin, Ch. ,1969: PhD dissertation. Yale University.
Morrison, L.V., 1979: Geophys. J.R. astr. Soc., 58, 349.
Morrison, L.V., Stephenson, F.R., 1981: Reference Coordinate Systems for Earth Dynamics. Ed. E.M. Gaposchkin & B. Kolaczec, 181.
Rosselló, G., Jordi, C., Salazar, A., 1989: Proceedings of IAU Colloquium (Belgrade, 1987). Celestial Mechanics (in press).

SYSTEMATIC ASTROMETRIC ERRORS IN PULSAR TIMING

L. FAIRHEAD
Bureau des Longitudes, 77 Avenue Denfert-Rochereau
75014 Paris, France
and
Department of Astronomy, University of California
Berkeley, CA 94720, USA

ABSTRACT. A new analysis of the timing data acquired on the fast pulsar PSR1937+214 is presented. Parameters are evaluated with various models based on two ephemerides, two atomic time scales and two TT−TB time transformations. Comparisons are carried out with results from other programs. We provide evidence that systematic errors induced by the model adopted are 5 to 10 times larger than the formal uncertainties calculated by the fitting procedure. Great care must thus be taken when using results from different millisecond pulsars timing programs for accurate astrometric purposes

1. Introduction.

Soon after the discovery of the millisecond pulsar PSR1937+214 (Backer *et al*, 1982), it was realized that beyond its exceptional significance for astrophysics, this new celestial object could also contribute significantly to solar system dynamics studies, to the long-term stability of atomic clocks and to astrometry. The interest grew when additional millisecond pulsars were discovered to form a new class of pulsars.

The use of millisecond pulsars for astrometry will be fully established with the combination of timing and VLBI observations of these objects. The unprecedented astrometric precision of ±0.2 milli-arcseconds for the equatorial coordinates of PSR1937+214, estimated by Rawley *et al* (1988) from timing observations, prompted us to assess possible sources of systematic errors.

We developed a new data analysis package adopting this astrometric approach and flexible enough to accomodate various Earth ephemerides, atomic time scales and TT−TB transformations for the assessment of systematic errors.

2. Observations and Reduction Method.

2.1 OBSERVATIONS

The data we have analysed were taken at the 305-m radio-telescope in Arecibo from November 1982 to October 1984 by a group from University of California at Berkeley and Princeton University (full detals are in Davis *et al*, 1985).

J. H. Lieske and V. K. Abalakin (eds.), Inertial Coordinate System on the Sky, 205–212.
© 1990 *IAU. Printed in the Netherlands.*

We were also provided by Dr. J.Taylor of Princeton with more precise data from Arecibo spanning seventeen months from November 1987 to March 1989. This data set was analysed using our pulsar timing package and compared to the results from the Princeton package.

Pulse arrival times are measured relative to the station's UT. For the analysis, it is necessary to transform this scale to a uniform time scale by first transforming UT to a national or international standard atomic time scale (TA), then to a terrestrial time (TT) and finally to a Barycentric Time scale (TB). TB will also be substituted for time in the ephemeris used to provide the coordinates of the Earth. Several national standards atomic time scales (NBS, NRC, USNO, PTB ...) as well as their combination by the BIPM (TAI, TTBIPM) (Guinot, 1988) are available. Discrepancies exist between these time scales such as an annual variation of $\approx 300-400$ ns between TA(USNO) and TA(PTB) (Guinot, private communication). The transformation from a national standard atomic time scale to the TT scale reduces to the addition of the last term of equation (6) in Guinot (1986) which depends on the Earth orbital velocity and the clock geocentric location. Finally, there are two approaches to convert from TT scale to the TB scale. One is based on a numerical "time ephemeris" as described by Hellings (1986) and the other is an analytical formula (e.g. Fairhead and Bretagnon, 1989). The consequences of these two different approaches will be discussed below.

It is important to realize that the TB scale constructed according to such a scheme will not be unique and should be designated by the name of the atomic time scale initially used in the time chain described in section 2, such as TB(USNO) or TB(BIPM).

2.2 DATA ANALYSIS

The analysis consists in calculating precisely the changes in the propagation time of an electro-magnetic signal from the pulsar to the radio-telescope. These changes depend on the time-dependent position and velocity of the telescope with respect to the centre of the Earth, the position and velocity of the Earth in a reference frame centered on the Solar System Barycentre (SSB), the position and proper motion of the pulsar with respect to the SSB as well as the delay on an electro-magnetic signal traveling in the solar system gravitational field, the interstellar dispersion and the transformation from terrestrial time to coordinate time. The effects of the pulsar strong gravitational field need not be included in this analysis of PSR1937+214 data as it is an isolated pulsar.

The computation of the propagation delay is carried out in the framework of General Relativity theory in isotropic coordinates. The corresponding metric can be found for example in Hellings (1986). The following simple formulation for the transformation from pulse arrival time to pulse emission time is not complete but will suit our illustration needs (the full formulation is in Hellings(1986)):

$$T_N = t_N^{obs} - \frac{R_0}{c} + \vec{k} \cdot \vec{r}/c + \Delta_{rel} + \Delta_{TB} + kDM/\nu^2 \qquad (1)$$

where t_N^{obs} is the arrival time of a pulse measured at the radio-telescope and T_N is its emission time at the pulsar. R_0 is the pulsar distance from the SSB, $\vec{k}$ is a unit vector from the SSB in the direction of the pulsar at time t_N^{obs}. $\vec{r}$ is the position of the radio-telescope with respect to the SSB at time t_N^{obs}. Δ_{rel} is the relativistic delay on the signal. Δ_{TB} is the transformation from Terrestrial Time, TT, to coordinate time, TB. The delay kDM/ν^2 is caused by interstellar medium dispersion, with ν the observation frequency Doppler-shifted to the SSB by taking the velocity of the telescope into account. Once the measured arrival times (t_N^{obs}) have been transformed into emission times (T_N) using the model of equation (1), one has:

$$T_N - T_0 = NP \qquad (2)$$

where T_0 is the emission time of an arbitrary pulse. N is the number of pulses emitted by the pulsar between T_0 and T_N and P is the period of the pulsar at T_N. P is modeled by:

$$P = P_0 + \frac{1}{2}\dot{P}_0(T_N - T_0) \tag{3}$$

where P_0 and $\dot{P}_0$ are the period and its first derivative with respect to time at T_0.

As the a priori values for all the parameters of the model are approximate, equation (2) will not be verified exactly and one has:

$$T_N - T_0 = (N + \delta)P \qquad -0.5 < \delta < 0.5 \tag{4}$$

The prefit residuals δP and the partial derivatives of $T_N - T_0$ with respect to the parameters of the model are then fitted using a linear least-square procedure to determine new parameters for the model, such as the period and period derivatives of the pulsar, its position, or the orbital parameters of the Earth.

2.3 REFERENCE SYSTEMS AND EPHEMERIDES

2.3.1 The Ecliptic Reference Frame Pulsar timing data are essentially sensitive to the Earth orbital motion and to the pulsar position as shown by the leading term $\frac{\vec{k}\cdot\vec{r}}{c}$ in equation (1). The ephemeris providing the Earth orbital motion also defines the ecliptic plane. Consequently, the natural coordinate system for the pulsar position and proper motion is the ecliptic reference frame of the ephemeris adopted. We have chosen this new astrometric approach, over the classical use of the equatorial system, to avoid a source of uncertainty which is the transformation from ecliptic to equatorial coordinates. Positions of the line of the equinox can differ by $0.1''$ and values of the obliquity by $0.03''$ as given by different ephemerides and recommended by the IAU (Standish 1982, Bretagnon private communication).

2.3.2 Ephemerides A high accuracy ephemeris is needed since timing measurement uncertainty for PSR1937+214 is at the $1\mu s$ level corresponding to 300 m on the orbit of the Earth. We have used two ephemerides in this analysis: the JPL ephemeris DE200 (Standish 1982) and an ephemeris developped at the BdL (LeGuyader, 1989). Finally, we compared our results to those obtained using the CfA ephemeris PEP740R.

The actual orbit of the Earth is unique but its representations by these three ephemerides will differ at some level. The superposition of the three representations of the Earth orbit will lead to possible non-alignments, rotations rates and linear drifts between the ephemerides reference frames. For example, there is a limit of 0.06 milli-arcsecond per year in the rotation between the DE200 and PEP740R reference frames (Standish as quoted by Rawley, Taylor and Davis (1988)).

3. Results

The results of several solutions on the 1982-84 data, combining two ephemerides (BDL, DE200), three atomic time scales (USNO, TAI, BIPM) along with the two TT−TB transformation formulæ are presented in tables **I** and **II**. The residuals $-P\delta$ were fitted by a weighted least square procedure for the following parameters: the period P_0 and its first derivative $\dot{P}_0$, the ecliptic coordinates, λ_0 and β_0, the proper motion, μ_{λ_0} and μ_{β_0}, at epoch J2000 and the epoch of origin T_0. Post-fit residuals are similar for each solution and were tested for normality by using the χ^2 test and for goodness of fit (Bevington 1969). In the result tables, the normalized χ^2 and the normalized goodness-of-fit are included and denoted by χ^2 and G, respectively. The root-mean-square (r.m.s.) for the post-fit residuals of each solutions is also indicated.The degree of freedom in our analysis is 188 such that $\chi^2 \approx .3$ corresponds to a high confidence level (larger than 99 %) for the gaussian distribution of the post fit residuals and G $\approx$ 1 indicates a good fit of our model to the data points. Hence, a priori measurements uncertainties used in the fit are realistic.

Parameter	BDL/BIPM	BDL/USNO	BDL/TAI
P_0 (ms)	1.557806448862(5)	1.55780644886354(5)	1.55780644886286(5)
$\dot{P}_0$ (10^{-19}s s^{-1})	1.05126(1)	1.05128(1)	1.05127(1)
λ_0 (J2000.0)	301°58′23″.7842(12)	301°58′23″.7843(12)	301°58′23″.7843(12)
β_0 (J2000.0)	42°17′48″.3145(2)	42°17′48″.3145(2)	42°17′48″.3146(2)
μ_{λ_0} (μas y^{-1})	$-381(52)$	$-372(52)$	$-395(52)$
μ_{β_0} (μas y^{-1})	$-676(121)$	$-672(121)$	$-685(121)$
T_0 (JD)	2445303.27316791	2445303.27316791	2445303.27316791
r.m.s. (μs)	0.970	0.985	0.974
χ^2	0.30	0.42	0.31
G	0.85	0.88	0.86

Table I : Parameters obtained for PSR1937+214 using the BdL ephemeris and 3 different atomic time scales (BIPM= atomic time from the BIPM (Guinot, 1988), USNO= atomic time for the US Naval Observatory (A1), TAI= Atomic time as given by the BIH). The analytical transformation from TT to TB was used in all 3 solutions.

Parameter	DE200/BIPM	DE200/USNO	DE200/USNO/JPL
P_0 (ms)	1.55780644886286(5)	1.55780644886354(5)	1.55780644886288(5)
$\dot{P}_0$ (10^{-19}s s^{-1})	1.05127(1)	1.05129(1)	1.05129(1)
λ_0 (J2000.0)	301°58′23″.6715(12)	301°58′23″.6717(12)	301°58′23″.6716(12)
β_0 (J2000.0)	42°17′48″.3371(2)	42°17′48″.3372(2)	42°17′48″.3372(2)
μ_{λ_0} (μas y^{-1})	$-359(52)$	$-349(52)$	$-375(52)$
μ_{β_0} (μas y^{-1})	$-669(121)$	$-664(121)$	$-683(121)$
T_0 (JD)	2445303.27316791	2445303.27316791	2445303.27316791
r.m.s. (μs)	0.969	0.984	0.974
χ^2	0.29	0.22	0.27
G	0.86	0.88	0.87

Table II : Parameters obtained for PSR1937+214 by using the JPL ephemeris DE200, two different atomic time scales (BIPM= atomic time from the BIPM (Guinot, 1988), USNO= atomic time for the US Naval Observatory (A1)) and the numerical TT to TB transformation (denoted by JPL in the column heading, the other 2 solutions used the analytical formula).

Parameter	Princeton	BdL	Berkeley
P_0 (ms)	1.557806467568084 (12)	1.557806467568084 (13)	1.557806467573990 (70)
$\dot{P}$ $(10^{-19}$s s$^{-1})$	1.051149 (33)	1.051178 (31)	1.050923 (35)
$\alpha(J2000.0)$	$19^h39^m38^s.560220(4)$	$19^h39^m38^s.560217(4)$	$19^h39^m38^s.562237(3)$
$\delta(J2000.0)$	$21°34'59''.14177(13)$	$21°34'59''.14184(12)$	$21°34'59''.25119(10)$
r.m.s. (μs)	0.381	0.377	0.444

Table III : Analysis of the 1987-1989 data set by the various timing packages. All parameters have been estimated for epoch= 2447362.67855354 JD.

Finally, we have analysed the 1987-89 data using the JPL ephemeris DE200 and the TAI time scale fitting for the following parameters: the period P_0 and its first derivative $\dot{P}_0$, the equatorial coordinates, α_0 and δ_0 at epoch J2000 and the epoch of origin T_0. Results from the Princeton timing package, the Berkeley timing package and ours on the same data are given in table III. The position determined with the Berkeley program was transformed to J2000. using the procedure described in Murray (1989). The correlation matrix from our analysis is in table **IV** and shows that none of the parameters are correlated.

	P_0	$\dot{P}$	α	δ
P_0	1.00	0.57	-0.02	0.41
$\dot{P}$	0.57	1.00	-0.11	0.76
α	-0.02	-0.11	1.00	-0.23
δ	0.41	0.76	-0.23	1.00

Table IV : Correlation matrix for the BdL fit on the 1987−89 dataset.

4 Discussion

The different solutions were compared with each other and table V gives the differences between the various parameters.

From the values of the r.m.s., the χ^2 and the goodness of fit given by tables I and II, we consider that all the solutions on the 1982− 84 data are statistically equivalent . The choice of ephemeris, atomic time scale or TT to TB time transformation does not affect the r.m.s. or the goodness of fit for this set of data.

	USNO/BIPM	BDL/DE200	PEP740R/DE200
ΔP_0 (fs)	0.69(14σ)	0.01(< 1σ)	6(50σ)
$\Delta\dot{P}(10^{-19}$ s s$^{-1})$	0.00002(2σ)	0.00001(1σ)	0.0002(6σ)
$\Delta\alpha(^s)$	0.000009(1σ)	0.000018(2σ)	0.002(500σ)
$\Delta\delta('')$	0.00001(< 1σ)	0.001(5σ)	0.11(1000σ)

Table V : Differences between pulsar parameters when using different models.

The only significant difference between Taylor's result and ours on the same data is a 1σ difference in the period derivative $\dot{P}$ whereas the only difference in the models used is the UTC time scale used (UTC(NIST) being used by Taylor and UTC by us). This difference in $\dot{P}$ would be expected to reveal a quadratic term in UTC(NIST)-UTC.Such a quadratic term is indeed present in UTC(NIST)-UTC over the time interval in question and with the right order of magnitude to explain the $\dot{P}$ difference ($1.2\mu s$ amplitude).

The r.m.s. given by Taylor on the 1987-1989 data is equivalent to ours but was calculated with totally independent software except for the ephemeris used. We can thus consider that the rms of $\approx .36\mu s$ truly characterises that data and that the precision may be limited by receive noise, instabilities in the atomic time or the ephemeris used. We have also determined that with these high quality data we are approaching the precision level of double precision real numbers on most computers and that special steps (such as treating julian dates as integer and fractionnal day numbers) have to be taken to ensure that this limit does not hinder us.

4.1 OBSERVED DIFFERENCES IN THE PERIOD

The use of different time scales (BIPM,USNO) in our solutions mainly affects the value of the pulsar period P as shown by table **VI**. The difference $\Delta P \approx 7 \times 10^{-16} s$ is not compatible with the formal error given by the least-square fit ($\sigma \approx 5 \times 10^{-17} s, \Delta P = 14\sigma$). This is explained by a linear drift of $4 \times 10^{-13} ss^{-1}$ between these two time scales, revealed by a direct comparison of these two time scales. This linear drift is explained by the difference in definition between the two time scales. The BIPM time scale is supposed to be exact and to realize the S.I. second at the 10^{-13} s level whereas the USNO timescale is required to be stable.

The difference in the parameters when using a numerical or an analytical $TT-TB$ transformation is insignificant on the $1982-84$ data as shown by the comparison of the solutions designated by DE200/USNO and DE200/USNO/JPL in table **II**. By making a direct comparison between the two procedures (Fairhead and Bretagnon, 1989) we have calculated that the periods determined with these two procedures will only differ by $1.5 \times 10^{-17} s$ which is the present formal error for the period (table **III**).

The differing values for the period obtained by the Berkeley program and ours can be explained by a Doppler effect on the period due to the relative velocity of the Earth orbit between one ephemeris and the other (DE200 and PEP740R). The 6×10^{-15} s difference in the values of the period corresponds to a relative velocity of 35 km y^{-1} in the pulsar direction.

This relative velocity is explained by the different values used in the ephemerides for the masses of the outer planets. Over short periods of time (such as the time interval spanned by the data analyzed), the variation of the Sun orbit around the barycentre of the solar system caused by the different values of the masses will appear as a linear drift in the position of the Earth between the two ephemerides. We have determined this linear velocity using a simple ephemeris of the Sun position with respect to the solar system barycenter and by substituting for the masses of the outer planets, thevalues recommende by the I.A.U. and those adopted in the JPL ephemrides. We thus find a linear drift in the motion of the Sun between the two ephemerides of 30 km y^{-1} in the direction $\alpha = 18^h 28^m, \delta = 0°$ for the two years spanned by the data (1982-1984). This leads to a linear drift in the direction of the pulsar, of 25 km y^{-1}. This is of the order of magnitude that is needed to explain the Doppler shift in the periods determined with respect to the two ephemerides.

4.2 OBSERVED DIFFERENCES IN PSR1937+214 COORDINATES.

The use of the different ephemerides BDL and DE200 (using the same atomic time scale) mainly affects the position of the pulsar. The differences in coordinates (0.000015^s in α and $0.001''$ in δ) are marginally larger than the formal uncertainties (2σ and 4σ, respectively). An annual periodic difference in the position of the Earth will be absorbed in the position of the pulsar by the least-square-fit procedure. This is because the position of the pulsar and the position of the Earth are correlated by the leading term $\frac{\vec{k} \cdot \vec{r}}{c}$ in equation (1). A difference of 700 m in $\vec{r}$ will correspond to the observed difference in the position of the pulsar of 0."001. The direct comparison of the two ephemerides, BDL and DE200, shows such a difference over the two years of data, figure (3). An absolute difference of 700 m corresponds to a relative difference of 10^{-9} in the position of the Earth. The eccentricity of the Earth which creates an annual term in the position of the Earth has a relative uncertainty of 10^{-9} in the BDL ephemeris (Bretagnon, private communication). This could explain the discrepancy in the two ephemerides.

The differences in PSR1937+214 positions between the Berkeley results and ours could be explained by a rigid rotation between the two reference systems in which the coordinates are determined. The reference frame used in the Berkeley package is the one defined by the PEP740R ephemeris which is aligned onto the reference frame of PEP311, a previous ephemeris developed at the CfA for epoch 1982. The reference frames of these two ephemerides drift by ~ 5 milliarcsecond y^{-1} because of the different values for the mean motion of the Earth in the two ephemerides (Chandler, private communication). The reference frame of our program is the one inherent in DE200 which has been aligned onto the J2000 reference frame (Standish, 1982).

Bartel *et al* (1985) and Backer *et al* (1985) have studied the rotations between the reference systems of the VLA and those of the different ephemerides available form the CfA and the JPL by comparing pulsar timing and interferometric positions. Thus, Bartel *et al* provide an epoch dependent rotation that best aligns PEP311 to DE118. We have used this rotation plus the rotation matrix between DE118 and DE200 given by Standish (1982) to transform the PEP740R coordinates for PSR1937+21 to the DE200 reference frame but differences of $\Delta\alpha = 0.0009^s, \Delta\delta = 0.0017''$ still remain. These differences in coordinates could be explained by a yearly periodic difference between the two ephemerides of ~ 10 km or, as seems more likely, by uncertainties in the rotations between the various reference frames and in the epoch-dependency of these rotations.

5. Conclusion

We have shown that the parameters determined from the analysis of timing data of one fast pulsar are very sensitive to the choice of ephemerides and atomic time scale used in the analysis. Important differences in the value of the period (14 σ and 45 σ) can be explained respectively by a linear drift of some 4×10^{-13} s/s between two atomic scales used and by a linear drift of 35 km y^{-1} in the Earth orbit between the two independent ephemerides (DE200 and PEP740R). Differences in the pulsar position could be explained by annual periodic differences of some 700 meters between two ephemerides and by a rotation (0.19" in that region of the sky) between the dynamical reference systems inherent to the two ephemerides. Furthermore, when comparing the Princeton results and ours, we have seen that the quality of the present data is so good that irregularities in the realisations of the UTC time scales used in the analysis are greater than the intrinsic precision of these timing data. One must therefore be very specific about which ephemeris or time scale is used in the analysis when using millisecond pulsar timing data for accurate astrometric applications. One can easily lose all the benefits of these high-precision data because of systematic errors.

The analysis of timing data from many fast pulsars should enable us to decorrelate all these effects and to better constrain atomic time scales on a long time basis (> 1 year). Furthermore, they will be useful to compare the different available ephemerides of the Earth and to determine the rotations between these ephemerides.

Acknowledgments

We are grateful to Drs R.W.Hellings, D.C.Backer for the donation of the earlier timing data of PSR1937+214 used in this analysis, to Dr. J.Taylor for the permission to use his latest data and results and to Drs. P.Bretagnon and J.F.Chandler for helpful discussions.

Bibliography

Backer D.C., Kulkarni S.R., Heiles C., Davis M.M. & Goss W.M., (1982), *Nature*, **300**, p.615.
Backer D.C., Fomalont E.B., Goss W.M., Taylor J.H. & Weisberg J.M., (1985), *Astron. J.*, **90**, p.2275.
Bartel N., Capallo R.J., Ratner M.I.,Rogers A.E.E., Shapiro I.I. & Whitney A.R., (1985), *Astron. J.*, **90**, p.318.
Bevington P.R., (1969), Data Reduction and Error Analysis for the Physical Sciences, McGraw-Hill Book Company.
Bretagnon P., (1982), *Astron. & Astrophys.*, **114**, p278
Davis M.M., Taylor J.H., Weisberg J.M. & Backer D.C., (1985), *Nature*, **315**, p.547.
Fairhead L. & Bretagnon P., (1989) *accepted for publication by Astron. & Astrophys.*
Guinot B., (1986), *Celestial Mech.*, **38**, p.155.
Hellings R.W., (1986), *Astron. J.*, **91**, p.650.
Le Guyader C. (1988) in preparation
Murray C.A., (1989), *Astron. & Astrophys.*, **218**, p.325
Rawley L.A., (1986), *Ph.D. Thesis*, Princeton.
Rawley L.A., Taylor J.H. & Davis M.M., (1988), *Astrophys. J*, **326**, p.947
Standish E.M., (1981), *Astron. & Astrophys.*, **101**, p.L17
Standish E.M., (1982), *Astron. & Astrophys.*, **114**, p.297

Discussion

MURRAY: (a) Were you not able to measure the parallax of PSR1937+21? (b) How far away do you think it is?

FAIRHEAD: The timing data from 1982 to 1984 was not precise enough to determine the parallax of PSR1937+21. Parallax measurment from timing data should be possible using the most recent data. (b) PSR1937+21 is 2 kpc away.

KOPEJKIN: I have two remarks.
(1) First, I think that you can delete in the right-hand-side of the equation for the timing model the term which depends on radial velocity of pulsar and one which depends on the transversal velocity quadratically. The radial velocity term is incorporated into the period of the pulsar and another term is incorporated into the pulsar's period derivative.
(2) As far as I understand, you consider TT scale as proper time at the geocenter, is that correct? (Author: Yes, it is). In my opinion it is better and more correct to consider the TT scale as the coordinate time of a geocentric reference system constructed by the procedure outlined in the previous report by Brumberg, Klioner and Kopejkin. Realization of TT is done on real surface of the Earth at the points of location of atomic clocks. More information about this question is contained in the paper of Brumberg and myself which will be published in the journal *Celestial Mechanics*.

PULSAR ASTROMETRY—AVAILABILITIES AND RELATIVISTIC ASPECTS OF A PULSAR REFERENCE FRAME

O.V. DOROSHENKO, Yu.P. ILYASOV
Lebedev Physical Institute
Leninskii pros. 53
117924 Moscow, USSR

S.M. KOPEJKIN, M.V. SAZHIN
Sternberg State Astronomical Institute
Universitetskii pros. 13
119899 Moscow, USSR

ABSTRACT. We propose a quasi-inertial four-dimensional reference frame in the solar system asymptotically-flat spacetime on the basis of pulsar astrometry techniques. The Pulsar Reference Frame (PRF) consists of the Pulsar Time Scale (PT) and the the Pulsar Reference System (PRS) on the sky .

1. Pulsar Time Scale

It was shown after 10 years timing observations of pulsars in the USSR and the USA that some pulsars have a high rotational stability— especially millisecond pulsars (Il'in et al. 1984; Sazhin 1989; Rawley et al. 1987). They can be more stable than the best atomic clock. The Pulsar Time Scale (PT) based on such pulsars was proposed as an independent new realisation of barycentric time TB (Il'in *et al.* 1984; Il'in *et al.* 1982). The PT is defined as a continuous sequence of time intervals between radiopulses of the pulsar in the barycentre:

$$t_N = t_0 + P_0 N + \frac{1}{2} P_0 \dot{P} N^2 + O(N^3) \tag{1}$$

where t_0 is the starting epoch of PT, P_0 is the pulsar period corrected for its radial velocity, $\dot{P}$ is the period derivative corrected for its radial acceleration and Shklovskii's term (Shklovskii 1969), and where N is the number of pulses.

A reduction from the topocentric arrival time τ_N to the barycentric one t_N should be made by:

$$c\,\tau_N = c(t_N + \Delta t_r) - \mathbf{k}_0 \cdot \mathbf{x}_N - \dot{\mathbf{k}}_0 \cdot \mathbf{x}_N \Delta t_N + \frac{1}{2R_0}(\mathbf{k}_0 \times \mathbf{x}_N)^2 +$$

$$+ \sum_B \frac{2GM_B}{c^2} \ln\left(\frac{r_{NB} + \mathbf{r}_{NB} \cdot \mathbf{k}_0}{2R_0}\right) + 10^{-2}\left(\frac{1}{f^2} - \frac{1}{f_0^2}\right)\frac{DM}{2.41}c \tag{2}$$

where

$$\mathbf{x}_N = \mathbf{x}_E + \mathbf{w}_N + O(c^{-2}) \qquad\qquad \mathbf{w}_N = \mathbf{P}^{-1}\mathbf{N}^{-1}\mathbf{S}^{-1}\mathbf{y}_N \tag{3}$$

J. H. Lieske and V. K. Abalakin (eds.), Inertial Coordinate System on the Sky, 213–216.
© 1990 *IAU. Printed in the Netherlands.*

where Δt_r is the relativistic time correction from terrestrial time TT (TAI) to barycentric time TB; k_o is the unit vector from the barycentre to the pulsar; $\dot{k}_o$ is the proper motion vector; R_o is the barycentre-pulsar distance; x_E is the barycentric vector to the Earth's centre-of-mass; y_N is the geographical vector to an observer; P, N, and S are matrices of precession, nutation and diurnal rotation of the Earth; DM is the dispersion measure; and

$$r_{NA} = \left[r^i_{NA} \cdot r^i_{NA}\right]^{\frac{1}{2}}$$

$$r^i_{NA} = x_N{}^i(t_N) - x_A{}^i(t_A)$$

w h e r e $x^i_A(t_A)$ represents the solar (or planet) coordinates at t_A, the moment of closest approach of radio pulse to the Sun (or planet) and f and f_o are the local-received and barycentric-received frequencies respectively.

2. Pulsar Reference System—PRS

The practical realisation of the Pulsar Reference System (PRS) is achieved by VLBI measurement of the pulsar coordinates (Fomalont et al. 1984). A reduction of the observations should be made with an equation which has a precision on the order of some picoseconds (Kopejkin 1990):

$$
\begin{aligned}
\tau_2 - \tau_1 = &\frac{1}{c}k_o\cdot b + \frac{1}{c}\left(\dot{k}_o\cdot b\right)\Delta t + \Delta \tau_r + \frac{1}{c^2}\left[v_E\cdot b - (k_o\cdot b)(v_E\cdot k_o) - (k_o\cdot b)(\dot{w}_2\cdot k_o)\right] \\
&- \frac{1}{c^3}(k_o\cdot b)\left[\frac{1}{2}v_E^2 + v_E\cdot\dot{w}_2 + \frac{1}{2}\dot{w}_2^2 + 2\bar{U}(x_E) + U_E(w_2) - (v_E\cdot k_o)^2 - 2(v_E\cdot k_o)(\dot{w}_2\cdot k_o) - (\dot{w}_2\cdot k_o)^2\right] \\
&- \frac{1}{c^3}\left[\frac{1}{2}(v_E\cdot b)(v_E\cdot k_o) + (v_E\cdot b)(\dot{w}_2\cdot k_o)\right] \\
&+ \frac{2GM_E}{c^3}\ln\frac{r_1 + (k_o\cdot w_1)}{r_2 + (k_o\cdot w_2)} + \sum_{A\neq E}\frac{2GM_A}{c^3}\frac{k_o\cdot b + N_{EA}\cdot b}{R_{EA}(1 + k_o\cdot N_{EA})}
\end{aligned}
\tag{4}
$$

where b is the baseline vector in the Geocentric nonrotating reference system (GRS), w_1 and w_2 are the first and the second antennae coordinates in the GRS, τ_1 and τ_2 are topocentric times of arrival of the same radio pulse to the first and the second antennae, $\Delta\tau_r$ is a relativistic correction for different time rates between observer clocks caused by distinct geographical positions of the radiotelescopes, v_E is the barycentric velocity of the Earth, w_2 is the velocity of the second radio telescope, U_E is the geopotential, and where $\bar{U}$ is the solar system gravitational potential excluding the Earth, $N^i_{EA} = r^i_{EA}/r_{EA}$.

3. Reference Pulsars

The parameters of 17 reference pulsars are compiled in a catalogue from Lyne et al. (1989). The pulsars of small $\dot{P}$ are chosen so as to obtain objects with a weak "activity" A (Cordes and Downes 1985). An error of measurement of pulse arrival times σ_{TN} depends on the signal/noise ratio. It was calculated under the conditions that the radiotelescope had 1000 m² effective area, 10 MHz bandwidth at 400 MHz, 100 K system noise temperature for 1 hour observations with a time-constant 0.1 of equivalent width of pulse. The RMS of timing errors in pulsar noise σ_N which depend on pulsar rotation fluctuations are taken from Lyne *et al.* (1989). The distribution of reference pulsars over the sky is good enough.

PSR	α(2000) h m s	δ(2000) ° ' "	P s.	Ṗ 10^{-15}s/s	σ_{TN} mks	σ_N mks	A	μ_α mas/yr	μ_δ mas/yr
0031–07	00 34 08.890	–07 22 01.	0.9429	0.4083	436.0	960.	–1.68	–	–
0355+54	03 58 53.62	+54 13 44.5	0.1564	4.3912	25.9	1111.	–1.38	5.3	6.2
0834+06	08 37 05.524	+06 10 13.70	1.2738	6.79918	37.2	500.	(–0.01)	1.9	51.1
0950+08	09 53 09.206	+07 55 35.81	0.2531	0.22915	3.8	770.	–1.85	14.9	30.6
1133+16	11 33 03.200	+15 51 00.85	1.1879	3.73273	8.2	450.	–2.07	–101.5	356.7
1237+25	12 39 40.450	+24 53 49.13	1.3824	0.95954	25.8	230.	–2.33	–106.0	42.1
1449–64	14 53 32.77	–64 13 15.0	0.1795	2.74754	4.5	–	(–0.65)	–	–
1556–44	15 59 41.497	–44 38 45.9	0.2570	1.01955	12.9	–	(–1.35)	–	–
1604+00	16 07 12.045	+00 32 40.13	0.4218	0.30607	51.9	290.	–2.07	–0.7	–.72
1821–24	18 24 31.948	–24 52 10.56	0.0031	0.00155	5.3	–	(–5.96)	–	–
1913+16	19 15 27.922	+16 06 27.41	0.0590	0.00864	89.7	–	(–4.74)	–	–
1919+21	19 21 44.710	+21 53 01.82	1.3373	1.34809	17.5	500.	(–1.16)	–	–
1933+16	19 35 47.739	+16 16 40.58	0.3587	6.00354	4.1	1730.	–1.49	–	–
1937+21	19 39 38.471	+21 34 59.25	0.0016	0.00001	0.2	–	(–6.77)	–0.3	–0.5
1957+20	19 57 22	+20 49	0.0016	<0.00003	1.0	–	–	–	–
2016+28	20 18 03.750	+28 39 54.25	0.5580	0.14936	18.1	1.5	–1.60	–	–
2217+47	22 19 48.068	+47 54 53.96	0.5385	2.76421	20.9	1.6	–1.46	–	–

4. Practical Applications

The construction of the PRF produces a possibility to define more precisely (up to 1 mas) the mutual orientation of the ecliptic and the equatorial reference frames, to measure the accurate position of the solar system barycentre, and to obtain corrections to the ephemerides of the Earth and planets.

One additional consequence of the PRF can be its application for investigation of the stochastic gravitational wave background (Sazhin 1978). The amplitude of gravitational waves which can be generated by some cosmological sources for instance, the domain stage of the early universe or cosmic strings, can reach 10^{-14} in the frequency band 10^{-7} to 10^{-8} Hz.

Thus, pulsar astrometry comparing timing and VLBI parameters of pulsars can essentially improve the inertial reference frame on the sky and can extend our knowledge in fundamental astrometry and astrophysics.

References

Cordes, J.M., Downes, G.S.: 1985, *Astron. J. Suppl. Ser.*, **59**, 343–382.
Fomalont, E.R., Goss, W.M., Lyne, A.G. *et al.*: 1984, *Monthly Notices Roy. Astron. Soc.*, **210**, 113–130.
Il'in, V.G., Ilyasov, Yu.P., Kuzmin, A.D., *et al.*: 1982, *Trans.IAU*. **XVIII**, 241.

Il'in, V.G., Ilyasov, Yu.P., Kuzmin, A.D., *et al.*: 1984, *Docladi Academy*, **275**, No. 4, 835–838 (in Russian).
Kopejkin, S.M.: 1990, *Astron. Zh.*, No. 1, in press (in Russian).
Lyne, A.G., Manchester, R.N., Taylor, J.H.: 1989, Pulsars catalogue. 1–42 (private communication).
Rawley, L.A., Taylor J.H., Davis M.M., *et al.*: 1987, *Science*, **238**, 761–765.
Sazhin, M.V.: 1978, *Astron. Circ.*, No. 1002 (in Russian).
Sazhin, M.V.: 1989, *Measur.Tech.*, No. 1, 27–29 (in Russian).
Shklovskii, I.S.: 1969, *Soviet Astr.*, **13**, 562.(in Russian).

Discussion

BASTIAN: Are the presently known millisecond pulsars bright enough to be observed by VLBI with the full precision nowadays attainable with this technique?
KOPEJKIN: Yes, some are. The pulsar PSR 1937+21 was observed by VLBI.

POINTS: A GLOBAL REFERENCE FRAME OPPORTUNITY

J.F. CHANDLER AND R.D. REASENBERG
Smithsonian Astrophysical Observatory
60 Garden Street
Cambridge, Massachusetts 02138
USA

ABSTRACT. POINTS is a space-based optical astrometric interferometer, capable of measuring the angular separation of two stars about 90° apart with 5-microarcsec (μas) nominal accuracy. During the intended ten-year mission, a repeated survey of a few hundred targets over the whole sky, including a few bright quasars, will establish a "rigid" reference grid with 0.5 μas position uncertainties. At that level, the grid will be free of regional biases and tied to the extra-Galactic frame that is our present best candidate for an inertial frame. POINTS will also determine parallaxes and annual proper motions at about the same level. Further, the planetary ephemeris frame will be tied through stellar aberration to the grid at about 300 μas. Additional targets of interest, to a limiting magnitude of greater than 20, will be observed relative to the grid, yielding determinations with uncertainties depending on the observing schedule. Measurement at the microarcsec/year level of the apparent relative velocities of quasars that are widely separated on the sky will severely test the assumption of cosmological quasar distances and may also constrain models of the early Universe.

1. Introduction

We consider a large-bandwidth optical astrometric interferometer to be operated in space. This interferometer (dubbed POINTS for Precision Optical INTerferometer in Space) would measure the angular separation of stars separated by about 90° on the sky with a nominal measurement accuracy of 5 microarcseconds (μas). The key to obtaining such accuracy is the use of stable materials in a thermally controlled environment, real-time metrology of critical instrument dimensions, and closure information from the astrometric data. We estimate the instrument would measure daily about 60 pairs of stars; a random set of such measurements, if suitably redundant, contains the closure information necessary to detect and correct time-dependent measurement biases to well below the nominal measurement accuracy. The 90° target separation yields direct observation of absolute parallax; we do not encounter the problem of finding zero-parallax reference stars which would be severe at the microarcsecond level.

The nominal limiting magnitude, m = 17, assumes moderate control of unmodeled changes in the instrument's angular acceleration. However, techniques exist for extending the range fainter by several magnitudes. The wide magnitude range permits an observing schedule that includes both bright stars and quasars. We have found that roughly one third of the observing time in a ten-year mission would allow the determination of relative positions, parallaxes, and annual proper motions below the 1-μas level for a set of several hundred reference objects. These well-observed stars and

J. H. Lieske and V. K. Abalakin (eds.), Inertial Coordinate System on the Sky, 217–228.
© 1990 *IAU. Printed in the Netherlands.*

quasars would form a "rigid frame" and would serve as references for all additional target stars, as well as being targets of primary scientific interest. Because of the quasars in the observation set, all POINTS observations would be in a frame which is the best available realization of an inertial system. The remaining two-thirds of the observing time would be available for a wide variety of scientific applications.

The European astrometric instrument HIPPARCOS has ushered in the age of space astrometry. Initial progress reports for this survey instrument, including discussions of possible follow-on missions, appear elsewhere in this volume. POINTS is the natural interferometric descendant of HIPPARCOS: both combine stars from well separated fields in a single measurement; both use 360° closure for self-calibration; both yield a global star map that gains rigidity from a truss-like measurement schedule.

It is now widely recognized that interferometric instruments will play a major role in many aspects of space-based optical astronomy. (See, for example: Proceedings of the Workshop on High Angular Resolution Optical Interferometry from Space, Baltimore, June 1984, *BAAS* **16**, 1984; Proceedings of the Colloquium on Kilometric Optical Arrays in Space, Cargèse, October 1984, ESA SP-226, 1985; Report of the Workshop on Imaging Interferometry in Space, Cambridge, October 1985, distributed by Battelle, 1987.) Results of major importance will come from imaging interferometers with higher resolution and more light-gathering power than the Hubble Space Telescope (HST). However, such instruments must be large to achieve their advantage over existing ones. POINTS, which is small, could open new areas of astrophysical research and change the nature of the questions being asked in some old areas. It could be the first of a new class of powerful instruments in space and could prove the technology for the larger members of that class to follow.

In Section 2, we discuss briefly the design of POINTS. In Section 3, we discuss the operation of a POINTS mission. In Section 4, we consider the data analysis task for POINTS, and in Section 5, we discuss some of the expected results from a POINTS mission, with emphasis on the reference-frame aspects.

2. Instrument design

The POINTS instrument consists of two starlight interferometers mounted at a nearly right angle and a metrology system. The instrument determines the angular separation between two widely separated stars by measuring the (adjustable) angle between the interferometers and, independently, the offsets of the target stars from their respective interferometer axes. With proper selection of target stars and only a small adjustment capability in the interferometer angle, both members of a target pair of stars can be simultaneously within one second of arc of their respective interferometer axes. Consequently, off-axis distortions and the attendant biases are essentially eliminated. Central to the instrument architecture is the real-time monitoring of the angle between the interferometers and the metrology along the starlight optical path, each of which uses a laser interferometer scheme based on currently available technology.

Figure 1 shows an artist's rendition of the preliminary design for POINTS mounted on the Multimission Modular Spacecraft; Figure 2 shows a rendition of POINTS mounted as a Space Station Attached Payload. Each interferometer has a 2-m baseline separating two afocal telescopes,

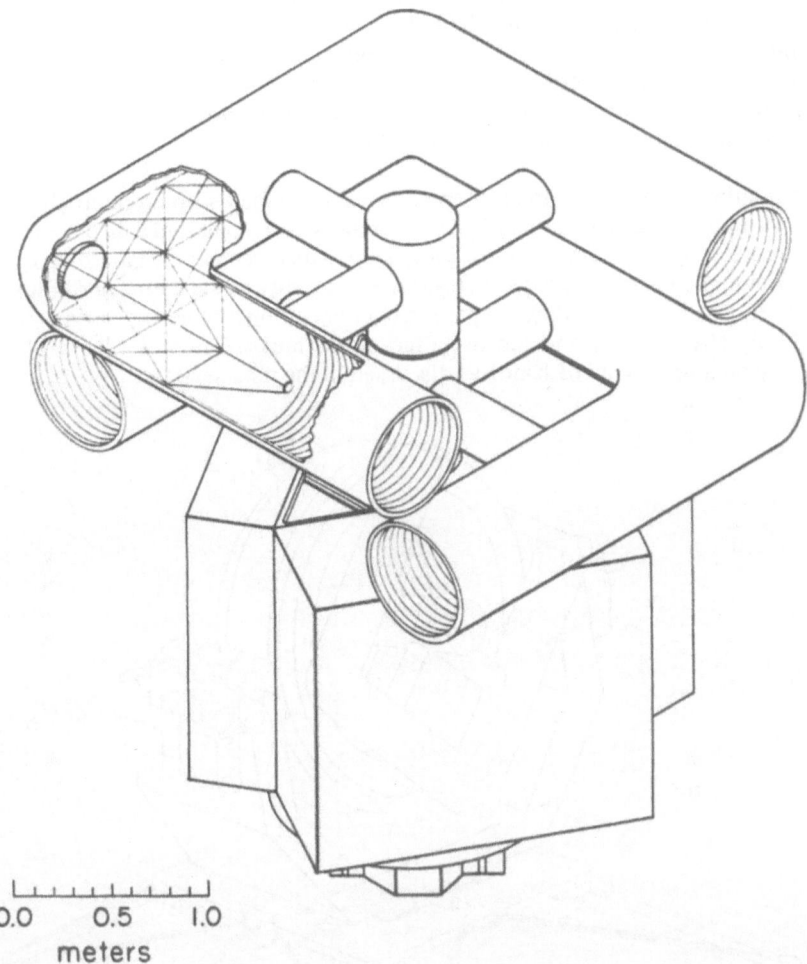

Figure 1. An artist's rendition of POINTS with 2-m separations between pairs of telescopes 25 cm in diameter. The instrument, shown mounted on the Multimission Modular Spacecraft, comprises two U-shaped interferometers joined by a bearing that permits the angle between the principal axes of the interferometer to vary by up to a few degrees from its nominal value of 90°.

each with a primary mirror 25 cm in diameter. The axes of the interferometers are separated by a roughly 90° angle adjustable within an articulation half-range presently estimated to be 3°. In each of the two interferometers, the afocal telescopes collect samples of the starlight and direct them toward fringe-forming and detecting assembly (not shown in either figure). After the compressed samples are mixed by a beamsplitter, the light is dispersed and focused to form a channeled spectrum, from which the astrometric information is extracted.

The determination of the interferometer angle is accomplished with an internal laser metrology system. For a 2-m baseline, the nominal 5-μas uncertainty corresponds to a displacement of one end of the interferometer by 50 picometer (pm). Since similar displacements of internal optical elements are also important, the instrument requires real-time metrology of the entire starlight optical path at the few pm level. This metrology does not pose an overwhelming problem because the accuracy need be achieved only after 100 sec and because a bias in the measurement is acceptable as long as it changes slowly on a time scale of hours, as discussed in the next section.

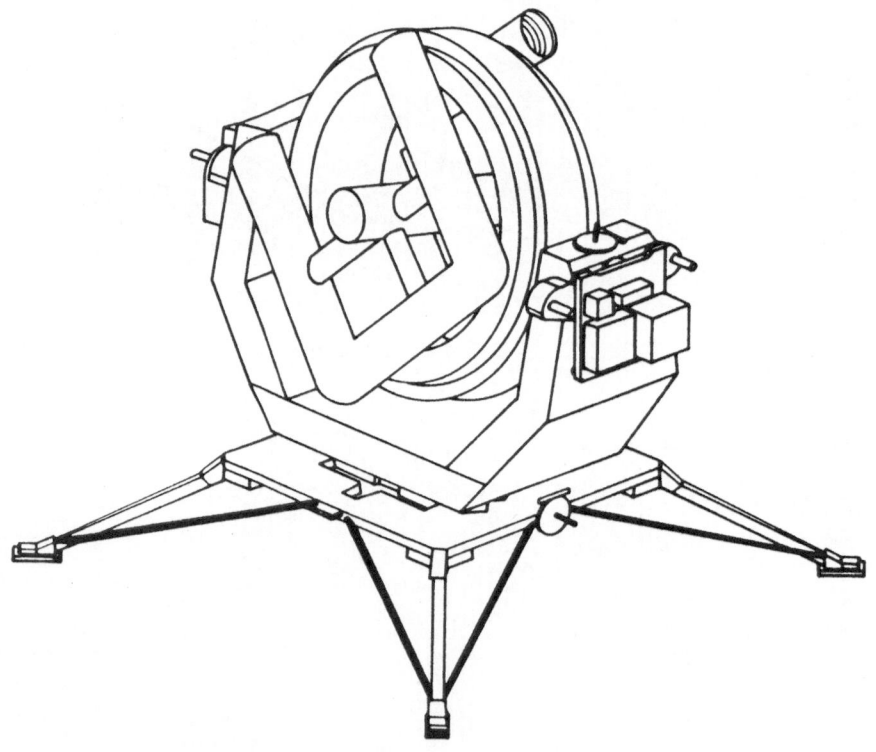

Figure 2. An artist's rendition of POINTS with the same specifications as in Figure 1, but mounted as an Attached Payload on Space Station. The attachment uses the Payload Pointing System (PPS) for coarse pointing. Fine pointing and vibration isolation are both provided by a magnetic suspension, which is shown as a ring mounted inside the PPS and holding POINTS.

3. Instrument operation

Our understanding of the characteristics of POINTS' operation is based on a series of covariance studies and other analyses. We have investigated several patterns for the distribution of target stars on the sky, including uniform Monte Carlo distributions, and have found no fundamental advantages for any particular pattern. We have, therefore, conducted most of the covariance studies with sets of uniformly distributed Monte Carlo stars.

Since POINTS would be a global instrument, it is natural to consider a global reference frame as the basis for interrelating POINTS observations. See Kovalevsky (1984) for a discussion of astrometric instruments in space and definitions of types of astrometry. This global approach, which is in sharp contrast to the use of local frames defined for classical small-field astrometric instruments, provides significant advantages. All observed objects, not just the stars of our primary reference grid, are potentially available to contribute to the stability of the reference frame used for each observation. Each object that contributes to the reference frame stability can be studied astrometrically and its motions modelled.

When an observation set has sufficient redundancy, it can be analyzed to yield a rigid frame; it serves to determine the angular separation of all pairs of observed stars, even those that were not observed simultaneously. The degree of redundancy is well characterized by M, the ratio of the number of observations to the number of stars observed. In a series of Monte Carlo covariance studies, we found that the minimum redundancy necessary for a rigid frame is about $M = 3.5$. With moderate redundancy, $M = 4.2$, the uncertainty in the separation of any star pair is about equal (on average) to the instrument measurement uncertainty. A solidly redundant observing schedule for 300 stars and 5 quasars (1500 star-star observations and 250 star-quasar observations to provide extra redundancy to compensate the quasar's higher magnitude), in which a roughly 30-day observation sequence would be repeated four times per year for ten years, would yield mean stellar parameter uncertainties of 0.6 µas in position, 0.4 µas/y in proper motion, and 0.4 µas in parallax (Reasenberg, 1986). Note that the parallax determination to 0.4 µas is better by nearly a factor of two than one would naively calculate from the coordinate uncertainties in a single series. The enhancement reflects POINTS' direct parallax observation. The star grid is free of regional biases by virtue of the global astrometry that determines it.

In other Monte Carlo covariance studies, we investigated the ten-year observing sequence mentioned above, but with fewer observations. From this study, it appears that if at least one of the observing series is complete (i.e., $M \geq 3.5$), then observations can be deleted from the other series by a variety of random or systematic procedures yielding an increase in the mean parameter-estimate uncertainty which depends only on the square root of the total number of observations. Further, additional stars can be added to the observation sequence with a minimal number (perhaps 20) of observations per star. Thus, if the instrument were run full-time on a ten-year star survey, the survey could comfortably include 7000 stars and would result in the knowledge of their positions, annual proper motions, and parallaxes at the 2 µas level. Independent of the target selection and scheduling, aberration would insure that the stellar grid would be connected to the Earth ephemeris frame with a 300-µas uncertainty.

Our studies have also showed that the 90° nominal interferometer angle is not crucial to the design of the instrument, but it minimizes the required articulation range. For POINTS to have plenty of reference stars available for each target, the product of the articulation range and the sine of the nominal angle should be as large as possible, since that product is proportional to the solid angle from which reference stars may be selected.

The metrology system mentioned in the previous section contains finite-size optical components, each of which will introduce a bias into the measurement of the interferometer angle. This bias will, of course, be time dependent at the microarcsec level. It is essential that we be able to determine and correct for the instrument bias, preferably without the introduction of additional hardware. The required determination and correction both naturally occur when the observations are combined in a least-squares estimate of the individual stellar coordinates (including proper motion and parallax), instrument model parameters, and the expected biases. In particular, our covariance studies have shown that even without the introduction of a special observing sequence, it is possible to estimate simultaneously the stellar coordinates and several instrument bias parameters per day without significantly degrading the stellar coordinate estimates. Thus, metrology biases and related errors can be allowed to change on a time scale of hours without significantly degrading the performance of the instrument.

This kind of calibration, which uses the astrometric data, extends even to the determination of the lengths of the interferometer baselines, which serve as the scale factors in relating fringe phases to angular offsets of the target stars from their respective axes. Observing the same pair of stars with successively different articulation angles allows the scale factors to be determined along with the total separation angle between the stars. The observation time scale of a few minutes largely decouples the baseline calibration from the rest of the data analysis, which must remove the effects of long-term drifts in the instrument.

Our performance analyses assume the target to be an isolated, unresolved point source, *i.e.*, one with no significant structure. In practice, this condition will not always be met. Star spots, flares, and the like will set a limit to the astrometric usefulness of the measurements of some targets. Of course, we may learn useful things about some stars from the motions of their centers of light. At 100 pc, the Sun has a magnitude of 10. A shift of the center of light by 1% of its diameter at this distance would cause an apparent astrometric shift of nearly 1 μas. Many of the targets are expected to be less than 20 pc away.

4. Data Analysis

Although the mission will support a large and diverse set of science objectives (Reasenberg *et al.*, 1988a), we picture the first stage of the data analysis as being done centrally because of the synergistic nature of the data: one investigator's target is everybody's reference star. The amount of effort required would depend on the degree to which we need to use spacecraft engineering data to elucidate systematic errors. In any case, it would not be a complex effort nor would it require special computing facilities. If the data were available today, the analysis could be done, for example, on a "super minicomputer" as a time-shared job or even on one of the high-power personal work stations now available. The software needed would be similar to that already in use for POINTS sensitivity studies, but with the more detailed instrument models that will be possible after we have a working instrument. The required algorithms are those of standard weighted least-squares parameter estimation and error analysis. Those algorithms and the equations of condition relating the measurements to the adjustable parameters are, of course, embodied in the existing software.

Early in the mission, the analysis would be limited to positions determined during observing periods of a few days to a month. Of course, the best *a priori* proper motions and parallaxes would be included in the analysis model, although these would have little effect. At about one half year, it would first become possible to estimate the full set of five parameters (two position, two proper motion, and one parallax) for each star. The stability of the solution should increase considerably during the following year. Because of this progressive refinement of the parameter set, the POINTS reference frame would, in effect, be available for experiments from the very beginning of the mission, despite the fact that the "final" reduction of data already taken would be delayed while the reference grid stabilized. Among the classes of early science from the mission would be a determination of the RR Lyrae and Cepheid distance scales, a study of the spiral arms of the Galaxy, a determination of the mass and mass distribution of the Galaxy from observations of the Magellanic clouds and of clusters, respectively, and a determination of stellar masses by parallax measurements of visual binaries.

After about two years of observation, the postfit residuals from the data reduction would be used to investigate irregularities in the motions of individual stars. These apparent motions would be analyzed both for their intrinsic scientific interest and to improve the stability of the resulting reference frame. At first, the position of each star would be determined with respect to the mean of the positions of all the stars. After the initial modeling of the irregularities in stellar proper motions, the position of each star would be determined with respect to the nominal modeled reference positions of all the stars. This iterative procedure is expected to converge quickly as long as the number of stars in the reference grid is large enough (*i.e.*, greater than 100).

One of the possible irregularities in stellar positions is the astrometric "wobble" due to planets orbiting the target stars. A series of numerical experiments has shown that suitably large planets could be detected through their astrometric signatures as seen by POINTS. These experiments involved a set of 100 nearby target stars, and in each experiment a subset of stars were chosen (blindly) at random to have a single planet each. The planet masses and orbital elements were all assigned with Monte Carlo distributions. Using the iterative procedure just described, planet-like signatures were sought in the Monte Carlo measurement noise over a ten-year simulated mission, and planet signatures with amplitudes above the noise level were invariably detected; this threshold is equivalent to 1% of the Sun's motion due to Jupiter as seen at a distance of 10 pc.

5. Applications

In this section, we discuss some of the scientific results that could be obtained from POINTS. They depend on the high astrometric accuracy and, in some cases, on the wide target separation that POINTS would provide. A brief list includes (a) a distance scale based on direct parallax determinations for a large number of Cepheids; (b) a determination of the masses of stars in binary systems and those close enough to apply the method of perspective acceleration; (c) parallax measurements yielding both absolute stellar magnitudes and, in conjunction with mass estimates and other data, a sharpened mass-color-luminosity relation; (d) a vastly improved global reference frame and a tie to existing ones; (e) the study of the mass distribution in the Galaxy; (f) a strictly geometric (*i.e.*, coordinate and parallax) determination of the membership of star clusters, especially useful in the case of "peculiar" stars; (g) a bound on or a measurement of quasar proper motions; (h) a light-deflection test of general relativity (Reasenberg *et al.*, 1988b); and (i) a search for other planetary systems.

No attempt has yet been made to develop a scientifically balanced allocation of the instrument observing time. The time required to perform all of the indicated studies far exceeds the time that would be available with a single ten-year mission. Further, additional uses are likely to be suggested when it becomes widely known that POINTS-type data will be available.

In any case, it is clear that, in one way or another, all the results hinge on the ability of a POINTS Mission to realize its goal of establishing a global reference frame. Defining a reference frame based on POINTS measurements requires that we select coordinate axes. Traditionally, these axes correspond to the mean equator and equinox of the Earth at some epoch, which seems a reasonable basis for a frame derived from observations from Earth's surface. However, POINTS represents a non-traditional approach to astrometry, freed from Earth's spin vector; the choice of axes can be made arbitrarily. For example, the coordinates and proper motions of all the reference stars could be tied by weak constraints to the values from a traditional catalog so that the ensemble of reference stars provides the link to classical optical catalogs. Similarly, the coordinates of the reference quasars could be tied by such constraints to the values from a radio-source catalog. Alternatively, three star coordinates, such as the right ascension and declination of one star and the position angle of a second star relative to the first, could be given defined values.

The new FK5 catalog, which is now the best of its type available, is characterized by statistical position errors of order 10 milliarcsec (mas) at the mean epoch of about 1940; errors in the proper motions are about 1 mas/y (Fricke *et al.*, 1988). Thus, the contemporary accuracy is ~50 mas. In addition to having these statistical errors, such catalogs are known to have systematic zonal effects resulting from instrumental flexures, the atmosphere, and the use of a variety of different instruments covering separate parts of the sky. These kinds of errors are discussed, for example, by Podobed (1965), and more theoretically by Eichhorn (1974). The most promising near-term effort to address the question of reference frames based upon quasars is the envisioned cooperative program involving HIPPARCOS, the Very Large Array (VLA), and the HST. HIPPARCOS is expected to produce a rigid framework of star positions requiring adjustments both to orient the system and to remove residual rotation rate. These adjustments are to be provided by VLA observations of radio stars observed optically by HIPPARCOS, and also by measurements of the relative separations of quasars and HIPPARCOS stars by HST.

Unfortunately, the expected HIPPARCOS proper-motion errors are as large as 3 mas/y because of the short time base of the observations and will be greater if the mission is cut short. These errors will degrade the HIPPARCOS catalog to the current level of the FK5 in an astrometrically brief period of under two decades. More important, this degradation will continue, making the HIPPARCOS catalog inferior to the FK5. The solution to this problem could be a second HIPPARCOS mission, as advocated by some members of the HIPPARCOS Team. POINTS would provide an independent check on the systematic errors of the HIPPARCOS catalog, including an immediate extragalactic link. The direct referral of star positions to an extragalactic reference by POINTS would furnish an extremely accurate grid of optical sources over a wide magnitude range, thus providing a practical approach to extending this reference to other objects. The use of a small POINTS catalog to enhance the much larger HIPPARCOS catalog needs to be investigated.

The radio and optical reference frames may be related by objects which emit detectable radiation in both spectral regions. There are discrepancies at the 0.3-arcsec level in the northern hemisphere between the FK4 optical reference frame and the radio reference frame (Johnston et al., 1985). These discrepancies are believed to be due to systematic errors in the optical frame. In the southern hemisphere, the optical frame is significantly less accurate than in the northern hemisphere. POINTS would establish an optical reference frame free of such systematic errors and accurate to a few μas over the entire celestial sphere, as well as tied to the radio frame by redundant observations of bright extragalactic radio sources.

There is a limited number of known compact optical extragalactic sources showing radio emission. See Table 1 for a list of 36 quasars and BL Lac objects of magnitude 15.5 or brighter, extracted primarily from Veron-Cetty and Veron (1987) and first presented in this form by Reasenberg et al. (1988a). Instruments such as the VLA and the HST will extend this number to more than 100. The radio reference frame should be accurate to approximately 0.1 mas. POINTS' optical positions of radio objects would relate the optical and radio reference frames to the accuracy of the radio positional measurements or to the extent that the radio and optical emission are coincident. The POINTS reference frame could be used to help in finding and removing systematic errors in the radio reference frame so that the reference frames may be related on a submilliarcsecond scale. This data set could also be used to examine the coincidence of the radio and optical emission at the milliarcsecond level in the reference objects.

In addition, POINTS could tie its new reference frame to the existing one of the planetary ephemerides. At present, the latter rests primarily on spacecraft Doppler and range tracking and Lunar laser ranging data, which are all measures of distances or line-of-sight velocities, but which, by the application of planetary dynamics, allow the construction of a geometric framework with angular consistency comparable to the relative uncertainties of the distances. The uncertainty in the planetary reference frame depends upon the object and the epoch chosen, but a typical value is 1 mas. A knowledge of the Earth ephemeris is, in fact, necessary for taking into account the effect of stellar aberration. Indeed, Earth orbital parameters would be included in the parameter set for the data analysis, and the results would include a possible refinement of the ephemeris and a tie to the POINTS reference frame at roughly the 300-μas level from the aberration alone. Direct observation of small solar-system bodies with POINTS could provide an additional tie of even higher precision.

In the process of tying its optical reference frame to that of extragalactic radio sources, POINTS would perform an important check on quasar distances. Quasars ought to have motion relative to the local comoving frame comparable to the approximately 100 km/s seen for closer objects. Thus, the cores of quasars ought not to show relative motion if they are at cosmological distances and are not complicated by source structure variations; these hypotheses would be cast into doubt if such motions were to be seen. Recent work at radio frequencies has placed a 20 μas/y bound on the relative motion of the cores of two quasars separated by about 30 arcmin (Bartel et al., 1986). With ten years of observing, POINTS should detect, at several standard deviations, an angular velocity of 5 μas/y for widely separated quasars. If, as generally expected, quasars are at cosmological distances, this angular velocity corresponds to relativistic relative velocities across the primitive universe. Indeed, although there is no consensus on the relevant cosmological deceleration constant, it seems possible that two observable quasars can be found that lie outside each other's event horizons. In the simplest case, two quasars each with a redshift Z=3 separated by 180° would lie just at each other's horizon (Field, 1990).

Table 1. Bright Quasars and BL Lac Objects

Name	Right Ascension (1987.5)	Declination	V (mag)	Radio Flux (6 cm) Jy
	h m s	° ′ ″		
III Zw 2	00 09 50.8	+10 54 07	15.4 (S)	0.42
PG0026+12	00 28 34.8	+13 11 56	14.8 (S)	0.002
TON S180	00 56 30.8	−22 26 00	14.4 (S)	
TON S120	01 21 14.7	−28 24 58	14.7	
Fairall 9	01 23 17.1	−58 52 16	13.2 (S)	
PKS0405−12	04 07 13.2	−12 13 34	14.6	1.990
0716+714	07 21 53.4	+71 20 36	13.2 (L)	1.121
1E0754+3928	07 57 09.5	+39 22 30	14.4	
PG0804+761	08 09 23.4	+76 04 56	15.1	
PG0844+349	08 46 55.4	+34 47 52	14.0	
OJ 287	08 54 48.9	+20 06 31	14.5 (L)	2.61
PG0953+415	09 56 06.5	+41 19 15	14.5	
PKS1004+13	10 06 45.9	+12 52 37	15.1	0.420
EX1059+730	11 01 45.7	+72 50 47	14.7	
PG1116+215	11 18 29.0	+21 23 24	15.1 (S)	
4C 29.45	11 58 51.8	+29 19 05	14.4	0.890
PG1211+143	12 13 39.4	+14 07 22	14.6	
Mk 205	12 21 11.6	+75 22 46	14.5	
W COM	12 21 31.7	+28 13 58	15.0 (L)	
3C 273	12 28 28.3	+02 07 17	12.9	43.4
OP−106	13 05 33.0	−10 33 20	15.2	1.28
B2 1308+32	13 09 52.0	+32 24 52	15.2 (L)	1.59
PG1351+640	13 52 53.4	+63 49 25	14.8	
PG1441+442	14 13 18.8	+44 03 43	15.0	
S4 1435+63	14 36 28.5	+63 39 52	15.0	1.240
AP LIB	15 16 55.8	−24 19 27	14.8 (L)	1.94
1519−65	15 21 45.6	−06 41 45	14.9	
PG1634+706	16 34 34.5	+70 33 04	14.9	
4C 39.49	16 53 52.2	+39 45 36	14.0 (L)	1.313
PG1700+518	17 01 07.0	+51 50 26	15.2	
3C 351.0	17 04 31.6	+60 45 31	15.3	1.21
PG1718+48	17 19 17.2	+48 04 59	14.7	
Mk 509	20 23 27.6	−10 47 02	13.2 (S)	0.004
PKS2155−304	21 58 06.9	−30 17 14	13.1 (L)	0.31
BL LAC	22 02 43.3	+42 16 40	14.7 (L)	2.96
4C 31.63	22 03 15.0	+31 45 38	14.5	0.298

Objects above marked with a (L) are BL Lacs, objects marked (S) are Seyfert 1's with bright stellar nuclei, all others are QSO's.

6. Discussion

We have described a small, novel astrometric instrument that could both perform several significant scientific studies and prove technologies which could eventually be useful for larger interferometers, both imaging and astrometric. A preliminary evaluation of the scientific uses of the instrument shows that the observing schedule for a ten-year mission could easily and usefully be filled. In fact, the large number of areas of research that POINTS would strongly impact would surely result in oversubscription of the observing time. Increased scientific throughput is possible by scaling the instrument. Serious problems with the scaling are not encountered for an instrument that fits in the Shuttle bay.

The principal technical challenge posed by POINTS is the control of systematic error. The architecture of the instrument has been developed around this problem, which we address at three levels: the use of highly stable materials and thermal control; real-time laser metrology of critical dimensions; and the correction of biases by means of the closure information content of the raw astrometric data.

POINTS would establish an optical global reference frame of unprecedented precision, tied to the frames of both the planetary ephemeris and extra-Galactic radio sources and free of regional biases at the level of precision. It would permit the evaluation of the radio frame for regional biases and would address the co-location of the centers of light and radio emission for several tens of targets.

Acknowledgement. This work was supported in part by NASA OSSA through grant NAGW-1647 from the Innovative Research Program and grant NAGW-1355 from the Planetary Division, and by the Smithsonian Institution, both directly and through its Scholarly Studies Program.

References

Bartel, N., Herring, T.A., Ratner, M.I., Shapiro, I.I., and Corey, B.E. (1986). VLBI limits on the proper motion of the 'core' of the superluminal quasar 3C345, *Nature* **319**, 733–738.

Eichhorn, H. (1974). *Astronomy of Star Positions* (Frederick Ungar Publishing, NY).

Field, G. (1990). Private communication.

Fricke, W. *et al.*, (1988). *Veröffentl. Astron. Rechen-Institut*, Heidelberg, N. 32.

Johnston, K.J., de Vegt, C., Florkowski, D.R., and Wade, C.M. (1985). *Astron. J.* **90**, 2390.

Kovalevsky, J. (1984). *Sp. Sci. Rev.* **39**, 1–63.

Podobed, V.V. (1965). *Fundamental Astrometry*, Translated from the Russian by A.N. Vyssotsky, The University of Chicago Press.

Reasenberg, R.D. *et al.* (1988a). *Astron. J.* **96**, 1731.

Reasenberg, R.D., Babcock, R.W., Chandler, J.F., and Shapiro, I.I. (1988b). In *Proceedings of the International Symposium on Experimental Gravitational Physics*, edited by Hu Enke and P. Michelson.

Veron-Cetty, M.-P., and Veron, P. (1987). "A Catalog of Quasars and Active Nuclei", *ESO Scientific Report* No. 5 (Garching:ESO).

Discussion

BASTIAN: Did you really say that this instrument should be operated mounted on the Space Station? It would constantly be shaken and jolted. No such high-precision work would probably be possible.

CHANDLER: The Space Station is only one of the possible options. The first configuration I showed on the transparencies was a free-flying one. If POINTS is operated as a Space Station Attached Payload, then a major effort will indeed be required to provide both fine pointing and vibration isolation.

HØG: What is the technical level of your study? I ask because I miss a realistic picture of a baffle to limit stray light from the Sun. In fact your drawings do not look different from what you presented in 1984.

CHANDLER: The study is still in an early stage, and many technical details (including the necessary internal and external baffles for shielding the detectors from stray light) remain to be worked out. The primary emphasis of recent work has been in the area of laser metrology and sensitivity studies.

RÉQUIÈME: At the present time we have a lot of problems with the optical structure of radio stars used to link the VLBI and optical reference frames at the level of 10 mas. We also have difficulties with the radio structure of extragalactic sources at the level of 1 mas. How will you select your stars in order to obtain the microsecond accuracy? Are you certain that the optical reference frame obtained with these stars will be coincident with the VLBI extragalactic frame, even at the level of 0.1 mas?

CHANDLER: We will naturally choose sources as compact and as stable as possible, but we cannot assure coincidence of the radio and optical positions at the microarcsecond level. By observing as many radio sources as possible, we can expect to *test* the level of coincidence and to provide a means of calibrating the global consistency of the VLBI reference system.

RELATIVISTIC REDUCTION OF ASTROMETRIC OBSERVATIONS AT POINTS LEVEL OF ACCURACY

V.A. Brumberg[1], S.A. Klioner[1], S.M. Kopejkin[2]
(1) Institute of Applied Astronomy, 197042 Leningrad
(2) Sternberg State Astronomical Institute, 119899 Moscow

ABSTRACT. The framework of general relativity theory (GRT) is applied to the problem of reduction of high precision astrometric observations of the order of one microarcsecond. The equations of geometric optics for the non-stationary gravitational field of the Solar system have been deduced. Integration of the equations of geometric optics results in the isotropic geodesic line connecting the source of emission (a star, a quasar) and an observer. This permits to calculate the effects of relativistic aberration of light due to monopole and quadrupole components of the gravitational field of the Sun and the planets taking into account their motions and rotation. Transformations between the reference systems are used to calculate the light aberration occurring when passing from the satellite system to the geocentric system and from the geocentric system to the barycentric system. The barycentric components of the observed position vector reduced to the flat space-time are corrected, if necessary, for parallax and proper motion of a celestial object using the classical techniques of Euclidean geometry.

1. Introduction

At present, the discussion of high precision measurements of light deflection and time delay of radio signals in the Solar system gravitational field has confirmed the effects of the post-Newtonian approximation of GRT within the accuracy of 1.5% and 0.1% respectively (Will, 1986). Meanwhile, the precision of measurement technique applied in astrometric observations is still fast increasing. In this respect, some specific programs are now in elaboration aimed to determine the nonlinear effects of the post-post-Newtonian approximation of GRT. A particular attention is paid to the project POINTS of a space interferometer to be placed on an Earth artificial satellite (Reasenberg and Shapiro, 1986; Reasenberg et al., 1988; Chandler and Reasenberg, 1990). Preliminary estimation enable one to conclude that the instrumental precision of measurement of angular distances between

J. H. Lieske and V. K. Abalakin (eds.), Inertial Coordinate System on the Sky, 229–240.
© 1990 IAU. Printed in the Netherlands.

celestial objects may be of the order of one microarcsecond. This is comparable with the magnitude of light deflection due to the post-post-Newtonian solar gravitational field as well as to the post-Newtonian corrections for oblateness, motion and rotation of the Sun and the planets. Beside these effects, the reduction of observations of space interferometer should involve the relativistic corrections for light aberration occurring in converting the observed positions of stars to the fixed barycentric reference system of the Solar system. Corrections for parallax and proper motion of observed objects should be also taken into account. Some relativistic effects for light deflection in the gravitational field of one fixed gravitating body have been determined in (Epstein and Shapiro, 1980; Fischbach and Freeman, 1980; Richter and Matzner, 1982, 1983; Sarmiento, 1982; Cowling, 1984; Brumberg, 1987). But these papers do not consider the relativistic effects due to the motions of the gravitating body and an observer with respect to the barycentric reference system and present no algorithm for taking into account parallax and proper motion of observed objects. The aim of the present paper is to develop the consistent relativistic approach for reduction of observations of a space interferometer with due regard to all effects of the order of one microarcsecond.

2. Reference systems

Neglecting the influence of the gravitational field of the Galaxy on the light propagation and motion of the gravitating bodies one may consider the Solar system as isolated. In GRT the characteristics of a reference (coordinate) system introduced in some space-time domain and the potentials of the gravitational field are described by a single object, i.e. the metric tensor $g_{\alpha\beta}$. This tensor is determined by solving the Einstein field equations with appropriately chosen boundary and (or) initial values. It is well known that one may impose on $g_{\alpha\beta}$ four arbitrary complementary conditions. We adopt harmonic conditions $\left(\sqrt{-g}\, g^{\alpha\beta}\right)_{,\beta}=0$. In all formulae used below the greek indices run through values 0,1,2,3; the small latin indices take values 1,2,3; each pair of repeated indices means summation; comma denotes ordinary partial derivative; raising and lowering of the latin indices are performed with the aid of the unit matrix.

The gravitating bodies of the Solar system are considered here as spheroids with constant spin vectors. Numerating the bodies by capital latin letters let us characterize body A by the quantities as follows: mass M_A, mean radius L_A, spin vector $\underline{S}_A$, and the oblateness parameter J_A. Let us introduce also barycentric velocity v_A of the body A, velocity v_{T} of the body's matter relative to its centre of mass, and distance D_A between body A and the nearest body. The independent small parameters of the problem at hand include J_A, $\varepsilon_A = v_A/c$, $\varepsilon_{\mathrm{T}} = v_{\mathrm{T}}/c$,

$\eta_A = GM_A/c^2 L_A$, and $\delta_A = L_A/D_A$. G and c are the gravitational constant and the light velocity respectively. Besides, there exists a small parameter a characterizing the ratio of the mass of the planets and their satellites to the mass of the Sun.

Barycentric reference system (BRS) of the Solar system is used to describe the light propagation from the observed celestial object to an observer and to study the motion of the bodies inside the Solar system. BRS serves as a global reference system. Its metric tensor is resulted from the Einstein field equations by using the post-Minkovskian approximation technique in parameter η_A (Damour, 1983; Blanchet and Damour, 1986,1987). The BRS origin coincides with the Solar system barycentre. Its spatial axes are dynamically non-rotating (Brumberg and Kopejkin, 1989a).

Geocentric reference system (GRS) is used to study the motion of the Earth satellite involved in space interferometry observations and to derive the transformation (reduction) formulae of the position vector components of the observed celestial object from GRS to BRS. GRS is constructed in the space domain restricted by the orbit of the Moon. In solving the Einstein field equations one applies therewith the post-Newtonian approximation technique in parameters ε_A and η_A (Kopejkin, 1988,1989a; Brumberg and Kopejkin, 1989a,b). The GRS origin coincide with the centre of mass of the Earth. Its spatial axes are dynamically non-rotating but they rotate kinematically relative to BRS with the velocity of relativistic precession.

Satellite reference system (SRS) represents a coordinate system with an observer (a space interferometer) at its origin. It is designed for the specific description of observed quantities and for the derivation of the transformation formulae for the position vector components of the observed object from the SRS (instrumental) axes to the GRS axes. SRS is constructed in the space domain restricted by the terrestrial surface. Construction of SRS is performed by solving the Einstein vacuum field equations using the post-Newtonian approximation technique (Brumberg and Kopejkin, 1989b; Kopejkin, 1989a). The world line of the satellite is assumed to be geodesic. Therefore, at the SRS origin the metric tensor reduces to the Minkovsky tensor and its first derivative vanish identically. The SRS spatial axes are dynamically non-rotating but are subjected to kinematical rotation with respect to GRS.

Let us denote the BRS, GRS and SRS coordinates by $x^\alpha = (ct, x^1)$, $w^\alpha = (cu, w^1)$, $\xi^\alpha = (c\tau, \xi^1)$ respectively. At the SRS origin the time τ represents the proper time of an observer and the spatial axes ξ^1 realize the instrumental triad of the observer's equipment (an interferometer). Up to constant factors the time scales u and t are equal to TT and TB respectively (Brumberg and Kopejkin,1989c). Transformations between BRS, GRS and SRS are given explicitly in (Kopejkin, 1988,1989a; Brumberg and Kopejkin, 1989a,b).

3. Geometric optics in the Solar system gravitational field.

In optical region the length of the electromagnetic waves is less than the Solar system space-time curvature by many orders. For this reason (Misner et al., 1973) the light propagation is governed by the geometric optics laws implying the motion of photons in null (isotropic) geodesics of the space-time.

The space-time is split up into three regions with the origin at the Solar system barycentre: 1) external region $R > R_0$, R_0 being the radius of the orbit of Pluto, 2) internal region $R < R_1$ and 3) buffer region $R_1 < R < R_0$. Let the light be emitted at the moment t_1 by a source far outside the Solar system and be received by the observer at the moment t_2. Let x_1^i and x_2^i be the coordinates of the source and the observer at the moments t_1 and t_2 respectively. Equation of light propagation is solved separately in the external and internal regions with subsequent matching of both solutions in the buffer region leading to the intermediate solution (Kopejkin, 1990).

The external region is dominated by the monopole component of the total gravitational field of the Solar system. Therefore, the equation of the null geodesics may be presented in the form

$$\ddot{x}^1 = -\frac{GM}{R^3} x^1 + c^{-2}\frac{GM}{R^3} \left(-\dot{x}^2 x^1 + 4(\underline{x}\,\dot{\underline{x}})\,\dot{x}^1\right) + O\left(\frac{GM}{R^2}\frac{L^2}{R^2}\right) \quad (1)$$

The remainder terms are due to the quadrupole component of the Solar system total gravitational field. To solve Eq. (6) one substitutes into its right-hand member the unperturbed solution $x_N^i(t) = x_1^i + ck^i(t-t_1)$ with $|\underline{k}| = (k^i k^i)^{1/2} = 1$. The resulting ordinary differential equation is solved under initial conditions: 1) $x^1(t_1) = x_1^1$ and 2) $\lim_{t \to -\infty} c^{-1}\dot{x}^1(t) = k^1$. These conditions mean physically that the light trajectory passes through the point of emission at moment t_1 and the BRS coordinate velocity of a photon at the infinite isotropic past is equal to the light velocity locally measured in SI units. The specific form of the external solution $x_E^i(t)$ is given, for example, in (Brumberg, 1972). The remainder terms of the obtained solution are proportional to $c^{-2}GM L^2/R^2$ with $L = R_0$ and $M = \sum_A M_A$.

The internal region is characterized by the gravitational fields of the individual attracting masses moving much slowly than a photon. Due to this and taking into account the smallness of the time interval of the light propagation through the internal region one may present the equations of null geodesics in the form

$$\ddot{x}^1 = F_1^1 + F_2^1 + F_3^1 + F_4^1 + O(\eta_A \varepsilon_A^2) + O(\eta_A^2 a) + O(\eta_A^2 \varepsilon_A) \quad (2)$$

with

$$F_1^1 = \sum_A \frac{GM_A}{R_A^3} \left(-R_A^1 + c^{-2} \left(-\dot{\underline{x}}^2 R_A^1 + 4 \left(\dot{\underline{x}} \, \underline{R}_A \right) \dot{\underline{x}}^1 \right) + \right.$$

$$\frac{4}{c} \left(\underline{v}_A \, \dot{\underline{x}} \right) R_A^1 - \frac{4}{c} \left(\underline{R}_A \, \dot{\underline{x}} \right) v_A^1 - \frac{2}{c} \left(\underline{v}_A \, \underline{R}_A \right) \dot{\underline{x}} - \tag{3}$$

$$c^{-4} \, 4 \, (\dot{\underline{x}} \, \underline{R}_A) \, (\underline{v}_A \, \dot{\underline{x}}) \, \dot{\underline{x}}^1),$$

$$F_2^1 = \frac{2}{c^4} \frac{G^2 M_S^2}{R_S^6} \, (\dot{\underline{x}} \, \underline{R}_S)^2 \, R_S^1 - \frac{2}{c^4} \frac{G^2 M_S^2}{R_S^4} \, (\dot{\underline{x}} \, \underline{R}_S) \, \dot{\underline{x}}^1, \tag{4}$$

$$F_3^1 = \frac{4}{c^2} \sum_A \frac{G}{R^3} \left((\underline{S}_A \wedge \dot{\underline{x}})^1 - \frac{3}{2} R_A^{-2} (\dot{\underline{x}} \, \underline{R}_A) \, (\underline{S}_A \wedge \underline{R}_A)^1 - \right.$$

$$\frac{3}{2} c^{-2} R_A^{-2} \left(\underline{R}_A \, (\underline{S}_A \wedge \dot{\underline{x}}) \right) \left(\dot{\underline{x}} \wedge (\underline{R}_A \wedge \dot{\underline{x}}) \right)^1) \tag{5}$$

$$F_4^1 = 6 \sum_A \frac{G}{R_A^5} \, I_A^{pq} \, R_A^q \, (\delta^{1p} - 2 \, \dot{x}^1 \dot{x}^p) -$$

$$15 \sum_A \frac{G}{R_A^7} \, I_A^{pq} \, R_A^q \, R_A^p \, (R_A^1 - 2 \, (\dot{\underline{x}} \, \underline{R}_A) \, \dot{\underline{x}}^1) \tag{6}$$

with $R_A^1 = x^1 - x_A^1$, $x_A^1(t)$ and $v_A^1 = dx_A^1/dt$ are respectively the BRS coordinates and velocity components of the centre of mass of body A. I_A^{ij} is the trace-free quadrupole moment of body A. The terms F_1^1 and F_2^1 entering into Eq. (7) describe respectively the post-Newtonian and post-post-Newtonian effects of the monopole components of the gravitational fields of the attracting bodies taking into account their motion. In calculating the post-post-Newtonian perturbations it is sufficient to consider only the terms depending on the mass of the Sun M_S. The terms F_3^1 and F_4^1 are due to the rotation and the quadrupole components respectively of the solar and planetary gravitational fields.

The internal solution $x_1^1(t)$ of Eq. (7) is looked up in the form

$$x_1^1(t) = x_N^1(t) + c^{-2} \, (B^1(t) + C^1(t) + D^1(t)) \tag{7}$$

with unperturbed solution $x_N^1(t) = x_2^1 + c\sigma^1(t - t_2)$. Functions B^1, C^1, D^1 satisfy the equations

$$\overset{..1}{B} = F_1^1 + F_2^1 , \qquad \overset{..1}{C} = F_3^1 , \qquad \overset{..1}{D} = F_4^1 \tag{8}$$

At the moment t_2 the trajectory of the internal solution should pass through the point of observation implying $x_1^1(t_2) = x_2^1$. Vector σ^1 being the arbitrary constant of integration is determined later in matching the internal and external solutions. In solving the first of Eqs. (13) the motion of the attracting bodies is assumed to be uniform and rectilinear, i.e.

$$x_A^1(t) = x_A^1(t_A) + v_A^1(t-t_A) + O(a_A (t-t_A)^2) \tag{9}$$

with $a_A^1 = dv_A^1/dt$ being the acceleration of the centre of mass of the body A and t_A being some fixed moment of time. In solving remaining two Eqs. (13) the centre of mass of any body A may be regarded as being at rest at moment t_A. The remainder terms in Eqs. (7) and (14) are responsible for the errors of the internal solution. By the suitable fixing t_A regarded as the parameter of the solution one can minimize the magnitude of the errors of the internal solution. It turns out that t_A corresponds to the moments of the closest approach with body A provided the latter is located between the light emitter and the observer. For this case the magnitude of the residuals is proportional to $c^{-4}GM_A a_A R_A ln(R_A r/d_A^{-2})$ where r is the distance between the observer and body A and d_A is the impact parameter of the light trajectory with respect to body A. If the observer is located between the light source and body A then t_A is to be coincident with the moment of observation t_2 and the magnitude of the residual errors of the internal solution is proportional therewith to $c^{-4}GM_A a_A R_A ln(R_A/r)$.

Solution of Eq. (13) is partly given in (Brumberg, 1987; Klioner,1989). In the complete form it should be published in our future paper. From the methodological point of view it is of interest to estimate the magnitude of the light deflection due to various factors in passing through the Solar system (see TABLE 1). These estimates demonstrate that within the microarcsecond accuracy one has to take into account the whole Solar system. Indeed, we have to consider under certain conditions the influences of three largest asteroids Ceres, Pallas, Vesta, the Galilean satellites of Jupiter, Titan, Triton and perhaps some other satellites of Saturn and Uranus whose physical properties are not well known yet. The values of δ_1 due to these bodies varies from 0.5 μas (Pallas) to 33 μas (Titan).

Matching of the external and internal solutions is performed in the buffer region at some moment t_* provided that coordinates $x_B^1(t_*)$

and $x_\parallel^1(t_*)$ coincide and the difference between tangent vectors $\dot{x}_\parallel^1(t_*)$ and $\dot{x}_\perp^1(t_*)$ is minimal. These conditions enable to find the matching radius R_* for any body A. More specifically, R_* is determined by either of two equations

$$M_A a_A \ln(R_* \, r \, d_A^{-2}) = c^2 ML^2 R_*^{-3} \; , \quad M_A a_A \ln(R_*/r) = c^2 ML^2 R_*^{-3} \; .$$

For any attracting body the matching radius exceeds the radius R_0 of external region and may be chosen for all bodies in common enabling the difference between the tangent vectors of the external and internal solutions to be less than one microarcsecond.

The matching procedure implies that $\sigma^1 = k^1 + O(c^{-2} GML^2 R_*^{-3})$. This procedure results in some intermediate solution coinciding with the internal solution for $R < R_*$ and identical to the external solution for $R > R_*$. Formally, one may regard the intermediate solution as coinciding with the internal solution in the whole space-time since outside the Solar system the internal solution differs from the external solution only by the terms $O(c^{-2} GML^2 R^{-2})$.

TABLE 1. Estimates of relativistic effects due to Solar
system bodies.

Body	δ_1	δ_2	δ_3	δ_4	δ_5	ψ_{max}
Sun	$1.75 \cdot 10^6$	2	0.1	0.8	11	180°
Mercury	83	0.06	0.02	–	–	$9'_\circ$
Venus	493	0.002	0.06	–	–	$4.5_\circ$
Earth	574	0.6	0.06	–	–	180°
Moon	26	0.002	0.003	–	–	8
Mars	116	0.2	0.01	–	–	$25'_\circ$
Jupiter	16300	240	0.8	0.2	0.001	$90_\circ$
Saturn	5800	95	0.2	0.04	–	16.5
Uranus	2100	25	0.05	0.007	–	1.2
Neptune	2600	10	0.05	0.006	–	$51'$
Pluto	$\simeq 20$	–	–	–	–	$0.5''$

The first five columns of the Table present the maximal values in μas of relativistic effects under study. These values have been estimated as follows: the post-Newtonian light deflection due to monopole field of the body $\delta_1 = 4GM/c^2 L$; the correction due to quadrupole field $\delta_2 = \delta_1 \cdot J$; the influence of the motion of the body $\delta_3 = \delta_1 \cdot v/c$; the effect of the rotation $\delta_4 = 4GS/c^3 L^2$; the post-post-Newtonian deflection due to monopole field $\delta_5 = 15\pi/4 \cdot G^2 M^2/c^4 L^2$. The absent values are less than 10^{-3}. The last column contains the maximal angular distance between a source and the body at which the influence of the body on the apparent source position is still to be taken into account.

4. Aberration of light

Let us denote the SRS, GRS and BRS spatial components of the tangent vector to the null geodesic by $s^1 = c^{-1} d\xi^1/d\tau$, $q^1 = c^{-1} dw^1/du$ and $p^1 = c^{-1} dx^1/dt$ respectively. Directly measurable quantities are the components of vector $-s^1$ directed oppositely to vector s^1.

Aberration of the light is caused by the transformations from one reference system to another system at the point of observation. Aberration relations between vectors s^1, q^1 and p^1 have the form

$$s^1 = (K_0^1 + K_j^1 q^j) / (K_0^0 + K_j^0 q^j), \quad q^1 = (\Lambda_0^1 + \Lambda_j^1 p^j) / (\Lambda_0^0 + \Lambda_j^0 p^j) \quad (10)$$

$K^\alpha_{\ \beta} = \partial \xi^\alpha / \partial w^\beta$ and $\Lambda^\alpha_{\ \beta} = \partial w^\alpha / \partial x^\beta$ being the transformation matrices of the coordinate bases at the point of observation. The length of vector s^1 is equal to unity since at the SRS origin the metric is flat. Using this condition one may calculate from (15) the lengths of vectors q^1 and p^1. Aberration relations between the unit vectors s^1, $m^1 = q^1/q$ and $n^1 = p^1/p$ derived from (15) have the form

$$
\begin{aligned}
s^1 =\ & \underline{m}^1 + c^{-1} (\underline{m} \wedge (\underline{m} \wedge \underline{v}_{\underline{1}}))^1 + c^{-2} \left(\frac{1}{2} (\underline{m}\ \underline{v}_{\underline{1}}) (\underline{m} \wedge (\underline{m} \wedge \underline{v}_{\underline{1}}))^1 - \right. \\
& \frac{1}{2} (\underline{m} \wedge \underline{v}_{\underline{1}})^2 \underline{m}^1 + R^{1j}\ \underline{m}^j) + c^{-3} (\frac{1}{2} (\underline{m}\ \underline{v}_{\underline{1}})^2 (\underline{m} \wedge (\underline{m} \wedge \underline{v}_{\underline{1}}))^1 - \\
& \frac{1}{2} (\underline{m}\ \underline{v}_{\underline{1}}) (\underline{m} \wedge \underline{v}_{\underline{1}})^2\ \underline{m}^1 + 2\ U_{\underline{K}}(\underline{w}_{\underline{1}}) (\underline{m} \wedge (\underline{m} \wedge \underline{v}_{\underline{1}}))^1 + \\
& R^{1j} (\underline{m} \wedge (\underline{m} \wedge \underline{v}_{\underline{1}}))^j + O(c^{-4})
\end{aligned}
\quad (11)
$$

$$
\begin{aligned}
m^1 =\ & \underline{n}^1 + c^{-1} (\underline{n} \wedge (\underline{n} \wedge \underline{v}_{\underline{K}}))^1 + c^{-2} (\frac{1}{2} (\underline{n}\ \underline{v}_{\underline{K}}) (\underline{n} \wedge (\underline{n} \wedge \underline{v}_{\underline{K}}))^1 - \\
& \frac{1}{2} (\underline{n} \wedge \underline{v}_{\underline{K}})^2\ \underline{n}^1 + (\underline{n} \wedge (\underline{R}_{\underline{K}} \wedge \underline{a}_{\underline{K}}))^1 + F^{1j} \underline{n}^j) + \\
& c^{-3} (\frac{1}{2} (\underline{n}\ \underline{v}_{\underline{K}})^2 (\underline{n} \wedge (\underline{n} \wedge \underline{v}_{\underline{K}}))^1 - \frac{1}{2} (\underline{n}\ \underline{v}_{\underline{K}}) (\underline{n} \wedge \underline{v}_{\underline{K}})^2\ \underline{n}^1 + \\
& 2\ U_{\underline{K}}(\underline{w}_{\underline{1}}) (\underline{n} \wedge (\underline{n} \wedge \underline{v}_{\underline{K}}))^1 + 2\ \bar{U}(\underline{x}_{\underline{K}}) (\underline{n} \wedge (\underline{n} \wedge \underline{v}_{\underline{K}}))^1 + \\
& (\underline{n}\ \underline{v}_{\underline{K}}) (\underline{n} \wedge (\underline{R}_{\underline{K}} \wedge \underline{a}_{\underline{K}}))^1 + (\underline{a}_{\underline{K}}\ \underline{R}_{\underline{K}}) (\underline{n} \wedge (\underline{n} \wedge \underline{v}_{\underline{K}}))^1 -
\end{aligned}
\quad (12)
$$

$$((\underline{R}_E \wedge \underline{a}_E) \ (\underline{n} \wedge \underline{v}_E)) \ n^1 + F^{1j} \ (\underline{n} \wedge (\underline{n} \wedge \underline{v}_E))^j + O(\ c^{-4})$$

Here U_E is the Earth potential, $\bar{U}$ is the potential of all Solar system bodies excluding the Earth, functions $F^{1j} = -F^{j1}$ characterize kinematical rotation of GRS with respect to BRS and functions $R^{1j} = -R^{j1}$ characterize kinematical rotation of SRS with respect to GRS (Brumberg and Kopejkin, 1989a; Kopejkin, 1989b).

Vector n^1 is related with the unit vector k^1 characterizing the direction of propagation from the source to the observer by means of expression resulted from the internal solution (12)

$$n^1 = k^1 + c^{-3} \ (\ (\underline{k} \wedge (\underline{\dot{B}} \wedge \underline{k}))^1 + (\underline{k} \wedge (\underline{\dot{C}} \wedge \underline{k}))^1 + (\underline{k} \wedge (\underline{\dot{D}} \wedge \underline{k}))^1) -$$

$$c^{-6} (\ \frac{1}{2} \ (\underline{k} \wedge \underline{\dot{B}})^2 \ k^1 + (\underline{k} \ \underline{\dot{B}})(\underline{k} \wedge (\underline{\dot{B}} \wedge \underline{k}))^1 \ ) + \qquad (13)$$

$$O \ (c^{-4} \ \delta_A) + O \ (c^{-4} \ J_A) + O \ (c^{-4} \ \alpha)$$

with dot denoting the differentiation with respect to time t.

5. Parallax and proper motion

For the sake of convenience let us re-designate the moment of emission by T, the moment of reception by t, the coordinates of the emitter at the moment T by R^1 and the coordinates of the observer at the moment t by x^1. Vector k^1 is expressed in terms of the coordinates of the emitter and the observer as follows (Brumberg, 1972,1987)

$$k^1 = (\ R^1(T) - x^1(t))/ \ |\underline{R}(T) - \underline{x}(t)| \ + O(c^{-2}) \qquad (14)$$

For a limited time interval the coordinates of the emitter may be presented in the form

$$R^1(T) = R^1_0(T_0) + V^1(T_0) \ \Delta T + \frac{1}{2} \ \dot{V}^1(T_0) \ \Delta T^2 + O(\ \Delta T^3) \qquad (15)$$

with $\Delta T = T - T_0$, T_0 being the initial epoch of emission. $V^1 = dR^1/dT$ and $\dot{V}^1 = d^2 R^1/dT^2$ are BRS velocity and acceleration of the emitter respectively.

The parallax may be taken into account by expanding the right-hand member of Eq. (19) in powers of the parallactic ratio $\varpi \simeq |\underline{x}|/R$

$$k^1 = R^{-1} \, R^1 - R^{-3} (\underline{R} \wedge (\underline{x} \wedge \underline{R}))^1 - \frac{1}{2} \, R^{-5} (\underline{R} \wedge \underline{x})^2 R^1 -$$

$$R^{-5} (\underline{R} \, \underline{x}) \, (\underline{R} \wedge (\underline{x} \wedge \underline{R}))^1 + O(\pi^3) \tag{16}$$

The proper motion of the light source is taken into account by expanding the right-hand member of (21) in powers of ΔT and using (20). Finally, one gets

$$k^1 = k_0^1 \, (\, 1 + (\underline{\pi} \, \underline{\mu}) \, \Delta T - \frac{1}{2} \, \underline{\pi}^2) + \mu^1 \Delta T \, (\, 1 + \frac{1}{R} \, (\underline{k}_0 \, \underline{x})) -$$

$$\pi^1 (\, 1 + \frac{1}{R_0} (\underline{k}_0 \, \underline{x}) - \frac{1}{R_0} (\underline{k}_0 \, \underline{V}) \Delta T) + \frac{1}{2} \, \dot{\mu}^1 \Delta T^2 + O(\pi^3) + O(\Delta T^3) \tag{17}$$

with $k_0^1 = R_0^1 / R_0$, $\dot{k}_0^1 = R_0^{-1} (V^1 - (\underline{k}_0 \underline{V}) k_0^1)$. $\mu^1 = (\underline{k}_0 \wedge (\dot{\underline{k}}_0 \wedge \underline{k}_0))^1$ is the vector of proper motion and $\pi^1 = R_0^{-1} (\underline{k}_0 \wedge (\underline{x} \wedge \underline{k}_0))^1$ is the vector of parallax.

BRS time interval ΔT is not directly measurable quantity. It should be expressed in terms of the BRS time interval $\Delta t = t - t_0$ at the point of observation (t_0 being the initial epoch of observation) by means of relation

$$\Delta T = (\, 1 + c^{-2} (\underline{k}_0 \underline{V}))^{-1} (\Delta t + c^{-1} (\underline{k}_0 \underline{x})) +$$

$$O(\, c^{-1} R_0 \, \pi^2) + O(\, c^{-1} |\underline{x}| \, \mu \, \Delta t) \tag{18}$$

The term $c^{-1} (\underline{k}_0 x)$ in (23) is of the sinusoidal form with the maximal amplitude of the order of 500 seconds and the period of one year. This term should be taken into account for stars with large proper motion. For example, for the Barnard star with $\mu \simeq 10''$ per year such term results to the change of the star coordinate by the order of 200 microarcseconds per year. Within the accuracy of the one microarcsecond this term may be easily detected.

6. Conclusion

This paper presents an algorithm of reduction of astrometric observations to be performed on an Earth satellite with the precision of one microarcsecond. To be short, this algorithm is reduced to Expressions (16)-(18) and (22),(23) for the transformation from the observed vector $-s^1$ to the BRS unit vector k_0^1. Two independent components of the latter yield the position of the source on the celestial sphere at epoch t_0. A set of vectors k_0^1 referred to one and the same initial epoch t_0 for sufficiently large number of sources determines an inertial reference system on the sky.

It may be noted that the relativistic precessions F^{ij} and R^{ij} need not to be known for reduction of observations performed on a satellite insofar one deals here with relative observations. The relativistic precession is wanted only to provide absoluteness to the inertial reference system constructed with the aid of satellite board observations.

References

Blanchet, L., Damour, T.(1986) Phil. Trans. Roy.Soc. London, A320, 379.
Blanchet, L., Damour, T. (1988) Phys. Rev., D37, 1410.
Brumberg, V.A. (1972) Relativistic Celestial Mechanics. Nauka, Moscow (in Russian).
Brumberg, V.A. (1987) Kin. Fiz. Nebes. Tel, 3, 8 (in Russian).
Brumberg, V.A., Kopejkin, S.M. (1989a) Nuovo Cim., 103B, 63.
Brumberg, V.A., Kopejkin, S.M. (1989b) in J. Kovalevsky, I.I. Mueller, B. Kolaczek (eds.), Reference Frames in Astronomy and Geophysics, Kluwer, Dordrecht, 115.
Brumberg, V.A., Kopejkin, S.M. (1989c) Cel. Mech. (in press).
Chandler, J.F., Reasenberg, R.D. (1989) this volume.
Cowling, S.A. (1984) MNRAS, 209, 415.
Damour, T. (1983) in N. Deruelle, T. Piran (eds.), Gravitational Radiation, Norht-Holland, Amsterdam, 59.
Epstein, R., Shapiro, I.I. (1980) Phys. Rev. D22, 2947.
Fischbach, E., Freeman B.S. (1980) Phys. Rev. D22, 2950.
Klioner, S.A. (1989) ' Propagation of the Light in the Barycentric Reference System considering the Motion of the Gravitating Masses ', Preprint No. 6, Inst. Appl. Astr. Acad. Sci. USSR, Leningrad (in Russian).
Kopejkin, S.M. (1988) Cel. Mech., 44, 87.
Kopejkin, S.M. (1989a) Astron. Zh., 66, 1069 (in Russian).
Kopejkin, S.M. (1989b) Astron. Zh., 66 (in press).
Kopejkin, S.M. (1990) Astron. Zh., 67 (in press).
Misner, C.W., Thorne, K.S., Wheeler, J.A. (1973) Gravitation, Freeman, San Francisco.
Reasenberg, R.D., Shapiro, I.I. (1986) in J. Kovalevsky, V.A. Brumberg (eds.), Relativity in Celestial Mechanics and Astrometry, Reidel, Dordrecht, 383.
Reasenberg, R.D., Babcock, R.W., Chandler, J.F. et al. (1988) Astron. J., 96, 1731.
Richter, G.W., Matzner, R.A. (1982) Phys. Rev. D26, 1219, 2549.
Richter, G.W., Matzner, R.A. (1983) Phys. Rev. D28, 3007.
Sarmiento, G.A.F. (1982) Gen. Rel. Grav., 14, 793.
Will, C.M. (1986) in J. Kovalevsky, V.A. Brumberg (eds.), Relativity in Celestial Mechanics and Astrometry, Reidel, Dordrecht, 355.

Discussion

HUGHES: This paper, with its complicated considerations, gives us a perfect example of the necessity of very carefully setting up a considered nomenclature for dealing with space/time. It becomes naive in the extreme to speak simply of some "coordinate system" or such, without a very careful specification of the underlying theory, approximations etc. in the "reference system" or whatever we finally call such a thing.

BASTIAN: Is POINTS able to detect low-frequency gravitational waves, of which the universe may be filled?

KOPEJKIN: A level of 0.1 mas would be expected, according to a study by Braginsky.

BASTIAN: So at least *some* challenge remains.

TURYSHEV: How do you define the region of matching of coordinate systems?

KOPEJKIN: The region of matching is initially bounded by the distance to the nearest attracting body. For example, for the matching of barycentric and geocentric reference frames, the region of matching is bounded initially by the distance from the Earth to the Moon. After the determination of the functions incorporated in the coordinate transformations, the region of matching can be extended to a greater distance, namely to the point in space where the determinant of the coordinate transformations is equal to zero.

ASTROMETRIC INTERFEROMETRY —
CAN IT ESTABLISH A FUNDAMENTAL SYSTEM?

G.H. KAPLAN
U. S. Naval Observatory
Washington, D.C. 20392
USA

ABSTRACT. The astrometric optical interferometer on Mt. Wilson is providing a new source of astrophysical and astrometric data on bright stars. The instrument, with 12-meter baselines, has been in operation since late 1986. The interferometer is capable of wide-angle astrometry, that is, the determination of very precise stellar positions within a reference frame defined by bright stars spread across a large area (of order one steradian) of the sky. This paper addresses the question of whether such an instrument can be used to establish a *fundamental* system — that is, one tied to the Earth in some well-defined way. Some astrometric data from this instrument are presented to illustrate the difficulties involved. Proposed means of addressing these problems in future instruments are discussed.

1. Introduction

This paper expands upon some ideas which followed from a discussion in August 1988, among those of us involved with the Mt. Wilson astrometric optical interferometer project, concerning the "precision" and "accuracy" of the astrometric results from the interferometer. It explores in some detail a concern expressed by Michael Shao about the validity of the astrometric results since the interferometer's baselines did not appear to co-rotate with the Earth. I am considering here only the overall geometry of the interferometer observations and what its implications are for that type of instrument. The results may apply to other types of Earth-based, "fundamental" instruments — transit circles, astrolabes, zenith tubes — but a more general discussion is beyond the current scope and my own expertise.

Here, I use the terms "absolute" or "fundamental" to describe star coordinates measured with respect to the Earth's instantaneous rotation axis: declinations and relative right ascensions. I will not deal with the origin of right ascension, i.e., the location of the equinox. I use the word "axis" frequently, and it should be considered synonymous with "angular velocity vector", that is, both an orientation and a magnitude (spin rate) are involved. Two axes are coincident only if both their orientation and magnitude are identical.

I wish to say at the outset that I do not mean to imply that observations which are not fundamental are not useful. In fact, astrometric observations from space (*e.g.*, HIPPARCOS) will not be fundamental in the usual sense. However, if we are planning to eventually replace transit circles with interferometers we need to consider all the ramifications. As we enter an era where the most accurate wide-angle astrometry may be non-fundamental, we will need to pay much more attention to exactly

J. H. Lieske and V. K. Abalakin (eds.), Inertial Coordinate System on the Sky, 241–250.
© 1990 *IAU. Printed in the Netherlands.*

how our coordinate systems are both defined and realized.

2. The Mt. Wilson Optical Interferometer

The optical interferometer on Mt. Wilson, near Los Angeles, is unfamiliar to many in the astrometric community. A significant amount of observational data has already been obtained from this unique instrument, which is essentially a long-baseline Michelson stellar interferometer. Stellar fringes were first detected in September 1986, and routine observational programs in stellar diameters and astrometry were begun in July 1987. The interferometer has been a joint project of the Smithsonian Astrophysical Observatory, the Naval Research Laboratory, the Massachusetts Institute of Technology, and the U. S. Naval Observatory. Key people involved in the development of this system include Michael Shao, Mark Colavita, Kenneth Johnston, Richard Simon, David Mozurkewich, Donald Hutter, James Hughes, John Hershey, John Pohlman, Braden Hines, and David Staelin.

The Mt. Wilson interferometer is a highly automated, rapid-response, wideband system which combines and interferes starlight from a pair of independently-moving 25-cm siderostats. Currently, the instrument can alternate between pairs of siderostats forming two baselines, each 12 meters long. The instrument actively tracks a star's "white light fringe" in real time and has been designed specifically for absolute astrometry. The astrometric capabilities of this instrument were demonstrated on a limited data set from observations taken in the autumn of 1986 when only one baseline was available (Mozurkewich *et al.* 1988). Since that time, the second baseline has become operational, a number of instrumental systems have been improved, and experiments have begun with two-color observations. The latter is a means for greatly reducing the effects of the Earth's atmosphere on the observations. More technical detail on this instrument may be found in Shao *et al.* (1988), Colavita *et al.* (1987), Shao (1988), and Kaplan *et al.* (1988).

When one-color data from a half dozen nights of observations are combined, the formal errors in position for each star have been generally in the range 15-25 milliarcseconds in declination and somewhat worse in right ascension. The two-color technique promises to reduce these formal uncertainties to perhaps a few milliarcseconds. The current instrument is limited to stars brighter than about visual magnitude 5 and sky coverage is restricted to declinations between +15 and +55 degrees. However, this instrument has been sufficiently successful that the U. S. Naval Observatory is currently involved in the design and construction of a much larger astrometric optical interferometer to be completed by 1994 and located in the coastal mountains of central California.

Clearly we must understand exactly what our star positions mean; specifically, in what reference frame are they expressed and how can they be related to more conventional systems, for example, the FK5? Up to this point we have essentially been treating the Mt. Wilson optical data like VLBI data as far as our data reduction procedures are concerned. However, our data are unlike VLBI data in two ways: (1) we can observe only over part of a day (less than half of one rotation of the Earth), and (2) our baseline components vary significantly over the course of the observations. We have been aware of these distinctions for some time, and each of us involved in the project has expressed the situation in slightly different ways. We have made statements to the effect that "we have different baselines for different stars" or "for any given star, the baseline is a function of hour angle", or something similar. Given these realities, I reconsider here the statement frequently made by interferometry people (radio or optical) that "our declinations and relative right ascensions are absolute". Under what conditions is that true?

3. Interferometers and the Earth's Rotation Axis

Unlike transit circles, PZTs, and astrolabes, interferometers have no tie to the local gravity field and therefore have no well-defined latitude. Viewed from inertial space, an interferometer baseline is simply a vector which rotates around some axis once per sidereal day (see Figure 1). The only actual tie to the Earth is the *assumption* that the axis about which the baseline rotates is coincident with that of the Earth. A lot rests on that assumption, both for the VLBI/ Earth rotation community and for those of us interested in interferometric astrometry.

Regardless of whether the interferometer's axis does co-incide with that of the Earth, one can always determine, from an appropriate set of observations, a number of parameters describing the baseline, **b**, and its apparent axis of rotation, **w**. The *a priori* star coordinates at the ob-serving epoch are assumed to define an inertial coordinate system (at some level of accuracy) for these measurements. There are a number of ways of parameterizing this geometric information, some more use-ful than others.

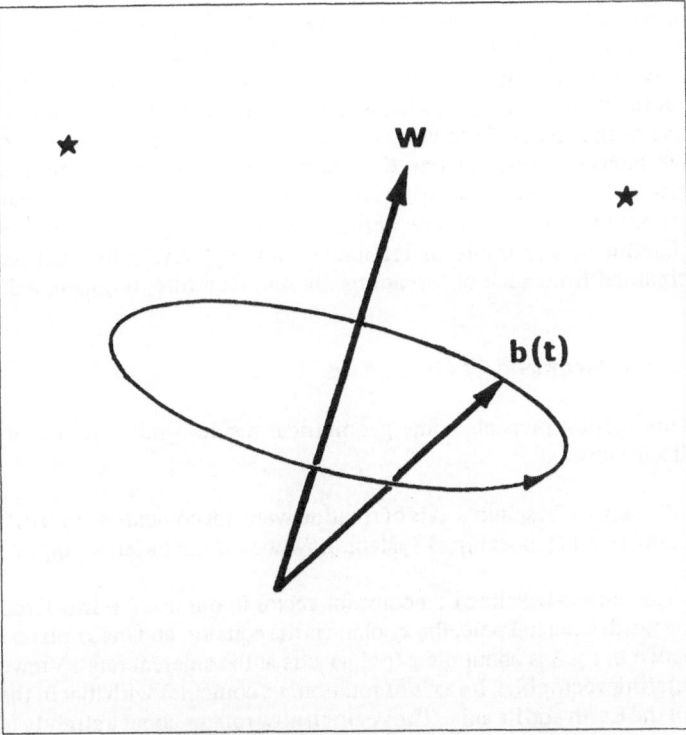

Figure 1. Diurnal motion of an interferometer baseline, **b**, about its apparent axis **w**.

The Earth rotation commu-nity, in giving very high weight to VLBI data, takes as axi-omatic the coincidence of any VLBI axis **w** with that of the Earth. In fact, a stronger assumption is made: VLBI baselines are tied rigidly to the crust of the Earth. The latter assumption is a sufficient but not a necessary condition for the coincidence of the axes.

But Earth rotation studies are a special case of interferometric astrometry, and for those of us interested primarily in star or radio source positions, there is no requirement that the baselines be tied to the Earth's crust. Indeed, there is no need that the baselines in any way remain constant from one night to the next or even that the same baseline be used. However, it *is* necessary to assume that during the course of observations, the baseline's axis of rotation is the same as that of the Earth. *The extent to which interferometrically determined star coordinates are fundamental rests entirely on the validity of the assumption that the baseline's apparent axis of rotation is coincident, in both orientation and magnitude, with that of the Earth.*

The data indicate that the VLBI community is on firm ground in stating that their source positions are fundamental in the sense I am using here. The optical interferometer data are quite different. We know that the Mt. Wilson baselines do not remain constant during our observations. In our current data reduction approach, we solve for rates of change of the rectangular baseline components during the observations. Occasionally we also solve for the corresponding accelerations. We cannot avoid some kind of procedure like this; the data demand it. The tacit assumption is that solving for the baseline motion effectively reduces the baseline to a constant vector within our local Earth-fixed coordinate system, thus maintaining the fundamental nature of our instrument. By the time we published our original astrometry paper (Mozurkewich *et. al.* 1988), we had begun to recognize the subtle complications of what we were doing, and we couched our results in appropriately conservative terms. In the following, I explore these subtleties more fully. My thesis here is that, as long as we have to solve for baseline motions from the observations themselves, the observations are effectively related to an arbitrary axis of rotation. Although we can, in fact, locate that axis of rotation within the reference frame defined by the catalog of *a priori* star positions, by this point we have abandoned any frame of reference with any physically well-defined anchor. Any star positions obtained from such observations are not, therefore, fundamental.

4. The Geometry of Offset Axes

This section presents some geometrical results which are useful background information for this discussion.

Suppose a baseline's axis of rotation were not coincident with that of the Earth. Viewed from a truly Earth-fixed (topocentric) system, how would the baseline appear to move?

Consider a baseline **b**, a constant vector in our usual Earth-fixed system: the z axis points toward the north celestial pole, the xy plane is the equator, and the xz plane is the local meridian. The system, hence **b**, rotates about the z (polar) axis at the sidereal rate. Viewed from an inertial frame, **b** is the rotating vector $\mathbf{b}(t)$. Its axis of rotation, **w**, coincides with that of the Earth. Now suppose we slightly tilt the Earth and its axis. The vector **b** now rotates about a slightly offset axis, **w**′, again at the sidereal rate. Viewed from the inertial frame, **b** is now the rotating vector $\mathbf{b}'(t)$. Form the difference vector $\mathbf{d}(t) = \mathbf{b}'(t) - \mathbf{b}(t)$ (see Figure 2). What does the locus of $\mathbf{d}(t)$ look like, back in the Earth-fixed system where **b** is a constant?

A little bit of algebra shows the following. In the Earth-fixed (topocentric) system the difference vector $\mathbf{d}(t)$ executes a small ellipse over the course of a day. The plane of the ellipse is orthogonal to **b** (as we would expect, since we are only dealing with rotations here and no change of length should result). The major axis of the ellipse is in the plane formed by **b** and the z (polar) axis. The length of the semimajor axis is simply the angular offset of **w**′ from **w** (in radians) times the length of **b**. The ratio of minor to major axis is identical to the ratio of the z (polar) component of the baseline to the total baseline length (that is, $\mathbf{b} \cdot \mathbf{w} / |\mathbf{b}||\mathbf{w}|$). The instantaneous phase angle of **d** as it traces out this small ellipse depends on both the local sidereal time and the relative orientation of the two axes. The vector **d**, does, however, trace out the ellipse in a counterclockwise manner as viewed from the origin, and completes the circuit in one sidereal day.

If, in addition, the rotation rates are slightly different for the two axes, the ellipse becomes distorted in the direction of its minor axis (that is, orthogonal to the rotation axis). In this case the ellipse will not close on itself after one day.

How would an offset rotation of one of the optical interferometer's baselines appear in the topocentric Mt. Wilson reference frame? Assume a quarter-arcsecond angular separation of the direction of rotation axis of the baseline from that of the Earth (we shall see in section 5 that this is not too unreasonable a number). As seen from Mt. Wilson, the baseline would appear to undergo an elliptical oscillation over the course of a sidereal day. As stated above, the plane of the ellipse would be orthogonal to the baseline.

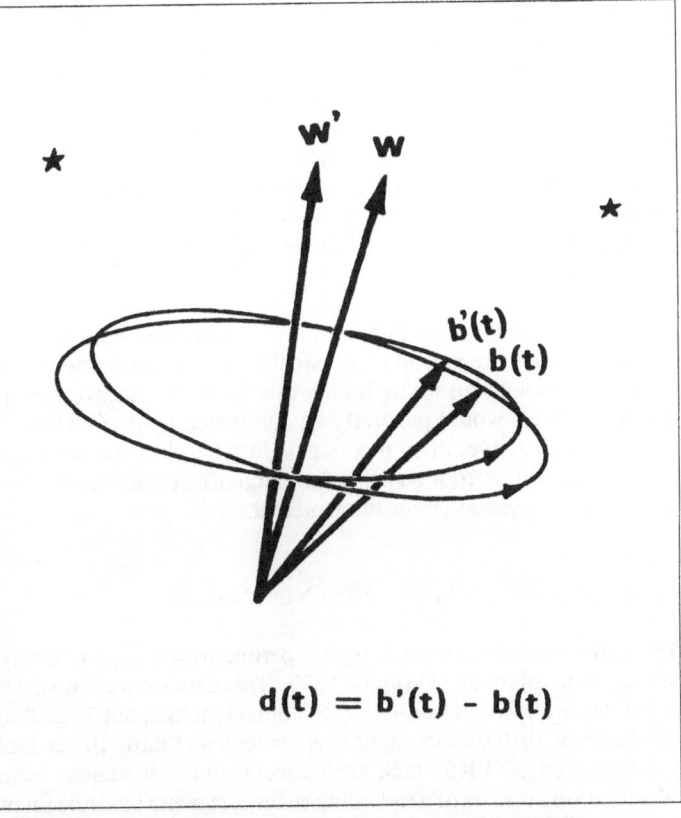

Figure 2. Comparison of diurnal motion of an interferometer baseline about two axes, w and **Figure 1.** Diurnal

For either of our 12-meter astrometric baselines the plane of the ellipse would obviously be vertical and, for the hypothesized quarter-arcsecond axis offset, the major axis would be 29 microns long. The oscillation would progress in a counterclockwise manner viewed from the south. For the south-north baseline, the plane of the ellipse would lie in a vertical east-west plane, with the major axis oriented vertically, and a minor-to-major axis ratio of 0.83 (eccentricity 0.56). For the south-east baseline, the ellipse would lie in a vertical plane oriented at an azimuth of 147 degrees, with the major axis tilted 51 degrees from vertical on a great circle passing through the celestial pole. The minor-to-major axis ratio would be 0.45 (eccentricity 0.89).

Suppose one of the baselines happened, by chance, to undergo the diurnal counterclockwise elliptical oscillation described above. That motion, viewed from an inertial frame, would be equivalent to rotation about an axis offset from that of the Earth by a quarter arcsecond. The exact orientation of the offset axis — the position of the apparent celestial pole for that baseline — would depend on the phase angle of the baseline as it traces out its ellipse as a function of local sidereal time.

Now the geometry of this elliptical motion is specified fairly tightly. The probability of any baseline undergoing this kind of motion by chance over the course of an entire day is minuscule. However, we have no way of knowing what our baselines do over the course of an entire day, and we do not need to know. The only information we have on baseline motion, and the only information that is important, is derived from our observations, which rarely take up more than 0.4 day. Furthermore, the size of the ellipse and the phase angle of the oscillation are arbitrary. Therefore, any arbitrary baseline motion might *simulate* an arc of the ellipse over the span of observations. The condition for this to occur is that the motion be dominated by periodic components with periods of about a day or longer. Another condition, that the motion preserve baseline length, is actually not required since a "stretch" or "shrinkage" of the baseline can be considered to be an independent (orthogonal) effect.

The baseline motions that we have experienced with the Mt. Wilson instrument are definitely dominated by long-period components; most often we assume them to be linear over the course of a night, although on many nights that is clearly an oversimplification. We actually expect that thermal effects would generally be dominated by periodicities of one solar day.

It seems possible, then, that our nightly baseline motion might mimic the motion that would be equivalent to that from an axis of rotation offset from that of the Earth. If that is indeed so, then our instrumental system is not fundamental.

5. An Experiment With a Real Night's Data

I have performed some numerical experiments with a particularly good set of one-color observations taken on the night of 21 August 1987. This data set was chosen because it spans the entire night, the residuals are low (4 microns RMS, equivalent to about 70 milliarcseconds on the sky), and because the baseline drift on that night was obvious and fairly linear-looking on both baselines. That night we observed 32 FK5 stars, each several times at various hour angles on both baselines. Each observation consists of a recording of time, star and baseline identifier, and the measured optical path length difference between the two active siderostats (in VLBI terminology, the "delay", expressed in units of length). We pass the observations through two programs, CALC and SOLVE. CALC computes the *a priori* geometry of the observation from conventional models of precession, nutation, aberration, Earth rotation, atmosphere, etc., an assumed constant baseline vector, and FK5 star positions (the models are documented in Kaplan et al. 1989). CALC compares its computed path length difference with that actually measured, producing path length (O-C)s. SOLVE is a standard least-squares program for extracting various kinds of geometric information, including star positions, from CALC's (O-C)s.

Figure 3(a) shows the 21 August data for the SN (south -> north) baseline, where the baseline is treated as a constant over the course of the night. Time increases downward, and the horizontal axis represents post-solution residuals in optical path length, expressed in microns. The gaps in the data represent periods of time when the SE (south -> east) baseline was in use. This is the simplest possible treatment of the data; star positions were not solved for (some of the scatter in the residuals is due to slightly incorrect star positions). Obviously treating the baseline as a constant is inadequate; the slope in the residuals amounts to about 40 microns over the course of the night for the SN baseline. The data for the SE baseline, which is not shown, show a slope of almost 20 microns (this was the "old" SE baseline, which was only 8 meters long).

Figure 3(b) shows the same data treated in the conventional way, that is, the three rectangular components of the baseline are assumed to vary linearly with time over the course of the night and the three rates of change for each baseline are solved for (the "linear-drift" model). The RMS of the residuals is 3.72 microns for the SN baseline data shown (4.22 for the SE baseline data).

Next, I added two parameters and their partials to the program SOLVE (all SOLVE parameters can be turned on or off at will). The parameters DELPSI and DELEPS are two small angles in the ecliptic system. (The fact that we have no sensitivity to the ecliptic is irrelevant for this purpose; we know well enough where it is on the sky and these are differential angles which we need to only about three significant digits.) These parameters provide the location of the baseline's apparent axis of rotation within the reference frame of the *a priori* star coordinates. The same parameters could yield information on precession/nutation modelling errors *if* the baselines were well-enough anchored in the Earth's crust so that we could assume that the baseline's apparent axis of rotation was coincident with that of the Earth. For the optical interferometer data, I cannot make that assumption; I simply want to know whether the baseline's motion — from whatever causes — can be modelled as equivalent to that from a slightly offset axis.

Figure 3(c) shows the results of my experiment with DELPSI and DELEPS. Rather than solving for the rates of change of the baseline components, I solve for DELPSI and DELEPS for each baseline. That is, I solve for the orientation of the axis about which each baseline appears to be rotating (the "offset-axis" model). I also solve for the rate of change of baseline length for the two baselines, so that the number of parameters in the solution is the same as for the run which produced Figure 3(b). The RMS of the residuals for this run is 3.98 microns for the SN data shown (4.18 for the SE data). Thus, the goodness of fit is about the same as for the conventional run. If I omit the change-in-length parameter the fit degenerates slightly, to an RMS of 4.30 microns. The values of DELPSI and DELEPS indicate axes a few tenths of an arcsecond offset from nominal but are not the same for the two baselines. The values of DELPSI and DELEPS are quite well determined — the worst case differs from zero by 2.6 times its mean error — and remain the same if the change-in-length parameter is omitted. The value of the change-in-length parameter corresponds to that which can be computed from the conventional linear-drift run. In short, the scheme works.

I have run a similar experiment with data taken on 8 October 1987. For this data set, the offset-axis model actually was slightly better for both baselines than the conventional linear-drift model. I should say that these data sets have not been pre- or post-selected in any way; they were chosen simply on the basis of the overall quality of the data and the strong signature of the baseline changes over the course of the night. I was actually quite surprised to see that, in practice, the two models were essentially equivalent, that is, for these data at least, modelling the baseline changes as a rotation about an offset axis is pretty much equivalent to modelling the changes as linear drifts with time. This is what we would expect, of course, from the geometry I have described in section 4, but I thought it important to verify the effect with real data. I do not claim that all optical interferometer data fit into this mold and I expect that neither model is entirely adequate.

What the offset-axis model provides is the baseline's axis of rotation within the reference frame of the star coordinates. If our baselines were tied to the Earth this would be quite valuable data, containing information both on large-scale rotations or distortions in the input star catalog and (accumulated over many years) on the validity of our models for Earth orientation and rotation. But without the assurance that the baseline's axis of rotation is that of the Earth, DELPSI and DELEPS do not provide us with any information that we can practically use. We have lost any well-defined

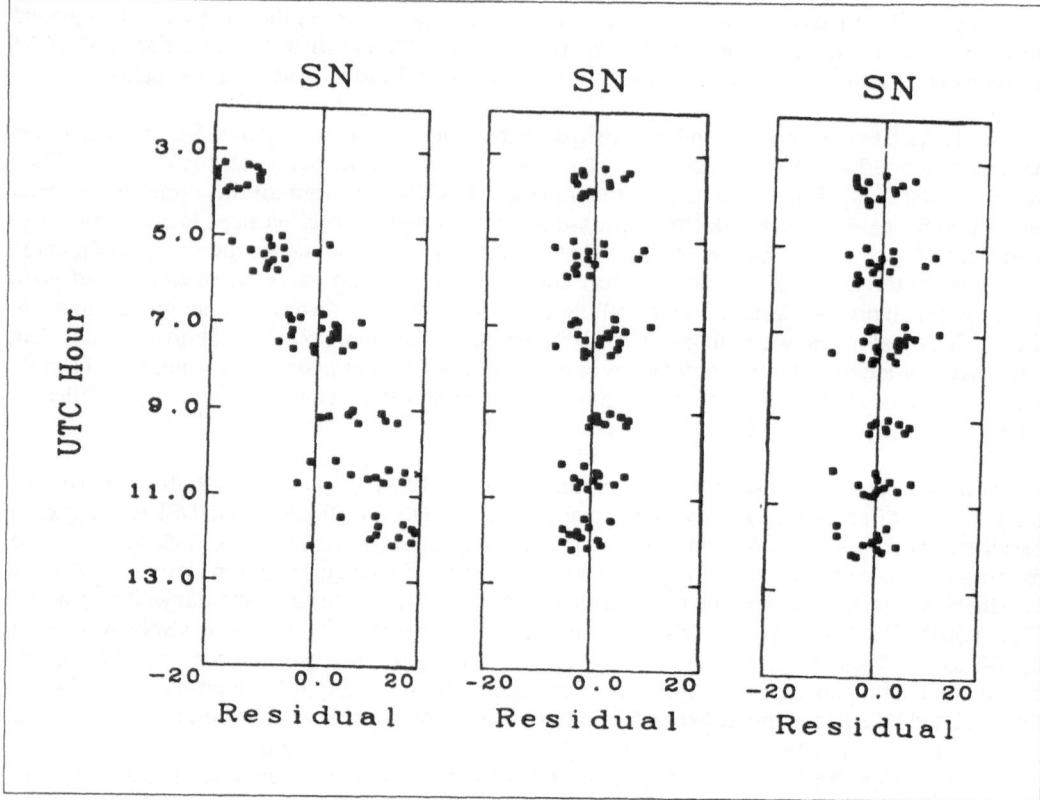

Figure 3. Interferometer path-length residuals, in microns, with various baseline motion models: (a) (left) no motion; (b) (center) linear drift of rectangular components; (c) (right) rotation about slightly offset axis

physical anchor for our coordinate system; we are reduced to trusting the input star catalog (the FK5) because our axis of rotation has no external significance.

We can always solve for star position corrections within the reference frame of the input star catalog. I put artificial errors in the catalog positions of three FK5 stars (numbers 52, 862, and 1619) by adding 0.20 arcsec to both coordinates of each. I can recover these errors from the 21 August data either from a solution based on the linear-drift model of baseline motion or a solution based on the offset-axis model. I then rotated the entire FK5 input catalog system (including the three star positions with the artificial errors) by 0.13 arcsecond and re-reduced the data. In the offset-axis model solution the DELPSI and DELEPS values adjusted themselves to reflect the additional rotation of the system. But the star positions corrections were unchanged, that is, they still indicated their original 0.20 arcsecond errors. Those are their errors *within the rotated FK5 system*, not their total errors. When I applied the linear-drift model solution, I obtained corresponding results — the rates of change of the baseline components adjusted themselves but the star position corrections were essentially unchanged. This is the point of this paper: *as long as we have to solve for baseline motion from the observations themselves, we can recover star position corrections only within the system of the input catalog. Our observations are not fundamental.*

6. Conclusion

In order to do fundamental astrometry, we must effectively "tie down" our baselines to the Earth in some well-defined way. Unfortunately, the siderostats and the piers they sit on must be exposed to many types of environmental influences. It is not realistic to expect baselines which are tens of meters long, exposed to the elements, to be stable to a few parts in 10^8 in three dimensions over many hours. Accepting the fact that there will be significant changes in the baseline components during our observations, we need to measure these changes with respect to either (1) the Earth itself or (2) an astronomical reference system known *a priori* to be fundamental to a high degree of precision.

The new Naval Observatory astrometric interferometer now being designed will incorporate a complex laser metrology system to continuously measure the siderostat positions with respect to Earth-fixed benchmarks. Guaranteeing that these benchmarks, called optical anchors, are stable at the micron level is a challenge, but one with which the geodetic community has had experience. Jim Hughes and Don Hutter are engaged in the design work for the metrology system, which will rest on thermally isolated invar tripods driven into bedrock.

The other approach is to use a small number of stars whose positions and proper motions are very precisely known within an established fundamental system, and measure the baseline motion from their observations alone. Quasars with VLBI positions would be the best candidates, but the interferometer would have to operate efficiently beyond 15th magnitude. We expect that the new Naval Observatory interferometer will be able to reach the brightest quasars, so we anticipate that this strategy will also be available to us.

We are entering an era when the most precise wide-angle astrometry — that performed from space platforms — will not be fundamental. It may be the case that the traditional Earth-based astronomical coordinates of right ascension and declination are actually not optimum for the very high precision astrometry of the future. There has already been much discussion in favor of abandoning the equinox as the origin of right ascension; perhaps this is only the beginning of the disassembly of the traditional fundamental astronomical reference system. Nevertheless, we should recognize that the need to relate terrestrial systems to astronomical systems will remain. The issues may be cast in a different terminology, but the basic scientific challenges remain. A small class of astronomical instruments —those which are capable of establishing what we now call fundamental systems — will remain the essential link between the sky and the Earth. With careful design, optical interferometers will soon be important members of this class.

References

Colavita, M.M., Shao, M., Staelin, D.H.: 1987, *Appl. Opt.* **26**, 4113.
Kaplan, G.H., Hughes, J.A., Seidelmann, P.K., Smith, C.A., Yallop, B.D.: 1989, *Astron. J.* **97**, 1197.
Kaplan, G.H., Hershey, J.L., Hughes, J.A., Hutter, D.J., Johnston, K.J., Mozurkewich, D., Simon, R.S., Colavita, M.M., Shao, M., Hines, B.E., Staelin, D.H.: 1988, in *Proceedings* of the

NOAO-ESO Conference on High Resolution Imaging by Interferometry, ed. F. Merkle (European Southern Observatory, Garching), p. 841.

Mozurkewich, D., Hutter, D.J., Johnston, K. ., Simon, R.S., Shao, M., Colavita, M.M., Staelin, D.H., Hines, B., Hershey, J.L., Hughes, J.A., Kaplan, G.H.: 1988, *Astron. J.* **95**, 1269.

Shao, M.: 1988, in *Proceedings* of the NOAO-ESO Conference on High Resolution Imaging by Interferometry, ed. F. Merkle (European Southern Observatory, Garching), p. 823.

Shao, M., Colavita, M.M., Hines, B.E., Staelin, D.H., Hutter, D. ., Johnston, K.J., Mozurkewich, D., Simon, R.S., Hershey, J.L., Hughes, J.A., Kaplan, G.H.: 1988, *Astron. Astrophys.* **193**, 357.

Discussion

KHARIN: (1) What is the diameter of the mirror at the siderostat and (2) what detectors were used?

KAPLAN: (1) The primary siderostat mirrors on Mt Wilson are 25 cm in diameter. However, the effective aperture of the instrument is only 8 cm. (2) We detect interference by modulating the instrumental optical path length by about one wavelength at a frequency of 500 Hz. We then slowly slew the delay line. When interference is achieved, it occurs only over a range in delay of about one wavelength because of the wide bandwidth. Therefore, we search for a 500 Hz modulation of the light in the combined beams, which indicates that the beams are interfering. We use standard cooled photomultiplier tubes for detection.

TREUHAFT: Is it practical to monitor the baseline motion as a function of temperature or by direct metrology?

KAPLAN: The baseline motion over the course of a night seems not to be generally repeatable from night to night, even though the Mt Wilson temperature function is. We have experimented with pier-to-pier metrology systems, but these monitor only distance (length), whereas our baseline motion is mostly in the form of a rotation.

MORRISON: (1) Is it your intention to publish a catalogue with the current Mt Wilson interferometer? (2) To what extent are the astrophysical and astrometric programmes complementary in the use of observation time? (3) How long will the current astrometric programme continue?

KAPLAN: (1) We are a little reluctant to publish a "catalog," as such, since (a) we can observe a relatively small number of stars; (b) we have only single-epoch positions; and (c) we do not yet fully understand our systematic errors. We have published (Mozurkewich *et al.* 1988) our observed offsets from FK5 positions based on one-color observations. Our two-color observations are much more precise, and we do plan to publish these results also.
(2) There is very little overlap—they are essentially independent observing programs.
(3) Undoubtedly until the new interferometer is built, in 1993.

BASTIAN: If among the stars reachable by your instrument there would be three radio objects, then you could check your results against independent measurements. Because there is nothing except radio data rivaling your accuracy, are there any such radio stars?

KAPLAN: I agree we are in the situation of having no good checks on our accuracy. The situation is somewhat like the early days of VLBI. HIPPARCOS observations would be *very* helpful. Our sky coverage and magnitude range is quite restricted with the current Mt Wilson interferometer, and there are not enough radio stars in our observing range to make a meaningful check.

Part 4
Realization and comparison of reference frames

Plate VIII. The original Pulkovo Observatory viewed from the northwest. The 15-inch
refractor can be seen in the open center turret. Nineteenth century photograph courtesy
of W.R. Dick.

Plate IX. The reconstructed Pulkovo Observatory viewed from the northwest in an aerial view. Photograph provided by Pulkovo Observatory.

THE DEFINITION AND STABILITY OF LOCAL INERTIAL REFERENCE FRAMES

R. N. Treuhaft and S. T. Lowe
Jet Propulsion Laboratory, California Institute of Technology
4800 Oak Grove Drive
Pasadena, California 91109, USA

ABSTRACT. Inertial reference frames spanning approximately $10°$-$30°$ square on the sky and capable of locating objects to few-hundred microarcsecond accuracies are useful for a broad class of astrometric measurements. Deep space tracking and general relativistic angular deflection experiments are examples of astrometric measurements which can profitably reference the positions and/or motions of objects to a field of radio sources in a local frame. A method for defining local inertial reference frames has been developed based on Very Long Baseline Interferometry (VLBI) measurements of extragalactic radio sources. By observing the radio emission from the object to be located in the frame, as well as that from about five radio sources which define the frame, dominant systematic astrometric errors can be minimized through parameter estimation. The entire reference frame measurement is of the order of 30 minutes including all the sources in a frame. The limiting error for single-epoch position determination in a local frame is the unknown structure of both target and reference objects. Structure can cause systematic milliarcsecond-level errors. The limiting error for epoch-to-epoch differential position measurements is tropospheric fluctuations, assuming that the radio source structures do not change from one epoch to the next. Preliminary results of an epoch-to-epoch measurement of relativistic gravitational deflection by Jupiter, in which the total deflection was about 600 microarcseconds, suggest that the local reference frame is stable at the 240-microarcsecond level over twelve days. Data have been taken at longer time intervals to determine the annual stability of the frames. At the time of preparation of these proceedings, those data have not yet been analyzed.

1. Defining a Local Inertial Reference Frame

A local reference frame in this report can extend over $10°$ to $30°$ square on the sky. It is inertial in the sense that positions and motions of an object referenced to the frame are attributable to forces *on* the object, as opposed to measurement errors associated with the object or the frame. The local inertial reference frame technique can be seen as an intermediary between the full sky radio frame and the few arcminute frame defined by a radio antenna beamwidth. In addition to being intermediate in spatial extent, its $\approx$200-microarcsecond (μas) precision is, on the average, better than global reference frame results, which are based on much larger data sets taken over years [Sovers, 1989, and Ma, 1989]; but the accuracy is poorer than the 5-10 μas achieved in the antenna beamwidth frame [Marcaide and Shapiro, 1983]. The accuracy attained in a given frame is a measure of how "inertial" the frame is. That is, it is a measure of the level at which positions and

J. H. Lieske and V. K. Abalakin (eds.), Inertial Coordinate System on the Sky, 253–260.
© 1990 IAU. Printed in the Netherlands.

motions measured in that frame can be attributed to forces acting on the object being measured.

Covariance studies have shown that dual frequency (for charged particle calibration) Very Long Baseline Interferometry (VLBI) measurements of five reference radio sources, with one source measured twice, define a frame which enables approximately 200-μas position measurements [Treuhaft, 1988]. This accuracy results by assuming perfectly known reference source positions, which implies that the structure of the reference and target sources is also perfectly known. In the presence of milliarcsecond structure [Charlot, 1989], a single local reference frame measurement can only produce few-milliarcsecond results. Unless all sources in the frame are either mapped, or their structures parameterized to much better than 1 milliarcsecond, the full potential of the local reference frame approach can only be realized in epoch-to-epoch differential measurements. If the epochs of the individual reference frame determinations are sufficiently close that the reference source structure effects do not vary at more than the 100-μas level, then differential accuracies of the order of 300-μas should be attainable on the differential measurement.* Epochs separated by less than a few months should satisfy that criterion [Porcas, 1987].

As described in detail elsewhere [Treuhaft, 1988], the seven observations in a minimal local reference frame observation set determine seven parameters. One parameter is the projection of the position of the target object onto the baseline vector at the time of the measurement. Two parameters are associated with clock behavior: epoch and rate. Two parameters are associated with orthogonal earth rotations, and two with static zenith tropospheric delays, one for each station. Assuming 20 picoseconds system noise on each VLBI group delay measurement implies 120-μas accuracy for the single-epoch measurement. Tropospheric and source structure/position uncertainty effects, when treated with a consider analysis [Bierman, 1977], yield the errors quoted above.

2. The Short Term Stability of Local Reference Frames

The stability of a reference frame will be defined as the rms scatter of repeated position determinations in that frame. A test of the short term stability of local reference frames was performed by measuring the relativistic gravitational deflection of the radiation from P 0201+113, which passed within 200 arcseconds of Jupiter on 21 March 1988. Two VLBI experiments were performed, one on 21 March 1988 and one on 2 April 1988. The Deep Space Network stations involved were DSS 13 and DSS 15, the 26-m and 34-m antennas at Goldstone, California, and DSS 43, the 70-m antenna near Canberra, Australia. Data were taken at 2.3 GHz and 8.5 GHz with spanned bandwidths of approximately 40 and 100 MHz for the low and high frequency band respectively. A total of 28 channels, each sampling at 4 Mbits/second were used. According to general relativity, the maximum difference in apparent position of the source, projected onto the baseline vector, between the two observation epochs was about 600 μas; this difference in apparent position is due to deflection by Jupiter's gravitational field on 21 March. The effect of Jovian gravitation on 2 April was down to 25 μas. Ten local reference frame measurements were obtained for each day. The frames used are described below.

In figure 1, preliminary results of the measured angular deflection of P 0201+113 are shown versus observation time. Only the DSS 13 to DSS 43 baseline is shown. The times given are the universal times of the observations of the target source on 2 April. The

---

* The differential accuracy is $\sqrt{2}$ times the "structure-free" single-epoch accuracy, because the measurement consists of a difference between two effectively independent measurements.

TABLE 1. Radio Sources Used in Local Reference Frame Stability Test

Source Name	Right Ascension hr min sec	Declination degrees min sec
P 0201+113	02 03 46.65734	11 34 45.3629
P 0019+058	00 22 32.44128	06 08 04.2715
P 0106+01	01 08 38.77113	01 35 00.3200
GC 0119+04	01 21 56.86177	04 22 24.7377
P 0202+14	02 04 50.41399	15 14 11.0450
CTD 20	02 37 52.40576	28 48 08.9919
GC 0235+16	02 38 38.93021	16 36 59.2753
OD 166	02 42 29.17103	11 01 00.7285
3C 454.3	22 53 57.74796	16 08 53.5614

Details of the data shown follow. The data in figure 1 are for DSS 13 to DSS 43 only. The solid curved line shows the expected gravitational deflection by Jupiter versus time, and the horizontal line shows the result if the experiment were totally insensitive to the gravitational deflection difference between the two epochs. The rms scatter of the differential results about the expected Jovian deflection is 336 μas, which implies a single-epoch stability of 240 μas. The reduced χ^2 values in the figure are calculated for the expected Jovian deflection between epochs, χ_J^2, no detectable change between epochs, χ_0^2, and a constant bias between epochs, χ_C^2. The probability that Jovian deflection would produce a χ_J^2 of the value quoted or greater is approximately 0.2. The probability that zero actual deflection would produce the χ_0^2 value shown or greater is much less than 0.005, and the corresponding probability for a constant bias is 0.008. Thus this preliminary result is more consistent with planetary gravitational deflection than with either a null detection or a constant bias between epochs. However, the value of χ_J^2 indicates that errors have been underestimated by about 20%, on the average. Actually, it is clear from inspection of the figure that underestimation of the error on only one or two points is a more likely explanation of the large χ^2. A possible explanation of the underestimate of the error on the first point is given in the next paragraph.

The results of figure 1 are preliminary in several respects. The reduced χ^2 values calculated do not include known correlations between measurements. These correlations can be substantial between small-net determinations and the large-net measurement from which earth rotation and troposphere parameters were derived. A precise treatment of these correlations is underway. The underestimation of errors is in part due to the failure to account for tropospheric fluctuation effects. Incorporating these effects in the formal error using a fluctuation model [Treuhaft and Lanyi, 1987] is also underway. For example, a calculation of the fluctuation effect on the first data point, which was taken at 10° elevation in Australia for P 0201+113, yields an adjusted error bar of approximately 500 μas, as opposed to the 240 μas bar shown. It should also be noted that in the last large network (the ninth point from the left), one of the seven observations failed, and seven parameters could not be estimated. Earth rotation parameters were therefore used from the previous large network observed two hours beforehand (the fourth point from the left). The errors introduced by this approach should be well below the formal error of the result, but this must be verified. There may also be effects due to errors in the cross correlation of the VLBI data which are currently under investigation. These errors could be of the order of the error bars, changing the conclusion substantially. Given those cautions, the

256

angular shifts shown are the difference in measured position between the two epochs at equal sidereal times. That is, the observation schedule of 21 March was exactly repeated on 2 April, except for one or two instrumental failures on reference frame sources. If a reference frame source failed at one epoch, the analogous scan was excluded from the analysis at the other epoch. Failure to do so—e.g. using four reference sources for a P 0201+113 position determination at one epoch and five at the other—resulted in errors as large as 1 milliarcsecond. These milliarcsecond errors are probably due to apriori source coordinate errors of that order.

As mentioned above, each point in figure 1 corresponds to a local reference frame measurement consisting of a target observation of P 0201+113, and one or more reference sources. Only the first, fourth, and ninth measurements consisted of the seven nominal observations needed for a complete parameter set determination. All other points were derived from smaller networks, in which either two or three observations were used. In these smaller networks, a clock epoch (for only two observations) or a clock epoch and rate (for three observations) were estimated along with the projection of the source vector onto the baseline. All other parameters in smaller networks were fixed at the values derived from the previous 7-observation measurements. A complete list of the target (first entry) and reference source J2000 positions [Sovers, 1988] is given in table 1 below. Reference sources for the larger networks differed slightly, depending on mutual visibility. The reference source used in the smaller networks was the one closest to P 0201+113, P 0202+14.

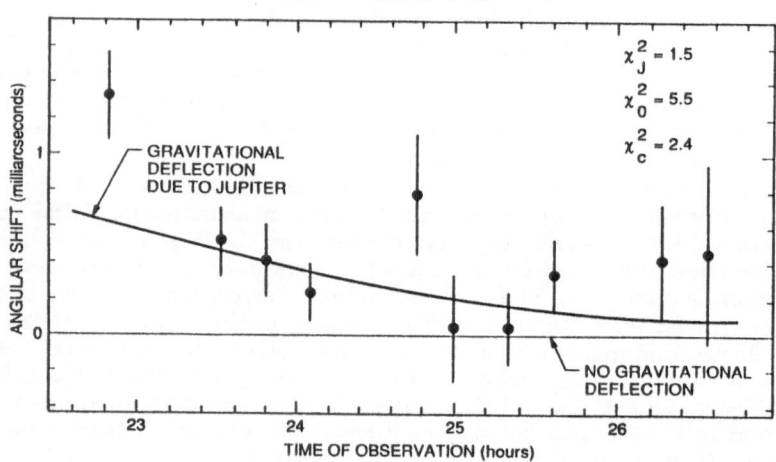

Figure 1. Differential angular deflections of P 0201+113, between observation epochs separated by 12 days, versus time of observation at second epoch. The deflections were measured over the DSS 13 to DSS 43 baseline.

data currently suggest that 10 repeated local reference frame differential measurements–i.e. 20 single-epoch measurements, of approximate accuracy 340 μas, were sensitive to the change in target position induced by planetary relativistic gravitational deflection over the course of twelve days. If this conclusion withstands further analysis and checks, this local reference frame measurement will be the first to detect planetary deflection; all other solar system deflections have been observed in the sun's gravitational field. It is worth noting that a single deflection measurement of the above accuracy 2° from the sun's center would enable a post-Newtonian γ parameter determination to 1 part in 1000, equaling the accuracy of the best determination [Reasenberg, et. al., 1979], if the phase fluctuations induced by the solar plasma could be calibrated. Repeated independent measurements could substantially improve the knowledge of γ.

3. Summary and Future Directions

A local inertial reference frame can be defined by six reference observations of five radio sources. Measuring a radio emitter's position within that frame requires one more observation of that object, totalling seven observations for a complete single-epoch measurement. The six reference source observations are used to solve for six parameters relating to the dominant errors associated with clock behavior, earth rotation, and static tropospheric delay. Single-epoch measurement accuracies are limited by radio source position uncertainties and source structure effects at the milliarcsecond level. They are also limited by tropospheric fluctuations. Differential measurements made between epochs over which radio source structures, although possibly unknown, do not change, are limited by tropospheric fluctuations. Differential measurements made over few-month time scales should therefore be insensitive to uncalibrated, but stationary, radio source structure or position uncertainty.

A test of the differential measurement technique, aimed at detecting relativistic gravitational deflection by Jupiter, currently suggests that over a 12-day interval, 240-μas single-epoch stabilities were observed. The differential measurement results imply that single-epoch measurements are much more stable than accurate, as expected. Assuming three different hypotheses, 1) the observed deflections arose from gravitational deflection, 2) they arose from zero actual deflection, and 3) they arose from a constant bias in the deflection measurements, yields probabilities of 0.2, much less than 0.005, and 0.008. These numbers are the probability of observing the quoted reduced χ^2 or greater for each hypothesis. The preliminary nature of these probabilities and the error analysis in the experiments was discussed.

In addition to further analysis of the data presented, analysis of the DSS 15 to DSS 43 data will also be completed. At the moment, the conclusions from that largely independent measurement are similar to those presented here. Data taken on the DSS 15 to DSS 45 (the 34-m antenna in Australia) in September of 1989 will have been processed by the time these proceedings are published. The same observing schedules were run at two week time intervals. Differential measurements of P 0201+113 between the two-week epochs will allow a further check of instrumentation and the differential technique. Because Jupiter was many tens of degrees away for both observations, the angular shifts should be consistent with a null effect over that time interval. Comparing the 1988 epochs to the 1989 epochs will allow an assessment of the level to which source structure fluctuations and uncalibrated instrumental changes corrupt the data on annual time scales.

Should experiment confirm that the single-epoch stability of local reference frames is at the 240-μas level, as suggested by the current data set and error analysis, reducing dominant error sources could then be addressed [Treuhaft, 1988]. Applying water vapor radiometer (WVR) calibrations to the data should reduce the tropospheric fluctuation error due to the wet component of the atmosphere. The level to which that component can be calibrated over 1 to 30 minute time scales is not determined as yet. If the 200-μas level of the wet fluctuation error is reduced, for example, by an order of magnitude, then system noise errors of the order of 100 μas will become dominant. Successful use of the VLBI phase delay observable, as opposed to the group delay observable on which the described technique has been based, would reduce the system noise component of the error to about 4 μas. Beyond using phase delays and WVRs, dry tropospheric fluctuations at about 50 μas might be calibrated with barometric arrays around each station of the baseline, in an attempt to increase the local reference frame stability to that of the antenna beamwidth technique. It is also conceivable that radio source structure fluctuations could be parameterized and estimated with additional reference frame observations, which would then allow single-epoch accuracies to be of the order of single-epoch stabilities. This approach has not yet been considered quantitatively.

4. Acknowledgment

The research described in this report was carried out at the Jet Propulsion Laboratory, California Institute of Technology, under a contract with NASA.

REFERENCES

Bierman, G. J., *Factorization Methods for Discrete Sequential Estimation*, Academic Press, New York, 169, 1977. Charlot, P., "Radio Source Structure in Astrometric and Geodynamic VLBI," submitted to *Astronomical Journal*, 1990.

Ma, C., "The Realization of an Inertial Reference Frame from VLBI," in proceedings of IAU Symposium 141, 1989.

Marcaide, J. M. and Shapiro, I. I., "High-Precision Astrometry via Very Long Baseline Radio Interferometry: Estimate of the Angular Separation Between the Quasars 1038+528A and B," *Astronomical Journal*, **88**, 1133, August 1983.

Porcas, R. W., "Summary of Known Superluminal Sources," in *Superluminal Radio Sources*, ed. Zensus, J. A., and Pearson, T. J., Cambridge University Press, Cambridge, p. 18, 1987.

Reasenberg, R. D., Shapiro, I. I., MacNeil, P. E., Goldstein, R. B., Breidenthal, J. C., Brenkle, J. P., Cain, D. L., Kaufman, T. M., Komarek, T. A., and Zygielbaum, A. I., "Viking Relativity Experiment: Verification of Signal Retardation by Solar Gravity," *Astrophysical Journal*, **234**, L219, December 1979.

Sovers, O. J., Edwards, C. D., Jacobs, C. S., Lanyi, G. E., Liewer, K. M., and Treuhaft, R. N., "Astrometric Results of 1978-1985 Deep Space Network Radio Interferometry: The JPL 1987-1 Extragalactic Source Catalog," *Astronomical Journal*, **95**, 1647, June 1988.

Sovers, O. J., "An Extragalactic Reference Frame from DSN VLBI Measurements–1989," in proceedings of IAU Symposium 141, 1989.

Treuhaft, R. N. and Lanyi, G. E., "The Effect of the Dynamic Wet Troposphere on Radio Interferometric Measurements," *Radio Science*, **22**, 251, March 1987.

Treuhaft, R. N., "Deep Space Tracking in Local Reference Frames," *Jet Propulsion Laboratory Telecommunications and Data Acquisition Progress Report*, **42-94**, 1, August 1988.

Discussion

YATSKIV: Some years ago, Dr Kovalevsky has proposed to use the terms "small field astrometry" and "global astrometry." You are using "local" and "inertial" frames. What is the reason for these changes?

TREUHAFT: I agree in principle with the terms "small field" and "global" astrometry. The only reason for the change is that "small field" usually applies to fields of the order of 1° x 1°, but in this report I am discussing fields of about 20° x 20°.

KHARIN: Is it possible to obtain radiointerferometric positions of the natural satellites and, if so, what is the accuracy of these observations?

TREUHAFT: Because the natural satellites are angularly broad, on the order of one arcsec, they can only be observed on short baselines. For example, observations of the Galilean satellites, with the Very Large Array in New Mexico (having a 30 km maximum baseline), have yielded approximate accuracies of 30–50 mas. This should be compared to the milliarcsecond or better accuracy attainable on compact extragalactic sources.

XU: How can you get enough reference sources in such a small field?

TREUHAFT: The density of current radio source catalogs (about 200 sources around the sky above −45° declination) is sufficient to guarantee approximately 5 reference sources in the 20° x 20° field used.

YE: What would be the approaches you would use in order to measure phase delay instead of group delay?

TREUHAFT: The phase delay is a higher accuracy data type than the group delay. However, much more accurate *apriori* information is required to employ this data type. We plan to use high precision group delays as the *apriori* information needed for employing the phase delays.

XU: Your results are from differential VLBI. Does that mean it is the internal precision?

TREUHAFT: The preliminary results shown in this paper are epoch-to-epoch differential radio source coordinate determinations. I have referred to the results in terms of "differential accuracy" but the term "precision" is also appropriate. Since two independent data sets were compared, however, I would not use the term "internal precision" which, to me, implies an internal consistency check on one data set.

ZHDANOV: What are the prospects for "measuring" the post-newtonian parameter γ?

TREUHAFT: If 200–300 microarcsecond precision data could be obtained on a raypath 2° from the Sun's center, then a γ measurement of about 2 parts in 1000 would result. This accuracy, which is comparable to the current knowledge of γ, could be obtained with a *single* pair of 30-minute measurements. Many such pairs of measurements could be made to statistically improve the result. The problem of solar plasma fluctuations is now being studied.

AN EXTRAGALACTIC REFERENCE FRAME FROM DSN VLBI MEASUREMENTS – 1989

O. J. SOVERS
Jet Propulsion Laboratory, California Institute of Technology
4800 Oak Grove Drive
Pasadena, California 91109

ABSTRACT. Assessment of the impact of recent improvements in Deep Space Network (DSN) instrumentation, as well as of joint data analyses, provide a prognosis for the accuracy level to be expected in future realizations of an inertial radio reference frame. Intercontinental dual-frequency radio interferometric measurements during 68 sessions (including two recent sessions employing Mark III instrumentation) from 1978 to 1989 using NASA's DSN stations in California, Spain, and Australia give 8900 pairs of delay and delay rate observations. Analysis yields a catalog of positions of 200 extragalactic radio sources north of −45° declination. The resulting source position formal uncertainty distributions peak below 1 milliarcsecond, with three fourths being smaller than 2 mas. Comparison with independent measurements shows some evidence for systematic errors at the milliarcsecond level.

1. Introduction

During the past decade, the field of radio interferometric measurements on intercontinental baselines has matured, and is now on the threshold of yielding determinations of angular position that are accurate to a milliarcsecond over the entire sky. The extragalactic radio sources which are observed with Very Long Baseline Interferometry (VLBI) are thus the leading candidates for establishing a reference frame for angular positioning at the 1 mas level. This paper is a report of recent progress in VLBI astrometric measurements with Deep Space Network antennas. Recent instrumentation improvements promise a substantial increase in measurement precision in the 1990s. The source position catalogs resulting from two independent astrometric programs are compared, and the effects of errors in the standard models of precession and nutation are considered in an attempt to establish a confidence level for the accuracy of VLBI reference frames.

2. Observations and Analysis

A total of 68 day-long VLBI observing sessions were carried out during 1978-89 on the two DSN intercontinental baselines: California to Spain and California to Australia. Until late 1988, the experiments employed Mark II data acquisition systems and a bandwidth synthesis technique with channels of 2 MHz bandwidth, spanning about 40 MHz at both

261

J. H. Lieske and V. K. Abalakin (eds.), Inertial Coordinate System on the Sky, 261–270.
© 1990 *IAU. Printed in the Netherlands.*

2.3 and 8.4 GHz. Recently, Mark III data acquisition systems were introduced, with spanned bandwidths of 100 to 400 MHz. 8578 pairs of Mark II delay (D) and delay rate (DR) observables were included in the analysis, along with 321 D+DR pairs from two Mark III sessions in 1989. All observations were dual-frequency, and employed H-maser frequency standards.

Modeling was performed in Solar System barycentric coordinates defined in terms of the mean equator of J2000.0, and used the Project MERIT standards (Melbourne et al., 1983) for astronomical constants and Earth models, with certain exceptions noted below. Tropospheric delay modeling employed the Lanyi (1984) mapping function, and included the effects of varying surface temperature. The usual reference source 3C 273 fixed the origin of right ascension. Universal time and polar motion values were taken from the uniform series BIH87C02, extended through 1989. A multiparameter diagonally weighted least-squares fit estimated session-specific clock and station location parameters, a troposphere zenith delay at each station for every 3-hour period, and the right ascension and declination of each source. Corrections to the present IAU model of nutation were estimated for each session in the form of daily nutation offsets in longitude and obliquity. This is imperative for the Mark III data, which show a 50% higher scatter without nutation corrections.

3. Results: 1989 DSN Source Catalogs

Several fits to DSN VLBI data during 1989 produced a number of source catalogs. Three of these are described in some detail in order to illustrate some points concerning effects of selection of data and alternative sets of estimated paramaters.

Two catalogs were generated in early 1989 for submittal to the International Earth Rotation Service (IERS). One, JPL 1989-3, was our standard analysis of all DSN data through 1988. This fit estimated a set of coordinates for each Spanish or Australian station, along with the position of the celestial pole, in each observing session. The latter results were expressed in terms of corrections to the 1980 IAU angles in longitude ($\Delta\psi$) and obliquity ($\Delta\varepsilon$). The data base for a second catalog, JPL 1989-2, also included all TEMPO (Time and Earth Motion Precision Observation) dual-frequency data collected at JPL from 1980 to 1988, an additional 5340 observations. The relatively short duration (<3 hr) of the TEMPO observing sessions prohibits estimation of station coordinates and celestial pole position for each session; hence the fit assumed the Minster-Jordan AM02 plate motion model, and estimated corrections to the 1980 IAU precession constant and nutation amplitudes. With the exception of a reference day, values of UT1 and the two components of polar motion were also estimated.

The third catalog, JPL 1989-5, is an exact analog of 1989-3, with the exception that 669 new observations were included. Half of the new observations were derived from analyses of the first two DSN VLBI sessions employing Mark III data acquisition systems. Table 1 summarizes the characteristics of the observations, modeling, and statistics of the resulting source position uncertainties for these three catalogs, while Figs. 1 and 2 show histograms of average observation epochs and uncertainties in arclengths in the RA and δ directions for JPL 1989-5.

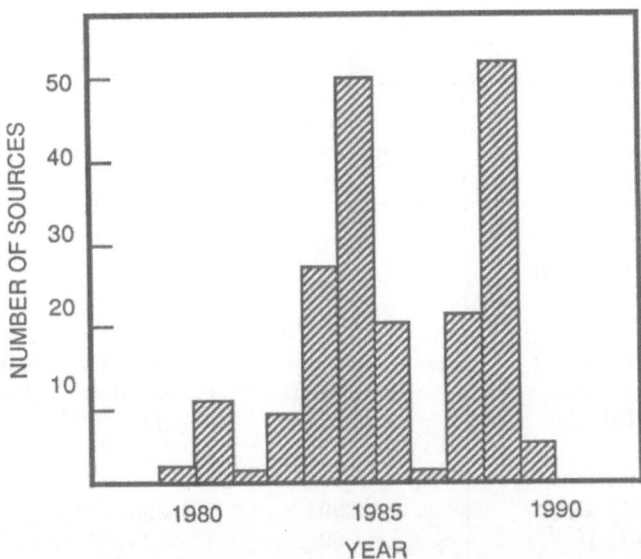

Figure 1. Average observation epochs for the 200 sources in the JPL 1989-5 catalog.

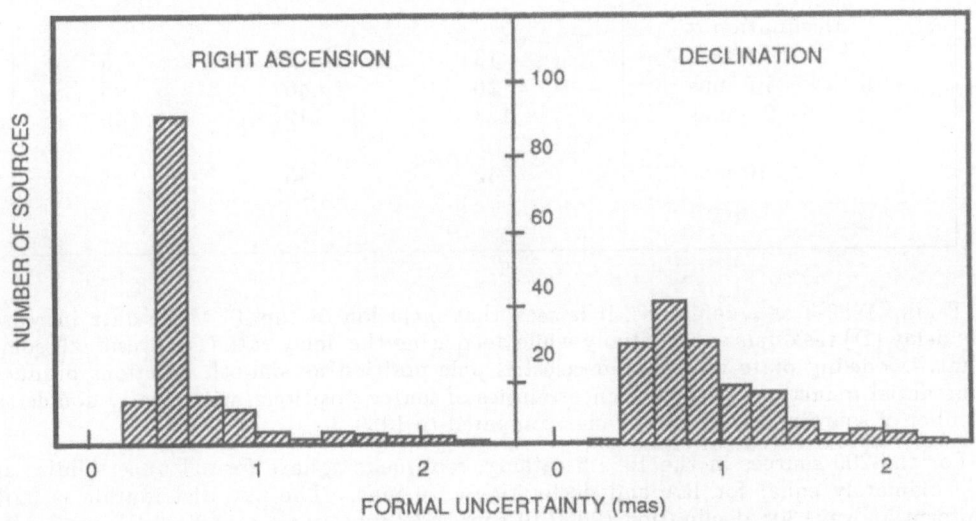

Figure 2. Histograms of formal uncertainties in RA (arclength) and δ for JPL 1989-5.

Table 1. Recent DSN Source Catalogs

Catalog Name	1989-2 TEMPO	1989-3 IERS	1989-5 Current
Obsvs.: Mk II	13570	8230	8578
Mk III	–	–	321
Modeling:			
Tectonics	AM0-2	Estimated	Estimated
UTPM	Estimated	Constant	Constant
Nutation	Amplitudes	Angles	Angles
Post-fit resids.:			
RMS D, ps	399	330	318
RMS DR, 10^{-13}	1.29	1.35	1.33
Number of sources	186	195	200
Declination σ			
<0.5 mas	13	1	5
<1.0 mas	76	46	95
<2.0 mas	123	112	149
>10 mas	42	45	21

Taking 1989-3 as a reference, it is seen that inclusion of the TEMPO data increases the delay (D) residuals substantially while decreasing the delay rate (DR) residuals somewhat. Modeling plate motion and celestial pole position as smooth functions produces substantial reductions in formal uncertainties of source positions, with nearly double the number of sources having $\sigma_\delta < 1$ mas compared to 1989-3.

For the 200 sources in the 1989-5 catalog, root-mean-square formal uncertainties are approximately equal for RA and declination: 1.6 mas. The sky distribution is fairly uniform above $-30°$ declination (20 ± 10 sources/sterad), and a factor of 5 sparser for $-30° > \delta > -45°$. Recent emphasis in our observing program has been on sources near the ecliptic plane, for potential use in navigating interplanetary spacecraft. This is reflected in some of the non-uniformity between $\delta = \pm23°$. Approximately 10% of the sources have $\sigma_\delta > 10$ mas as a consequence of being recent additions to the observing program. Mean source observation epochs have a wide range, with peaks of ≈50 each in 1984.0−1985.0 and 1988.0−1989.0.

3a. Impact of Mark III Data

The impact of improved data quality is seen by examining the residuals in the fit producing the JPL 1989-5 catalog (Table 2). Mark III residuals are nearly a factor of 8 smaller than Mark II residuals, with a less dramatic improvement in delay rates. The instrumentation is so stable that data from one of the two Mark III observing sessions are adequately fit with a single (two-parameter) linear clock model over the entire 24 hours. At this level of tens of picoseconds, source structure effects are also becoming obvious. Comparing 1989-5 with 1989-3 shows that adding 321 Mark III and 348 Mark II observations has doubled the number of sources with $\sigma_\delta < 1$ mas, and increased the number with $\sigma_\delta < 2$ mas by 1/3. Expectations for JPL source catalogs in the near future are thus ≈ 200 sources with formal declination uncertainties below a milliarcsecond.

Table 2. 1978-89 Fit Residuals: Impact of Mark III Data

Data	Number of observations	RMS delay (ps)	RMS rate (fs/s)
All	8899	318	133
Mk II	8578	324	134
Mk III	321	43	112

3b. Nutation and Precession Corrections

To summarize the corrections to the 1980 IAU models of precession and nutation implied by DSN VLBI measurements, Table 3 reports the results of one such fit. The data base and parametrization were identical to those used to generate the 1989-5 catalog, with the exception that the precession constant, as well as the amplitudes of both in- and out-of-phase semiannual, annual, and 18.6-year nutation terms were estimated. These results are essentially in agreement with post-fit analyses of $\Delta\psi$, $\Delta\varepsilon$ values from the 1989-5 fit. Here the annual and semiannual nutations are terms #10 and #9, respectively, in the 1980 IAU series (Seidelmann, 1982). With the exception of the 18.6-year nutation in obliquity, the out-of-phase corrections are not significant. In contrast, all corrections to in-phase terms exceed 2σ, and the precession correction is also highly significant. These results further underline the necessity of correctly modeling the motion of the celestial pole in analyses of VLBI observations.

Table 3. Precession and Nutation Amplitudes from 1978-89 DSN VLBI Data

Term	In phase	Out of phase
Precession, mas/yr	-1.96±0.13	...
18-year ψ, mas	-3.45±2.12	0.07±1.00
ε	0.44 0.17	0.59 0.22
Annual ψ, mas	3.98±0.35	0.03±0.37
ε	1.57 0.13	0.10 0.11
Semiannual ψ, mas	1.27±0.28	0.38±0.36
ε	-0.51 0.12	0.17 0.13

It must be emphasized that the 12-year extent of data is not sufficient to separate precession from the 18-year nutation. Magnitudes of correlation coefficients between precession and any 18-year term do not exceed 0.79, however. The dominant problem is the high (0.93) correlation between in- and out-of-phase 18-yr nutation terms.

4. Accuracy Assessment: Source Coordinate Comparisons

One traditional way of assessing accuracy has been to perform comparisons of independently determined source coordinates. This was done for the 98 sources common to JPL 1989-5 and a recent catalog (GSFC 1989) based on CDP and IRIS data (Ma et al., 1989). After removal of a rotation (-2.1, 0.6, 5.1 mas about the x, y, z axes respectively), three sources (P 0420$-$01, P 1055$+$01, 3C 279) are noted as outliers ($> 4\sigma$ differences in RA). The comparison is summarized in Table 4, where it may be seen that the only effect of omission of the three outliers is to reduce the normalized χ^2 for RA differences. Figure 3 shows differences in RA and δ (arclengths), as well as the differences normalized by the root-sum-squared errors (note that the unmodified formal uncertainties are used for both catalogs, and the three outliers are included). With the exception of the RA outliers noted above, the normalized differences appear to be normally distributed. Examination of the position dependence of JPL-GSFC differences reveals no apparent systematic differences.

Root-mean-square differences in RA (arclength) and δ are 2.0 and 1.3 mas, respectively. If the outliers are omitted, both are nearly normally distributed ($\chi^2_\nu \approx 1$). The RMS arclength difference for 4465 common arcs between sources is 2.5 mas, with no discernible trend vs. arc length over the range of 0° to 180°. The normalized χ^2 for arclength differences, however, is 1.5 (2.7 if the three outliers are included). The sizable x and y rotations point to the need for reconciling nutation modeling (see Sec. 5 below), while the z rotation is exactly equal to the rotation that was applied in the process of generating GSFC 1989 to align it with the FK5 system. The magnitudes of the differences indicate that realistic errors are probably close to the formal uncertainties, with the exception of the three sources noted above.

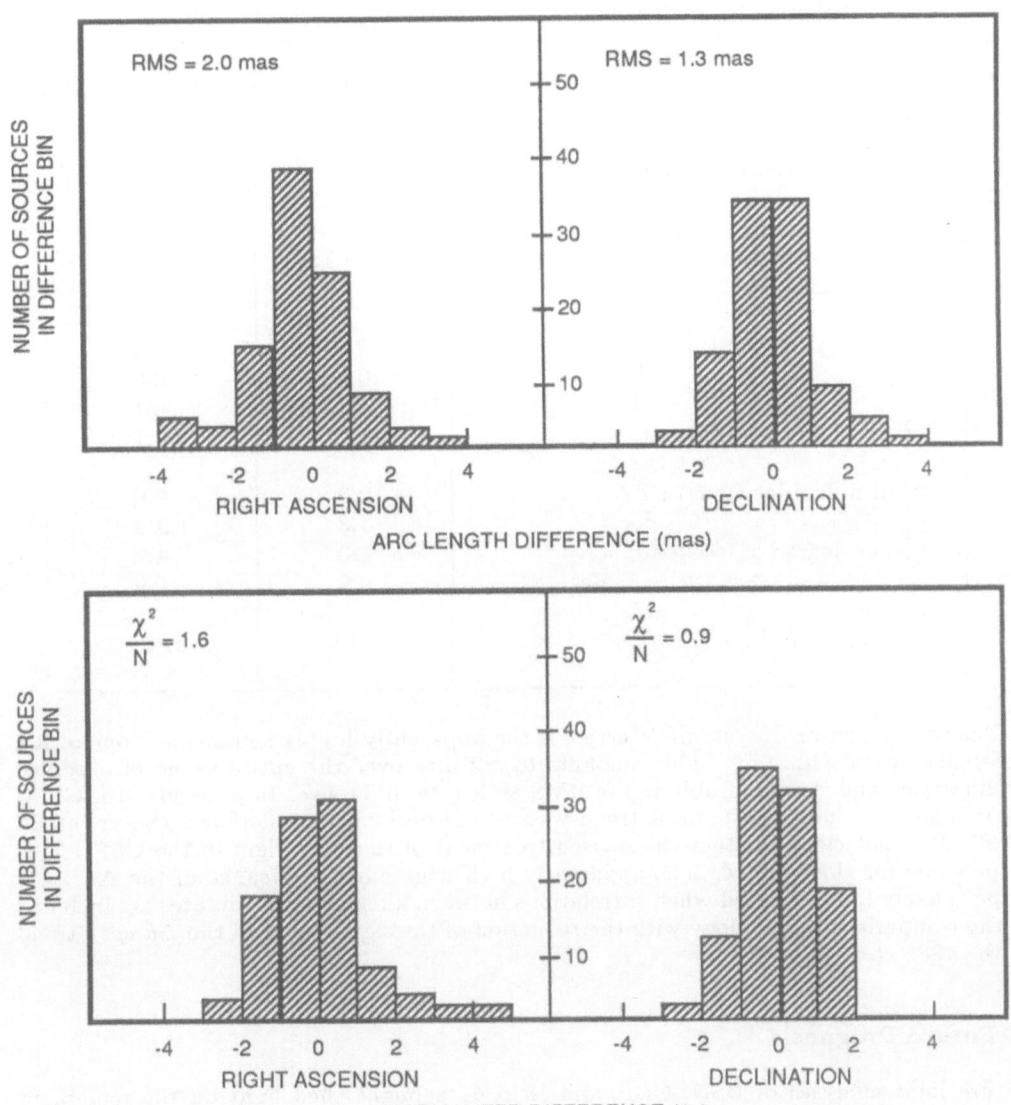

Figure 3. Histograms of differences between coordinates of 98 sources common to the JPL 1989-5 and GSFC 1989 catalogs. For the two lower plots, the differences are normalized by root-sum-squared formal uncertainties.

Table 4. Catalog Comparison: JPL 1989-5 vs. GSFC 1989

	All	Omit 3
Number of common sources	98	95
RMS uncertainty for common sources		
JPL 1989-5 : RA, dec. (mas)	1.7 1.7	1.7 1.7
GSFC 1989 : RA, dec. (mas)	1.2 0.6	1.2 0.6
Rotational offsets (mas) x	−2.1	−2.1
y	0.6	0.6
z	5.1	5.1
χ^2 per degree of freedom	1.7	1.1
RMS difference (mas) : RA	2.0	2.0
dec.	1.3	1.3
χ^2 per degree of freedom : RA	1.6	1.2
dec.	0.9	0.9
Δ(RA) vs. δ slope (μas/deg)	19±4	19±4

The sole indicator of systematic errors is the apparently highly significant slope of RA differences vs. declination. This amounts to ≈2 mas over the entire range of observed declinations, and is comparable to the RMS arclength difference. In previous JPL-GSFC comparisons, a similar systematic trend was seen in declination differences (Sovers et al., 1988). It is not clear whether the revised treatment of the RA origin in the GSFC fit is responsible for this change. The apparently high degree of significance of the $\Delta\alpha$ vs. δ slope is likely to be reduced when correlations between all source coordinates are included in the comparison, by analogy with the reduction of the significance of the $\Delta\delta$ vs. δ trend in the 1988 comparison.

5. Future Prognosis

Future joint analyses of DSN, CDP, and IRIS data might shed light on the remaining discrepancies between magnitudes of formal uncertainties and catalog differences. Analyses of 1984-1989 IRIS data, as well as of a considerable fraction of the CDP observations of astrometric quality dating back to 1979, are nearly complete at JPL. This is being done with independent parametrization and independent software (Masterfit: Sovers and Fanselow, 1987). Results of the new southern astrometric program (Russell and Johnston, 1989) will densify sky coverage and contribute significant numbers of source positions in the previously unexplored region of the celestial sphere below −45° declination. Plans are to eventually combine the several hundred thousand VLBI observations from all these programs to generate a definitive extragalactic reference frame.

Preliminary work exposes one danger of such an undertaking: if the position of a source is very different in fits to two individual data sets, the discrepancy may enter either the fit residuals or bias other estimated parameters in the combined fit. Only to the extent that it is practical to examine such details of large-scale combined fits will it be possible to have complete confidence in the resulting uncertainty estimates.

Uncertainties due to defects in the current IAU nutation model may misalign the entire VLBI reference frame by several milliarcseconds, as was seen in the presence of GSFC-JPL rotational offsets in Table 4. That such rotations are entirely due to errors in nutation modeling was demonstrated by adding the CDP session of 1980/10/17 (the GSFC nutation reference day) to the fit producing JPL 1989-5. The x and y rotational offsets between this modified JPL 1989-5 and GSFC 1989 are reduced to below 0.5 mas. Naturally, errors in celestial pole coordinates on the reference day will be directly reflected in misalignment of the radio source catalog.

Fixing the origin of right ascension is the second aspect of correctly orienting a radio source catalog. There are two parts to this problem: connecting the radio frame to the optical FK5 frame on the one hand, and to the frame of the planetary ephemerides on the other hand. Hipparcos data and observations of radio stars will contribute to the former, while analyses of planetary probe orbits, planetary occultation measurements, and VLBI and timing experiments on pulsars will yield ties to the ephemeris frame. VLBI data themselves have some sensitivity to absolute right ascensions (*e.g.*, through the Sun's gravitational bending and aberration effects). It appears, however, that such internal ties will not be possible at the several mas level, due to insufficient sensitivity.

6. Acknowledgments

Colleagues who scheduled experiments and analyzed results during the late 1980s include R. P. Branson, R. J. Dewey, J. S. Gotshalk, C. S. Jacobs, S. T. Lowe, J. S. Ulvestad, and T. J. Vesperini. The research described in this report was carried out by the Jet Propulsion Laboratory, California Institute of Technology, under a contract with NASA.

7. References

Lanyi, G. E. (1984). *International Symposium on Space Techniques for Geodynamics, Proceedings 2* (Research Institute of the Hungarian Academy of Sciences, Sopron, Hungary),184; *JPL/NASA TDA Prog. Rep. 42-78*, 152.

Ma, C., Shaffer, D. B., de Vegt, C., Johnston, K. J., and Russell, J. (1989). Astron. J., in press.

Melbourne, W., Anderle, R., Feissel, M., King, R., McCarthy, D., Smith, D., Tapley, B., and Vicente, R. (1983). USNO Circular No. 167, Washington, D.C.

Minster, J. B. and Jordan, T. H. (1978). J. Geophys. Res. **83**, 5331.

Russell, J. L. and Johnston, K. J. (1989). Paper presented at this meeting.

Seidelmann, P. K. (1982). Celest. Mech. **27**, 79.

Sovers, O. J., Edwards, C. D., Jacobs, C. S.. Lanyi, G. E., Liewer, K. M., and Treuhaft, R. N., Astron. J., **95**, 1647.

Sovers, O. J., and Fanselow, J. L. (1987). *JPL/NASA Publ. 83-39, Rev. 3*.

Discussion

CHANDLER: You mentioned the Phobos "lander" data—are these the data from the Soviet Phobos mission in Mars orbit (rather than something from the future)?

SOVERS: I am not directly involved in this work, but my information is that analysis of tracking data for the Phobos mission through mid-1989 is underway.

XU: What is the origin of RA you have used for your VLBI catalogue?

SOVERS: It was the usual value of the right ascension of 3C273, determined by Hazard *et al.* from lunar occultation measurements in the 1970s. This has been used as an RA reference point in all VLBI analyses until very recently.

REALIZATION OF AN INERTIAL REFERENCE FRAME FROM MARK III VLBI

C. MA
Goddard Space Flight Center
Geodynamics Branch
Greenbelt, Maryland 20771
USA

ABSTRACT. Over 350 000 dual frequency Mark III VLBI observations from several geodetic and astrometric observing programs have been used to realize an inertial reference frame through the positions of 325 compact extragalactic radio sources uniformly distributed over the sky with standard errors typically under 1 milliarcsecond (mas). Internal and external tests indicate that the reference frame defined by the relative positions of these radio sources should be accurate and stable at the 1-2 mas level. Because the conventional precession and nutation models are adjusted in the estimation of the source positions, the positions and relative angles are not degraded over the interval of observations or at epochs away from the reference epoch.

1. Introduction

Compact extragalactic radio sources can be used to define a kinematically fixed reference frame against which precise measurements of celestial and terrestrial motions can be made. This paper describes the results obtained from the analysis of ten years of Mark III VLBI data acquired by several geodetic and astrometric observing programs. The important features of the Mark III VLBI system applicable to high precision astrometry are reviewed. The implications of the currently used method of analysis are discussed. In addition to describing the current status of catalog distribution and precision, several sources of systematic error are considered. The future prospects of the radio reference frame are forecast.

There are several papers developing precise radio catalogs from VLBI observations (Fanselow *et al.*, 1984; Ma *et al.*, 1986; Robertson *et al.*, 1986; Sovers *et al.*, 1988; Ma *et al.*, in press; Sovers, this volume). This paper extends the previous catalogs through new observing programs designed to fill gaps in the surveyed sky and through further extension of the geodetic VLBI data base. This paper is, however, preliminary in nature, and the details of the new data will be described elsewhere (Russell *et al.*, in preparation; Ma *et al.*, in preparation).

2. Technique

VLBI observations for high precision geodesy and astrometry rely on certain features of the Mark III VLBI system which are not commonly used in astronomical VLBI. These are given in Table 1.

There are two simultaneous observing frequencies to allow calibration of the ionosphere. The 400 MHz receiver bandwidth at the primary observing frequency allows precise determination of the

J. H. Lieske and V. K. Abalakin (eds.), Inertial Coordinate System on the Sky, 271–280.
© 1990 *IAU. Printed in the Netherlands.*

Table 1. Mark III VLBI System

Receiver	Record	Clock	Correlators
8.4 Ghz - 400 MHz	28 tracks	Maser	Haystack
2.3 GHz - 80 MHz	2 MHz/track		Bonn
	12 passes/tape		Washington
			Kashima

group delay observable. The recording of 14 simultaneous tracks (reduced from the 28 possible for operational simplicity) permits both a good delay resolution function and high sensitivity. There have been some experiments carried out where both the receiver bandwidth and the recorded bit rate were doubled. Twelve passes per tape permits considerable savings of tape and transportation costs. The source of time and frequency is always a hydrogen maser. There are four correlator facilities which have contributed to the data set described in this paper. Geodetic observations commonly involve networks of four to seven stations with baselines as short as 1 km but more commonly 2000 to 6000 km. Some baselines involving South Africa and Australia may be as long as 11000 km. Astrometric networks are generally two or three stations with 3000 to 10000 km baselines. During a normal 24-hour experiment 20 to 40 sources are observed. Counting an observation as the delay and delay rate pair from a single baseline looking at a single source for 100 to 800 seconds, geodetic sessions usually generate 400 to 1000 observations. The number of observations from astrometric sessions is generally smaller.

3. Observing Programs and Data

The geodetic and astrometric observing programs whose data are included in this paper are summarized in Table 2. It should be mentioned that there is another astrometric program run by the Jet Propulsion Laboratory (JPL) using the Deep Space Network stations which parallels certain aspects of the programs used here (Sovers, this volume). Table 2 shows the observing programs, sponsoring organization(s), starting dates, and number of observations, sessions, sources, and sites to 1989.5. The objective of the Crustal Dynamics Project (CDP) is to measure global and regional site velocities while IRIS and NavNet make frequent, evenly spaced measurements of polar motion and UT1. Approximately 50 of the geodetic sources have been used regularly and have at least several hundred observations. The data are stored in the Mark III VLBI data base at Goddard Space Flight Center after being made available by the sponsoring organization.

4. Data Analysis

These data were analyzed using the CALC/SOLVE software developed at Goddard. The a priori models include a theoretical VLBI delay (Robertson, 1975), IAU-sanctioned precession and nutation, general relativity (Shapiro, 1967), and tides (both solid Earth and ocean loading). The ionosphere and atmosphere are calibrated using real-time information. Adjusted parameters include source positions, site positions, daily Earth orientation parameters, daily nutation offsets and

Table 2. Observing Programs and Data to 1989.5

Geodetic:	*Data*:
Crustal Dynamics Project (NASA-1979)	340000 observations
IRIS (NGS-1980)	840 sessions
NavNet (USNO-1987)	70 sources
	30 sites
Astrometric:	
Crustal Dynamics Project (NASA-1980)	12000 observations
Reference Frame (NRL-1987)	40 sessions
South (NRL, CSIRO, HRAO, CRL-1988)	

NASA - National Aeronautics and Space Administration
IRIS - International Radio Interferometric Surveying
NGS - National Geodetic Survey
USNO - United States Naval Observatory
NRL - Naval Research Laboratory
CSIRO - Commonwealth Scientific and Industrial Research Organization (Australia)
HRAO - Hartebeesthoek Radio Astronomy Observatory (South Africa)
CRL - Communication Research Laboratory (Japan)

nuisance parameters such as the clocks and residual atmospheres. See Ma *et al.* (in press) for further details. The post-fit weighted rms residual from a solution using all the data is 42 ps while the astrometric data alone fit at 56 ps.

It should be emphasized that the adjustment of nutation offsets in longitude and obliquity for each day of data is a departure from conventional practice that is necessary because the intrinsic accuracy of the VLBI data is better than that of the standard precession and nutation models. Herring *et al.* (1986) demonstrated that the IAU 1980 nutation series has errors at the 1-2 mas level while Sovers *et al.* (1988) suggested that the precession constant might be in error by as much as 1.6 mas/yr. If the standard precession and nutation models were used without adjustment, the positions of sources observed at only a few, scattered epochs in one year would be affected by the short term nutation errors while the positions of sources observed continually over a long interval or greatly separated in time would be blurred by the errors in the long period nutation terms and the precession constant.

The current practice in VLBI analysis is to adopt the standard IAU precession and nutation models for a reference day, preferably one on which the network is strong and the data of high quality. These models, along with an arbitrary choice of right ascension zero point, define the orientation of the catalog coordinate system with respect to the extragalactic objects. For each other day, a pair of offsets in longitude and obliquity is estimated. The data from different days are combined through the overlap of sources observed, and the actual celestial pole of the catalog is that of the reference day. The mean epoch of the observations therefore loses its conventional meaning. Implicit in this method of analysis are the assumptions that extragalactic objects have negligible proper motions because of their great distance and that source structure is insignificant. Figure 1 shows the adjustments in longitude (scaled by $\sin \epsilon$) and obliquity from 1979–88. The overall slope in longitude

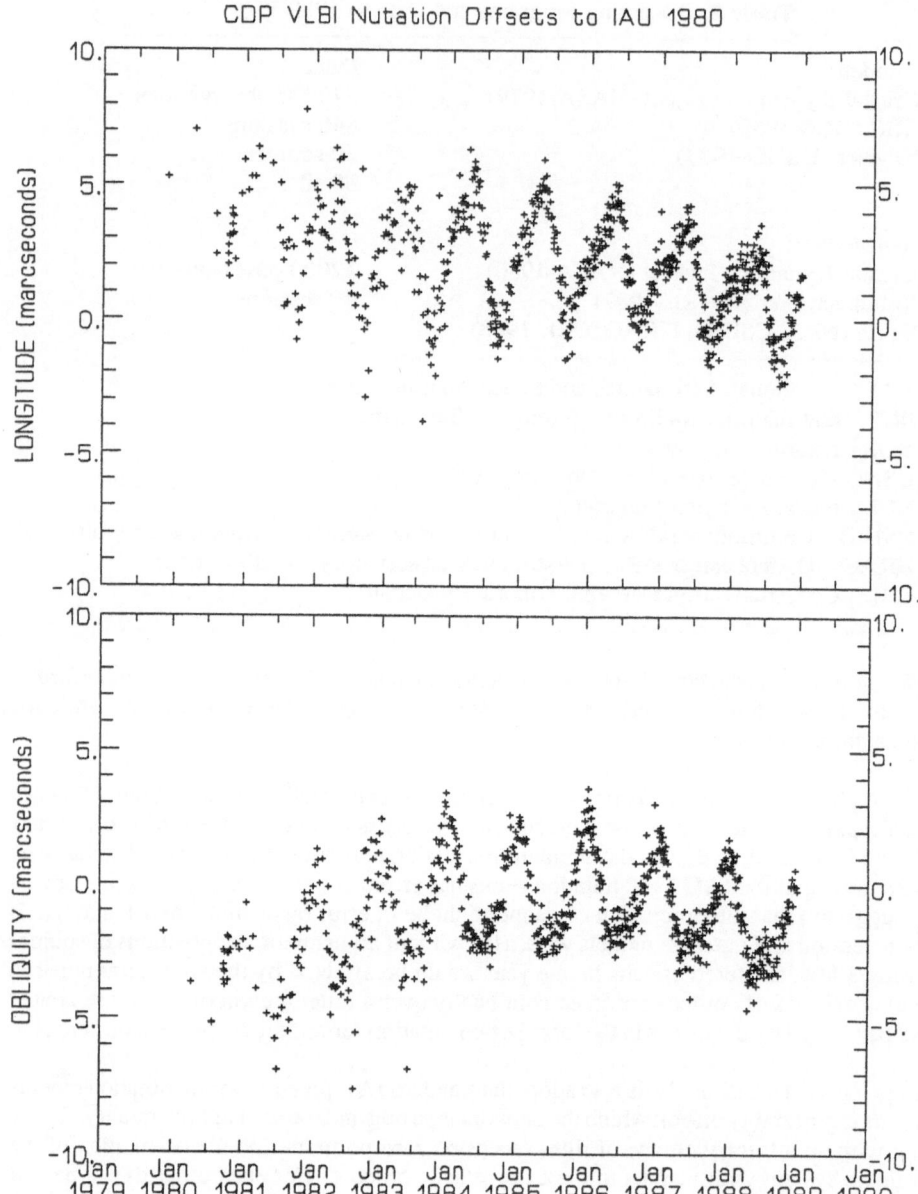

Figure 1. Daily nutation offsets estimated from CDP/IRIS VLBI (1979–88).

is indicative of the error in precession while the curvature in obliquity reflects the error in the principal nutation term. It can be seen that the amplitude of the errors is much larger than the internal error of the VLBI positions (section 5). The adjustment of the precession/nutation model also prevents the degradation of the relative positions at epochs distant from the observations.

5. Astrometric Results

The astrometric results include 325 sources with at least one good observation. There are 305 sources with formal standard errors in right ascension and declination less than 5 mas and 224 sources with formal standard errors less than 1 mas. The distribution of the sources is shown in Figure 2 on an equal area projection. Because of the nature and scope of the geodetic programs, most of the 50 best observed sources are in the north although there are some as far south as −30°. The sources below −45° were observed with the Australia-South Africa baseline while the sources between −45° and −30° were observed with the Australia-Japan-Hawaii network. The distribution of sources by right ascension sector and declination band is shown in Figures 3 and 4. Figures 5 and 6 show the histograms of formal errors $\sigma(\alpha)\cos\delta$ and $\sigma(\delta)$, respectively, in 0.5 mas bins. The peaks at the right are the remaining outliers. It can be seen that the overall distribution of sources overall is quite uniform and that the distribution of errors is concentrated below 1 mas.

6. Errors

While the formal errors are quite small it is instructive to examine the data in different ways to understand the stability of the adjusted positions and the possible systematic errors. Table 3 shows a comparison of the positions from a catalog generated from many years of data with the positions generated from annual solutions. Except for using the same reference day, the annual solutions were completely independent. The same reference day was also used for the multiyear solution.

Table 3. Comparison of 80-87 Catalog with Annual Solutions

year	number sources	number observations	X mas	Y mas	Z mas	sin δ mas	Δδ mas
80	50	13000	−0.1	0.2	−1.3	−4.0	3.8
81	49	9000	0.5	0.7	−1.5	−0.5	1.0
82	39	13000	0.5	−0.1	−0.7	−0.3	0.6
83	28	15000	0.3	−0.3	−0.6	1.0	−0.4
84	53	34000	0.4	−0.0	−0.2	0.1	0.1
85	34	47000	0.1	0.6	−0.5	0.3	−0.2
86	46	52000	−0.1	−0.1	0.5	−0.1	−0.1
87	145	67000	0.3	−0.3	−0.1	0.2	−0.3

X, Y, and Z are rotations about orthogonal axes oriented toward $(\alpha,\delta) = (0,0)$, $(6hr,0)$ and $(0,90°)$, respectively, between the annual catalog and the multiyear catalog. The last two columns are a systematic difference in declination as a function of declination and an offset in declination, respectively. It can be seen that in the later years, which contribute most to the overall solution, there

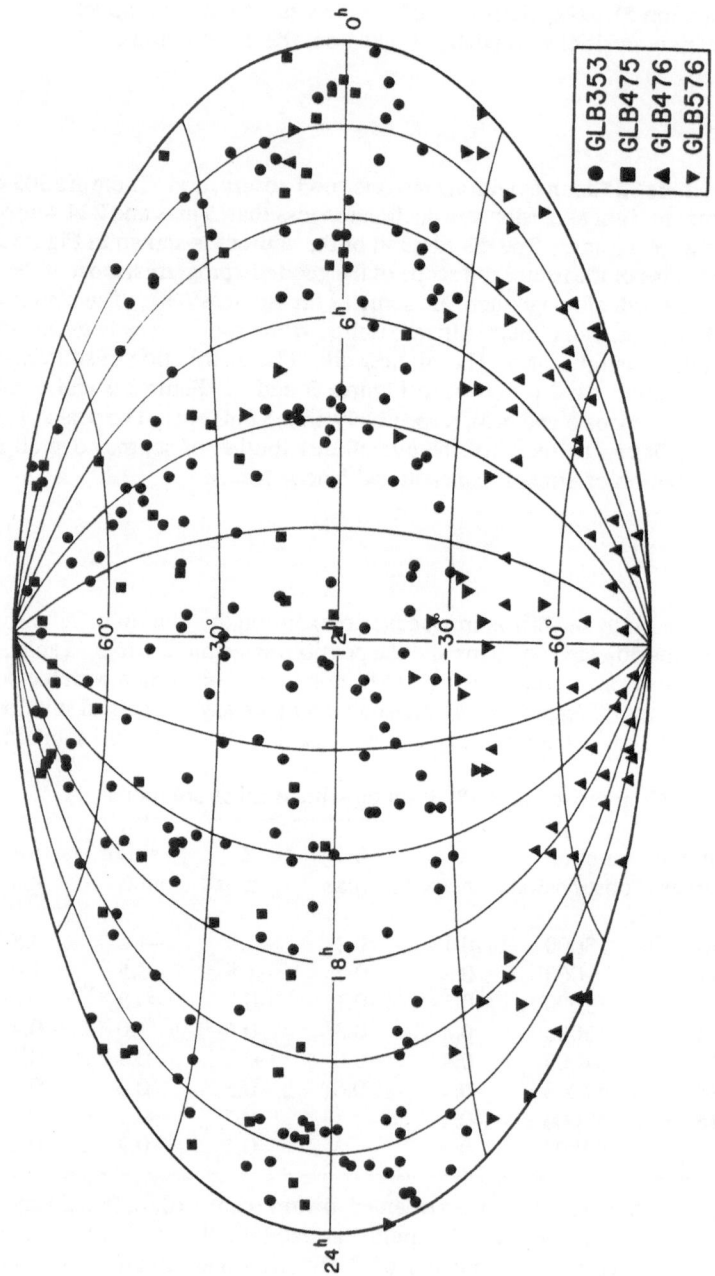

Figure 2. Equal area projection showing 325 radio sources. GLB353 - CDP/IRIS; GLB475 - NRL Reference Frame; GLB476 - South (Australia-South Africa); GLB576 - South (Australia-Japan-Hawaii).

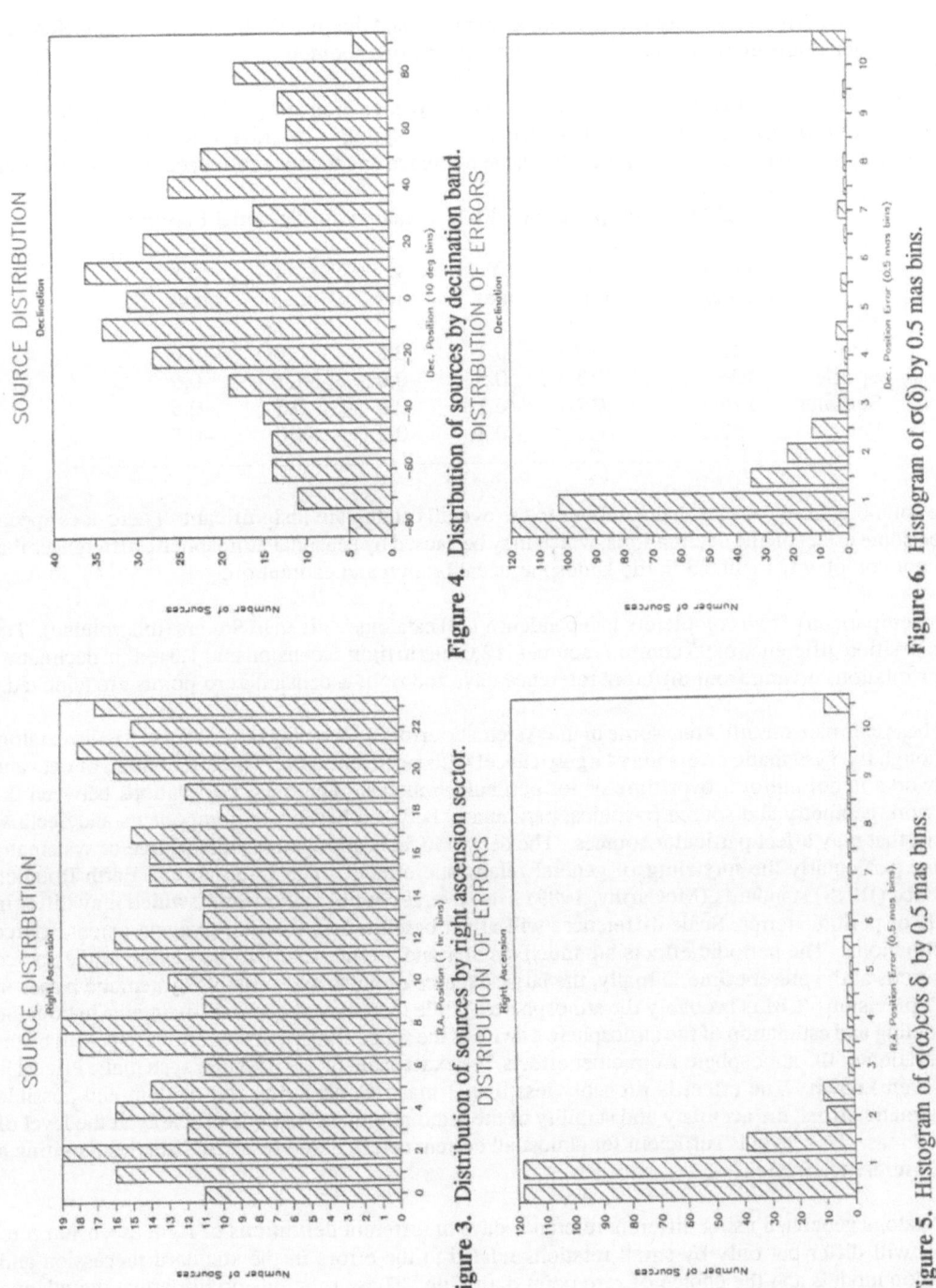

277

SOURCE DISTRIBUTION
Declination

Figure 4. Distribution of sources by declination band.

DISTRIBUTION OF ERRORS
Declination

Figure 6. Histogram of $\sigma(\delta)$ by 0.5 mas bins.

SOURCE DISTRIBUTION
Right Ascension

Figure 3. Distribution of sources by right ascension sector.

DISTRIBUTION OF ERRORS
Right Ascension

Figure 5. Histogram of $\sigma(\alpha)\cos\delta$ by 0.5 mas bins.

are very small differences between the annual solutions and the full catalog. The larger differences in the earlier years are probably caused by limited network geometry.

Table 4 shows a comparison between the same multiyear catalog and positions from solutions that use data from specific seasons (winter — Dec, Jan, Feb; spring — Mar, Apr, May; summer — Jun, Jul, Aug; fall — Sep, Oct, Nov). Again the same reference day was used for each seasonal solution.

Table 4. Comparison of 80-87 Catalog with Seasonal Solutions

	number sources	X mas	Y mas	Z mas	$\sin \delta$ mas	$\Delta\delta$ mas
Winter	120	0.0	−0.1	0.2	0.7	−0.6
Spring	106	−0.3	0.5	−0.0	−1.1	0.9
Summer	116	0.2	−0.1	0.0	−0.5	−0.5
Fall	37	−0.2	−0.3	−0.2	−1.0	−0.8

The rotations from the seasonal solutions to the overall catalog are insignificant. There does appear to be some effect on the declinations, which may be caused by seasonal atmospheric differences that are not completely included in the atmospheric calibration and estimation.

A comparison of two completely independent VLBI catalogs is given in Sovers (this volume). The rms position difference of 95 common sources is 2.0 mas in right ascension and 1.3 mas in declination after rotations arising from different reference days and right ascension zero points are removed.

These comparisons illustrate some of the systematic errors which may be present in a radio catalog, although the systematic errors may largely cancel with sufficient data. The distribution of data and networks is not uniform over time or for particular sources. There are correlations between the network geometry and source positions, particularly between baseline z-components and declinations, that may affect particular sources. The delay model is another possible source of systematic error, particularly the modeling of general relativistic effects. The International Earth Rotation Service (IERS) standards (McCarthy, 1989) lists several possible algorithms, which may differ in scale or periodic terms. Scale differences will affect baseline lengths and, to some extent, source declinations. The periodic effects are much smaller and are not likely to be visible in the source positions at the present time. Finally, the largest source of noise and possible systematic biases in high precision VLBI is probably the atmosphere. While there have been improvements in both the modeling and estimation of the atmosphere as well as the scheduling of observations to permit better separation of the atmosphere from other effects, the exact scale of the possible systematic effects is not well known. The effect is probably less than 1 mas. Considering the random and possible systematic errors, the accuracy and stability of the radio reference frame is probably at the level of 1 to 2 mas. This level is sufficient for almost all current celestial and terrestrial studies requiring a fixed reference frame.

Catalogs generated using different reference days or different definitions of right ascension zero point will differ but only by small rotations related to the errors in the standard precession and nutation models and the choice of zero point definition. These rotations do not affect the relative angles between sources.

The connection to the optical frame is another, very complicated problem. The FK5 positions of some 40 optical counterparts of extragalactic radio sources have been determined from recent plates. These right ascensions can be used to set the zero point of the radio coordinate system. After the zero point is set, a group of well-behaved, compact radio sources (the IERS suggests about 20) can be used to maintain the stability of the zero point for further analyses and as more data are added. It should be emphasized, however, that the radio zero point is independent of the optical equinox origin and is set to the FK5 origin only to facilitate radio/optical comparisons. Likewise the pole of the radio catalog is completely independent of the pole of FK5.

7. Future Prospects

The current state of the radio reference frame is quite good, and the future looks even brighter. The geodetic programs organized by NASA, NGS, and USNO will continue, the latter two perhaps indefinitely because of their functions, so that a core set of sources will be continually measured. The goal of the NRL reference frame and southern observing programs is to locate a well distributed set of 400 radio sources with precision approaching 1 mas for maintaining the radio reference frame and for linking the optical FK5 frame to the radio frame. These observations will continue for several years using such networks that are appropriate and available. Coming into play in the next few years will be the Very Long Baseline Array, a set of 10 identical stations as far south and west as Hawaii, as far north as the state of Washington, and as far east as the Virgin Islands. While this facility is designed primarily for study of radio source structure, it will be available for astrometry and will also require astrometric calibration. It has the potential for becoming the primary, ongoing supplier of high quality astrometric data. The astrometric data base will continue to expand and must be preserved just as carefully as plates. It will then be possible to combine even more observing programs and to improve the analysis over time. Our group and the group at JPL are undertaking a cooperative effort in this direction. We have already included JPL Mark II data in some of our analyses, although not here, and JPL has begun to analyze IRIS and CDP data. We have compared our analysis software and will continue these comparisons as the models are improved. These efforts will give more refined limits on possible systematic and analysis errors.

8. References

Fanselow, J.L. *et al.* (1984) *Astron. J.* **89**, 987.
Herring, T.A. *et al.* (1986) *J. Geophys. Res.* **91**, 4745.
Ma, C. *et al.* (1986) *Astron. J.* **92**, 1020.
Ma, C. *et al.* (1990) *Astron. J.*, in press.
McCarthy, D.D. ed. (1989) *IERS Standards (1989)*, Paris Observatory, Paris.
Robertson, D.S. (1975) Ph.D. thesis, Massachusetts Inst. of Technology.
Robertson, D.S. *et al.* (1986) *Astron. J.* **91**, 1456.
Shapiro, I.I. (1967) *Science* **157**, 806.
Sovers, O.J. *et al.* (1988) *Astron. J.* **95**, 1647.
Sovers, O.J. (1990), in IAU Symposium 141, this volume.

Discussion

YATSKIV: How many estimates of the zenith atmospheric delay per day are you including in your solution?

MA: At the very first stage only a single zenith atmospheric delay is employed. The final analysis is done using between 8 and 24 residual atmosphere parameters per station per day.

KAPLAN: Does your astrometric program include any galactic objects?

MA: Our astrometric program as currently conceived does not include any galactic objects. We would consider using suggested galactic objects, but such sources would need to be observable (at both 8.4 GHz and 2.3 GHz) on our astrometric networks, which are generally considerably less sensitive than the astronomical VLBI networks.

WALTER: GSFC has published several catalogues of VLBI positions since the middle of the seventies. If new observations become available will then all observations be lumped together for setting up a successor catalogue, or do you confine yourself to the more recent data? Work on astronomical constants would be facilitated by the latter case which would for instance allow a better distinction of epochs of observation.

MA: The normal procedure of the Goddard VLBI group is to use all the data (both old and new) of acceptable quality available at the time to make each new source position catalog. In practice this means essentially all of our Mark III data (from fixed stations). There are also Mark I data from 1972–78, but these are not used because they lack dual frequencies for ionospheric calibration. It is simple to make catalogs using only data from specific time intervals, and these catalogs could be generated for studies of astronomical constants. The Mark I data, if somehow properly calibrated, might provide a useful extension of the time base for directly estimating the long-period astronomical constants, but they would not give much additional information for the source position catalog. It should be noted that source catalogs generated only from our geodetic data are much smaller than catalogs generated from our geodetic and astronomical data together.

A PROGRESS REPORT ON THE ESTABLISHMENT OF THE RADIO / OPTICAL REFERENCE FRAME

J. L. Russell and K. J. Johnston
U. S. Naval Research Lab, Code 4030
Washington, D.C. 20375 USA

C. de Vegt and N. Zacharias
Hamburger Sternwarte, Gojenbergsweg 112
D-2050 Hamburg 60
Germany, Fed. Republic

C. Ma and D. Shaffer
Crustal Dynamics Project
Goddard Space Flight Center, Code 621.9
Greenbelt, MD 20771 USA

D.L. Jauncey and G.L. White
Australia Telescope National Facility
CSIRO Division of Radiophysics
Sydney, Australia

J.E. Reynolds
Mount Stromlo and Siding Spring Obs.
Australian National University
Woden, ACT 2606, Australia

G. Nicolson and A. Kemball
Hartebeesthoek Radio Astronomy Observatory
Meiring Naude Rd
Pretoria, South Africa

R. Hindsley and J. Hughes
U. S. Naval Observatory
Washington, D.C. 20392

1. Introduction

In 1987 we began a 5-year program to establish a reference system of at least 400 extragalactic sources which are compact and flat spectrum in the radio and which also display optical emission. This reference frame is to be global with about one source per 100 sq deg (Johnston, et al., 1988). The program incorporates some data which had been obtained previously for other purposes. Altogether, in one or another aspect of the program, 489 sources have been considered so far.

This progress report summarizes the program as of September 1989. The published radio and optical positions are in Ma, et al. (1990) and Russell, et al. (1990a,b).

J. H. Lieske and V. K. Abalakin (eds.), Inertial Coordinate System on the Sky, 281–284.
© 1990 *IAU. Printed in the Netherlands.*

2. Radio Observations

To date we have obtained S/X dual frequency VLBI observations of 347 sources, 325 of which have at least one good observation and about 300 of which are suitable for inclusion in the radio reference frame. About two-thirds of these are in the northern hemisphere.

The distribution of observed radio sources is shown in Figure 1. The symbols in the figure indicate whether the sources were first observed as part of the NASA Crustal Dynamics Project (source list from GLOBL solution GBL353; various baselines), the new northern hemisphere observations (GBL475; Fairbanks-Hatcreek-Maryland Point), the mid-southern observations (GBL576; Kauai-Kashima-Tidbinbilla), or the far southern observations (GBL476; Tidbinbilla-Hartebeesthoek). The largest gaps in the distribution are in the galactic plane. The sources observed to date include 198 of the 233 recommended by the IAU working group on the Radio/Optical Reference Frame (Argue, et al, 1984). The positions for the southern sources in general are not as accurate as the northern because of differences in the number of observations; the CDP sources are especially well-observed and accurate.

The radio results are adopted as the positions of the sources and establish the reference frame. More details of the formation of the radio reference frame from VLBI are given by Ma (1990) and Ma et al. (1990). We note that the RA zero point is fixed using positions of optical counterparts of 28 quasars which have been derived on the system of the FK5.

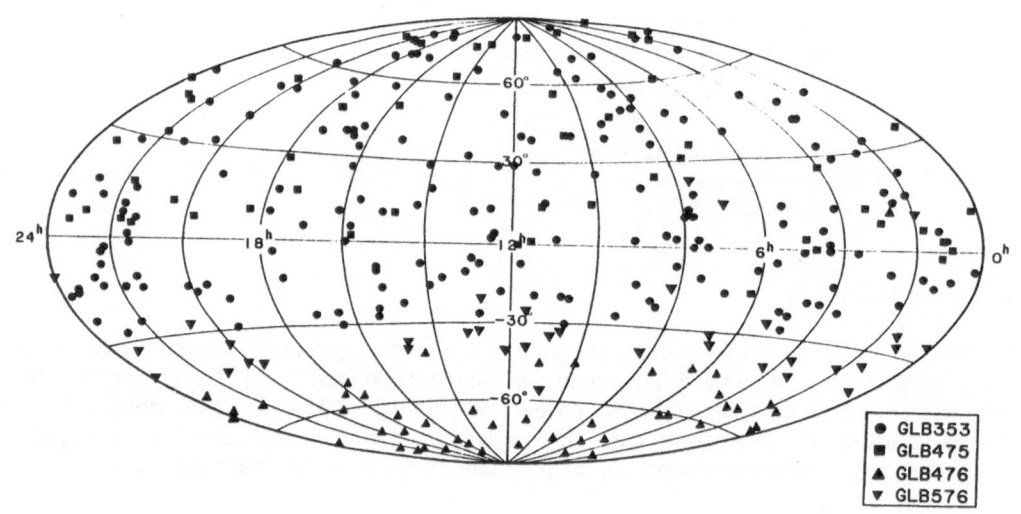

Figure 1. Distribution of sources which have been observed with VLBI.

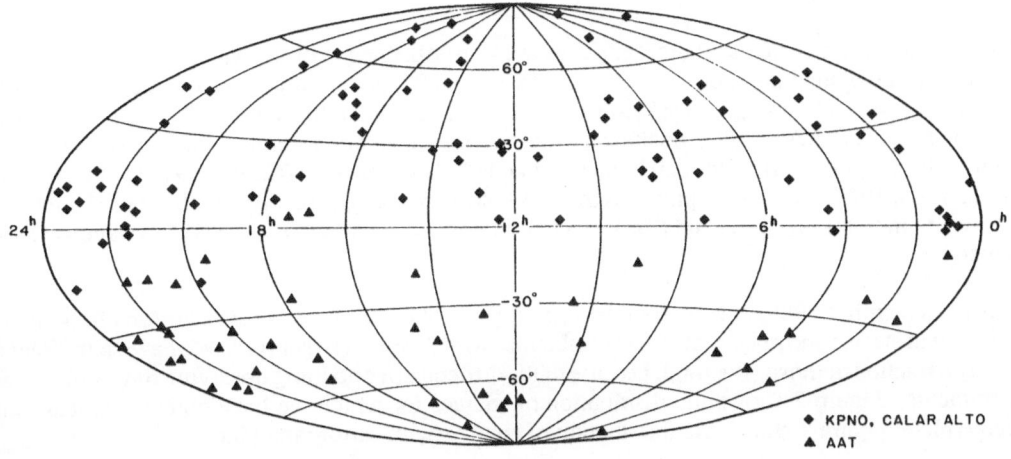

Figure 2. Distribution of sources which have source plates.

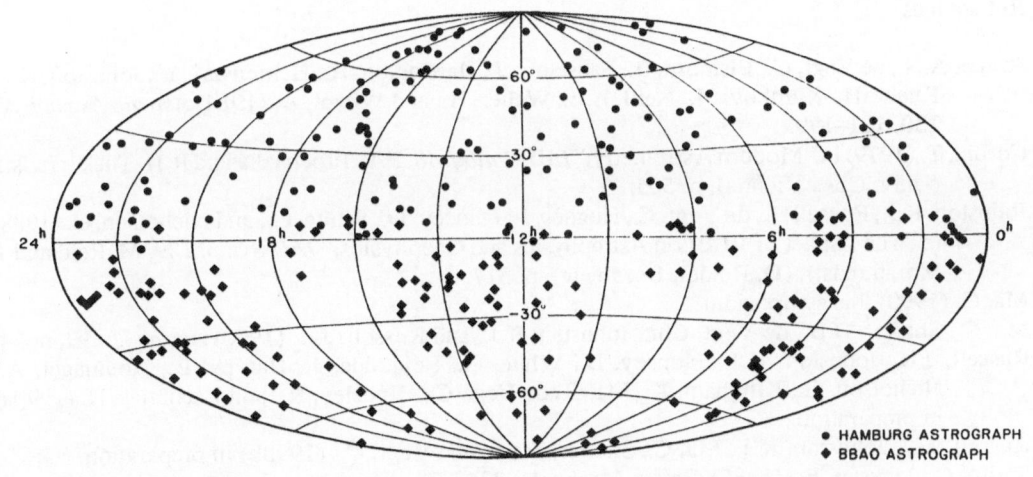

Figure 3. Distribution of sources which have astrograph plates.

284

3. Optical Observations

The optical positions of the radio/optical sources are obtained with a two-step procedure. The position of the optical counterpart, generally $18 < m_v < 22$, is determined using deep prime focus photographs or CCD images from 4-m class telescopes, so far the KPNO 4-m, the Calar Alto telescope and the AAT. The position is referred to secondary reference stars of intermediate magnitude ($12 < m_v < 16$) which can be measured relative to AGK3RN (Corbin, 1979) or SRS (Smith and Jackson, 1985) stars using plates from wide field astrographs, specifically those of Hamburger Sternwarte in Hamburg, FRG and Black Birch Astronomical Observatory, the USNO station in New Zealand.

As of September 1989, at least four astrograph plates each have been obtained for 170 northern sources and at least one plate each has been obtained for 173 southern sources. We have source plates of only a fraction of those, 126 total, because of the difficulty of obtaining observing time on the large instruments. Figure 2 shows the distribution of sources for which we have obtained at least one source plate; Figure 3 shows the distribution of sources with astrograph plates.

4 Summary

We plan to reobserve the sources annually with VLBI to improve the radio reference system and to continue acquisition of optical data. We will also be observing new sources to increase the density of the source distribution in the reference frame and to replace some of the marginal sources currently included.

References

Argue, A.N., de Vegt, C., Elsmore, B., Fanselow, J., Harrington, R., Hemenway, P., Johnston, K.J., Kuehr, H., Kumkova, I., Neill, A.E., Walter, H., and Witzel, A. (1984) *Astron. Astrophys.*, **130**, 191–199.

Corbin, T. (1979) in "Modern Astrometry," *IAU Colloq. 48*, F.V. Prochazka and R.H. Tucker (eds.), (Univ. Obs., Vienna), p. 505.

Johnston, K.J., Russell, J., de Vegt, C., Hughes, J., Jauncey, D., White, G., and Nicholson, G. (1988) in "The Impact of VLBI on Astrophysics and Geophysics," *IAU Symp. 129*, M. Reid and J. Moran (eds.), (D. Reidel, Dordrecht), p. 317.

Ma, C. (1990) this symposium

Ma, C., Shaffer, D.B., de Vegt, Chr., Johnston, K.J., and Russell, J.L. (1990) *Astron. J.* **99**, no. 4.

Russell, J.L., Johnston, K.J., Jauncey, D., White, G., Reynolds, J., Harvey, B., Nothnagel, A., Nicholson, G., Kingham, K., Ma, C., de Vegt, C., Hindsley, R., and Zacharias, N. (1990a) in preparation.

Russell, J.L., Johnston, K.J., Ma, C., Shaffer, D., and de Vegt, C. (1990b) in preparation.

Smith, C., Jackson, E., (1985) *Celest. Mechanics* **37**, 277.

CORRECTIONS TO THE LUNI-SOLAR PRECESSION FROM
RADIO POSITIONS OF EXTRAGALACTIC OBJECTS

H. G. WALTER
Astronomisches Rechen-Institut
Moenchhofstr. 12-14
D-6900 Heidelberg
Fed. Rep. of Germany

ABSTRACT. In an attempt of a realization of the radio reference frame
a compilation catalogue of positions is derived from independent
observation catalogues of extragalactic objects the coordinates of
which had been determined by means of Very Long Baseline Interferometry.
The compilation catalogue comprises 209 objects which are divided into
a core consisting of 50 objects having mean positional accuracies of
0.5 milliseconds of arc (mas) and an extension with positional
accuracies better than 2 mas. Comparison of this catalogue with an
independent compilation catalogue led to confidence limits at the 1 mas
level. - The compilation catalogue is supposed to represent a static
reference frame of fixed extragalactic points. As the epochs of the
contributing observations span nearly 10 years it was tried to
interpret the apparent motion of the fixed points recognizable in the
observation catalogues as an effect of luni-solar precession. The pilot
study points at a reduction of the conventional value of about 2 mas
per year.

1. INTRODUCTION

In recent years Very Long Baseline Interferometry (VLBI) has given rise
to a steadily increasing number of observation catalogues of positions
at the level of a few milliarcseconds (mas). The accumulated material
suggests the establishment of a compilation catalogue (CC) constituting
a radio reference frame of extraglactic objects. Assuming an isotropic
model of the universe these objects, at cosmological distances, are not
affected by proper motions and, thus, approximate a static reference
frame. Tentatively, this frame may play the role of a reference for
comparing observation catalogues of different origins. It may also be
instrumental to the detection and interpretation of apparent motions of
the extragalactic objects.
    Below the construction of the compilation catalogue is outlined
and compared with the independently derived catalogue of extragalactic
radio sources of IERS (Arias et al., 1988 a). Another application of

J. H. Lieske and V. K. Abalakin (eds.), Inertial Coordinate System on the Sky, 285–292.
© 1990 IAU. Printed in the Netherlands.

the CC concentrates on the apparent motions which are manifest in the observation catalogues. Stimulated by the findings of Fanselow et al. (1984) and, likewise, of Sovers and Edwards (1988) about a possible inconsistency of the IAU 1976 model of precession it is attempted to interpret these apparent motions as an effect of luni-solar precession. The observation equations are set up accordingly, and solved by a least squares fit. Several case studies are performed corroberating the hypothesis of a slightly too large value of precession.

2. CONSTRUCTION OF THE COMPILATION CATALOGUE

A catalogue of positions of extragalactic radio sources has been compiled from individual observation catalogues obtained during the last decade by the Goddard Space Flight Centre (GSFC), the Jet Propulsion Laboratory (JPL) and the U.S. National Geodetic Survey (NGS). In a previous paper (Walter, 1989 a) the method of catalogue construction is treated followed by a preliminary version of the catalogue. On setting up the system of the compilation catalogue, the line of thought was pursued that it is defined as the weighted mean of the instrumental systems of the observation catalogues, i.e. the coordinate determination is constrained to

$$\sum_{i=1}^{n} w_i \, \Delta c_i = 0, \tag{1}$$

and the observation equations read

$$\Delta b + \Delta c_i = b_i - b_o, \qquad i = 1, \ldots, n. \tag{2}$$

The meaning of the notation is:

Δb: correction to some approximate position b_o
Δc_i: systematic correction of the i-th catalogue
w_i: weight of the i-th observation catalogue
n : number of observation catalogues containing the respective radio source.

Eqs. (1) and (2) form a system of linear equations which is solved for the unknowns Δb and Δc_i.

With the advent of substantial new observations since 1988 an up-dated version of the compilation catalogue has been set up subdivided into a core and an extension (Walter, 1989 b). Account has been taken of the following observation catalogues: Fanselow et al. (1984) - Q17; Ma et al. (1986) - Q19; Robertson et al. (1986) - Q20; Sovers et al. (1988) - Q22; Ma (1988) - Q23; Carter et al. (1988) - Q26. The core consists of 50 objects the positions of which are determined from at least 4 observation catalogues, and the extension is comprised of 159 objects. The extension arises simply by applying the mean systematic corrections found in the core solution to the coordinates of the respective observation catalogues. Subsequently, weighted mean positions are calculated from these modified positions of the observation catalogues for all those objects which are not treated in the core. Table 1 summarizes the properties of the two sets and their union.

TABLE 1.   Properties of the compilation catalogue:
            Core, extension and core + extension

	Average internal accuracy (mas)		Number of objects	Range of declination (degree)
	RA	Dec		
Core	0.34	0.52	50	−30, +74
Extension	1.6	1.79	159	−45, +85
Core + Extension	1.30	1.49	209	−45, +85

It is intended to use the compilation catalogue as a basis for comparing not only the observation catalogues having contributed to the CC but also any other existing or forthcoming catalogues irrespective of their nature of being an observation or compilation catalogue. Furthermore, the basis is suitable for studying long-term effects in the context with astronomical constants such as precession. The sections below exemplify the usage of this basis for comparing compilation catalogues and studying the luni-solar precession.

3. COMPARISON OF TWO COMPILATION CATALOGUES

By mere coincidence of activities Arias et al. (1988 b) and Walter (1987) have set up compilation catalogues along different lines of thought. It is interesting to compare these catalogues, i.e. CC1 (Arias et al., 1988 a) and CC2 (Walter, 1989), in order to get some feeling for the external accuracy.

A total of 200 objects is common to both catalogues. Their right ascension and declination differences have been plotted versus RA and Dec. Typical for the differences in RA is a shift of about 1 mas. The differences in Dec with respect to RA (see Fig. 1) are dominated by a distinct sinusoidal function of 24 hours period with an amplitude of 3 mas, whereas the declination differences with respect to declination seem to be balanced showing the customary pattern of decreasing accuracy around declination zero.

When the comparison is limited to either the 50 core objects or the 159 objects of the extension the basic features of the above plots remain unaltered, only the error budget changes. Table 2 presents the weighted mean coordinate differences, their standard deviations, the weighted RMS differences, and the number of objects subjected to each comparison.

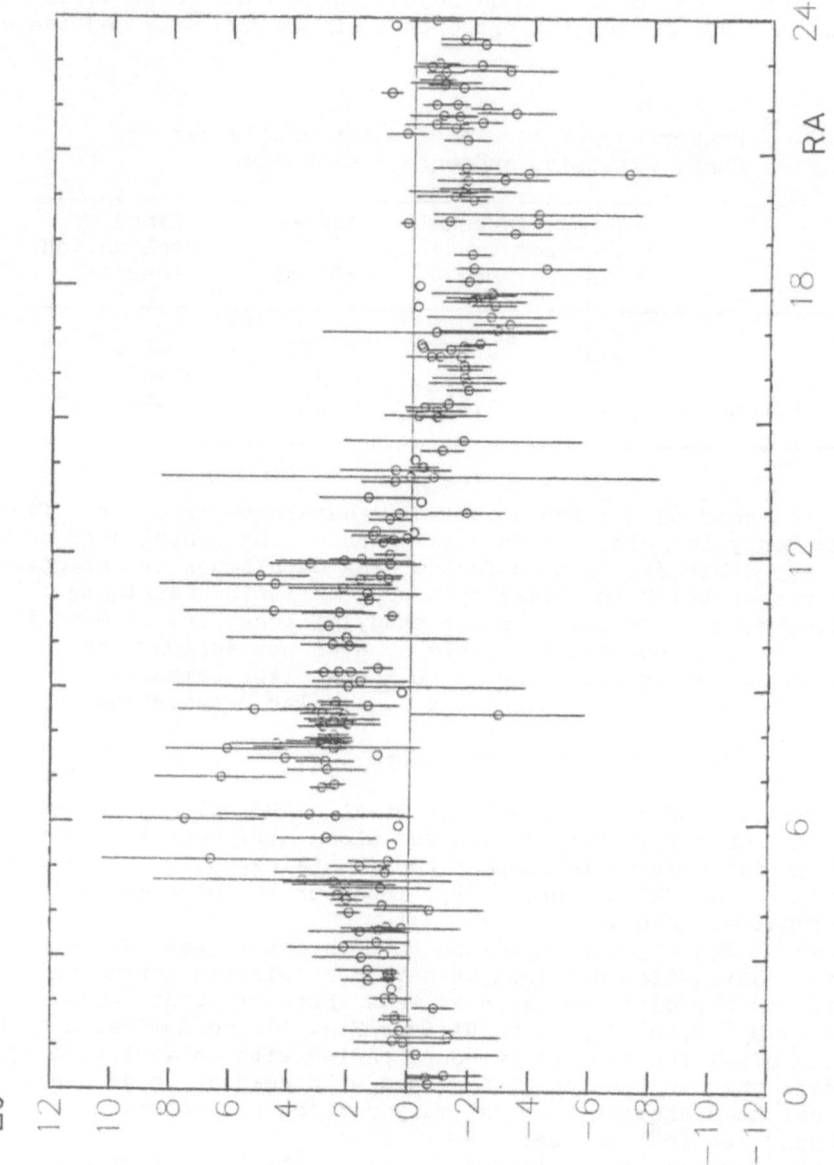

Fig. 1. CC1-CC2: Declination differences versus right ascension,
units: 0.001 arcsec, hours

TABLE 2.   Error budget of catalogue comparisons:
           CC1 (Arias et al., 1988 a) minus CC2 (Walter, 1989 b)

	Weighted mean differences (mas)		RMS differences (mas)		Number of common objects
	RA	Dec	RA	Dec	
Entire CC2	0.57 ± 0.04	0.26 ± 0.08	0.81	1.13	200
Core	0.53 ± 0.07	0.36 ± 0.10	0.70	0.80	50
Extension	0.65 ± 0.06	0.12 ± 0.12	0.98	1.45	150

Using a 5 % significance number the significance test of the differences between CC1 and the entire CC2 yields 19 (10 %) rejections in RA and 12 (6 %) in Dec of the hypothesis of equal mean values. If the test is restricted to the core the figures become 12 (24 %) and 4 (8 %), respectively; for the extension one gets 6 (4 %) and 8 (5 %).

At present the offsets between the two compilation catalogues are not fully understood. One reason may be ascribed to the employed methods or to the selection of observation catalogues which is not exactly identical for CC1 and CC2. Another reason could be a statistical effect to the extent that 4 or 5 observation catalogues are insufficient for setting up a mean system.

4. A PILOT STUDY INTO THE POSITION DIFFERENCES

Since extragalactic objects are assumed fixed in inertial space one would expect that position differences between observation catalogues uniformly referred to J2000.0 are randomly distributed due to accidental errors. If, however, the value of precession employed to reduce a position at observation epoch to the fundamental epoch J2000.0 would be in error, a secular effect should be manifested in these differences.

TABLE 3.   Variation of general precession in Dec (Δn_α, Δn_δ), and corrections $\Delta\psi$ of the luni-solar precession derived from Δn_α and Δn_δ. Units: 0.001 arcsec/y = 1 mas/y.

Origin of selected observations	Number of observations	Δn_α	$\Delta\psi$	Δn_δ	$\Delta\psi$
Qn ∩ Core 1) n = 17,19B 22A,23,26	217	−0.39 ±0.14	−0.98 ±0.35	−0.43 ±0.09	−1.11 ±0.22
2) n = 17,22A, 26	124	−1.10 ±0.29	−2.77 ±0.72	−0.83 ±0.15	−2.12 ±0.39

Reportedly the IAU 1976 value of precession has been rigorously applied to reduce the data in deriving the observation catalogues. Therefore, also CC2 is marked by this precession value. The difference between the position in each observation catalogue and CC2 is formed for each object. Bearing in mind that these position differences are associated with the differences between the observation epoch and the mean epoch of observation in CC2 one can write the observation equations below for all sources in the observation catalogues which are represented in CC2. The following designations are used:

t —   epoch of observation
$\overline{t}$ —   mean epoch of observation in CC2
T —   fundamental epoch (J2000.0)
p_α , p_δ —   variation in RA (α) and Dec (δ) due to general precession, i.e.
     $p_\alpha = m + n \sin \alpha \tan \delta$
     $p_\delta = n \cos \alpha$
     m — general precession in RA
     n — general precession in Dec.

To first approximation the object coordinates α, δ at observation epoch corrected by a precessional term are equal to the coordinates $\overline{\alpha}$, $\overline{\delta}$ in CC2 at mean epoch also corrected by a precessional term, i.e.

$$\alpha(t,T) + \Delta p_\alpha(T-t) = \overline{\alpha}(\overline{t},T) + \Delta p_\alpha(T-\overline{t}), \tag{3}$$
$$\delta(t,T) + \Delta p_\delta(T-t) = \overline{\delta}(\overline{t},T) + \Delta p_\delta(T-\overline{t}), \tag{4}$$

or, formulated as observation equations,

$$(\Delta m + \Delta n_\alpha \sin \alpha \tan \delta) \, (t-\overline{t}) = \alpha - \overline{\alpha} \tag{5}$$
$$\Delta n_\delta \cos \alpha \, (t-\overline{t}) = \delta - \overline{\delta}, \tag{6}$$

where the right-hand sides are supplied by the observation catalogues and CC2.

Since right ascensions are not measured directly but, instead, right ascension differences are given with respect to the reference source 3C273B, eq. (5) undergoes a modification. It has the effect that the coefficient of Δm vanishes and a correction term is added due to the precessional variations of the reference source. After some arithmetic eq. (5) becomes

$$\Delta n_\alpha \left((-\sin \alpha_R \tan \delta_R) + \sin \alpha \tan \delta \right) (t-\overline{t}) = \alpha - \overline{\alpha} \tag{5a}$$

where the coordinates of 3C273B are denoted α_R, δ_R.

The unknowns Δn_α and Δn_δ are solved by weighted least squares fits applied to eqs. (5a) and (6), and to a variety of observational data. Surprisingly consistent results in RA and Dec have been found for the data sets treated. For two of them the possible corrections of the precessional values are listed in Table 3.

5. CONCLUSIONS

From present day VLBI observation catalogues of extragalactic radio sources it is possible to build a compilation catalogue having internal position accuracies better than 0.7 mas for the core with objects of at least four observations, and better than 2.5 mas for the extension with objects of not more than three observations. Its external error is assessed by comparison with an independently constructed compilation catalogue giving 0.1 mas for the error of the weighted mean differences of position, which lies well within the limits of the average standard deviations. An offset of about 1 mas is found in right ascension on top of which a periodic variation of 24 hours with an amplitude of 3 mas is added. The significance test for the two catalogues yields notable values only for the core.

The catalogues agree at the level of 1 mas. Since they avail of almost the same observational data any deviations seem to be related to the mode of compilation rather than to source structure and baseline configuration.

From analysing right ascension and declination residuals follow corrections $\Delta\psi$ of the luni-solar precession ranging from -1.1 ± 0.2 mas/y to -2.8 ± 0.7 mas/y depending on the selected data.

ACKNOWLEDGEMENTS

Gratitude is expressed to Drs. C. Ma (GSFC) and O.J. Sovers (JPL) for making available their catalogues prior to publication.

REFERENCES

Arias, E.F., Feissel, M., Lestrade, J.-F. (1988a) 'An extragalactic celestial reference frame consistent with the BIH terrestrial system (1987)', Bureau International de l'Heure, Annual Report for 1987, Paris, D-113.

Arias, E.F., Feissel, M., Lestrade, J.-F. (1988b) 'Comparison of VLBI celestial reference frames', Astron.Astrophys. 199, 357.

Carter, W.E., Robertson, D.S., Fallon, F.W. (1988) ' Polar motion and UT1 time series derived from VLBI observations', Bureau International de l'Heure, Annual Report for 1987, Paris, D-23.

Fanselow, J.L., Sovers, O.J., Thomas, J.B., Purcell Jr., Cohen, E.G., Rogstad, D.H., Skjerve, L.J., Spitzmesser, D.J. (1984) 'Radiointerferometric determination of source positions utilising Deep Space Network antennas - 1971 to 1980', Astron. J. 89, 987.

Ma, C., Clark, T.A., Ryan, J.W, Herring, T.A., Shapiro, I.I., Corey, B.E., Hinteregger, H.F., Rogers, A.E.E., Whitney, A.R., Knight, C.A., Lundqvist, G.L., Shaffer, D.B., Vandenberg, N.R., Pigg, J.C., Schupler, B.R., Rönnäng, B.O. (1986) 'Radio-source positions from VLBI', Astron. J. 92, 1020.

Ma, C. (1988) private communication

Robertson, D.S., Fallon, F.W., Carter, W.E. (1986) 'Celestial reference coordinate systems: submilliarcsecond precision demonstrated with VLBI observations', Astron. J. 91, 1456.

Sovers, O.J., Edwards, C.D. (1988) 'Precession and long-term nutation from VLBI observations', Bureau International de l'Heure, Annual Report for 1987, Paris, D-109.

Sovers, O.J., Edwards, C.D., Jacobs, C.S., Lanyi, G.E., Treuhaft, R.N. (1988) 'The JPL 1988-2 extragalactic radio source catalog', Bureau International de l'Heure, Annual Report for 1987, Paris, D-17; see also Astrometric Results of 1978-1985, Deep Space Network Interferometry: 'The JPL 1987-1 extragalactic source catalog', Report of the Jet Propulsion Laboratory, 31 p.

Walter, H.G. (1987) 'Evaluation of catalogues of extragalactic radio sources', in Proceedings of the IAU Colloquium 100, "Fundamentals of Astrometry", H.K. Eichhorn (ed.), Kluwer Acad. Publ., Dordrecht, in press.

Walter, H.G. (1989a) 'A compilation catalogue of positions of extragalactic radio sources', Astron. Astrophys. 210, 455.

Walter, H.G. (1989b) 'A celestial reference frame based on extra-galactic radio sources', Astron. Astrophys. Suppl. Ser. 79, 283.

Discussion

MURRAY: (1) Can you be sure that all the catalogues which you used have been reduced with the same set of precessional constants?

(2) The declination differences between your compilation catalogues and IERS could show a secular change of obliquity if the epochs were different.

WALTER: (1) The forewords to the catalogues used here state that the reductions proceeded along the lines recommended by IAU or by the MERIT standards.

(2) As both compilation catalogues are set up from almost the same observation catalogues, the mean catalogue epochs should not differ more than 1 or 2 years, which seems too short for explaining the 24 hours' periodicity of the declination differences.

VITYAZEV: Have you found any evidence of the equinox motion by evaluating Δn from RA and from declination?

WALTER: So far the goal of the analysis was to show that a secular effect in terms of Δn is detectable in the residuals of RA and Dec. A more refined study will have to demonstrate whether Δn contains a component which should be ascribed to the equinox motion.

YATSKIV: We have constructed a compilation catalogue of extragalactic objects using 16 primary sources (See the paper by Yatskiv and Kuryakova in this volume). It seems to me that our results agree with those of Walter.

DEDICATED SOVIET VLBI-NETWORK "QUASAR"

A.M. FINKELSTEIN, G.S. GOLUBCHIN, V.M. GORODETSKY, V.G. GRACHEV,
V.S. GUBANOV, A.V. IPATOV, M.N. KAIDANOVSKY, E.I. KORKIN,
E.I. NIKOLAEV, S.G. SMOLENZEV, A.A. STOTSKY, N.D. UMARBAEVA
AND YA.S. YATSKIV
Institute of Applied Astronomy
8 Zhdanovskaya ul
197042 Leningrad
USSR

ABSTRACT. The radiointerferometrical network "QUASAR", composed of six dedicated VLBI stations linked via a geostationary satellite channel with the Center of Operations, is under construction in the Soviet Union. It is proposed to construct "QUASAR"-stations abroad, in China, Bulgaria and India. A short review of basic scientific and technical features of the project is given in this paper.

In 1988 the USSR Academy of Sciences made a decision to construct a dedicated VLBI network of six radiotelescopes situated over the territory of the Soviet Union (near Leningrad, at Ukraina, North Caucasus, near Ashkhabad, at Lake Baikal, and at Kamchatka peninsula) and linked with the Center of Operations, which is under construction in Leningrad, via a special geostationary satellite channel. It is also considering the possibility of locating other "QUASAR"-stations abroad: in China, Bulgaria and India. The system is named "QUASAR". For the realization and further exploitation of this system, the Institute of Applied Astronomy has been organized by the USSR Academy of Sciences and it involved in this project different groups of scientists and engineers from Leningrad and Moscow.

The network will provide data for precise determination of inertial, dynamical and terrestrial coordinate systems and their mutual orientation as well as for high resolution mapping of cosmic radiosources.

The network will be operated in two modes:

- "Off-Line": using digital magnetic tape recorder with band of 144 MHz per station and
- "On-Line": with the transmission of radiointerferometrical signal from stations to the Center of Operations via a satellite channel with the speed of 4.5 Mbit per second per station.

The technical specifications of the network ate given in Table 1.

Four stations (near Leningrad, at North Caucasus, near Lake Baikal and Ashkhabad) and the Center of Operations are under construction at the present time. The first three stations of the network together with the Center of Operations will be operational in the beginning of 1992 and all six stations in 1994. During this period we are intending to construct and to introduce into the activity some "QUASAR"-stations abroad. The deadline for project realization is 1995.

J. H. Lieske and V. K. Abalakin (eds.), Inertial Coordinate System on the Sky, 293–294.
© 1990 *IAU. Printed in the Netherlands.*

Table 1. "QUASAR"-network specifications

1.1 Geometrical characteristics

Network	Maximum baseline	Longitude coverage	Latitude coverage
National	6700 km	119°	23°
International	7300 km	119°	48°

1.2 Antenna system

Antenna for:	Number	Diameter
observations of radiosources	8	32
observation of navigation satellites	8	1.3
Transmission signal via geostationary satellite:		
Far East, India, China, Center of Operations	4	12
other stations	5	4
monitoring of troposphere electrical characteristics	8	1.5
control of RT32 surface by radioholography	2	0.5

1.3 Receiving system. Radiometers for:

	Wavelength
observations of radiosources	0.7, 1.35, 3.5, 6, 13, 18/21
observation of navigation satellites	19
monitoring of troposphere	1.5, 1.0
control of RT32 surface	2.5

1.4 Time-frequency system

H-maser standard	$10^{-14} - 10^{-15}$
Primary time synchronization	20 ns via GLONASS

1.5 Data transmission system

"Off-Line"	magnetic tapes, 144 MHz per station
"On-Line"	satellite channel, 4.5×8 or 9.0×4 Mbit/s

1.6 Control and monitoring system

Central site computer	CM-1425
Number of workstations	15
"On-Line" system via satellite channel	64 Kbaud
Digital telephone line via satellite channel	64 Kbaud

1.7 Satellite: Geostatonary satellite GORIZONT with channel

bandwith	36 MHz
up-link / down-link frequencies	14 GHz / 11 GHz

1.8 Processing system

	1 Step	2 Step
Correlator (Mark-III format)		
number of stations	3	10
bandwidth per station	120 Mbit/s	288 Mbit/s
bus type	CAMAC	VME
input data	magnetic tapes and satellite channels	
Mainframe	VAX-6320 cluster	
total RAM	128 Mb	
disk memory	20 Gb	
Workstations	VS-3100, BESTA-88	
Software	VAX/VMS, UNIX	

1.9 Collocation

Laser ranging systems, gravimeters, seismic and meteorlogical data stations

A NEW APPROACH TO THE CONSTRUCTION OF A COMPILED CATALOGUE OF POSITIONS OF EXTRAGALACTIC RADIO SOURCES

YA.S. YATSKIV AND A.N. KURYANOVA
Main Astronomical Observatory
Ukrainian Academy of Sciences
252127 Kiev
USSR

ABSTRACT. An attempt of the realisation of the radio reference frame was undertaken. For this purpose a new approach to the construction of a compiled catalogue from observation catalogues of extragalactic radio sources was used. As a numerical example a compilation catalogue of primary radio sources based on the three individual catalogues (GSFC, JPL,and NGS) was constructed. Comparison of this catalogue with recent compiled catalogues led to the conclusion that their r.m.s. differences were about 1 mas.

1. Introduction

At present, two compiled catalogues of extragalactic radio sources are known (Arias et al., 1988; Walter, 1989). The r.m.s. differences of coordinates in these catalogues are about 2 mas. For the construction of these catalogues two different methods of compilation were used. Therefore it sounds reasonable to study different realisations of radio reference frame (RRF) and to apply a new approach to the construction of a compiled catalogue. We shall distinguish two problems of realisation of an optimum RRF:
  a) construction of primary radio reference frame (PRRF);
  b) extension of the PRRF to secondary radio sources.
This paper describes the construction of the PRRF.

2. Method of construction of PRRF

The PRRF is constructed under the general principle that the positions of the sources common to all individual RRF are used for combination solution and under the general assumption that individual catalogues of positions of radio sources (RS) are independent of each other.
   The compiled catalogue of the positions of extragalactic radio sources RSC(GAO UA) 89 C 01 was constructed in following steps:
  a) calculation of the lengths of arcs between RS: S_{ij}^k;
  b) comparison of arcs in different catalogues: (S_{ij}^k , S_{ij}^l);
  c) estimation of mean value of $\bar{S}_{ij}$ and residuals : $\Delta S_{ij}^k = S_{ij}^k - \bar{S}_{ij}$
  d) construction of individual reference frames (RF) defined by two selected RS;

J. H. Lieske and V. K. Abalakin (eds.), Inertial Coordinate System on the Sky, 295–296.
© 1990 *IAU. Printed in the Netherlands.*

e) construction of compiled RF defined by two RS;

f) construction of compiled RF under the conditions: "not net rotation and mean displacements of RS";

The purpose of the steps a, b, c, is to construct the net of arcs between the radio sources which is free from the RRF orientation and to estimate the weights of individual catalogues. For the construction of the RF defined by two RS the positions of the RS 0234+285 and RS 0851+202 were used.

3. Numerical example

For the construction of the compiled catalogue RSC(GAO UA) 89 C 01 we have used the coordinates of 16 common RS given in the individual catalogues of GSFC, JPL and NGS (BIH Annual Report for 1987). The relative orientations (A1, A2, A3) and r.m.s. differences of coordinates ($\sigma_\alpha \cos \delta$, σ_δ) between our catalogue and well known compiled catalogues RSC(IERS) 88 C 01, RSC(IERS) 89 C 01 (IERS Annual Report for 1988) and CC (H.G. Walter) are given in the Table.

Table. Relative orientations and R.M.S. differences between compiled catalogues under consideration. Unit: mas.

Compiled catalogues	A1	A2	A3	$\sigma_\alpha \cos \delta$	σ_δ
RSC(IERS) 88 C 01	−0.06	+0.52	−0.19	0.072	0.202
RSC(IERS) 89 C 01	−0.89	+0.63	−0.11	0.426	1.203
CC(H.G.Walter)	+0.03	+1.74	+0.63	0.332	0.575

References

Arias, E.F., Feissel,M., Lestrade, J.-F. (1988), *BIH Annual Report for 1987*, p.D-113.
Walter, H.G. (1989), *Astron. Astrophys.* **210**, 455–461.
IERS Annual Report for 1988, p.II–34.

Discussion

WALTER: At present it seems premature to state which compilation catalogue is the more reliable one. Indeed, the discrepancies of catalogue comparisons draw one's attention to the original observation catalogues which need careful examination with respect to identity, data reduction methods used, and the epochs. Then a sound basis for comparison can be established.

YATSKIV: I agree.

ARIAS: In our opinion, the differences shown in the plots between the Yatskiv and Walter compilation catalogues and the IERS Celestial Reference Frame partly arise from the different approaches used. In IERS the combination is realized on the basis of the relative orientations between frames and regional effects are not considered.

YATSKIV: I agree. In our approach we have taken into account both the orientation and the regional errors.

IN-ORBIT STATUS OF THE HIPPARCOS ASTROMETRY MISSION

M.A.C. PERRYMAN
Astrophysics Division
ESTEC, Postbus 299
2200 AG Noordwijk
The Netherlands

ABSTRACT. The Hipparcos satellite was successfully launched by Ariane 4 flight 33, from French Guyana, on 8 August 1989. However, the apogee boost motor, designed to place the satellite into its intended geostationary orbit, failed to function, and the satellite was left in a highly elliptical transfer orbit. After a perigee raising manoeuvre, using the hydrazine propulsion system, the orbital parameters are approximately: apogee 35900 km, perigee 526 km, eccentricity 0.72, inclination 6.8°, orbital period $10^h \ 40^m$.

Operational procedures have had to be revised during the first weeks of satellite operations, due to the non-nominal orbit. Commissioning and calibrations have taken longer than foreseen because of the operational difficulties and observational overheads, but routine observations commenced on 27 November. The essential differences between the nominal and the present missions are that observations can only be conducted for about 30-40 per cent of real time in the present mission, and the fact that the actual lifetime of the mission is still rather uncertain, and possibly somewhat shorter than anticipated. This paper gives a summary of the present satellite status, measurement results obtained so far, and in indication of the scientific goals that may be achievable with the present mission.

1. Introduction

Hipparcos, the European Space Agency's astrometry mission, has been fully described in the ESA publication SP-1111 (June 1989). Launched by an Ariane 4 from French Guyana on 8 August 1989, the failure of the apogee boost motor to place the satellite in its planned geostationary orbit appeared at first to strongly degrade the expected scientific return of the mission. However, apart from the failure of the apogee boost motor, the satellite behaviour has been entirely nominal, and the degradation of the solar arrays due to the particle radiation environment (the principal cause of the expected reduced mission lifetime) has not been as significant as had originally been predicted.

The perigee altitude was raised to 500 km on 7-8 September, the solar arrays and fill-in antenna were successfully deployed on 12 September, the baffles were opened on 25 September, the first star mapper measurements were made on 26 September, and 'initial star pattern recognition' (the term given to the initial determination of the satellite's three-axis attitude), with the satellite in sun-pointing mode, was successfully completed on

J. H. Lieske and V. K. Abalakin (eds.), Inertial Coordinate System on the Sky, 297–305.
© 1990 *IAU. Printed in the Netherlands.*

8 October. All six on-board detectors, the prime and redundant image dissector tubes, and the prime and redundant pairs of star mapper photomultipliers, are functioning nominally, as are the on-board mechanisms—the shutters, the switching mirror, and the refocusing mechanism.

All further instrumental calibrations were completed by 25 November, and the nominal data collection phase of the mission started on 27 November.

2. Orbital Parameters and Mission Lifetime

The satellite required hydrazine as reaction control fuel only for the spinning phase in transfer and drift orbit. The tanks were filled to their full capability of 32 kg, and a surplus of 26 kg, corresponding to a ΔV of about 47 ms^{-1}, was available for orbit manoeuvres. This was sufficient to raise the perigee to about 500 km to avoid atmospheric drag and atomic oxygen problems.

It was also possible to adjust the orbital period to a multiple of the great circle scan period. Such a scan-synchronous orbit allows the minimisation of occultations and avoids earth albedo light into the payload. The revised orbit maneouvres were completed on 18 September 1989, and the orbit characteristics for that epoch are:

Perigee height	526 km
Apogee height	35900 km
Orbital period	640 min
Eccentricity	0.72
Inclination	6.8°
Ascending node	105°
Argument of perigee	214°

3. Eclipses

Long (> 60 min) eclipses will occur in February/March 1990 and again in July 1990. The eclipse durations are not only longer than in the nominal orbit (100 min maximum, compared to 72 min for the nominal mission), but the ratio of the eclipse period to the orbital period becomes significantly larger. It is presently expected that the payload thermal control will have to be switched off during these intervals, probably making the resulting data unusable during this time interval.

The revised orbit results in an eclipse duration profile very different from that of a geostationary orbit. The eclipses will probably determine distinct operational phases:

- (a) first operational phase (2 months) lasting from the beginning of end November 1989, the planned completion date of payload calibration, to 2 February 1990 when the eclipse duration increases rapidly;
- (b) hibernation phase (2 months) lasting from 2 February to 9 April 1990, the end of the first eclipse season;
- (c) second operational phase (3 months) lasting from 9 April to 13 July 1990, the beginning of the second eclipse season;
- (d) third operational phase (13 months) lasting from 13 July to mid August when the eclipse duration increases rapidly once again.

4. Ground Station Coverage

In the nominal mission, the Odenwald ground station (FRG) would have been in continuous coverage with the satellite, and the total mission deadtime (due to occultations, calibrations, etc) would have amounted to less than 10 per cent. The present orbit, however, is covered by Odenwald for only 32 per cent of the time. The Perth and Kourou ground stations are included in the revised mission, resulting in about 81 per cent data acquisition time compared to full coverage. As a result of other operational and observational overheads, useful data collection occurs for about 30 per cent of the time.

However there are still non-visibility periods of more than 10 hours, a few times per week. It is under investigation to have a fourth ground station which would improve the coverage to 95 per cent and, more importantly, would reduce the non-visibility periods to less than 1 hour. The fraction of time for which useful data could be accumulated would also rise to about 50 per cent.

5. The Satellite Performance

5.1. POWER BUDGET

In the revised orbit the satellite power budget has to take into account three new aspects:

(1) degradation of solar cells due to increased dose from the radiation belts;
(2) temperature increase of the solar cells at perigee due to earth albedo light;
(3) worsened battery charge/discharge ratio due to the reduced orbital period and increased eclipse duration.

The effect of the solar array degradation due to the trapped high energy proton flux, is illustrated in Figure 1. Measurements of the in-orbit degradation, through measurements of the short-circuit current and the open-circuit voltage of the solar arrays, is presently in progress. Current lifetime prediction due to the solar array degradation is in the range 18–30 months. This prediction will be refined as the mission proceeds.

5.2. THE ATTITUDE CONTROL AND DETERMINATION

The attitude control and determination has to cope with three new problems:

(1) high disturbance torques at perigee;
(2) high star mapper rejection rates in the radiation belts;
(3) long non-visibility periods without reference stars.

5.2.1. *High Disturbance Torques.* The scientific measurements are based on an extremely smooth motion of the satellite leading to a control concept which leaves the satellite basically in free drift with occasional small thruster firings to keep the attitude within an allocated band around the nominal scanning law.

This concept is no longer valid around perigee due to the high disturbance torques caused by atmospheric drag, gravity gradient and magnetic torques. More frequent and more powerful thruster firings are required to counteract the high disturbance torques. As a consequence below 10 000 km altitude the control parameters have to be changed to provide a shorter 'look ahead' time and to allow for a longer thruster on-times. This not only reduces the value of the scientific measurements but also causes a problem for a hardwired back-

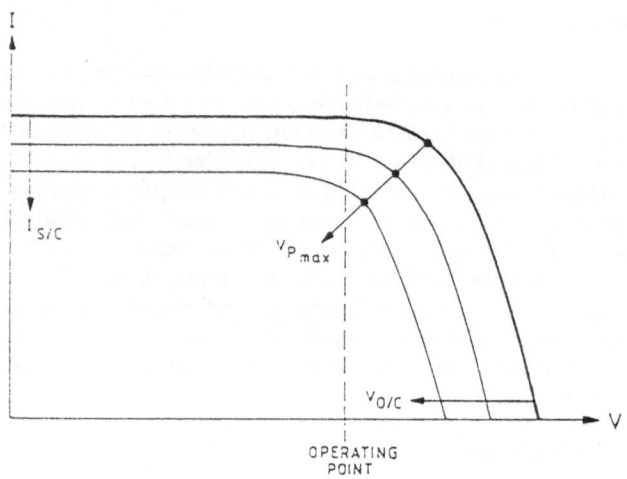

Figure 1. Radiation effects on the solar cells. The effect of the degradation is to decrease the short-circuit current and open-circuit voltage, such that the voltage at maximum power (V_{Pmax}) also decreases.

up function, referred to as the thrust impulse monitoring. The thrust impulse monitoring monitors the length and frequency of the thruster firings and switches automatically to the back-up branch of the attitude control subsystem if the hardwired thresholds are exceeded. For low altitude operations the thrust impulse monitoring therefore has to be disabled.

This new control concept has been already implemented with two control parameter sets controlled by time-tagged commands according to the predicted attitude. It has already worked successfully through many orbits. The cold gas consumption is only slightly higher than had been expected for the nominal mission.

5.2.2. *High Star Mapper Rejection Rate.* The background count rate of the star mapper increases drastically in the radiation belts. Consequently for a period of 2 to 3 hours around perigee all reference stars, except maybe very bright ones, will be rejected by the star mapper filtering process. Nevertheless, apart from those times affected by large solar flares, the attitude determination generally re-converges after perigee passages.

5.2.3. *Long Non-Visibility Periods.* The Hipparcos system requires a continuous uplink of the programme star file which contains the programme stars being measured and the reference stars needed to update the attitude determination and to guide the satellite along the nominal scanning law. The on-board programme star file buffer is sized to cover a time span of 1 to 2 hours. It is, however, insufficient to cover non-visibility periods of up to 12 hours. The solution envisaged is to use a special 'sparse programme star file' which contains only reference stars at time intervals just sufficient to support the attitude determination task.

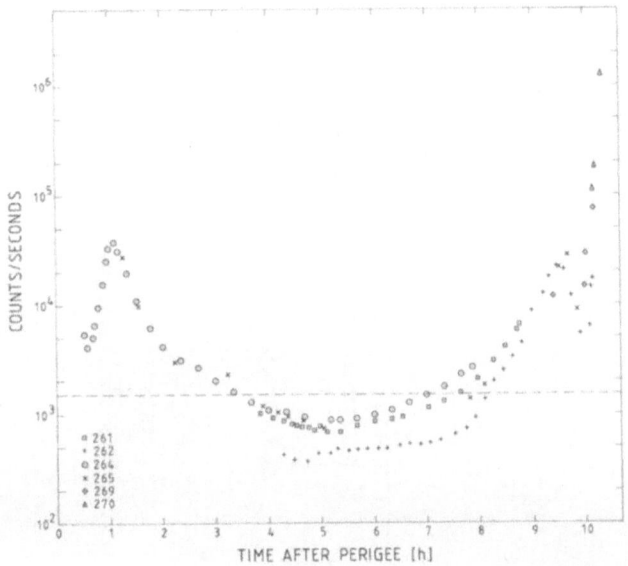

Figure 2. Photomultiplier tube background count rate. The numbers associated with the symbols indicate the day of the year 1989.

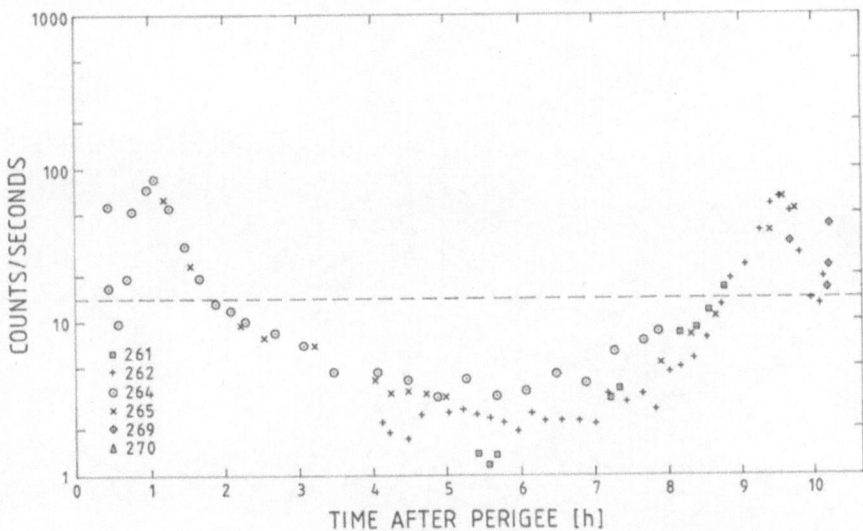

Figure 3. Image dissector tube background count rate. The numbers associated with the symbols indicate the day of the year 1989.

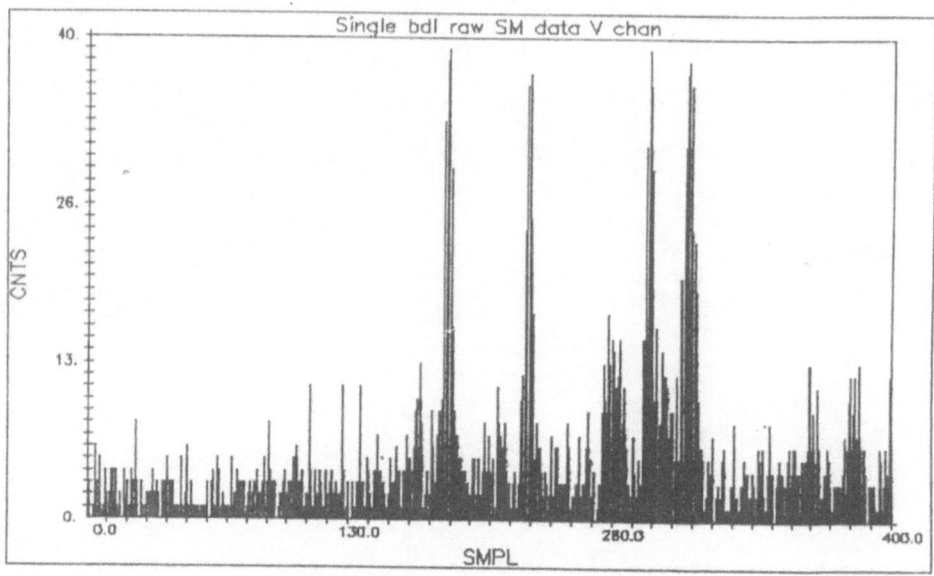

Figure 4. First results from the star mapper *V* channel.

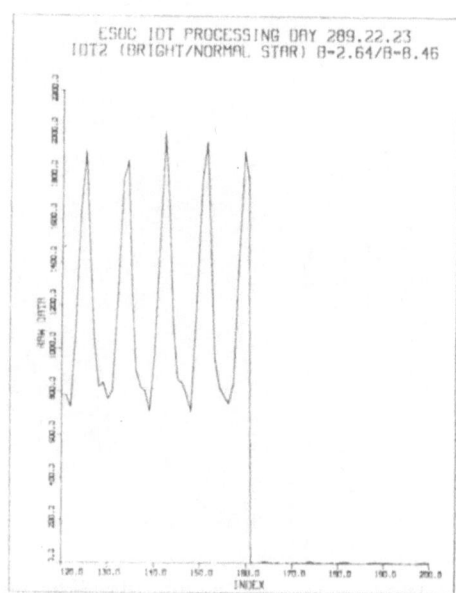

Figure 5. First results from the image dissector tube detector. The two stars observed are of magnitude 2.6 (at the left) and 8.5 (at the right, where the modulation is just visible).

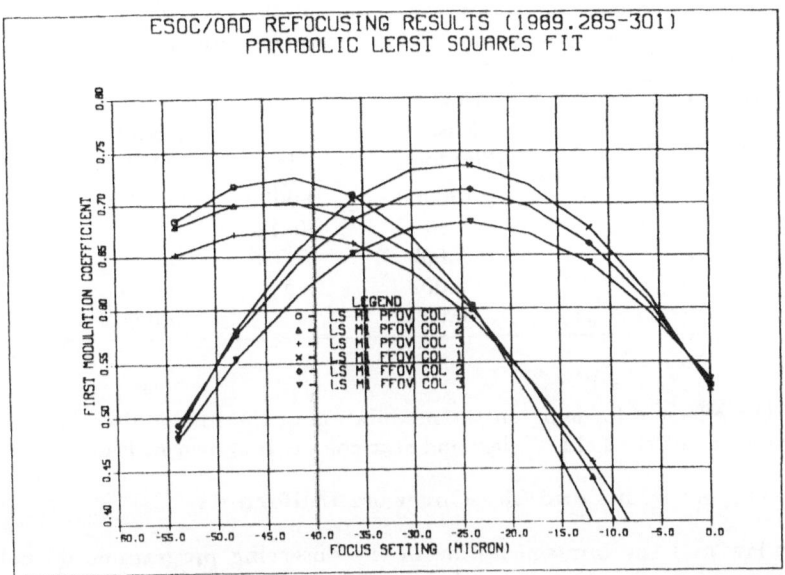

Figure 6. The effect of focusing on the modulation of the image dissector tube photon counts, as a function of the field of view and star colour. A good focus for both fields of view is achieved for a focus setting of −35 microns.

6. The Payload Detectors

The main payload related problem in the revised mission orbit is the increased background count rate due to Cerenkov and luminescence light caused by the high particle flux in the radiation belts. Some results are given in Figure 2 (background count rate of photomultiplier B) and Figure 3 (background count rate of the image dissector tube). The dotted lines indicate the background count rate assumed for accuracy analysis of the nominal mission. For the image dissector tube the background count rate most of the time is well below the nominal background level. In the 4 hour period around perigee (i.e. below an attitude of 23 000 km) the levels are less than a factor of 10 higher than nominal, and thus becomes comparable to straylight levels. It therefore can be expected that the limiting magnitude of observable stars is slightly reduced—from 12.5 to perhaps 12 mag in this region. The situation for the photomultiplier B is less favorable. The excess above the nominal background level is almost 1.5 orders of magnitudes for the outer belt and 3 orders of magnitude for the inner belt. Due to this high background the use of the star mapper for attitude determination within 3 hours around perigee is very doubtful.

The absolute background count rate levels are however well below levels affecting the detector hardware characteristics. It is therefore not necessary to switch off the detector high voltage supplies during radiation belt passages.

First results from the star mapper detector and the image dissector tube detector, showing the characteristic responses of the star transits, are shown in Figures 4 and 5

Table 1. Expected accuracies for various durations of the revised mission. The table is based on a nominal operational capability of the the revised mission.

Mission Product	Nominal Mission (30 months)	Revised Mission Duration (months)		
		(6)	(18)	(30)
No. of stars	120000	80000	110000	120000
Observations per star	80	0–10	30	50
Positions (arcsec)	0.002	0.015	0.006	0.003
Parallaxes (arcsec)	0.002	–	0.006	0.004
Proper motions (arcsec/yr)	0.002	–	0.008	0.005

respectively. The effect of focusing on the modulation of the image dissector tube photon counts, as a function of the field of view and star colour, is shown in Figure 6.

7. Scientific Observations and Sky Coverage Uniformity

The scanning law and the implementation of the observing programme do not have to be modified compared with the nominal mission, which means that the original Input Catalogue will still form the basis of the observing programme.

The number of observations for a particular star increases roughly in proportion to the mission lifetime. Sky coverage uniformity also improves with mission duration. However, the ability to decouple the determination of the positions, parallaxes and proper motions of the stars is also a function of the mission duration. In the nominal mission, a lifetime of at least 18 months was required to provide (a) full sky coverage, and (b) allow the astrometric parameter determination. Simulations by the Data Reduction Consortia have confirmed that, for the revised mission, a lifetime of at least 18 months is required for a fully self-consistent set of Hipparcos positions, parallaxes and proper motions to be obtained.

8. Predicted Accuracies

Estimates of the expected acuracy achievable by the present mission (and hence the scientific return) are critically dependent on the lifetime of the satellite. Present predictions indicate that, just as for the nominal mission, if the lifetime is less than 18 months then the scientific returns are strongly degraded compared with expectations—a substantial proportion of the scientific programmes would not be satisfied. However, a mission lifetime extending beyond 24 months improves the achievable accuracy, although the results will still represent some degradation with respect to original expectations. The assessment of the scientific impacts of this degradation are ongoing. A better indication of the scientific returns will be possible when the operational difficulties, which affect the total amount of useful data that can be acquired, have been fully evaluated. Table 1 provides an indication of the accuracies that may be achievable for various durations of the revised mission, on the assumptions that the operations can be conducted without significant loss of observing time, or without a significant degradation of the data quality. These assumptions will be confirmed over the coming weeks.

Discussion

Editor's note: The oral paper was presented by E. Høg, who responded to the comments and remarks.

MIYAMOTO: Prior to going to the discussion about the resolution, I would like to tell you that I, on behalf of IAU Commission 8, wrote a letter on 11 September to Prof. Lüst of ESA, in which I requested ESA to give earnest consideration to a new mission of HIPPARCOS as early as possible. Then, I received a reply of 22 September from Prof. Lüst, in which he states:
"I certainly anticipate activity in the scientific circles in Europe on the matter of Hipparcos 2, especially among the scientists who will have to accept that their projects be delayed, but you can be assured of my determination to look at this issue in depth."

HøG: Let us take this as an encouragement.

MORRISON: Was any consideration given to changing the observing strategy in a limited mission: e.g., by limiting the observations to brighter stars?

HøG: These questions were considered at the first Science Team meeting one week after the launch. The conclusion was quickly reached that nothing could be gained by changes in the scanning law, input catalogue, or observation strategy.

RATNATUNGA: Could you please comment on the prospect of a second HIPPARCOS and the continuity of observations with the current mission in the event of a decision for a replacement satellite which is so vital for astronomy?

HøG: I hope that all possible scientific observations will be continued with the present satellite for, say, 2 years, and I believe that most scientists will share this hope. Careful scientific arguments must of course be presented for such a continuation because the operational cost is far from negligible.

HUGHES: Are the two new additonal downlink stations now operational?

HøG: The one in Australia is operational, but I am not certain about the one in Kourou in French Guiana.

Note added in proof:

Updated information on 1 February 1990: since this paper was accepted for publication, progress with the HIPPARCOS operations have continued to improve the overall mission outlook in three main areas:
(1) indications are now that a satellite lifetime of 30 months or more may now be achievable;
(2) the long eclipse durations are likely to have a much smaller scientific impact than originally expected;
(3) the fraction of useful scientific data acqusition is now in the range 55-60 per cent, and should increase further once the NASA Goldstone ground station is incorporated into the network during March 1990. The first great circle reductions have been completed by the data reduction teams, and have confirmed the very high quality of the scientific data. If a mission lifetime of 36 months is reached, astrometric parameters close to the orginal 2 milli-arcsec targets might still be achievable.

The TYCHO Project on-board the HIPPARCOS Satellite

E. HØG
Copenhagen University Observatory
Østervoldgade 3
1350 Copenhagen K
Denmark

ABSTRACT. The Hipparcos mission had to be revised because the satellite did not reach the circular geostationary orbit. Observations from the elliptical transfer orbit will be degraded in the sense that good quality observations are expected, but only during less than 50 percent of the total time, and the mission duration will probably be less than 2.5 years. If the mission lasts 12 months a Tycho catalogue is expected containing at least 200 000 stars with typical accuracies of 0.10 arcsec for positions and 0.10 mag for B and V magnitudes.

Introduction

In the Tycho data analysis, see Høg (1989), the photon counts from the Hipparcos star mapper are processed to detect slit transits exceeding a certain signal-to-noise ratio. These detections or transits, collected throughout the mission, are identified with stars contained in a Tycho Input Catalogue. A limiting magnitude of about $B = 11$ mag, depending on star colour, was expected in the nominal mission of 2.5 years within small areas of 40 arcsec diameter centred on each Tycho Input Catalogue position, corresponding to a final Tycho Catalogue of at least 400 000 stars. The typical astrometric and photometric accuracy of a mean value in the Tycho Catalogue was expected to be 0.03 arcsec and 0.03 mag, respectively, for the majority of faint stars. Double stars with separations larger than about 2 arcsec would also be resolved.

A revised mission is now foreseen because the satellite could not be placed in the planned geostationary orbit due to failure of the apogee-boost-motor. Observations will be obtained from the elliptical orbit, as explained by Perryman (1989). The performance in this orbit is not yet known with any certainty, but probably the mission will last at least 12 months and result in a Tycho catalogue containing at least 200 000 stars with typical accuracies of 0.10 arcsec and 0.10 mag.

Observations and Data Analysis

The primary purpose of the Hipparcos star mapper is to observe the transit

J. H. Lieske and V. K. Abalakin (eds.), Inertial Coordinate System on the Sky, 307–310.
© 1990 *IAU. Printed in the Netherlands.*

time of bright stars of known position when they cross the slit system. The slit system consists of two groups of narrow slits, each of 40 arcmin length. The four slits in one group is perpendicular to the motion of stars in the field, the other group is inclined by 45 degrees, as has been illustrated and discussed by Høg (1986). By means of known star positions and observed transit times, the attitude of the satellite is determined. The satellite attitude must be known with a precision of about 1 arcsec during the mission in order to point the light sensitive area of the main detector at the individual programme stars as they cross the main field of view. A good attitude knowledge is also required later on to achieve the best astrometric precision in the data analysis of the programme stars, and here the attitude must be known with a precision of 0.1 arcsec.

Since stars used for the real-time attitude determination are a subset of the main mission programme stars, which are in turn a subset of all stars brighter than the star mapper detection limit, it became evident during the development of the Hipparcos project that observations of transit times for many stars, other than those required for attitude determinations, can be exploited to derive the positions of these stars. The photometric results for the stars are obtained from the analysis of the stellar photon fluxes at the slit transits.

The raw photon counts from the star mapper will undergo a sequence of processing steps, see Høg (1989). Firstly, the 'detection' process will be used to detect slit transits above a certain signal-to-noise threshold, and to estimate the epoch, amplitude and background associated with each such transit. Each transit is identified or associated with a star by means of a series of processes: 'prediction', 'identification' and 'recognition'.

These later processes are carried out at the three institutes Astronomisches Rechen-Institut in Heidelberg, Astronomisches Institut der Universität Tübingen, and Observatoire de Strasbourg, respectively. The 'detection' and 'photometry' are also carried out at Tübingen and the 'astrometry' process at Copenhagen University Observatory.

The Tycho Input Catalogue containing 2000 000 stars with an accuracy of positions about 1 arcsec at the epoch 1990 has been compiled at Strasbourg from the INCA data base and the Space Telescope Guide Star Catalogue, see Egret et al. (1989).

The Revised Mission

The accuracy obtained for a given star with Tycho is proportional to the inverse square root of the number of observations because photon statistics is the dominating error source. Since each star mapper crossing gives a practically complete determination of position and magnitudes it is much simpler to estimate the accuracy from a revised Tycho mission than from a revised Hipparcos main mission where many scans over a long interval of time is required before the position, parallax and proper motion components can be solved for. For the same reason the sky coverage during a certain mission length will be more uniform for Tycho than for the main mission, and the non-uniformity for the Tycho mission will not be considered here.

We do not expect difficulties in the data reductions from the fact that the observations are split up in shorter stretches. The number of reference stars should still be sufficient for all photometric and astrometric calibrations, provided the instrument stability is not degraded.

The average number of observations per star is proportional to the "useful observing time", and for the revised mission we must presently assume a useful observing time of only 25% of real time, compared to nearly 100% for the nominal mission. We know for certain that the observing time is less than 47% due to limited visibility from 3 ground stations and because the dark count of the photomultipliers is too high (>2500 counts per second) less than 2.25 hours from the perigee, due to Cherenkov radiation in the optics. A further decrease to one half must at present be assumed due to frequent attitude re-acquisitions and other presently unpredictable operations, thus leaving a useful Tycho observing time about 25% of real time.

The Table shows that the accuracy from a revised mission is much inferior to that of a nominal mission, but even 12 months provide an accuracy unavailable from the ground for such a large number of stars.

For shorter durations than 12 months the number of recognized stars will be further reduced due to too few crossings and to non-uniform sky coverage.

A severe reduction of scientific value is expected with respect to detection of new variable stars from photometry. This is due to the smaller number of observations in a given interval of time, and to the predicted shorter mission length.

Table 1. Number of Tycho stars and typical accuracy for the faint ones

Mission	Nominal	Revised	Revised
Duration (months)	30	12	30
Number of stars	400000	200000	400000
Astrometry (arcsec)	0.03	0.10	0.06
Photometry (mag)	0.03	0.10	0.06

References

Høg, E. (1989) 'Overview of the Tycho data analysis', in M. A. C. Perryman et al., The HIPPARCOS Mission, Volume III, The Data Reductions, ESA SP-1111, pp. 183-194.

Perryman, M. A. C. (1989) 'In-orbit status of the Hipparcos astrometry mission', in the present volume.

Høg, E. (1986) 'TYCHO astrometry and photometry', in H. K. Eichhorn and R. J. Leacock (eds.), Astrometric Techniques, IAU Symposium No. 109, pp. 625-635.

Egret, D., Didelon, P., McLean, B. J. (1989) 'The Tycho Input Catalogue', in M. A. C. Perryman et al., The HIPPARCOS Mission, Volume III, The Data Reductions, ESA SP-1111, pp. 427-436.

Discussion

MORRISON: You quoted the mean error for an automatic meridian circle to be 0.15 arcsec per observation. However, by repeated observations, the asymptotic mean error of a published position is 0.10 arcsec or better. This is the figure that should be compared with the TYCHO accuracy.

HøG: This means that 200 000 positions expected from a 1-year TYCHO mission is in the same sense worth as much as 10 years of observations with the Carlsberg Meridian Circle!

CHUBEY: What are the moments of inertia of the spacecraft in the present situation?

HøG: The moments of inertia are considerably larger due to the 500 kg of solid fuel, and this is taken into account in the attitude control.

WALTER: Is a weekly grid calibration sufficient bearing in mind that the stability of the grid may be impaired by the varying Earth albedo heating owing to the highly eccentric orbit of HIPPARCOS?

HøG: The varying Earth albedo will be minimized by phasing the satellite rotation so that the instruments point away from the Earth at perigee. The still remaining thermal effects will be studied.

HIPPARCOS AND REFERENCE SYSTEMS

F. MIGNARD

O.C.A. / CERGA

Av. Copernic

06130 Grasse (France)

ABSTRACT.

The Sphere determined from the observations carried out by the ESA satellite Hipparcos will have a residual rotation with respect to distant extragalactic sources. The alignement of the two systems of reference will be made at the mission completion. The procedure to be applied is discussed along with its expected accuracy.

1. INTRODUCTION

The main objective of global astrometry is to build a reference system materialized by the position and proper motion of as many luminous sources as possible (Kovalevsky, 1984). HIPPARCOS will provide such a system through the sphere solution. The HIPPARCOS solution will ultimately give for each star

i) a position as seen from the barycenter of the solar system at a reference epoch t_0 (for example 1991 January 1, 12h TB)

ii) the components of the proper motion at the same epoch. For a typical Hipparcos star (V= 9 mag) it is expected that the positions at the epoch and proper motion will be determined respectively with an r.m.s. error smaller than 2 mas and 2 mas per year, provided the mission lifetime is at least3 years.

2. THE HIPPARCOS AND VLBI REFERENCE FRAMES

However the Hipparcos measuring system will only provide a perfectly consistent system of angular arcs between the ~115000 entries of its stellar catalogue and not their absolute position on the sky. The mathematical structure of the equations relating the photons counts

311

J. H. Lieske and V. K. Abalakin (eds.), Inertial Coordinate System on the Sky, 311–321.
© 1990 *IAU. Printed in the Netherlands.*

recorded by the detectors to the astronomical coordinates of the stars are invariant by any 3-D rotation. Accordingly two catalogues differing by a mere rotation are equally acceptable solutions of the observation equations. Thus an arbitrary convention is necessary to finalize the Hipparcos catalogue. Put in other words the normal equations possess a global symmetry which let them unchanged by any transformation of the group SO(3), and so are the solutions of these equations. From a more physical point of view one can say that Hipparcos fits a very large number of arcs drawn on a sphere between more than 100000 points. The relative position of the points does not depend on the precise orientation of the sphere, which therefore will have to be arbitrarily fixed at a later stage.

A question arises naturally : among all the possible conventions to define the coordinate system for the Hipparcos catalogue, are there some more compelling than others ? If stars were only used as milestones to track the displacement of moving objects, without reference to their dynamics, taking one convention rather than another would not make any difference. But since the time of Newton, physicists and astronomers have singled out the inertial systems, with respect to which the laws of nature look more simple. Thus it is a desirable feature that the reference system to which proper motions are referred be inertial, or in the terminology of general relativity, quasi-inertial . Since it is doubtless that the Hipparcos sphere will rotate with respect to any materialization of the inertial frame, either dynamical or geometrical, additonal work will be needed to align it with the extragalactic frame. This is a prerequisite to any progress in dynamical astronomy based on the Hipparcos catalogue.

It is now anticipated that a new conventional inertial system attached to radio sources will shortly supersede the FK4 system. These radio sources are either quasi-stellar objects (quasars) or galactic nuclei. The possibility of defining a quasi-inertial reference system with extragalactic radio sources is based on the assumption that remote galaxies and quasars have no global rotational motion. In fact the actual assumption is somewhat stronger than that, since taken for granted that quasars are at cosmological distances, they have no significant angular motion detectable within the current capability of astrometry. Kovalevsky (1989) points out that in the most defavorable case a quasar at 100 Mpc with a transverse velocity reaching the speed of light, the proper motion would not be larger than 0.7 mas/year. A more realistic limit is obtained by matching the transverse velocity to the recession which yields a proper motion of 0.01 mas/year (Argue, 1989). A still more realistic constraint is placed by the measurement of the relative displacement of the two close-by (in angle) sources NRAO512 and 3C345 wich proves to be less than 0.02 mas/year (Bartel et al., 1986).

Unfortunately no quasar but one (3C 273 with a magnitude of 13) is optically bright enough to be seen by Hipparcos. Intermediate objects and observations will have to be used to realize the connection between the two reference frames.

The practical realization of the inertial frame is handed over to the Very Long Baseline Interferometers which, rotating with the earth determine the sources declination with respect to the instantaneous Earth's axis of rotation and their right ascension by reference to a selected fiducial source or some weighted average over many sources. At present the accuracy of radio sources position allows to materialize the extragalactic frame with a typical accuracy of 1 mas for about 100 objects, only for declinations above -40° (Sovers et al. 1988, Ma et al. 1986). Though the number of sources is rather small, the system possesses the remarkable property of not deteriorating with time because the error in the proper motion is negligibly small.

The alignement of the Hipparcos system to the VLBI's will permit to increase considerably the accessibility to the inertial frame for optical astronomy. The number of objects in the optical catalogue will be about 1000 times larger than the content in radiosources in the VLBI catalogue. With an average of 2 to 3 such stars per square degree a Schmidt plate will contain about 70 to100 Hipparcos stars for the determination of the plate constants.

3 . LINKING HIPPARCOS TO THE INERTIAL FRAME

3 . 1. General Principle

Several methods must be developed according as the link is between the Hipparcos frame and a stellar frame or if one searches to make the connection with the dynamical frame. The case of comparing the Hipparcos sphere to the dynamical reference frame will no be discussed in this paper. Interested readers may find valuable information in Söderhjelm and Lindegren (1982), Lindegren (1987) and Morando (1987).

Let us restrict to the first case ; it splits into two broad categories :

- Some stars belong to the two catalogues, but are not referred to a common reference system. This will be the case for a handful of radiostars observed by both Hipparcos and VLBI. In the latter case they are referred to the VLBI extragalactic reference frame mentioned in Sec.2 . Positions and proper motions are assumed to be available, along with a covariant matrix. The situation applies also to stars belonging to the proper motions survey of Lick and Pulkovo (Argue , 1989).

- No object appear in common in the two catalogues. This case pertains to the quasars in the VLBI system, not seen by Hipparcos. In that case one must resort to dedicated additional observations capable of giving the position of Hipparcos stars relative to quasars (Froeschlé and Kovalevsky 1982).

3 . 2 . Mathematical Procedure

Consider the link by means of radio stars, which is independant of the availability of observations with the Hubble Space Telescope.

Let X_i and V_i be respectively the position and velocity vector of the i th common object given in the HIPPARCOS catalogue the components of which are projected in the HIPPARCOS frame. In the present context the velocity vector V does not take into account the radial velocity ; this is the velocity vector associated to the transverse displacement of the star. Let Y_i and W_i be the same for the radio stars in the extragalactic frame.

Note that there is a subtle difference on how one should deal with the position vectors and the velocity vectors:

- The two position vectors noted X_i and Y_i stand for the same mathematical vector and strictly speaking should be written with the same letter. The two notations are necessary to recall that they are projected on two different basis.

- The velocity vectors refer to two different vectors, velocity of an object measured from two different reference systems, which are projected on two coordinates systems. We have then good reasons to use specific notations in that case.

Model assumption : in the following the two reference systems are assumed to differ only by a global rotation, possibly time dependant. In short, one neglects any kind of regional error that might be present in either system, that is to say each catalogue is supposed to provide a perfectly consistent set of positions and proper motions. This assumption will be checked at the end of the link from a careful examination of the residuals, to detect possible inadequacy of the model. In case the assumption fails the only way to remedy would be to consider that the global rotation is only the first term of an otherwise more complex expansion.

Let R be the rotation matrix between the two frames at the time t ; thus we have

$$X_i = [R] Y_i$$

$$V_i = [R] W_i + [\dot{R}] Y_i$$

The first equation accounts for the change of reference system while the latter allows for the change of frame and coordinate system. The rotation matrix depends on only three parameters, namely the axis of rotation and the angle of rotation, while $[\dot{R}]$ needs three other parameters to be defined.

Clearly one can restrict to infinitesimal rotations and neglect any term of second order in [R] and [$\dot{R}$]. The most general such rotation matrix has the following form

$$R = \begin{bmatrix} 1 & \gamma & -\beta \\ -\gamma & 1 & \alpha \\ \beta & -\alpha & 1 \end{bmatrix}$$

and is associated to the three direct infinitesimal rotations α, β, γ respectively about the x, y, z axes. The sign convention is chosen so that a rotation of the triad is meant and not that of the vectors. The sphere is slowly rotating with respect to the inertial frame in such a way that one can expand α, β and γ in the vicinity of the reference time t_0 of the HIPPARCOS catalogue as

$$\alpha = \alpha_0 + \dot{\alpha}(t - t_0) \quad ; \quad \beta = \beta_0 + \dot{\beta}(t - t_0) \quad ; \quad \gamma = \gamma_0 + \dot{\gamma}(t - t_0)$$

Finally the observation equations for a particular star become

$$X_i = \begin{bmatrix} 1 & \gamma_0 & -\beta_0 \\ -\gamma_0 & 1 & \alpha_0 \\ \beta_0 & -\alpha_0 & 1 \end{bmatrix} Y_i \tag{1}$$

for the position in the two catalogues and

$$V_i = \begin{bmatrix} 1 & \gamma_0 & -\beta_0 \\ -\gamma_0 & 1 & \alpha_0 \\ \beta_0 & -\alpha_0 & 1 \end{bmatrix} W_i + \begin{bmatrix} 0 & \dot{\gamma} & -\dot{\beta} \\ -\dot{\gamma} & 0 & \dot{\alpha} \\ \dot{\beta} & -\dot{\alpha} & 0 \end{bmatrix} Y_i \tag{2}$$

for the velocity vectors. The second matrix in (2) accounts for the induced velocity brought about by the time-dependant rotation.

In the above equations the subscript stands for the stars to be compared ; The X_i, Y_i, V_i and W_i are known (within observational errors) and the six rotation parameters are searched for. From a set of objects known in the two systems a least squares fitting yields the six unknowns of the rotation. However the above presentation is somewhat misleading, as it conveys the idea that the each stars will provide two vector equations amounting to six scalar observation equations. In fact only two independant scalar equations are availbale in (1) and another two in (2) . Hence it is useful to rewrite the observations equations in slight different way so as to make the linear structure more apparent and retain only independant equations . Let l and b be the longitude and latitude of a star, μ_l and μ_b the proper motion in longitude ($\mu_l = \dfrac{dl}{dt}\cos b$) and latitude. The choice of a particular set of spherical coordinates is irrelvant in the following discussion ; right ascension and declination can be substituted straightforwardly for the longitude and latitude. We obtain after some algebraic transformations of Eqts. 1 and 2 the design matrix as,

$$
\begin{bmatrix}
\Delta l\cos b \\[4pt]
\Delta b \\[4pt]
\Delta\mu_l\cos b \\[4pt]
\Delta\mu_b\cos b
\end{bmatrix}
=
\begin{bmatrix}
\cos l\sin b & \sin l\sin b & -\cos b & 0 & 0 & 0 \\[4pt]
-\sin l & \cos l & 0 & 0 & 0 & 0 \\[4pt]
\mu_b\cos l & \mu_b\sin l & 0 & \tfrac{1}{2}\cos l\sin 2b & \tfrac{1}{2}\sin l\sin 2b & -\cos^2 b \\[4pt]
-\mu_l\cos l & -\mu_l\sin l & 0 & -\sin l\cos b & \cos l\cos b & 0
\end{bmatrix}
\begin{bmatrix}
\alpha_0 \\[4pt]
\beta_0 \\[4pt]
\gamma_0 \\[4pt]
d\alpha/dt \\[4pt]
d\beta/dt \\[4pt]
d\gamma/dt
\end{bmatrix}
$$

where $\Delta l = l_{vlbi} - l_{hip}$ and accordingly for the other variables. If the observations were perfect in both catalogues, the link would be achievable with one and a half star only : six relationhips for six unknowns. In practice about twenty stars will be available to carry out the adjustement . The observation matrix will be a 4n x 6 matrix obtained by repeating the above matrix at different l_i and b_i. One should note that the parameter γ is determined only through Δl (the latitude remains unchanged by a rotation about the z-axis) and $d\gamma/dt$ affects only the proper motion in longitude. As stated before the signs are such that we go from the Hipparcos frame to the extragalactic frame by direct rotations about the coordinate axes.

The left-hand side column contains the random vector of the observations. It appears clearly that the noise in each equations is a combination of the errors in both the VLBI and Hipparcos catalogue. Typically we have, for example in the first equation,

$$\sigma^2(\Delta l\cos b) = \sigma^2_{hip} + \sigma^2_{vlbi} \tag{3}$$

which shows that the two catalogues play a symmetric part in the achievement of the accuracy of the link. Because of the quadratic nature of this combination the final accuracy is determined primarily by the less accurate catalogue. This will be critical for the determination of the motion of the Hipparcos sphere with respect to the inertial frame, as it can only be obtained through the comparison of proper motions. One should remark that not every piece of information need be used to determine the matrix of rotation and its first derivative. Good positional information can be processed separately with the first two equations (for each star) of the design matrix to determine the rotational shift between the two systems at the epoch. Afterwards one will select stars with reliable proper motions to solve for the derivatives $\dot{\alpha}, \dot{\beta}, \dot{\gamma}$. These proper motions could originate from various sources, in addition to radio stars program.

With the help of the design matrix it is easy to evaluate the accuracy to be obtained in the estimation, at least for a regular distribution of the linking stars. By neglecting the off-diagonal terms in the normal matrix one obtains from the variances when N stars are used for the link,

$$\sigma(\alpha) = \sigma(\beta) = \sigma(\gamma) = \frac{\sigma}{N^{1/2}} \cdot \frac{1}{<\cos^2 b>^{1/2}} \sim \frac{1.4\,\sigma}{N^{1/2}}$$

where σ is the standard deviation of $\Delta l \cos b$ and Δb (Eq.3). For the derivatives we have

$$\sigma(\dot{\alpha}) = \sigma(\dot{\beta}) = \sigma(\dot{\gamma}) = \frac{\sigma'}{N^{1/2}} \cdot \frac{1}{<\cos^4 b>^{1/2}} \sim \frac{1.7\,\sigma'}{N^{1/2}}$$

where σ' is the standard deviation of $\Delta\mu_l \cos b$ and $\Delta\mu_b \cos b$. This crude evaluation is in perfect agreement with the monte carlo simulation presented in the next section.

3 . 3 . Preliminary Results

A simulation has been set up by Froeschlé and Kovalevsky (1982) with the nominal accuracy expected from the Hipparcos phase study. They have solved the equations taken in the form (1) and (2), apparently without introducing the fact that $|\mathbf{X}| = |\mathbf{Y}| = 1$. It is difficult to assess how the use of redondant equations affects their results, although I am confident that their figures are representative of what would be obtained with indepedant equations. The formal standard deviations are probably slightly underestimated.

For radio stars observed in both systems they have found with the Hipparcos and radiosources astrometry of comparable accuracy ($\sim 0\rlap{.}{''}002$ in position) that the initial rotation is properly recovered with an accuracy of

$$\sigma = 0\rlap{.}{''}0035 \; /N^{1/2}$$

where N is the number of stars in common. As for the rotational motion of the Hipparcos sphere the error is given by

$$\sigma = 0\rlap{.}{''}005 \; /N^{1/2} \quad \text{per year.}$$

The reader is referred to their paper for a more detailed discussion of the results and the extension of the procedure to the link with the HST from the observations of quasars. To conclude this section it appears that a set twenty well distributed stars, with good VLBI positions and proper motions ($\sigma \leq 1$ mas and 1 mas/yr) , and free of regional errors should be sufficient to perform the link at the millisecond level.

4 . LINKING STARS

The objects to be used to link the Hipparcos catalogue to the extragalactic frame are derived from proposals made by individuals and accepted by a Selection Committee. The VLBI frame was initially thought purely as a radio frame with no particular attention to optical observation (Argue, 1989). The first report of possibe detection of radio-emission by stellar sources was made in the mid-60's for Betelgeuse. A recent compilation given by Wendker (1987) contains about 300 radio stars covering a wide variety of emitters, either steady thermal or highly variable non thermal. The spectral type and optical properties are also very different.

A working group was appointed by IAU in 1978 to select objects suitable for both optical and radio observations. The first list contained 234 sources and was published by Argue (1984). Only object brighter than V = 17 mag were retained, a convenient limit for the HST, but not for Hipparcos whose limitation is about V = 12.5 mag.

Most objects are Galactic radio stars having a very small angulardiameter. Typical stars are RS CVn binaries, UV Cet, early-type emission stars and water and SiO masers. About 200 such stars in existing catalogues would be suitable provided that accurate VLBI position and proper motions could be obtained. Several campaigns were undertaken to detect radio stars with VLBI and ascertain their position at different epochs. Table 1 below summarizes the attempts made until 1986 to catch the stars ; the last two colums show strikingly that pointing at a known radio star is not the same as recording it.

Table 1. Mark III VLBI campaigns of radio stars, (Lestrade et al. 1987a).

Years	Frequency (GHz)	Number of Stations	Number of stars observed	Number of stars detected
1982 Dec	8.4	4	12	5
1983 Feb	1.7	3	2	2
1983 Mar	8.4	5	15	5
1983 May	2.3 / 8.4	5	12	3
1983 Jul	5.0	6	12	7
1983 Oct	1.7	6	12	4
1984 Mar	5.0	4	1	1
1984 Oct	5.0	5	1	1
1985 Mar	8.4	3	6	0
1985 Jun	5.0	6	1	1
1986 Mar	10.8	4	3	2

At present ten radiostars have been monitored by VLBI during their radio outburst with the standard VLBI technique of bandwidth synthesis with sensitivity10 to100 mJy used for astrometry on extragalactic sources. Positions at a first epoch and multiple epochs for some starshave been reported with various accuracies by Lestrade et al. (1988a and 1988b). (see Table 2). The accuracy is mainly noise limited owing to the low flux density produced by radiostars compared to extragalactic sources. The best observed meet the precision requirement necessary to achieve the link without loss of accuracy.

Table 2 . VLBI position of ten radio stars

Star	Observation epoch	α (J 2000) deg.	RMS 0.″001	δ(J 2000) deg.	RMS 0.″001
LSI61303	1983.789	40.1320250	120	60.2293222	70
Algol	1983.567	47.0422117	4.5	40.9556553	5
UX Ari	1983.567	51.6472396	3	28.7155628	5
HR 1099	1983.215	54.1971833	3	0.5885167	20
HR 5110	1983.567	203.6987054	3	37.1824608	2
σ Cr B	1983.567	243.6716713	10	33.8590194	10
Cyg X1	1983.567	299.5903350	6	35.2016353	9
SZ Psc	1983.567	348.3490188	8	2.6753639	40
HD26337	1987.01	62.4202667	70	−7.8932333	7
HD77137	1987.03	134.9281875	30	−27.8161417	30

A new and very sensitive VLBI technique based on phase reference to a powerful and angularly close extragalactic source has been developed lately (Lestrade et al. 1987b). It has yielded a relative position of Algol with an accuracy of 0.5 mas, obtained during its quiescent phase of emission when the density flux is as low as 3 mJy. This technique is now routinely used at the millisecond level to monitor positions of radiostars and multiple epoch positons are now available for five stars. It is expected that additional positions and proper motions of such stars will be measured with VLBI by the time the Hipparcos catalogue is completed.

In term of sensitivity this new method allows VLBI measurements to match closely that of the VLA. However with the VLA faint radio stars are observed with an accuracy of 0.01 arcsec at the moment, not sufficent to carry out a valuable link.Further modifications in the VLA configuration and programs will hopefully give by 1994 proper motions and parallaxes of 20 stars with declination >30 deg to the milliarcsecond level (Johnston et al. 1988).

For the link through HST and quasars, 173 stars near 92 extragalactic objects (QSOs, BL Lacs and a few AGNs) have been identified. These stars are bright enough to be observed by Hipparcos and are included with a high priority flag in the Input Catalogue (Hemenway et al. 1988). The relative postion of the star with respect to the fixed quasar will be monitored at different time so that the proper motion can be determined. The difference between the HST and Hipparcos proper motion is the result of the time dependant rotation of the Hipparcos sphere relative to the extragalactic frame. The expected accuracy of 2 mas rms, is an order of magnitude better than the current accuracies obtained by ground-based optical techniques (Argue, 1989)

Closely connected to this work is the comparison between the radio positions and the optical positions obtained from meridian observations. VLBI positions are found to be comparable to optical positions although the differences are not significant below 0."05, the best precision obtainable with the automatic meridian circle of Bordeaux (Requième, 1973). An important point is to assess by how much the optical and radio photocenter are apart from each other. As the emission mechanism is poorly known it is of nearly no help in this regards ; however VLBI mappings indicate that the source size is of the order of 1 mas which is also the upper bound of the separation between the radio and optical photocenter.

CONCLUSION

Hipparcos was launched on 8 August 1989 from Kuru in French Guiana, but the failure of the apogee boost motor prevented from putting the satellite into its geostationary orbit. A degraded and probably shortened mission has started anyway. We do not know at the time of writing what the mission lifetime will be, as it is influenced by such unpredictable parameters like the number of solar flares and the induced changes in the Van Allen belts. A link with the extragalactic frame would be valuable provided the accuracy in the retrieval of star positions and proper motions by Hipparcos is sufficiently close to the nominal accuracy of few mas ; otherwise one should limit the ambitions to an adjustment to the FK5 frame, with a definition of 10 mas.

ACKNOWLEDGMENTS

J.F. Lestrade is gratefully acknowledged for his thorough reading of a preliminary version of this paper . Many of his remarks on the VLBI measurements have been incorporated in the final version.

REFERENCES

Argue A.N. , 1989. In the *Hipparcos Mission* , **Vol. II,** Chap . 16, ESA SP 1111.
Bartel et al. 1986. Nature, **319**, 733.

Froeschlé M., Kovalevsky J. , 1982. *Astronom. & Astrophys.* **116** , 89.

Hemenway P.D., Benedict G.F., Jeffreys W.H., Shelus P.J., Duncombe R.L., 1988. Proc. Sitges Coll. "*Scientific Aspects of the Input Catalogue Preparation*" **P. 461**. Torra and Turon C. eds.

Johnston K.J., Florkowski D. ,Vegt C. de, 1988. Proc. Sitges Coll. "*Scientific Aspects of the Input Catalogue Preparation*" **P. 447.** Torra and Turon C. eds.

Kovalevsky J. , (1984) . *Space Sci. Rev.* **39**, 1.

Kovalevsky J., 1989. In *Reference Frames in Astronomy and Geophysics*, **chap. 1**. Kovalevsky , Mueller & Kolaczek eds, Kluwer.

Kovalevsky J., Lestrade J.F., Preston R.A. .,1989. In the *Hipparcos Mission* , **Vol. III**, Chap . 17, ESA SP 1111.

Lestrade J.F., Preston R.A., Requième Y., Rappaport M., Mutel R.L., 1985. Proc. Aussois Coll. "*Scientific Aspects of the Input Catalogue Preparation*" **P. 251**. Perryman M. & Turon C. eds.

Lestrade J. F. , Preston R. A. , Niell A. E. , . 1987a, in *Proc. of the Third F.A.S.T. Thinkshop*, **P. 383**. Bernacca P.L. & Kovalevsky J. eds.

Lestrade J. F., Rogers A. E. E. , Niell A. E. , Preston R. A. . 1987b, in *The Impact of VLBI on Astrophysics and Geophysics* . IAU Sympos. 129, Cambridge, Mass. May 10-15.

Lestrade J.F. , White G.L. , Jauncey D. L., Preston R.A. , 1988b. Proc. Sitges Coll. "*Scientific Aspects of the Input Catalogue Preparation*" **P. 481**. Torra and Turon C. eds.

Lestrade J. F., Niell A. E. , Preston R. A., Mutel R. L. . 1988a, Astron. Jal. , **96**, 1746.

Lindegren L. 1986 . In *Proc. of the Third F.A.S.T. Thinkshop*, **P. 285**. Bernacca P.L. & Kovalevsky J. eds.

Ma C. et al. 1986. *Astron. Jal.* , **92**,1020.

Requième Y., 1973. *Astronom. & Astrophys.* , **23**, 453.

Söderhelm S. , Lindegren L., 1982 . *Astronom. & Astrophys.* , **110** , 156.

Sovers O.J. , Edwards C. P. , Jacobs C.S., Liewer K.M., Treuhaft R. M. , 1988. *Astron. Jal.*, In press.

Wendker H. J. , 1987. *Astronom. & Astrophys. Sup. Ser.* , **69**, 87.

A STRATEGY FOR LINKING HIPPARCOS TO AN EXTRA-GALACTIC REFERENCE FRAME

A.N. Argue[1]
Institute of Astronomy,
University of Cambridge,
Madingley Road, Cambridge CB3 0HA, UK.

ABSTRACT. The strategy for linking the HIPPARCOS Catalogue to extragalactic objects originally devised for the geostationary orbit, is no longer adequate for the elliptical orbit to which the satellite is now confined, but useful improvements in the rotation matrix may be obtained by incorporating proper motions from certain other catalogues and from ground based absolute proper motions.

1 Introduction

The locking of HIPPARCOS out of its geostationary orbit has created the urgent need for a drastic reappraisal of the strategy for the extragalactic link.

At the time of writing, there is an estimated 50% probability of the life time exceeding 18 months, but this is highly uncertain since it not estabilshed that it will be possible to reactivate the payload following the periods of prolonged eclipse. The following sigmas as a function of life expectancy have been predicted from simulations by Lindegren:

Life time	position $\sigma_{\alpha,\delta}$	parallax σ_π	proper motion σ_μ
1.5yr	0″.006	0″.0060	0″.008yr^{-1}
2.0	0″.003	0″.0045	0″.006yr^{-1}
2.5	0″.003	0″.0035	0″.005yr^{-1}

[1]JANET: ANA1@PHX.CAM.AC.UK

J. H. Lieske and V. K. Abalakin (eds.), Inertial Coordinate System on the Sky, 323–328.
© 1990 IAU. Printed in the Netherlands.

(For the geostationary mission originally planned, the duration would have been $2.5yr$ and each $\sigma = 0''.002$.) The estimates that follow here are based on a $2.5yr$ life time.

If a reconstitution of the celestial sphere can be performed successfully from the satellite data, there will result a system of positions, trigonometric parallaxes and proper motions for about 100,000 stars on a homogeneous, rigid instrumental system. The zero points for position and proper motions will be arbitrary in the first instance but will be aligned to FK5 as an off-line stage in the reduction. Lastly, the HIPPARCOS Catalogue will be aligned to the VLBI frame.

2 The Positional Frame

The positional frame will be greatly superior to any optical frame we now possess. In comparing it to FK5, it will be possible to smooth out small zonal errors known to exist in FK5 [1]. Then, linked to VLBI, there will be, for the first time, a homogeneous frame unified for optical and radio astrometry. This is urgently needed for the accurate correlation of optical and radio features in sources.

3 The Inertial Frame

For the solid body rotation for the frame, however, the rescue mission will not give any significant strengthening of FK5 beyond what is known already, *when the methods are confined to those envisaged by the FAST/NDAC link for the original mission* [2] *namely, using radio stars and Hubble Space Telescope* [3,11]. This is because both the numbers of objects and the predicted accuracy have become insufficient (the results now to be expected are given in the first line of Table 2). Although smaller than the σ_μ for one star ($0''.005yr^{-1}$), nevertheless much better values are needed for dealing with groups of stars, for instance in studies of galactic rotation. We shall show that improvements may be obtained by incorporating proper motions from other catalogues.

4 Proper Motions from other Catalogues

We propose that, following the transformation of the HIPPARCOS proper motions to FK5, the catalogues listed in Table 1 be then transformed to this HIPPARCOS-FK5 system. This could be done accurately because of the large overlap, as specified by the numbers in the third column. The FAST/NDAC procedures are then applied using these transformed catalogue proper motions in place of the HIPPARCOS instrumental values as originally planned for the geostationary mission. The likely numbers of stars in the FAST/NDAC proposals, selected for use with Space Telescope (HST) and radio stars respectively, are given in the fourth and fifth columns of Table 1. The numbers given for CAMC/Bordeaux are actual counts; the remaining numbers are estimates based on the various catalogue limits. For AC, the magnitude cut-off is assumed to be $B = 11^m$ and for IRS+NPZT and ACRS, $V = 10^m.5$. The declination limit for CAMC is $-45°$. The last five lines refer to "absolute proper motion" catalogues discussed in Section 7.

5 Accuracy of Rotation Link

We have used the simulations by Froeschlé and Kovalevsky [4] to estimate the uncertainty $\sigma R'$ in the time derivative R' of the fixed matrix rotation R for the first five catalogues in Table 1.

The results are in Table 2. Again we have used the limits of the various catalogues to estimate the numbers of objects, except for CAMC/Bordeaux: in the cases of IRS+NPZT and ACRS these estimates are very imprecise because we have not attempted to identify which particular stars are involved. For AC and CAMC/Bordeaux we have assumed 40 radio stars, this being the number likely to have accurately measured radio proper motions available at the end of the mission. For the VLBI accuracy we have assumed $0''.002$.

6 Conclusion

For the radio star link, AC and CAMC/Bordeaux in particular shew useful improvements over the rescue mission in the top line of Table 2. Indeed a straight mean of AC and CAMC/Bordeaux would give results very close to those expected for the original mission ($0''.0020 yr^{-1}$ for HST and $0''.0008 yr^{-1}$ for radio stars).

7 Ground based Optical Proper Motions

The last four lines in Table 1 refer to surveys that are directly tied to extragalactic objects. The values anticipated for $\sigma R'$ are (with the corresponding number of radio sources in parenthesis):

Lick/Yale	$0''.0002 yr^{-1}$	(-)
Potsdam	$0''.0014$	(10)
Kiev	$0''.0007$	(93)
Bonn	$0''.0003$	(20)
Hamburg/La Palma/CSIRO	$0''.0005$	(230)

Table 1: Surveys of Proper Motion: the third column gives the overlap between the catalogue and HIPPARCOS, the fourth and fifth the number of HIPPARCOS stars from the original proposals. Not all of the radio stars may have radio astrometry available at the end of the mission.

The mean errors for AC are from Brosche & Geffert [5]; ACRS (AC Reference Star Catalogue) from Corbin & Urban [8]; CAMC from Morrison *et al.* [6]; PPM from Röser and Bastian; CPC2 from de Vegt *et al.* [9]; Kiev from Yatsenko *et al.* [10] and the remainder from *The HIPPARCOS Mission II* [2].

Catalogue	Mean Error α,δ ($arcs.yr^{-1}$)	No. in Mission	No. in HST Link	No. in Radio Stars
HIPPARCOS	0″.005	100,000	147	157
AC	0″.003	$\sim 90,000$	90	142
PPM+CPC2	0″.004	$\sim 55,000$	92	~ 60
IRS+NPZT	0″.005	$\sim 38,000$	~ 30	~ 60
ACRS	0″.004	$\sim 90,000$	~ 80	~ 142
CAMC/Bordeaux	0″.003	$\sim 38,000$	77	40/61
Lick/Yale	0″.005	20,000		
Potsdam	0″.003	120		
Kiev	0″.003	340		
Bonn	0″.002	200		
Hamburg/La Palma/ CSIRO	0″.004	300		

Table 2: Uncertainty $\sigma R'$ in the frame rotation R' for six surveys of Table 1. In parenthesis: the number of radio sources assumed.

Catalogue	$\sigma R'$ HST Link		$\sigma R'$ Radio Stars	
HIPPARCOS	$0''.0040 yr^{-1}$	(78)	$0''.0021 yr^{-1}$	(40)
AC	$0''.0028 yr^{-1}$	(64)	$0''.0014 yr^{-1}$	(40)
PPM+CPC2	$0''.0036$	(60)	$0''.0020$	(30)
IRS+NPZT	$0''.0061$	(25)	$0''.0034$	(15)
ACRS	$0''.0036$	(60)	$0''.0022$	(36)
CAMC/Bordeaux	$0''.0027$	(77)	$0''.0014$	(40)

Again, $\sigma R'$ is significantly better than the proper motion of an individual star as measured by the rescue mission. These absolute proper motion surveys will provide a valuable supplement to the methods of Section 5. The original plan was to relegate these ground based methods to an inferior status of 'private projects' and use only the FAST/NDAC link for the HIPPARCOS Catalogue. Now it is urged that these methods have a *much more vital* rôle to play.

References

1. L. V. Morrison, P. Gibbs, L. Helmer, C. Fabricius, O. Einicke, Y. Réquième and M. Rapaport, 1990. *Evidence of systematic errors in FK5*. In 'Fundamentals of Astrometry', Proc. IAU Coll. 100, H. Eichhorn (ed.), in preparation.
2. A. N. Argue, 1989. *The Link to Extragalactic Objects*. In 'The HIPPARCOS Mission II', M. A. C. Perryman and C. Turon (eds.), ESA SP-1111 p. 199.
3. P. D. Hemenway, G. F. Benedict, W. H. Jefferys, P. J. Shelus and R. L. Duncombe, 1988. *The Extragalactic Link. Operational Preparations for the Hubble Space Telescope Observations*. In 'HIPPARCOS Scientific Aspects of the Input Catalogue Preparation', J. Torra and C. Turon (eds.), CIRIT, p. 461.
4. M. Froeschlé and J. Kovalevsky, 1982. *The Connection of a Catalogue of Stars with an Extragalactic Reference Frame*, Astron. Astrophys. 116, 89.
5. P. Brosche and M. Geffert, 1988. In 'Mapping the Sky', Proc. IAU Symp. 133, S. Débarbat *et al.* (eds.), p. 403.
6. L. V. Morrison, L. Helmer and L. Quijano, 1988. *Mapping the Sky with the Carlsberg Automatic Meridian Circle*, Ibid. p. 369.
7. S. Röser and U. Bastian, 1989. *Compilation of the PPM Catalogue*. In 'Star Catalogues: A Centennial Tribute to A. N. Vyssotsky', A. G. Philip and A. R. Upgren (eds.), p. 31.
8. T. E. Corbin and S. E. Urban, 1898. *Proper Motions of the Northern Astrographic Catalogue Reference Stars*, Ibid. p. 59.
9. Ch. de Vegt, N. Zacharias, C. A. Murray and M. J. Penston, 1989. *A Progress Report on the Second Cape Astrographic Catalog*, Ibid. p. 45.
10. A. I. Yatsenko, S. P. Rybka and R.-D. Scholz, 1987. *The Connection of the HIPPARCOS reference system to extragalactic objects by photographic astrometry*, Astron. Nachr. 308, 319.
11. F. Mignard, 1989. *HIPPARCOS and Reference Systems*. In this symposium.

Discussion

STRAND: It seems the value of 0.0002 arcsec for the Lick/Yale program is small. Can you explain this?

ARGUE: Yes, by the very large overlap between the Lick/Yale and the HIPPARCOS programmes.

YATSKIV: Why didn't you take into account the proper motion data from the Pulkovo program (i.e., the Tashkent, Kiev, and Pulkovo catalogues)?

ARGUE: I hope to receive more information on this programme at the joint meeting of IAU Commission 24 Working Group on Radio-Optical Objects and the HIPPARCOS INCA WG 2130, to be held here tomorrow.

THE REFERENCE FRAME DETERMINED FROM THE OBSERVATION
OF MINOR PLANETS BY HIPPARCOS

B. MORANDO & A. BEC-BORSENBERGER

Bureau des Longitudes

77, Avenue Denfert-Rochereau

75014 Paris · France

ABSTRACT. The observation of minor planets by Hipparcos offers the opportunity to obtain high precision positions for some minor planets. About fifty minor planets are on the programme. Their ephemerides had to be improved in order to reach a precision of 1 arsec and occultations by the Earth and the Moon had to be predicted.

From the position of a minor planet on reference great circles at different times better values of the initial position and velocity will be deduced but the reduction of the observations of the minor planets have to take into account the displacement of the photocentre relative to the centre which is due to the shape, the phase effect and the scattering properties of the surface. For some very small planets considered as star like this diplacement will be small and the precise positions obtained will allow to position the dynamical reference system relative to the Hipparcos system. For the bigger minor planets the observations by Hipparcos may give informations on the shape and scattering properties of the surface.

1. List of the minor planets on the programme

The times and the number of apparitions of the minor planets in the fields of view of the Hipparcos telescope has been calculated for each minor planet. Estimations of the maximum number of possible observations have been made following different values of the predicted Hipparcos magnitude. Predicted occultations by the Earth and the Moon have been studied by the FAST data analysis consortium at Cerga, and have been taken in account to obtain the windows of observability of the minor planets. The final selection of minor planets to be observed by Hipparcos has been made after consideration of the number of transits of minor planets for the period September 1989 to March 1992, the predicted magnitudes during these transits, and a possible degradation of the transparency of the telescope optics during the mission. Presently, new parameters have been introduced in the simulation, for estimating the number of transits which can be

329

J. H. Lieske and V. K. Abalakin (eds.), Inertial Coordinate System on the Sky, 329–336.
© 1990 *IAU. Printed in the Netherlands.*

obtained during successive sequences of observations from november 1989 and taking into account the increased number of predicted occultations by the Earth and the Moon due to the elliptical orbit of the satellite (Bec-Borsenberger A. 1989).

The following list gives the numbers and names of the minor planets retained after consideration of their frequency and magnitude at the time of the Hipparcos observations.

No	Name	No	Name	No	Name
1	Ceres	18	Melpomene	63	Ausonia
2	Pallas	19	Fortuna	88	Thisbe
3	Juno	20	Massalia	115	Thyra
4	Vesta	22	Kalliope	129	Antigone
5	Astraea	23	Thalia	192	Nausikaa
6	Hebe	27	Euterpe	196	Philomela
7	Iris	28	Bellona	216	Kleopatra
8	Flora	29	Amphitrite	230	Athamantis
10	Hygiea	31	Euphrosyne	349	Dembowska
11	Parthenope	37	Fides	354	Eleonora
12	Victoria	39	Laetitia	451	Patientia
13	Egeria	40	Harmonia	471	Papagena
14	Irene	42	Isis	511	Davida
15	Eunomia	44	Nysa	532	Herculina
16	Psyche	51	Nemausa	704	Interamnia

2. Improvement of the orbital elements

Because of the importance of having good ephemerides for the Hipparcos project, the improvement of the orbital elements of the Hipparcos minor planets has been realized in parallel by A. Bec-Borsenberger and J. Calaf (Bec-Borsenberger A. , Calaf J. 1989) so that there is a full and independant cross check of the two sets of improved orbital elements.

The main points of the determination of these elements by the analysis of the residuals between the observations and the ephemerides are recalled here. From initial conditions which have been taken in the ephemerides for minor planets (ITA), equatorial positions have been computed by numerical integration for each date of the available observations: those collected at the Minor Planet Center (Cincinatti) and, since 1983, about 12700 observations made specifically for the Hipparcos project (Bordeaux, Fabra, La Palma, San Fernando observatories). To reduce the residuals between the observed place and the computed place, the partial derivatives with respect to the initial conditions are computed and the system of equations of condition thus obtained has been solved by the least square method; the variations of the elements given by the solution furnished new elements. As, for the Hipparcos project, ephemerides have to be referred to the J2000 frame, the improved elements and the observations have been referred to this frame; then new improvements, including the last available observations made during 1987 and 1988, were made. Based on these ephemerides of the minor planets, a test of the Hipparcos mission over three years, from the 1st of April 1989, has been carried out at the Besancon observatory.

3. Determination of accurate positions from Hipparcos

An observation by Hipparcos gives the position of a minor planet along a given reference great circle for a given time. From a certain number of these observations the position and velocity at a given time, considered as beeing an initial time, can be determined.

On figure 1 RGC is the reference great circle at a given time and P its pole. N being the ascending node of the RGC on the the ecliptic of J2000.0, Hipparcos gives us only arc λ equal to NM", where M" is the projection on the RGC of the position M of the minor planet on the celestial sphere.

Let β be arc M"M and l and b the ecliptic coordinates of M. The two vectors :

$$\begin{bmatrix} X \\ Y \\ Z \end{bmatrix} = \begin{bmatrix} \cos \beta \cos \lambda \\ \cos \beta \sin \lambda \\ \sin \beta \end{bmatrix} \quad ; \quad \begin{bmatrix} x \\ y \\ z \end{bmatrix} = \begin{bmatrix} \cos b \cos l \\ \cos b \sin l \\ \sin b \end{bmatrix}$$

are such that :

$$\begin{bmatrix} X \\ Y \\ Z \end{bmatrix} = R \begin{bmatrix} x \\ y \\ z \end{bmatrix}$$

R is a rotation matrix which can be obtained for each observation as the ecliptic coordinates L and B of the pole P of the RGC are given. If :

$$i = \pi/2 - B \qquad u = \pi/2 + L$$

then :

$$R = \begin{bmatrix} 1 & 0 & 0 \\ 0 & \cos i & \sin i \\ 0 & -\sin i & \cos i \end{bmatrix} \cdot \begin{bmatrix} \cos u & \sin u & 0 \\ -\sin u & \cos u & 0 \\ 0 & 0 & 1 \end{bmatrix}$$

$$= \begin{bmatrix} \cos u & \sin u & 0 \\ -\sin u \cos i & \cos u \cos i & \sin i \\ \sin u \sin i & -\cos u \sin i & \cos i \end{bmatrix}$$

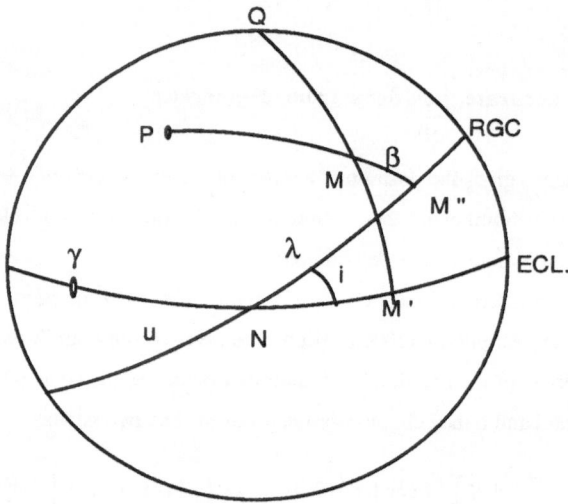

Figure 1.

The calculated coordinates x_c, y_c, z_c are functions of time t and of the initial conditions q_i (i = 1...6), the observed coordinates x_o, y_o, z_o are functions of time t and of $q_i + dq_i$, where the dq_i are corrections to he q_i

$$x_c = x(q_i,t) \quad ; \quad x_0 = x(q_i + dq_i,t)$$

....

....

with analogous expressions for the y and z. A Taylor expansion to the first order gives :

$$x_0 - x_c = \frac{\partial x}{\partial q_i}\, dq_i$$

....

....

But the corresponding variations of the coordinates X, Y, Z in the RGC reference system are given by :

$$\begin{bmatrix} dX \\ dY \\ dZ \end{bmatrix} = R \cdot \begin{bmatrix} \dfrac{\partial x}{\partial q_i} \\[6pt] \dfrac{\partial y}{\partial q_i} \\[6pt] \dfrac{\partial z}{\partial q_i} \end{bmatrix} = \begin{bmatrix} A_i\, dq_i \\ B_i\, dq_i \\ C_i\, dq_i \end{bmatrix}$$

This gives, in terms of λ, β, $d\lambda$, $d\beta$:

$$- \sin \beta \cos \lambda \, d\beta - \cos \beta \sin \lambda \, d\lambda = A_i \, dq_i$$
$$- \sin \beta \sin \lambda \, d\beta + \cos \beta \cos \lambda \, d\lambda = B_i \, dq_i$$
$$\cos \beta \, d\beta = C_i \, dq_i$$

(The second members are, of course, summed over i.)

Hipparcos only yields values for $d\lambda$, elimination of $d\beta$ between the first two equations gives the fundamental equation of the problem :

$$\boxed{\cos \beta \, d\lambda = \left(B_i \cos \lambda - A_i \sin \lambda \right) dq_i}$$

Many such equations for a given planet can be combined and solved for the dq_i by the least square method. The combination of many such positions will give the position of the dynamical reference system relative to the Hipparcos system (Söderhjelm S. , Lindegren L. 1982).

4. Displacement of the photocentre

The ephemerides give the position of the centre of the minor planet but the observations give the position of the photocentre. The situation is explained on figure 2.

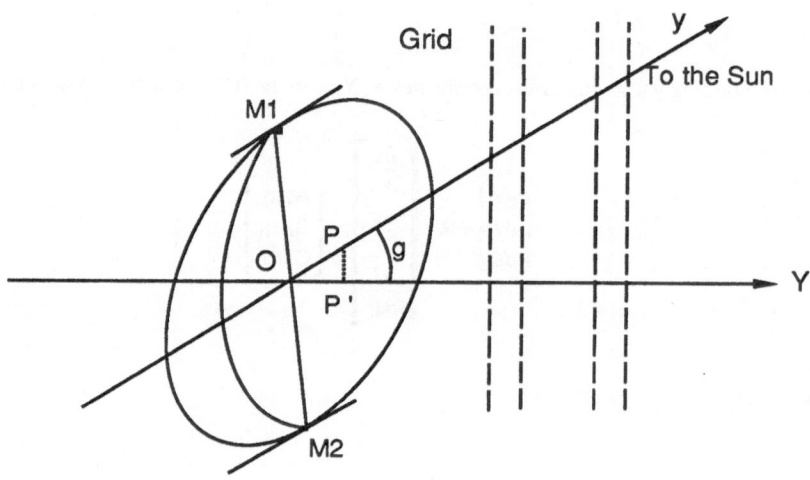

Figure 2.

This figure shows the focal plane of the Hipparcos telescope. Axis OY is perpendicular to the grid slits, Oy is directed towards the Sun. Assuming the minor planet to be a three axes ellipsoid, its image in the focal plane is an ellipse centered at O and the terminator is a half ellipse also centered at O and tangent to the first ellipse at M_1 and M_2. Hipparcos measures the abcissa along OY of the projection P'on OY of the photocentre P.

This peculiarity, which makes the observation of minor planets different from the observation of stars, has been studied by both reduction consortia, NDAC (Lindegren L. 1986) and FAST (Morando B. 1985, 1986). The position of P at any time may be calculated (Morando B. 1985) but depends strongly on the scattering properties of the surface of the minor planet. It is shown, for instance, that for a uniformly lit

Ceres the displacement of the photocentre is 0.004 arcsec for a phase angle of 20° but it is 0.02 arcsec if Lambert'law is true (Morando B. 1986). Unfortunately there are only few cases where shape-albedo models are known. However 17 of the minor planets on the programme are small enough to be considered as star like so that their positions will reach the nominal precision (Morando B. , Lindegren L. 1989).

5. Conclusion

Launched on the 9th of August 1989 Hipparcos could not be put on a geostationnary orbit so that the nominal mission will not be fulfilled. This means, as far as minor planets are concerned, that less observations will be made because occultations by the Earth and eclipses of the Sun as seen from the satellite will be more numerous than expected. Yet it is hoped that a good coverage made of observations to a few hundred arcseconds will be obtained.

REFERENCES

- Bec-Borsenberger A. (1989) ' Minor planets and other planetary system objects', The Hipparcos Book edited by M. Perryman ESA, vol.2, p.214
- Bec-Borsenberger A. and Calaf J. (1989) ' Improvement of orbital elements and high precision ephemerides for Hipparcos minor planets', in 'Asteroids, Comets Meteors', vol.3 (University of Uppsala) (under press)
- Lindegren L. (1986) ' Hipparcos observations of minor planets and natural satellites ', Proceedings of the third FAST thinkshop in Bari (Italy). Edited by P. L. Bernacca & J. Kovalevsky, p. 285
- Morando B. (1985) ' The modulation curve of a minor planets', Proccedings of the second FAST thinkshop in Marseilles (France). Edited by J. Kovalevsky, CERGA, p.125
- Morando B. (1986) ' A more elaborate model for the modulation curve of a minor planet ', Proceedings of the third FAST thinkshop in Bari (Italy). Edited by P. L. Bernacca & J. Kovalevsky, p. 277
- Morando B. and Lindegren L. (1989) ' The treatment of minor planets and planetary satellites', The Hipparcos Book edited by M. Perryman ESA, vol.3, p. 269
- Söderhjelm S. and Lindegren L. (1982) 'Inertial frame determination using minor planets. A simulation of Hipparcos observations', Astron. Astrophys. 110, p.156

Discussion

MORRISON: Regarding the problem of the accurate calculation of the phase effect due to the scattering law, I presume that you are aware of the series of papers based on photometric results obtained with the CAMC?

MORANDO: Of course, I forgot to mention that we have a lot of information on the physical properties of the minor planets on the programme. In some cases they may be insufficient owing to the high precision of the observations by HIPPARCOS.

ON THE IMPROVEMENT OF THE SYSTEM OF FUNDAMENTAL CATALOGUES BASED ON ASTEROID OBSERVATIONS

D.P. DUMA AND P.M. FEDIJ
Main Astronomical Observatory
Ukrainian Academy of Sciences
252127 Kiev
USSR

ABSTRACT. The question concerning the influence of the consideration of asteroid mutual perturbations on the calculated positions of 20 selected minor planets as well as on the fundamental catalogues systematic corrections is discussed.

1. Materials and methods

The equation of motions of 20 selected minor planets (SMP) were integrated within 1948–2000 interval. The epoch of the initial elements was 1980, December 27 (JD 2444600.5). The list of perturbing asteroids included Ceres (1), Pallas (2), Juno (3) and Vesta(4); the perturbations caused by major planets were excluded.

To estimate the mutual perturbations of SMP the geocentric RA and Dec of a minor planet derived on the bases of disturbed and non-disturbed orbits were compared every 50 days.

It is important to estimate the methodical error of the calculated positions of minor planets caused by the neglect of their mutual perturbation (mathematical model error). It can be done by comparing the most probable disturbed and non-disturbed orbits derived on the basis of the same set of observations. For minor planets (1)–(4), (39), and (40) with significant perturbing terms the estimates were found using the fictitious observations while the perturbation of corresponding geocentric coordinates served as differences O–C. The residuals after improvement of the orbital elements in the case of this model were interpreted as a first approximation of the method error to be found [1].

The influence of mutual perturbations of minor planets on the estimates of systematic corrections to the fundamental catalogue was studied using fictitious observations of Ceres, Pallas, Juno and Vesta [2]. The unknown corrections to the equinox ΔA, the equator $\Delta\delta_0$, and 6 orbital elements for every minor planet as well as to the 5 Earth orbital elements were included in the equations of condition. The equations were solved for every particular planet and for the four planets in common. The derived estimates of ΔA and $\Delta\delta_0$ present the errors of the corresponding catalogue corrections due to neglected perturbations.

At the last stage the residuals after combined solving were averaged for $3^h \times 15°$ equatorial areas.

J. H. Lieske and V. K. Abalakin (eds.), Inertial Coordinate System on the Sky, 337–338.
© 1990 *IAU. Printed in the Netherlands.*

2. Results and discussion

Limited values of perturbations in geocentric α and δ of SMP for the period mentioned above are given in the Table 1, when the combined influence of four massive asteroids was taken into account.

The maximum values of methodical errors of calculated geocentric positions within the whole interval reached $\Delta\alpha_m = \pm 0''.86$, $\Delta\delta_m = +0''.39$ for Vesta, $\Delta\alpha_m = -0''.33$, $\Delta\delta_m = -0''.10$ for Pallas and $\Delta\alpha_m = +0''.23$, $\Delta\delta_m = +0''.15$ for Harmonia. For other planets they are smaller. For some cases the error exceeds $0''.1$ near the epoch of elements. These errors are of systematical character and should be taken into account if high precision reduction of observations of SMP is carried out.

The analysis of solutions for catalogue zero-points corrections showed that the consideration of asteroid perturbations did not practically influence ΔA and $\Delta\delta_o$, especially in the combined solution. At the same time the neglected mutual perturbations may give rise to estimates of regional catalogue corrections up to $0''.05$-$0''.07$ which are of the same order as the systematic errors of the fundamental catalogue. That is why mutual perturbations of asteroids should be taken into account while systematic errors of fundamental catalogues are studied by ground-based and space-based observations of minor planets.

Table 1. The maximum values of perturbations of SMP caused by the first four asteroids within the 1948–2000 interval.

Planet	$\Delta\alpha_m$	$\Delta\delta_m$	Planet	$\Delta\alpha_m$	$\Delta\delta_m$
Ceres	$-0''.29$	$+0''.09$	Phocaea	$+0''.21$	$+0''.06$
Pallas	-0.72	-0.14	Gallia	$+0.04$	$+0.01$
Juno	$+0.35$	$+0.07$	Industria	-0.23	$+0.06$
Vesta	$+2.00$	$+0.40$	Hansa	-0.14	$+0.03$
Hebe	$+0.07$	-0.02	Herculine	-0.41	$+0.15$
Iris	$+0.35$	$+0.10$	Cheruskia	$+0.17$	$+0.16$
Parthenope	-0.28	-0.06	Olimpia	$+0.32$	$+0.06$
Melpomene	$+0.46$	-0.13	Mireille	$+0.68$	$+0.09$
Laetitia	-0.65	-0.12	Interamnia	$+0.64$	$+0.20$
Harmonia	$+1.14$	-0.33	Ivonne	-0.56	-0.10

References

1. Fedij, P.M. (1988) 'On the precision of the ephemerides of selected minor planets derived by not taking into account their mutual perturbations", *Kinematika i fizika neb. tel*, **4**, 6, p.86–88.
2. Fedij, P.M. (1989) 'The influence of mutual perturbations of minor planets on systematic corrections of star catalogues', *Kinematika i fizika neb. tel*, **5**, 1, p.94–96.

EXPECTATIONS FOR ASTROMETRY WITH THE HUBBLE SPACE TELESCOPE

R.L. DUNCOMBE
Center For Space Research
University of Texas at Austin
Austin, TX 78712

and

W.H. JEFFERYS, G.F. BENEDICT, P.D. HEMENWAY AND P.J. SHELUS
Department of Astronomy
University of Texas at Austin
Austin, TX 78712

ABSTRACT. The Hubble Space Telescope, a large optical instrument having an aperture of 2.4 meters and a length of 8.8 meters has been developed by the U.S. National Aeronautics and Space Administration in cooperation with the European Space Agency. The Space Shuttle will be used to place the telescope in orbit. The primary astrometric instrument will be one of the three Fine Guidance Sensors which have the capability of measuring the position of one object with respect to another to an accuracy of ±0".002 . To facilitate use of the Hubble Space Telescope, observers will be provided with the Astrometric Data Reduction Software package. The variety of astrometric problems and the several modes of operation are mentioned as well as the cooperative program with the European astrometric satellite project HIPPARCOS.

1. Hubble Space Telescope Astrometry

The Hubble Space Telescope (HST) will have the capability to measure the relative positions of stars within the field of view of a Fine Guidance Sensor (FGS) with an accuracy of ±0".002 rms. The magnitude range of HST astrometry using an FGS is from 9 to 17, which may be extended to 4th magnitude with the use of filters. While two FGSs are used to control the pointing of the HST, the third FGS may be used for astrometric measurements. FGS #2 has been designated as the prime FGS for astrometry because of guidance considerations.

Each FGS views a ninety degree sector of an annulus which comprises the outer portion of the HST focal plane (Figure 1). Each sector of the annulus (hereafter referred to as a "pickle") is four arcminutes wide and has a maximum chord length of about eighteen arcminutes. The measurement accuracy quoted above refers to the relative positions of two objects within the central twenty square arcminutes of the same pickle.

Figure 2 is a schematic diagram of an FGS. Each FGS includes an optical beam splitter, two orthogonal Koester's prism interferometers (one for each axis), and their associated photomultiplier tubes. Light from an object anywhere in the pickle is brought into the five arcsecond square aperture

339

J. H. Lieske and V. K. Abalakin (eds.), Inertial Coordinate System on the Sky, 339–346.
© 1990 *IAU. Printed in the Netherlands.*

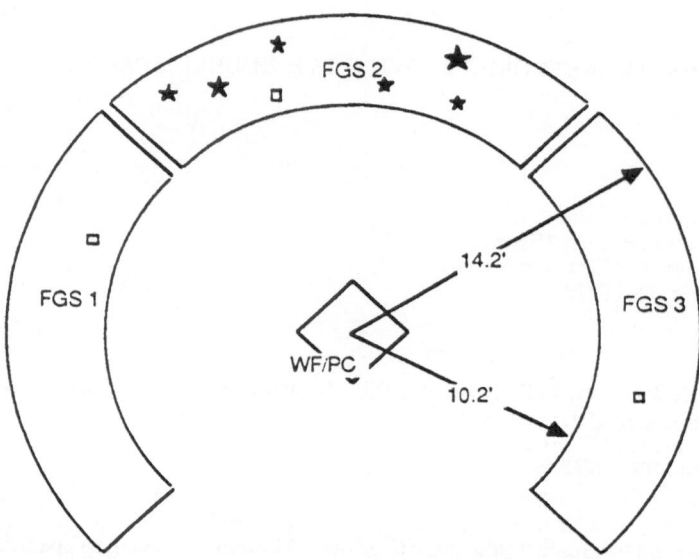

Figure 1. HST focal plane, with locations of FGS fields of view.

of the FGS detector assembly by the rotation of two beam deflectors called "star selectors." If the object is directly on the axis of an interferometer, the signal from each of the two photomultipliers associated with that axis will be equal, and their difference will be zero. If the object is off the axis, the difference of the two photomultiplier readings will provide an error signal that indicates how far off-axis the object is. FGS circuitry then nulls the error signal by repositioning the star selectors to bring the object back onto the interferometer axis. The position of one object with respect to another in the same pickle can be calculated from the angles θ_A, θ_B and $\Delta\theta_A$, $\Delta\theta_B$ produced by the star selector positions as shown in Figure 3.

The FGS has several modes of operation, when used as an astrometric instrument. In the "lock-on" mode, the interferometers are nulled successively on each object being measured within the same pickle. In the "multiple star" mode, the FGS aperture is moved diagonally across the target. The error signal in each axis varies as the aperture moves across the target, tracing out the instrumental transfer function (Figure 4). The transfer function can then be analyzed to detect duplicity of stars and to measure the relative positions of the two components. In the "moving target" mode, which is used for tracking targets such as asteroids, the FGS is kept locked on the object and its position is recorded periodically.

The "transit circle" mode will be used to observe objects fainter than the 17th magnitude. In this mode the Wide Field Planetary Camera (WFPC) will be used in conjunction with the two guiding

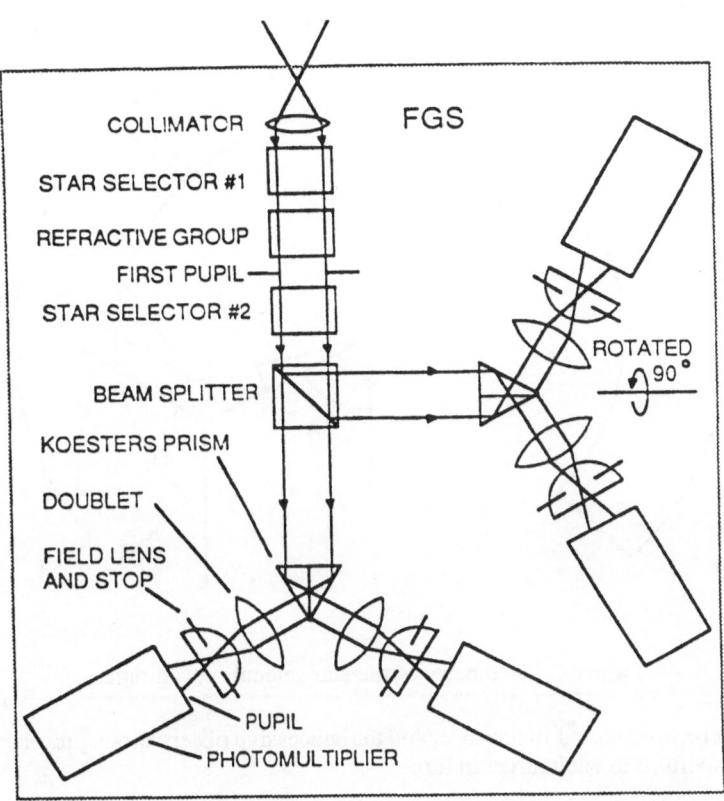

Figure 2. Schematic diagram of the FGS.

FGS units. The WFPC will be used to observe the two faint target objects in turn, while the HST is positioned so that the same pair of guide stars can be used for guiding during both observations. The position of the first faint object is measured with respect to the WFPC itself. Then the HST is repositioned to measure the second object with respect to the WFPC. The relative positions of the two target objects is the difference between the centroided images in the WFPC frame plus the difference between the FGS pointings on the same pair of guide stars. These observations are expected to produce results only slightly less accurate than those made by the astrometric FGS alone.

It can be seen from the foregoing that there are some restrictions on the use of an FGS for astrometry. Since two of the FGS are required for HST guidance, only one FGS is available for astrometric observations. Further, the FGS is not an imaging instrument; it can not see an object but can only sense its presence after moving to the object's approximate location. For this reason the

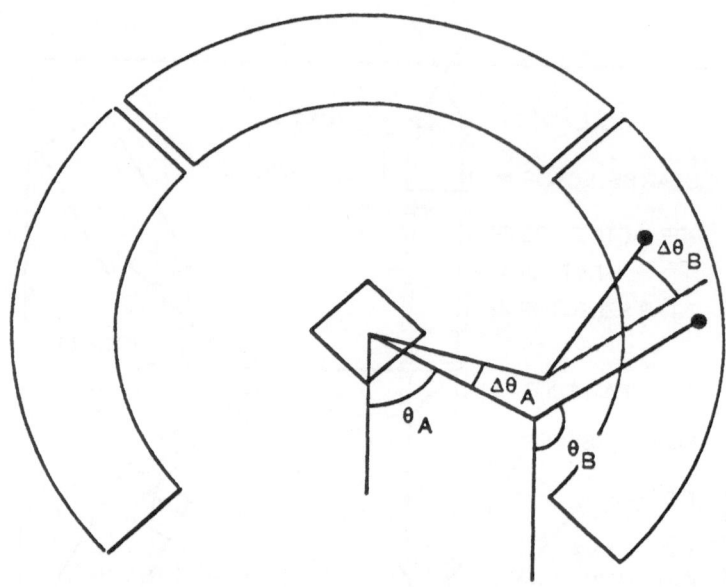

Figure 3. Geometry of the star selector coordinates.

pickle area must be inventoried in advance, and the successive observations pre-planned in order to direct the FGS aperture to each target in turn.

The ultimate determination of HST capabilities and limitations for astrometry in space will be made during the periods of Orbital Verification and Science Verification following launch. A star field, well-calibrated by ground observations, will be used to determine the zero-order plate scale and field distortion of the FGS. Since no star field is available which contains relative positions accurate to ±0".002 over a range of 18 arcminutes, an overlapping plate technique will be used in conjunction with the Optical Field Angle Distortion (OFAD) program. Then a moving target, such as an asteroid, will be tracked across an FGS field to measure the scale factor within the measuring accuracy of the FGS. Finally, a sample of known single and multiple stars will be observed to determine the FGS instrumental transfer function.

2. Software for HST Data Reduction

A library of astrometric data reduction software (ADRS) will be available to HST astrometric observers within the science data analysis system (SDAS) at the Space Telescope Science Institute (STScI). The ADRS user will pick from a modular list those software units to be applied to the data. Each module performs its defined operation on the data file, producing a new data file on which

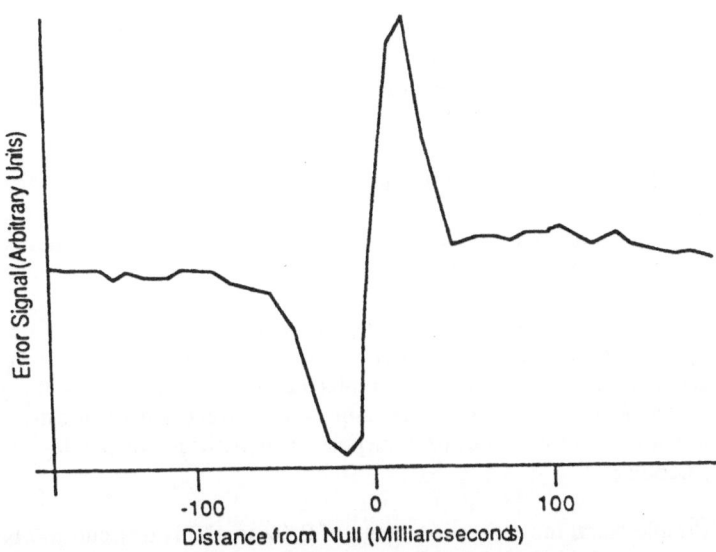

Figure 4. FGS transfer function.

further operations may be performed. The user may select from among standard operations which are available with any operating system (such as, creating and deleting files, editing, copying and combining files, saving and renaming files), and special modules created for astrometric purposes. The computer operating system allows the user to define and name a sequence of operations in which one software unit follows another. An on-line "help" program is provided to assist the users.

Numerous special astrometric software modules have been written specifically for HST data reduction. Included are modules for conversion among various coordinate systems, for the determination of the mean position of an object from a sequence of encoder readings (centroiding), for correction of the data for various instrumental effects (optical distortion, tilt, color terms), and to correct for physical effects such as differential aberration due to vehicle motion as well as the motion of the Earth. A special module for the analysis of transfer function data has been prepared at the Lowell Observatory capable of providing position angle and separation of double star components which have separations greater than about 0".03.

A basic unit of ADRS is GaussFit. This program has its own computer language, which is designed to facilitate the specification of complex data reduction models. The programming language of GaussFit provides a straightforward and easy way to formulate problems in nonlinear estimation, problems with correlated observations, problems where an equation of condition may contain more than one observation, problems involving exact constraints among the parameters, and problems in which the model can only be expressed algorithmically and not as a closed form expression. GaussFit uses orthogonal transformations instead of normal equations to solve the basic least squares problem, and it also allows the user to specify a robust estimation method that is resistant to outliers in the data. A GaussFit user's manual is available and further information may be obtained from W. H. Jefferys.

3. Proposed HST Astrometric Research Programs

HST astrometry, with accuracies from two to ten times greater than conventional ground based observations, will provide possibilities for new and exciting areas of research.

It will be possible to measure stellar parallaxes with an accuracy of ±0".001 or better, providing for extension of the trigonometric parallax method to star clusters and RR Lyrae stars. It will also allow improvement in the parallaxes of such fundamental distance calibrators as the Hyades. This improved measurement accuracy will make it possible to investigate systematic errors in parallaxes derived from ground observations. It may become possible to measure the parallaxes of central stars of some planetary nebulae.

In the study of double stars, the separation range of 0".02 to 1".0 is difficult to observe from the ground. Consequently, the statistics of double stars in this range are not well known. Statistics can be improved, however, if a scan of a star in the astrometric FGS is made routinely every time the telescope is pointed at some primary target for observation by one of the other instruments. Parallel observations such as this are feasible but their frequency is an operational matter that remains to be determined. In the case of known spectroscopic binaries, it may be possible to obtain visual orbits, thus providing more information about stellar masses.

It is proposed to utilize HST WFPC observations to detect the motions of stars within clusters, and to measure the effects of tidal forces on the cluster and of mass segregation within the cluster. Proper motions of stars will be detectable in a much shorter time than required for ground-based observations, and it may even be possible to measure proper motions in nearby galaxies.

HST observations of solar system objects will furnish new data on the gravitational fields of the planets and satellites. The accuracy of HST measurement of intersatellite angular distances will provide data good to 5 km at the distance of Jupiter and 10 km at the distance of Saturn. Longer term observations by HST of the satellite systems first observed by Voyager will help to solve some of the dynamical questions raised by the short duration data obtained on the Voyager fly-by.

Other proposed research programs include the determination of the gravitational deflection of light by Jupiter, the search for planets of nearby stars and the measurement of optical proper motions in quasars.

Observations with the FGS units of HST will provide a major contribution to fundamental astronomy in a cooperative program with the ESA astrometric satellite HIPPARCOS. The HIPPARCOS project was planned to produce a set of stellar positions, proper motions and parallaxes by essentially observing chords on the celestial sphere. The failure of HIPPARCOS I to achieve its planned orbit may prevent it from obtaining the approximately 100,000 stellar positions and parallaxes with the expected accuracy of ±0".001. Whatever the accuracy ultimately obtained, the reference frame for these data will be the HIPPARCOS Instrumental System which will have an unknown zero point and an unknown "solid body" rotation with respect to a non-rotating reference frame. It remains then, to determine the rotation of the HIPPARCOS Instrumental System and to tie it to some fundamental or absolute reference frame. Because of its capability to measure precise angular distances between objects of disparate magnitude, the HST FGS can be used to tie the HIPPARCOS Instrumental System to (a) very distant and hence relatively motionless objects such as quasars; (b) an absolute coordinate system derived from radio interferometric observations using radio sources which have discrete optical counterparts, and (c) a dynamical system such as that defined by the motions of selected asteroids. With respect to projects (a) and (b) plans have been made to observe the positions and motions of 160 HIPPARCOS catalog stars with respect to 90 extragalactic objects using the HST FGS or in some cases for fainter objects, the WFPC in conjunction with the guiding FGS units. With regard to project (c), plans are made to relate the orbits of selected minor planets, particularly at apparent crossing points, to stars in the HIPPARCOS catalog. Eight thousand candidate stars were selected and of these two thousand were accepted for the HIPPARCOS catalog.

Although HIPPARCOS I may not achieve all of its planned objectives, the HST Astrometry Team intends to carry through all of the planned link observations mentioned in (a), (b) and (c) that can be accommodated within the constraints of the HST ground system. These observations will be linked not only to the observations of HIPPARCOS I but also to on-going ground-based surveys by transit circle observations in both the northern and southern hemispheres.

We feel that the scope and intent of the program planned by the HIPPARCOS Project is of fundamental importance to all of astronomy and we whole-heartedly support the dedicated HIPPARCOS Team in their efforts to realize the ultimate goals of the original HIPPARCOS project.

The HST is now scheduled for launch in the Spring of 1990. We are eagerly awaiting this event to determine its astrometric capability.

4. Acknowledgements

The authors are pleased to acknowledge the cooperation of the HIPPARCOS Input Catalog group at Meudon, colleagues at Lowell Observatory, Georgia State University and Imperial College London for Speckle observations, and at Cambridge and the Astronomisches Rechen-Institut for finding charts.

We wish also to thank our colleagues on the Space Telescope Astrometry Team, Laurence Fredrick, William van Altena and Otto Franz for their cooperation on this project. We gratefully acknowledge support from the National Aeronautics and Space Administration under contract NAS8-32906.

5. References

1. M. Froeschlé and J. Kovalevsky, *Astron. and Astrophys.*,**116**, 89, 1982.
2. P.D. Hemenway, R.L. Duncombe, W.H. Jefferys and P.J. Shelus, "Using Space Telescope to Tie the HIPPARCOS and Extragalactic Reference Frames Together", in HIPPARCOS: Scientific Aspects of the Input Catalogue Preparation, T.D. Guyenne and J. Hunt, eds., European Space Agency Special Publication ESA SP-234, p.261, 1985.
3. P.D. Hemenway and R.L. Duncombe, "The Use of Space Telescope to Tie the HIPPARCOS Reference Frame to an Extragalactic Reference Frame", in *Astrometric Techniques*, IAU Symposium #109, H.K. Eichhorn and R.J. Leacock eds. p. 613, 1986.
4. W.H. Jefferys, G.F. Benedict, P.D. Hemenway, P.J. Shelus and R.L. Duncombe, "Prospects for Astrometry with the Hubble Space Telescope", *Celestial Mechanics*, **37**, 299, 1985.
5. W.H. Jefferys, M.J. Fitzpatrick and B.E. McArthur, "GaussFit: A System for Least Squares and Robust Estimation" *Celestial Mechanics*, **41**, 39, 1988.

Discussion

KLIONER: You said that the scientific program includes the measurement of the relativistic deflection due to Jupiter. Why only due to Jupiter? For example, the relativistic deflection of the light passing Saturn is 0.006 arcsec. What relativistic corrections do you take into account in your software?

DUNCOMBE: At present the only proposal in the program is the measurement of the relativistic deflection of light due to Jupiter. Other experiments may be proposed later.

STRAND: You mentioned that an accuracy of 0.002 arcsec will be attainable with the Hubble Space Telescope. Is that from a single observation or is it the result of several? Parallaxes of 0.001 arcsec mean error have already been obtained from ground based observations. With respect to investigation of systematic errors of stellar parallaxes, such a program is already in progress at the Flagstaff Statin of the U.S. Naval Observatory by observations of Quasars.

DUNCOMBE: It is the result of several observations. It should be noted that observational aims of the Hubble Space Telescope, established ten years ago, are now close to realization by ground-based techniques due to the rapid advancements in new technology.

KOPEJKIN: How long should we observe stars in the Large Magellanic Cloud for determination of proper motion of stars in this galaxy?

DUNCOMBE: The observations should be conducted over three years at least and over five years if possible, depending on the lifetime of the satellite.

SPACE ASTROMETRY AND THE HST WIDE FIELD/PLANETARY CAMERA

P. K. SEIDELMANN
U S NAVAL OBSERVATORY

INTRODUCTION

The launch of the Hipparcos spacecraft marked the beginning of space astrometry. Hopefully, this will be followed in the near future by the launch of the Hubble Space Telescope, which is not primarily an astrometric instrument, but has astrometric capabilities which will be described in this paper. In addition, there are plans and proposals for future astrometric spacecraft. These include the launch of a radio antenna, which combined with Earth-based antennae would provide a very, very long base line interferometer (Levy, 1986, 1988). There are proposals for launching optical interferometers, such as POINTS (Reasenberg et al 1988). There are also proposals by York and Gatewood (Gatewood et al., 1986; Gatewood 1987, 1989) for launching astrometric instruments using gratings and detectors. Thus, the future holds the prospects for a whole new capability in the field of astrometry.

GROUND-BASED ASTROMETRY

The field of ground-based astrometry can basically be divided into three categories: measurements of large angles, medium angles and small angles. Large angle measurements have been primarily restricted to brighter objects, radio sources and solar system bodies and to providing a hemisphere-wide reference system. Medium angle measurements have provided the tie between the brighter stars and fainter stars by means of astrographic catalogues. They have also provided the capability to search for unknown objects. Small angle astrometry has primarily addressed itself to parallaxes, the search for low mass companions and planetary objects and cluster dynamics.

Ground-based astrometry has been limited by the presence of an atmosphere. This controls the wavelengths that can be observed. It limits the resolution, the accuracy of the observation and generates scattered light around bright planets.

Speckle interferometry has provided the capability to observe relatively bright objects without the necessity of integrating over the atmospheric effects (McAlister 1986, 1987, Lu, et al. 1988, McCalister et al., 1987, 1988). The introduction of the charge coupled devices as detectors to replace the eye and photographic plate has increased the dynamic range, greatly reduced the observing time and provided a detector with a built-in measuring machine. To a large extent the CCD detector was introduced, at least in my case, in anticipation of its use in space. It provides the capability of providing improved accuracies in parallaxes (Monet and Dahn 1983, Dahn et al. 1988); observing capabilities for solar system objects that are not observable by other means (Seidelmann et al. 1981, Pascu et al 1983, 1987); the potential of an improved detector on a transit circle (Gehrels et al. 1986), and the potential for establishing an inertial coordinate system (Mao, et al, 1989).

347

J. H. Lieske and V. K. Abalakin (eds.), Inertial Coordinate System on the Sky, 347–354.
© 1990 IAU. Printed in the Netherlands.

An advantage of the CCD as a detector is demonstrated in the satellite observing program. With photographic plates the faint librational satellites would require long exposures and the images would then be lost in the scattered light of the planet. With a CCD the exposure lengths are reduced and, since the exposures are recorded digitally, the planetary scattered light can be subtracted based on a model. Thus, these satellites can be observed with a CCD, but not with photographic plates. Similarly, Nereid could be observed with a CCD from Flagstaff with six minute exposures that would have required with a photographic plate over an hour, a length of time that could not have been achieved with Neptune so far to the south.

SPACE ASTROMETRY

The primary advantage of doing astrometry from space is the absence of an atmosphere. Thus, the limitations imposed by the atmosphere on wavelength, resolution, scattered light, and accuracy are relieved. However, new limitations are imposed. These include the knowledge of the spacecraft orbit, the stability of the spacecraft, limitations on data rate, calibration requirements and the inability to tinker with the instrument. The accuracy and resolution limitations from space will not be imposed by the environment, but rather determined by the cleverness of the designer and the capabilities of the available hardware.

A number of papers at this meeting discuss the planned capabilities of the Hipparcos spacecraft to observe parallaxes, minor planets and triple stars. The Hipparcos spacecraft is an example of a large angle measuring device, although it is done in a different manner than from the ground (Kovalevsky 1986). The Hubble Space Telescope (HST), on the other hand, more closely resembles a ground-based instrument in space. General information about the HST has been published in a number of places in the literature (e.g. The Space Telescope Observatory, 1982). For astrometric purposes, HST includes the fine guidance sensors (FGS) that will be used to guide the telescope and can be used for astrometric positional observations. Their capability is described in another paper.

The HST has two imaging instruments. One is the Faint Object Camera which has extremely limited astrometric capabilities due to its very small field of view (maximum 45"). The other instrument, the Wide Field/Planetary Camera (WF/PC), has a large field of view compared to the faint object camera but a very small field of view compared to ground-based telescopes. The planetary camera provides a field of view of 68 arcseconds on a side with 4 CCD detectors. The Wide Field camera has a field of 160 arcseconds on a side. For astrometric purposes it is necessary to trace the path of light which enters the WF/PC by means of a pick-off mirror in the center of the field of view of the HST. The light enters through a window and passes through a filter, which is one of 48 possible selections. The light then encounters the four-faceted reflective pyramid that divides the light into four separate beams, which are directed through reimaging optics, through a field flattener, and onto a CCD. The choice between the Planetary and Wide Field modes is made by the selection of the position of the pyramid.

While the pyramid is essentially a reflecting optical component, it contains a spot with a nonreflecting coating approximately 1.23 arcsecond in diameter which is located in the middle of the field of view of one chip (PC8) of the planetary camera. It is toward the central corner of the field of view of CCD WF4. This nonreflecting Baum spot attenuates the light of an image falling on that spot. The pyramid also has very small spots along the ridges of the pyramid that are cleared of reflective coating. The pyramid can be back lit so that the light coming through one of these spots on a ridge of the pyramid appears on both of the adjoining CCD chips. This provides a means of tying the four images together in a mosaic.

CALIBRATION

The spots, called the Kelsall spots, provide an independent means of calibration. An exposure of the Kelsall spots can: 1) confirm that the pyramid is completely rotated into the correct position; 2) by measuring the positions of the Kelsall spots, detect whether an individual chip or reimager has moved, tilted or shifted in focus. Thus, the Kelsall spots provide an internal monitor on the WF/PC. Additional calibrations will be made by using either Omega Centauri, or NGC 6752.

The WF/PC is known to have pin cushion distortion from the reimaging systems. This effect can reach a magnitude of 1 1/2 pixels or 22 microns near the edge of the chip. The WF/PC also has the potential for distortions due to what is called the "potato chip effect." Since the CCD chips are thin devices, the chip may not be flat. The field flattener lens can displace the beam by 0.1 radian near the edge. The pupil is approximately 75 millimeters in front of the focal plane, as are the WF and PC field flatteners. The angle to the corner of the chip is then 0.11 radians and to the center of the edge of the chip 0.08 radians. The astrometric effects in the corner are 2.7 microns (18 milliarcseconds (mas) for WF and 8 mas for PC) and at the center of the edge 2.0 microns (13 mas for WF and 6 mas for PC).

Since the thinned chips are supported at the edges, the ripples go to zero where the effects of the field flattener are a maximum. The result of the combination of the field flattener and the ripple could be anticipated to be 1.5 micron (10 mas for the WF and 4 mas for the PC).

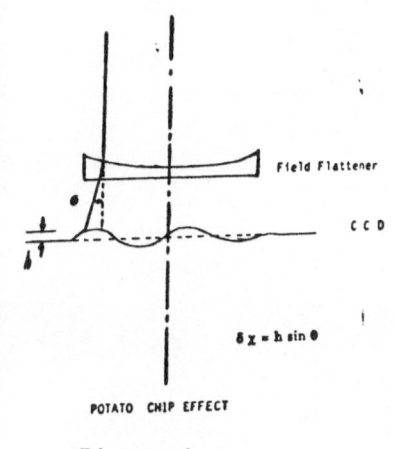

Figure 1

It is possible that the chip can change its shape relative to a flat surface as a function of temperature. Since there is a field flattener in front of the chip, this lack of flatness can cause a displacement in position of the image (Fig 1). A typical half amplitude of the ripple would be 25 microns and half the scale length about 100 pixels.

There is also a potential distortion of position due to the fact that there are defects in the CCD's. If an image falls on a bad pixel or a bad column, the image will be affected and the position accordingly affected. Therefore, small shifts in the pointing can be used to detect and average out shifts due to defects. These shifts also provide multiple exposures that improve the resulting accuracy as a function of the square root of the number thereof. Shifted positions also provide a means of isolating the cosmic ray events that will be present.

In addition to the changes in the shapes or flatness of the chips due to the changes in temperature, it is also possible that there will be shifts in the positions of the chips as a function of temperature. While the optical bench is made out of graphite epoxy for stability, there is a possibility of the motion of the chips due to the temperature of the cameras.

To investigate the distortions of the WF/PC, a series of observations were made during thermal vacuum test No. 6 using slant bars. Measurements of the slant bars provided a means of evaluating the third order model required for pin cushion distortion for real data. The measured displacements as determined from the differences between the measured positions of the slant bars and the positions of uniformly spaced, constant slope, slant bars is shown in Fig. 2. When the measurements are corrected by the third order model, the remaining residuls due to the potato chip effect and chip defects are shown in Fig. 3. Thus, it is expected that accurate positional results will require correcting the measured positions with a third order model for pin cushion distortion and a mapping of the remaining effects.

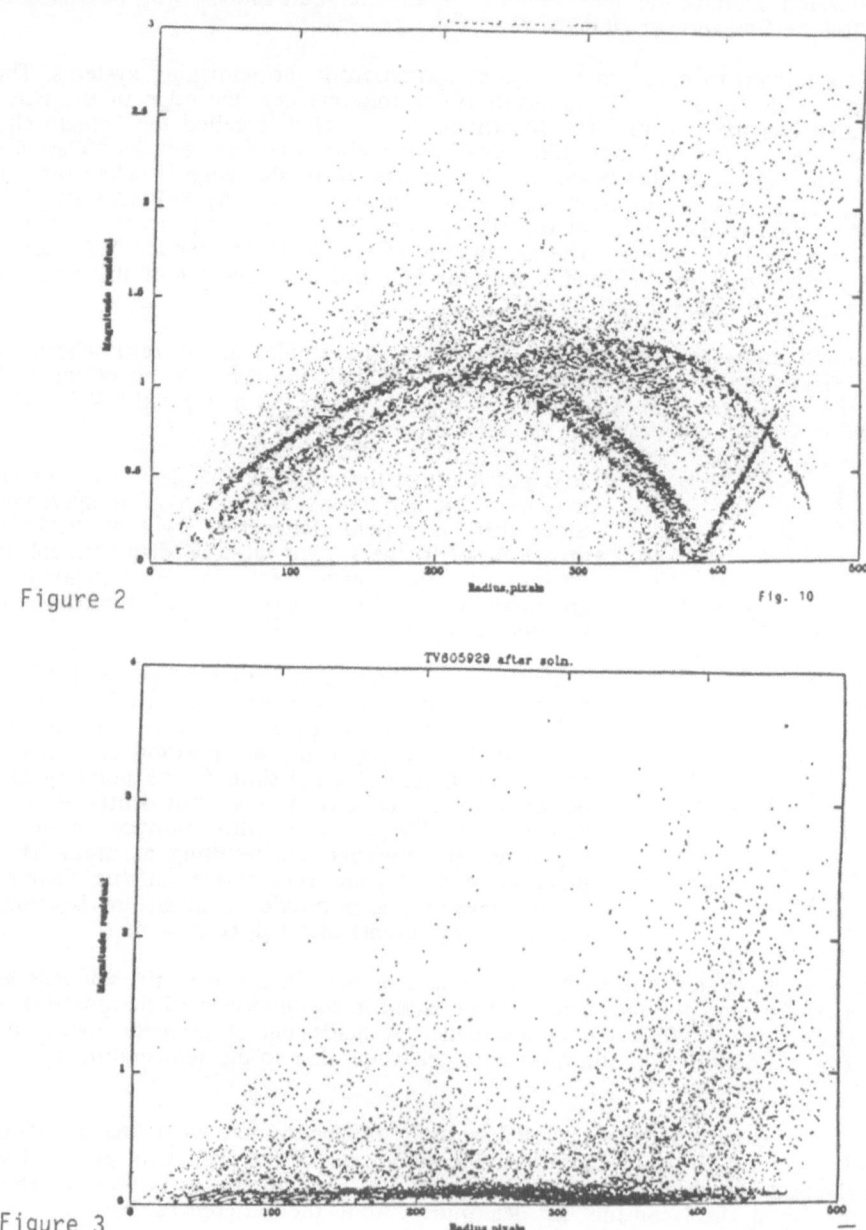

Figure 2

Fig. 10

Figure 3

Fig. 11

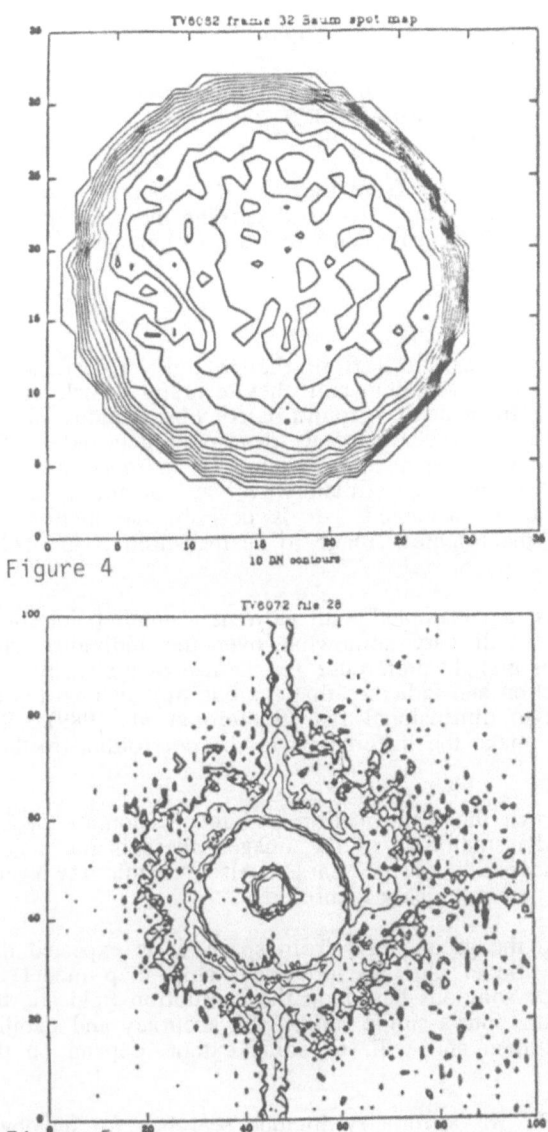

Figure 4

Figure 5

In flight, it is planned that images will be taken of a field of Omega Centauri or NGC 6752 which has a sufficient number of stars to represent the third order distortion. Shifted positions of the Omega Centauri field are designed for investigations of the potato chip effect, and changes thereof. In order to evaluate the potato chip effect observations must be made with a large number of stars covering all parts of each of the chips.

The Baum spot has a reduced reflectance that is a function of wavelength. A contour plot of the floor of the Baum spot is shown in Fig. 4. The axes are in pixels from (407, 402). The maximum DN is 250. A bright spot at (7, 17) is 40 DN above the floor. The reflectance of the spot was measured by placing a pinhole target spot on and off of the Baum spot in thermal vacuum tests. Preliminary results indicate that with filter F555W, the Baum spot reduces the reflectance by 6.156+0.064 magnitudes. Figure 5 shows the contour plot of the pinhole target in the center of the Baum spot with the diffraction spikes from the reimaging system. The diffracted light outside the spot reaches approximately the same DN level as the target pinhole under the spot.

Calibration of the WF/PC for astrometric purposes requires that we know the following information:

1. Distortions due to each reimaging system.
2. Distortions of each chip.
3. Thermal stability of each chip.
4. Flat field for each chip for each filter.
5. Angle between surface of chip and optical axis.
6. Point spread function for different places on chip.

7.   "Plate constants" for each chip.
8.   Location of the different chips with respect to each other.
9.   Stability of chip locations.
10.   Location of WF/PC optical field with respect to FGS and other instruments.
11.   Stability of WF/PC with respect to other instruments.
12.   Attenuation characteristics of Baum spot.

While some of these measurements can be made from thermal vacuum tests made prior to launch, the measurements must also be made in space to determine the real conditions. All stability characteristics require continued measurements to learn of changes.

REDUCTION METHODS

CCD's have been used for astrometric observations from the ground. Experience has accumulated concerning the methods of centroiding and the accuracy which can be achieved in measuring the images. In general, ground-based observations can be measured with a one dimensional fit to a Gaussian to determine centroids. Two dimensional fits do not significantly improve the resulting accuracies in most cases. As long as the signal-to-noise ratio exceeds approximately 5, an accuracy in the neighborhood of 1/40th of a pixel can be achieved. It is desirable to increase the exposure such that the DN number of the brightest image to be measured is just below the beginning of saturation.

On the HST the Wide Field Camera is undersampled with extremely small point spread functions. The point spread function will vary somewhat over the individual chip. Experience with simulated data indicates a slight preference for the use of a Cauchy function as opposed to a Gaussian function and indicates that the one dimensional fits are approximately as satisfactory as the two dimensional fits (Santoro et al. 1989). The errors due to the distortions, however, make the differences in the centroiding methods insignificant.

Experiences with simulations indicate that the identification and recognition of the star fields is extremely difficult. Due to the small size of the images, there is not a good visual identification of the relative magnitude of the stars. Also, cosmic ray events contribute many "volunteer stars" which are not easily identified.

Based on calibration observations from the ground and from space, it is expected that there will be a third order distortion model plus an individual chip map model for correcting the distortions. There will be solutions based on the calibration field and the Kelsall spots for combining the four chips into a single field. The accuracy and stability of that combining procedure is not known and will, to some extent, depend on the stability of the chips within the WF/PC.

The present plans for using the WF/PC for astrometry include searches for unknown satellites, observations of extremely faint satellites that cannot be observed from the ground; observations of Pluto and its satellite; observations of the rings of Jupiter, Uranus and Neptune; searches for low mass companions or planetary objects around selected stars; observations of minor planets; and observations of radio sources or quasars with respect to reference stars. There are plans for direct observations of low mass companions, but the characteristics of the HST mirrors and the instruments make the success of such observations unlikely.

CONCLUSIONS

We are on the threshold of astrometry from space. This offers the potential for achieving observations with significantly improved accuracy and resolution. Exciting science can be expected from the improved accuracies and resolutions to be achieved from space. There will be great competition, first for funding for astrometric instruments in space; and second, for observing time for astrometric observations on more general spacecraft, such as the HST. A strong case is going to have to be made for observing from space. Hopefully, significant improvements can be achieved, both from the ground and from space with the new techniques currently available. It will be necessary to perform astrometry where it can be done most efficiently, economically and effectively.

ACKNOWLEDGEMENTS

It is my pleasure to thank Earnest Santoro, Richard Schmidt, Stephen Panossian and David Monet, who contributed material used in this paper.

REFERENCES

Bandermann, L., Bareket, N., and Metheny, W., (1982) "Comparative Feasibility Study of Two Concepts for a Space-Based Astrometric Satellite." NASA - CR-166403.

Bernstein, H. H., Hering, R., and Walter, H. G. (1988) "Astrometric Parameters of Visual Double Stars Derived from Simulated Hipparcos Measurements." Astrophys. & Space Sci. 142, 161.

Dahn, C.C., Harrington, R. S. Kallarakal, V. V., Guetter, H. H., Luginbuhl, C. B., Riepe, B. Y., Walker, R. L., Pier, J.R., Vrba, F. J., Monet, D. G and Ables, H. D. (1988) "U. S. N. O. Parallaxes of Faint Stars." Astron. J., 95, 237.

Gatewood, G., (1987) "The Multichannel Astrometric Photometer and Atmospheric Limitations in the Measurements of Relative Positions." Astron. J. 94, 213.

Gatewood, G., (1989) "MAP Determinations of the Parallaxes of Stars in the Regions of HD 2665, BD +68 DEG 946, and Lambda Ophiuchi." Astron. J., 97, 1189.

Gatewood, G., Stein, J., Kiewiet de Jonge, J., Faste, D., and Breakiron, L (1986) "A New Astrometric System." Astrometric Techniques: IAU Symposium 109, Florida, 341.

Gehrels, T., Marsden, B. G., McMillan, R.S., and Scott, J. V. (1986) "Astrometry With A Scanning CCD." Astron. J. 91, 1242.

(1988) "Hipparcos' Final Test." Sky & Telescope, Vol. 75, No. 4, Apr. p. 358.

(1989) "Hubble Space Telescope Wide Field and Planetary Camera Instrument Handbook" Version 2. Space Telescope Science Institute.

Kovalevsky, J., (1986) "Hipparcos Satellite and the Organization of the Project." Astrometric Techniques: IAU Symposium 109, Florida, 581.

Levy, E. H., Gatewoood, G. D., Stein, J. W., and McMillan, R. S. (1986) "Astrometric Telescope of 10 Microarcsecond Accuracy on the Space Station." Advanced Technology Optical Telescopes (Proc. SPIE, Int Soc, Opt Eng 628) III, 181.

Levy, G. S., (1986) "First Successful Very Long Baseline Interferometry Observations Using an Orbiting Telescope." Preprint, Nobeyama Radio Observatory Report No. 109.

Levy, G.S., (1986) "Status of the Very Long Baseline Interferometry VLBI/Demonstration Using the Tracking and Data Relay Satellite System." Preprint, Nobeyama Radio Observatory Report No. 116.

Levy, G. S., (1988) "VLBI Using a Telescope in Earth Orbit - The Observations." JPL Astrophysics Preprint.

354

Lu, P. K., Demarque, P., Van Altena, W., McAlister, H. and Hartkopf, W., (1987) "ICCD Speckle Observations of Binary Stars. III. A Survey for Duplicity among High Velocity Stars." Astron. J., 94, 1318.

Mao Wei, Guo Xinjian, Xu Shui, Wu Guangjie, and Lu Ruwei, (1989) "Construction of an Inertial Coordinate System Using a CCD," Astron. Astrophysics 215, 190.

McAlister, H. A., (1986) "Speckle Interferometry in Astrometry." Astrometric Techniques: IAU Symposium 109, Florida, 293.

McAlister, H. A., (1987) "The Future of High Angular Resolution Astronomy: Seeing the Unseen." Vistas in Astronomy, 30, 27.

McAlister, H. A., Hartkopf, W. I. and Gutter, D. J. (1987) "ICCD Speckle Observations of Binary Stars. II: Measurements During 1982 - 1985 from the Kitt Peak 4m Telescope." Astron. J., 93, 688.

McAlister, H. A., Hartkopf, W. J., Bagnuolo, W. G., Sowell, J. R., Franz, O. G. and Evans, D. S. (1988) "Binary Star Orbits From Speckle Interferometry - I. The Hyades Binary Finsen 342 (70 Tauri)." Astron. J. 96, 1431.

Monet, D. G. and Dahn, C.C. (1983) "CCD Astrometry. I. Preliminary Results from the KPNO 4-m/CCD Parallax Program" Astron. J., 88, 1489.

Pascu, D., Seidelmann, P. K., Baum, W. A., and Schmidt, R.E. (1983) "Observations of Faint Planetary Satellites with Charge-Coupled Device."In The Motion of Planets and Natural and Artificial Satellites, edited by S. Ferraz-Mello and P. E. Nacozy, (Universidade de Sao Paulo, Sao Paulo, Brazil), 253.

Pascu, D., Seidelmann, P. K., Schmidt, R.E., Santoro, E. J., and Hershey, J. L. (1987), "Astrometric CCD Observations of Miranda: 1981 - 1985." Astron. J., 93, 963 - 968.

Reasenberg, R. D., Babcock, R. W., Chandler, J. F., Gorenstein, M. V., Huchra, J. P., Pearlman, M. R., Shapiro, I. I., Taylor, R. S., Bender, P., Buffington, A., Carney, B., Hughes, J. A., Johnston, K. J., Jones, B. F., and Matson, L. E., (1988) "Microarcsecond Optical Astrometry: An Instrument and its Astrophysical Applications," Astron. J. 96, 1731.

Santoro, E. J., Schmidt, R. E., Seidelmann, P. K., and Kristian, J., (1989) Centroid Analysis of Space Telescope Widefield Camera Point Spread Function Images" In Errors, Bias, and Uncertainties in Astronomy, Strasbourg, France.

Seidelmann, P. K. , Harrington, R. S., Pascu, D., Baum, W. A., Currie, D. G., Westphal, J. A., and Danielson, G. E. (1981) Saturn Satellite Observations and Orbits from the 1980 Ring Plane Crossing. Icarus, 47, 282.

"The Space Telescope Observatory," (1982) edited by N. B. Hall, Space Telescope Science Institute.

THE LOMONOSOV PROJECT FOR SPACE ASTROMETRY

V.V. NESTEROV, A.A. OVCHINNIKOV, A.M. CHEREPASHCHUK, AND
E.K. SHEFFER
Sternberg State Astronomical Institute
Universitetskij Prospekt 13
119899 Moscow
USSR

ABSTRACT. The LOMONOSOV project is aimed at developing a high-accuracy coordinate system of the entire sky to be used for sufficiently long period of time (30 to 50 years) in order to ensure the solution of a variety of applied and basic scientific tasks. This goal can be feasible as a result of comprehensive work, the basis of which is the space experiment, i.e., observations of stars with a telescope on board the Earth satellite. This method can be instrumental in overcoming the distortions characteristic of terrestrial astrometric observations performed through the atmosphere, and to achieve a high degree of efficiency. The other part of the LOMONOSOV project involves international backing of the space experiment through preparation of the input catalogue for 400 000 stars, and organization of high-precision ground observations of stars and other celestial objects in accordance with special programs.

Introduction

The contemporary state of astronomy is such that any serious fundamental scientific results can be obtained only with drawing up a catalogue containing information on several hundred thousand stars in the entire sky with their coordinates' accuracy throughout several dozens of years around $0''005$ to $0''0010$, as well as with color photometric data ensuring accuracy equal to 0.05 mag. To make up such a catalogue one should perform observations with even higher precision and to repeat them again in some time.

More than a century's experience of classical astronomy proves practical impossibility of achieving the accuracy of the ground-based measurements about $0''10$. Among the factors that put the limit of accuracy are local atmospheric fluctuations, insufficient stability of the selected directions which set up zero points, and technical imperfection of the measuring instruments operating under the impact of gravitation. It should be stressed that with a certain limit attained, greater observation frequency does not, in practical terms, result in a higher accuracy. Thus, for the past 300 years a well-known Polaris has been observed thousands of times but its coordinates are known to us with the accuracy of merely several hundredths of an arc second.

Of all conceivable technological innovations and the latest exploration techniques the astrometric satellite alone can, for the first time in history, be instrumental in constructing a homogeneous frame of reference for the entire celestial sphere with one or two orders of magnitude better accuracy

355

J. H. Lieske and V. K. Abalakin (eds.), Inertial Coordinate System on the Sky, 355–360.
© 1990 *IAU. Printed in the Netherlands.*

than any of the available systems (such as the international catalogue FK5 containing information on 3.5 thousand stars with 0".03 to 0".10 accuracy). This system promises to be devoid of the currently existing local nonuniformities and global differences between the two hemispheres of the sky. The totality of absolutely new data obtained by means of this technique will largely render obsolete the labor-consuming work of the past centuries and foster the unheard of progress in many astronomical research projects.

Program of Observation and Expected Results

The LOMONOSOV observation program covers:

- all stars up to 10.0 mag totalling some 400 000 and ensuring thereby, availability of about ten stars per square degree of the sphere;
- fainter stars (up to 13.0 mag) numbering some 8 000 and already selected for the ESA HIPPARCOS program as presenting a special interest for astrophysics and stellar astronomy;
- some 30 of the brightest extragalactic sources;
- some 40 Solar system bodies (planets and asteroids).

Specific results of the LOMONOSOV project involving a space experiment on measuring the angular separations between the above-mentioned objects and their four-color photometry by means of a telescope on board a spacecraft, and ground observations of the selected celestial objects aimed at absolutization of the future catalogue, will provide a possibility to make up a catalogue for scientists and practical workers containing 400 000 stars and complete up to 10.0 mag, covering the entire sky and accurate up to 0".002 to 0".010 with regard to positions, proper motions and parallaxes. This catalogue will remain fairly accurate for 30 to 50 years, while its first version can be ready by 1996.

Method of Observations and Reductions

The proposed experiment is reduced, basically, to the following.

A Cassegrain telescope with a 50 m equivalent focal length, 1 m main mirror diameter (with a 4 m focal length) and a nonaberrational field of view equal to 6 minutes of arc or 90 mm will be mounted aboard a spacecraft. The system of aperture mirror focuses in one field of view the images of two stars, or rather of two sections of the celestial sphere, which are divided in the sky by a 90° angular separation. Due to their reciprocal position the aperture mirrors should form a highly stable reference angle, while the differences of the true angular separation between the stars and the reference value are to be measured in the course of the experiment.

CCD matrices consisting of 800 x 800 elements are suggested to be employed as receiving and recording equipment with each element's dimensions of 15 x 15 μm corresponding to 0".06. Analysis of the matrix signals by means of special digital algorithms makes it possible to determine the distance between stars in the field of view with the accuracy of up to 0.3 μm, this value corresponding to nearly 0".001.

The experiment's strategy involves keeping the spacecraft pointed to the selected star (referred to as the "reference star"), located near the antisolar direction, with subsequent turns of the spacecraft relative to this direction and during the fine stabilization phase measurements of separations from the reference star to all other stars spaced from the former by 90° (referred to as "program stars"). Then the spacecraft is redirected to another reference star, and the separations from this reference star to another set of program stars are measured. During the rough stabilization phase photometric and spectrophotometric measurements of the program and reference stars, respectively, are performed.

The basic requirement to the spacecraft design involves the possibility of its fast reorientation with subsequent tri-axial stabilization. To ensure realization of the above-described measuring techniques it is necessary that following the gradual increase of the speed around the assigned axis it would be equal to, approximately, 0.5 degrees per second. The precision of the spacecraft's orientation after the turn should not be below several arc minutes. Upon achieving the stabilization, the residual angular velocities of the spacecraft should not exceed 0''.4 per time second, while the higher degree of stabilization (0''.01) required in the process of measurements should be ensured through utilization of a tracking mirror in the optical feedback circuit. Assuming that following each measurement the spacecraft must, on the average make a 1'.25 turn, it is possible to take 2 to 3 measurements each minute, or some 1.2 million measurements during one year's time.

The telescope capacity and characteristics of CCD matrices allow one to accumulate on the matrix during one second approximately 4×10^4 electrons from an A0 star of 10.0 mag. Assuming that a star image is distributed over 4×4 elements of the matrix, we find that to ensure 1% photometric accuracy in observing faint stars (10.0 mag) an exposure of up to 4 seconds is required. The maximum exposure for the faintest objects (such as quasars) can reach hundreds of seconds. The exposure time for the program stars is selected automatically by the computer aboard the spacecraft.

As has been mentioned above, the current technology will make it possible to take measurements at a rate not exceeding 2 to 3 measurements per minute, or 3000 daily. This allows one to assess the information content of the experiment with allowance for the fact that in measuring the distance between two stars it is necessary to analyse the matrix sections of 32×32 pixels, while a reading from each pixel is recorded by 14 information bits:

$$32 \times 32 \times 14 \times 2 \times 3000 = 84 \text{ Mbit/day}.$$

The remaining information, including photometric measurements, reference angle control, readouts of the dark field, etc. is evaluated at 25 Mbit/day. Thus, the experiment's information content makes up about 110 Mbit/day. With normal functioning of the onboard computer, this information will be recorded in memory and transmitted daily to the Earth.

Selection of the spacecraft's orbit is required by the necessity to minimize various interferences. Light interferences from the Earth and Moon, interferences from the Earth's radiation belts affecting the CCD matrices, as well as the desire to keep the spacecraft as long as possible in the useful portion of the orbit, require one to choose a 48-hour orbit with the apogee of some 12 000 km. The inclination of the orbit's plane to the ecliptic should be 50° to 60° in order to reduce the seasonal influence of the magnetospheric tail which creates additional interferences at the matrices. The experience with the ASTRON satellite prove that from a 200 000 km distance the Earth has a brightness of −19 mag, and measurements can be performed not closer than 30° to the luminary's edge.

To accomplish the entire program of observations in the optimal mode one should calculate in advance all the turning angles of the spacecraft, or in other words, have a list of all the stars to be included in the future catalogue with their approximate coordinates (the so-called *input catalogue*). Making up of such a catalogue is an important and labor-consuming task which can be realized on the basis of photographic observations carried out by the USSR observatories in both hemispheres, and the date from the *Carte du Ciel* astrographic catalogue.

The input catalogue has a direct bearing on the planning and optimization of the space experiment which are necessary for (1) collection of the maximum number of independent measurements in the shortest possible period, and (2) deriving the best-defined set of equations for the final stage of reconstructing the coordinates by the distances measured. The optimal plan should meet the following basic requirements:

- selection of some 3 000 reference stars within a ±35° band around the ecliptic. These stars should be single, invariable, relatively bright (7-9 mag), well-recognizable against the background of other stars, and have, as far as possible, precise coordinates;

- selection for each of the reference stars a 90°-space band with the width equal to the telescope's field of view, considering all the stars in this band as program stars;

- ensuring coverage of the entire sky with these bands;

- ensuring that each program star was observed at least with two reference stars so that the separation between the latter was close to 90° in order to allow the effective reconstruction of coordinates by the separations;

- ensuring the two-stage observation of each star at a half-year interval in order to determine its parallax;

- separation of proper motions and parallaxes;

- checking the angular separation from the Sun, Earth and Moon to the observed section of the sky; checking in the telescope's field of view the coverage of asteroids and major planets;

- possibility of selecting out of the entire totality of measurements only those which allow one to make up the final catalogue on the limited number of stars (approximately 20 000).

Following completion of all the measurements and prior to preparation of the final catalogue it will be necessary to bring all observations into a uniform system of coordinates and time. For correct application of the reduction formulae, the following information will be required:

- equatorial coordinates of the observed objects accurate to 1";
- component velocities of the Earth and spacecraft not worse than 20 cm/sec;
- spacecraft's coordinates in the geocentric system not worse than 1 500 m;
- Earth's coordinates in the heliocentric system not worse than 1 000 km;
- coordination of all time scales not worse than 0.01 sec.

Making up of the final catalogue is reduced to solving by one or another method the system of linear equations in which each equation links a specific measurement with ten unknown ones (five for each star). The normal system's matrix possesses dimensionality of 5 multiplied by the number of stars in the program; and since only relative angular measurements are considered, it has the deficiency of rank 6, corresponding to an unknown rotation in the coordinate system and its changes in time.

Solution of the system of normal equations can be achieved with various methods — iterative, for one. Another method is a two-stage solution. In this case one has to select from the best astrometric catalogues a limited number of stars (10 000 to 40 000) with relatively well-known coordinates, and consider only those equations which link them. As a result it is possible to find a solution adjusting the totality of these stars' coordinates inside themselves with zero points which correspond to the system of originally-accepted coordinates. Coordinates of the remaining stars in the experiment's program will be determined by the differences of their coordinates with those of the first-stage stars.

The final stage of deriving the coordinate system, i.e., its absolutization through connecting with various physical bodies' systems will be realized upon completion of the entire LOMONOSOV project.
Numerous astrometric and photometric measurements with a telescope aboard the spacecraft are also planned within the HIPPARCOS project under development by the European Space Agency since 1975.

The LOMONOSOV and HIPPARCOS projects employ different techniques of observing the celestial sphere and totally different methods of registering the star positions in the focal plane. Without doubt both projects will be complementary since there is a chance to find and elinminate possible systematic errors and thus increase the reliability of the data obtained. This task will require coordination of effort between Soviet specialists and their West European colleagues.

Discussion

ANONYMOUS: Could you describe the onboard and ground computers?
CHEREPASHCHUK: The onboard computer is under consideration. Probably our version of a computer will be of the PDP type. We would like to have a powerful ground computer. We hope for international cooperation on this question.

KLIONER: I have two questions. If I understand you correctly, the equipment of the space station will contain the onboard computer with quite complicated software. The first question is what are the principal characteristics of the onboard computer. The second question: if your project is supported, who—what organization—will develop the software for the onboard and ground-based control computers? I do believe that it is a very important question because the loss of one spacecraft was caused by the imperfection of the software in my opinion.
CHEREPASHCHUK: These questions are under study. We have no final decision and we hope we will collaborate with the international community.

Høg: It seems that you only need the two entrance mirrors forming a penta mirror which will give a stable 90° reference angle. So why do you have the third plane mirror at all?

CHEREPASHCHUK: This is only one of the possible schemes. The final plans will be adopted after calculation of thermal qualities.

CHUBEY: Did you resolve the problem of adjustment of large sets of arcs which are near 90° without having the connection between the common objects of the observational program?

CHEREPASHCHUK: Numerical experiments with the SAO star catalogue showed that this problem was solved. Final optimal strategy should be developed after the compilation of an input catalogue.

Høg: Do you obtain scientific data only when the satellite is visible from a ground station?

CHEREPASHCHUK: Yes, we have onboard memory and we transmit the scientific data once per day.

HUGHES: As a follow-up to Dr Høg's question: Did I understand you to say that you receive data (downlink) once per day? I ask this question because of the obvious related requirements for onboard memory, data format, data rates etc.

CHEREPASHCHUK: Yes, and appropriate equipment, both onboard and ground, will be installed.

REGATTA–ASTRO PROJECT:
ASTROMETRIC STUDIES FROM SMALL SPACE LABORATORY

G. AVANESOV, V. VAVAEV, Ya. ZIMAN, A. KOGAN, V. KOSTENKO,
V. KRASIKOV, V. FEDOTOV, V. HEIFETS AND Yu. CHESNOKOV
Space Research Institute
USSR Academy of Sciences
Profsojusnaja, 84/32
117810 Moscow
USSR

ABSTRACT. The REGATTA–ASTRO project provides a great set of astrometric and photometric measurements from Small Space Laboratory (SSL). SSL is a special spacecraft being designed at the Space Research Institute, USSR Academy of Sciences. The main feature of this spacecraft is the attitude control by solar light pressure. The main objective of the project is to compile precise global catalogues of star positions, parallaxes and proper motion containing stars with magnitude up to 8-9 and a position accuracy of $0''.01$. The astrometric concept of the project is based on a highly deterministic spacecraft angular motion and numerous observations of each star using wide-angle TV cameras with CCD detectors. The SSL computer provides measurement data pre-processing, photometric referencing and data compression. Ground-based computers process the data statistically. The estimated vector includes star coordinates, SSL angular motion parameters and generalized distortion of the instruments used.

1. REGATTA Project

Milestones

- Design, manufacturing and testing of Small Space Laboratory (SSL)
- Interplanetary plasma studies based on SSL (REGATTA–PLASMA project)
- Astrometrical investigation by SSL onboard instruments (REGATTA–ASTRO project)
- Remote sensing of minor bodies of the Solar System (long-term plans)

2. Small Space Laboratory

- Attitude and (sometimes) orbit control by solar light pressure
- No jet propulsion units
- No pressurised compartment
- Passive thermal control

Return

- Ecologic cleanliness
- Payload/total mass ratio up to 50%
- Highly deterministic angular motion and hence accurate instruments' pointing data

J. H. Lieske and V. K. Abalakin (eds.), Inertial Coordinate System on the Sky, 361–366.
© 1990 *IAU. Printed in the Netherlands.*

3. REGATTA–PLASMA Project

3.1. PRIMARY GOALS
- Solar activity mechanisms
- Ways of transmission of the solar influence via the interplanetary medium
- Response of near-planetary medium to the solar perturbations

3.2. STUDY METHOD
- Satellite net for multi-probe high temporal and space resolution measurements in collaboration with CLUSTER and SOHO spacecraft

3.3. DESIGN PECULIARITIES
- Deployable booms carrying electric and magnetic field sensors
- Spinning platform with autonomous power supply, thermal control and data acquisition subsystems

3.4. PROJECT IMPLEMENTATION
- REGATTA–E (1993). Studies in near-equatorial region of the Earth magnetosphere. Orbit inclination is 15°, $R_{MIN} = 5\ R_E$, $R_{MAX} = 12\ R_E$
- REGATTA–A (1995). Investigation of the auroral regions of the Earth magnetosphere and near tail; interaction with CLUSTER project. Polar orbit, $R_{MIN} = 3\ R_E$, $R_{MAX} = 20\ R_E$
- REGATTA–D (1996). Studies in "middle tail" of magnetosphere. Near-ecliptical orbit, R < 70 R_E, resonant with periodic Moon flybys
- REGATTA–B (1996). Unperturbed solar wind investigations. Halo orbit around L point of the Sun-Earth system
- REGATTA–C (1996). Distant region of magnetosphere. Halo orbit around L point of the Sun-Earth system

4. REGATTA–ASTRO Project

4.1. PRIMARY GOALS
- Measurements of the star positions, proper motions, parallaxes and magnitudes in UBVRI spectral bands
- Catalogue compilation

4.2. INSTRUMENTATION
- Astrometric TV cameras
- Slit photometers

4.3. SCANNING MODE
- Longitudinal axis of SSL (Z-axis) pointing to the Sun
- Spacecraft rotation around Z-axis with angular rate of 1 rev/day
- Auxiliary cameras inclined by an angle of 160 deg with respect to Z-axis

4.4. METROLOGIC CONCEPT OF THE EXPERIMENT
- Multiple observations of every star and following statistical filtering
- Highly deterministic spacecraft angular motion
- High thermal stability

4.5. Design Peculiarities
- No moveable parts
- Dynamic symmetry
- Optimal geometrical and optical parameters of the solar stabilizer to minimize the influence of random external perturbations

4.6. Orbit
- Quasi-satellite in the Sun-Earth system
- Maximum geocentric distance   12 million km
- Synodic period   11 months
- Ecliptical inclination   10 deg

4.7. Registration of stars by TV cameras
- Number of exposures during the star passage across a field of view   600
- Minimum number of consecutive passages per camera   5

4.8. Onboard Processing
- Data correction based on ground and onboard calibration
- Star image extraction
- Calculation of averaged moments of brightness distribution
- Star identification
- Calculation of star track parameters

4.9. Ground Processing
- Data correction to chromatic aberrations
- Computation of coordinate residuals
- Statistical appraisals of star coordinates
- Double star decomposition and magnitude determination
- Determination of parallaxes and proper motions
- Reduction of star positions to the unique epoch
- Transfer to the standard coordinate frame

4.10. Possible ways of future ASTRO Project development
- Observation of quasars to determine star catalogue attitude with respect to inertial frame
- Observations in different spectral bands, including IR and microwave

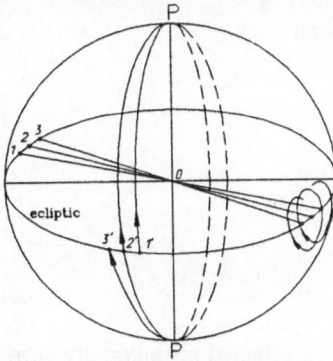

Figure 1. Mode of celestial sphere scanning

Table 1. Comparative parameters of REGATTA–ASTRO and HIPPARCOS projects.

	REGATTA–ASTRO	HIPPARCOS
Launch year	1993–1994	1989
Star position accuracy (after filtering)	0."01	0."002
Magnitudes	8-9	up to 13
Spectral bands	UBVRI	U
f/D, *meters*	0.1 / 0.07	1.4 / 0.29
Field of view, *deg*	5.3 × 8.0	0.9
Duration of total sphere survey, *years*	0.5	> 2
Number of observations of a specified star	> 10000	650
Single measurement error (μm / *arc sec*)	0.3 / 0.6	0.3 / 0.05
Number of stars being under observation simultaneously	up to 300	up to 4
Variations of angular position of an instrument with respect to the Sun direction, *deg*	0.1	40
Thermal stability of an instrument, *deg*	0.1	?

Table 2. Onboard Astrometric Complex (Main parameters)

TV cameras:	
Number of cameras	4 (astrometry) + 2 (photometry)
Focal length, *mm*	100
D / f ratio	1 : 1.4
Mean number of stars inside field-of-view	80
Lens transparency in 480-800 *nm* band	> 70%
Detector CCD array	
Number of pixels	520 × 580
Pixel size, μm	18 × 24
Exposure, sec	0.01-2.0
Shooting frequency, 1/*sec*	0.2-0.25
Photometers:	
Number of photometers	2
Photometer type	slit camera
Action mode	continuous
Optic axes	directed orthogonally with respect to the spin axis

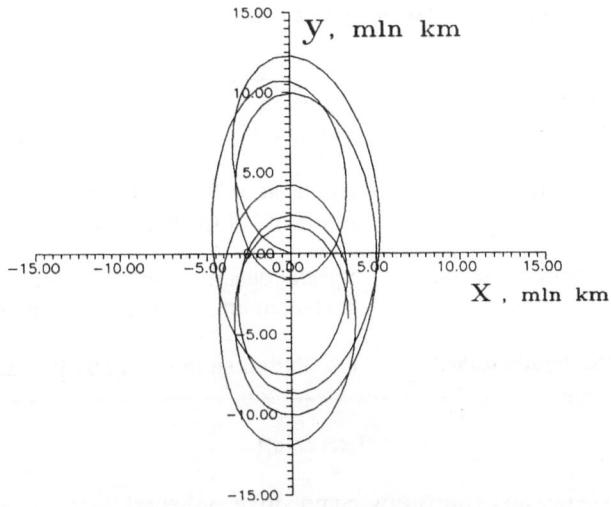

Figure 2. Quasi-satellite orbit of the REGATTA–ASTRO spacecraft. Synodical frame.

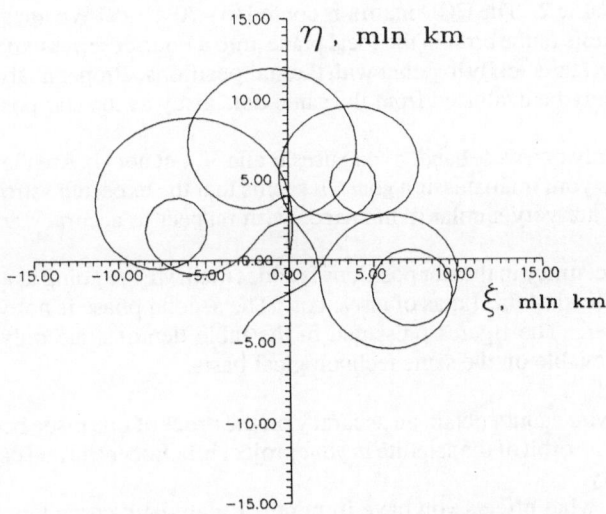

Figure 3. Quasi-satellite orbit of the REGATTA–ASTRO spacecraft. Inertial frame.

Table 3. Main sources of measurement errors

Source	Compensation method
Asymmetry of geometric and inertial parameters	Addition of asymmetry parameters to the list of quantities to be statistically determined
Instrumentation errors	Addition of "generalized distortion" parameters to the list of quantities to be statistically determined
Solar wind pressure variations	Special design of solar stabilizer. Onboard measurement of solar wind parameters
Micro-meteoric bombardment	Oscillation suppressing by hydraulic damper

Discussion

Röser: Are REGATTA-ASTRO and LOMONOSOV competitive projects?
Kogan: No, I hope not.

Hughes: (1) You have a very large field of view. Can you describe the optical system in a little more detail? (2) With a complex system it will be necessary to very carefully evaluate the systematic optical effects if the accuracy you mention is to be achieved.
Kogan: (1) The lens transparency in the 480–800 nm band is greater than 70%. The other parameters are given in Table 2. The CCD matrix is cooled to –70°C. (2) We are going to describe the optical distortion of the error in the focal plane into a Fourier series expansion and determine its coefficients statistically together with the star positions. Proper analysis shows that these coefficients may be evaluated from the same data array as the star positions.

Høg: HIPPARCOS uses only one wide band. TYCHO uses B and V, but not U. Are the photometric results in five colours your main mission goal? It seems that the expected astrometric results in the second phase are very similar to HIPPARCOS with respect to accuracy and to the number of stars.
Kogan: As far as the accuracy in the star positions is concerned, we are going to achieve the accuracy of 10 mas, instead of the 1 mas of HIPPARCOS. The second phase is not yet developed to the necessary level. The figures presented in the table demonstrate only the upper limit of accuracy achievable on the same technological basis.

Kopejkin: I think that you cannot obtain an accuracy on the order of one msec because of relativistic corrections. The orbit of the satellite in your project is heliocentric and cannot be determined very accurately.
Kogan: It's not clear what effects you have in mind. Relativistic corrections are susceptible to detailed calculations and that is being done for quite some time while defining the position of the spacecraft in deep space.

Xu: If the total duration of observations of your project will be only one-half year, it would seem to be too short for determining the parallax and proper motions of stars.

ON THE ROLE OF STAR CATALOGUES FOR AUTONOMOUS SPACE NAVIGATION

V. V. IVASHKIN
Keldysh Institute of Applied Mathematics
Miusskaya Sq., 4 , 125047 Moscow
USSR

ABSTRACT. The questions connected with constructing modern systems of autonomous navigation for spacecraft are considered. Increasing importance of these systems is shown, which is due to: a sharp increase in a number of spacecraft launches and congestion of on-ground control centers; and a necessity to provide more accurate and timely determination of motion parameters and motion control of spacecraft during their flights to the Solar System´s celestial bodies. Importance of the star sighting and angular measurements connected with stars for adequate solution of space navigation problems is shown. Thus the star catalogues play a significant role for constructing navigational algorithms.

1. INTRODUCTION. An autonomous system of space navigation allows some (optical, for example) on-board measurements of observed parameters, their statistical processing and determining the spacecraft motion parameters. Importance of these systems was already noted by pioneers of cosmonautics. With development of practical cosmonautics such systems were implemented in spacecraft designs since they sometimes allowed determining the spacecraft motion and orientation more timely and accurately, especially in situations of emergency and in transfers to the Solar System planets. For example, for decelerating the Luna-9 station to make a soft landing onto the Moon, by using an autonomous optical system the station velocity vector direction was very accurately determined and then reserved. Autonomous determination of the spacecraft orbit was provided in the Apollo Project. With time the autonomous navigation systems become more vital, even for artificial satellites.

2. STELLAR MEASUREMENTS IN AUTONOMOUS NAVIGATION

Astronomical navigation is a most important form of autonomous navigation. It is based on optical measurements of some parameters. As a rule, stars are main objects for observation here. They participate in navigation either directly or indirectly - allowing the construction of an onboard inertial coordinate system (ICS), in which the directions to

J. H. Lieske and V. K. Abalakin (eds.), Inertial Coordinate System on the Sky, 367–368.
© 1990 IAU. Printed in the Netherlands.

other celestial bodies (planets, the Sun) are then determined.

Now, the optical autonomous methods are gaining wide acceptance even in the near-earth orbital flights because potentially they allow constructing a fully autonomous and sufficiently accurate navigation system, which can help to relieve onground control centers. It is most promising to use the measurements of such parameters as angles of star elevations over horizon, the times of star set and rise above the planet horizon, angles of star refraction in the planetary atmosphere, angles of sighting line orientation relatively ICS for landmarks on the planet's surface as is shown in [1-3]. Such navigation allows determining the spacecraft position relatively a planet within 1 km or better.

In interplanetary transfers the main autonomous measurements are the angles of the orientation of sighting lines of closest celestial bodies relatively ICS. For distances about 100 million km to celestial bodies and accuracies of angular measurements about 1 arcsec the spacecraft position can be determined within a few hundreds kilometers, which is a satisfactory back-up to the ground measurements (especially in the emergency) for the middle part of flight to other planet.

The autonomous optical navigation is particularly important in the mission of rendezvous with a remote planet due to large time lags in communication with the spacecraft and an uncertainty in knowledge of the planet motion. In the framework of the Soviet-French Vesta Project in [4] the study of such navigation was carried out for transfer to an asteroid. Optical-TV sighting of the planet against the star background and computer processing of obtained measurements, characterizing the orientation of the planet's sighting line in ICS, allow good determination of the spacecraft's and landing probe's motion relatively the asteroid (within 10 to 25 km) and correction of knowledge on the asteroid position.

3. CONCLUSIONS. In modern cosmonautics the autonomous optical measurements using the stars are of great importance, and in future, as the problems to be solved will grow in complexity, their role will become even more and more significant. Therefore, the star catalogues which give a basis for using the star characteristics in navigational algorithms are very important for guidance, navigation and control of spacecraft.

4. REFERENCES

1. Battin, R.H. (1964) Astronautical Guidance, McGraw-Hill Book Company, Inc., New York.
2. Gounley, R., White, R., and Gai, E. (1984) 'Autonomous satellite navigation by stellar refraction ', Journal of Guidance, Control and Dynamics 2, 129-134.
3. Levine, G.M. (1966) 'A method of orbital navigation using optical sighting to unknown landmarks', AIAA Journal 4, 1928-1931.
4. Ivashkin, V.V. (1988) 'Navigation analysis of the space probe for investigation of the Vesta asteroid ', Preprint, Keldysh Institute of Applied Mathematics, USSR Ac. Sci., N152.

THE USNO (FLAGSTAFF STATION) CCD TRANSIT TELESCOPE AND STAR POSITIONS MEASURED FROM EXTRAGALACTIC SOURCES

RONALD C. STONE AND DAVID G. MONET
U. S. Naval Observatory, Flagstaff Station
P.O. Box 1149
Flagstaff, AZ 86002

ABSTRACT. The USNO (Flagstaff Station) is in the process of modernizing its 8-inch transit telescope. The upgraded instrument will use a CCD detector to measure the positions of both bright and faint stars directly with respect to extragalactic sources. Moreover, it will be able to observe large numbers of stars very rapidly, and improvements in accuracy are expected with further upgrades (e.g. CCD circle scanners, an interferometric telescope monitoring system, networks of environmental sensors). Many research opportunities related to coordinate systems exist for this telescope.

1. Star Positions from an Extragalactic Reference Frame

The U.S. Naval Observatory at Flagstaff is converting its 8-inch transit circle into a CCD scanning telescope that will be able to measure star positions differentially from extragalactic sources (QSO's, compact galaxies, and BL Lac objects) and radio stars with good radio positions. These objects can define an inertial reference frame.

The observing will be done in scan mode. The telescope will be set at a given declination and will scan the sky continuously by clocking the charge image of the sky across the CCD. The field of view of the current CCD chip is 20' by 20', and the telescope has the capacity to determine the centers of all the images in this area that are unblended and within the usable range of magnitudes. This is often more than 50 star images per field, and star positions should be measurable at a rate of 2000 images per hour. The limiting magnitude of the telescope is V $\approx$ 17.5 mag, and with a combination of short exposures and the use of screens for magnitude attenuation, stars as bright as V $\approx$ 5.0 mag should also be observable. Magnitudes with accuracies under ±0.05 mag can also be determined.

Koo, Kron, and Cudworth (1986) give counts of QSO's as a function of apparent magnitude. By using their statistics, the 8-inch TT is expected to observe 10 anonymous QSO's in each hour of scanning, except of course at low Galactic latitudes. Unfortunately, very few of these objects have been identified and even less have well determined radio positions. According to de Vegt (1986),

J. H. Lieske and V. K. Abalakin (eds.), Inertial Coordinate System on the Sky, 369–370.
© *1990 IAU. Printed in the Netherlands.*

about 233 extragalactic objects have good radio positions, of which the 8-inch should be able to observe 77 (or 33%) of them. All reductions will be made differentially, thereby minimizing many of the sources of systematic error (e.g. refraction, flexure, and clock errors).

2. The Refurishment of the 8-inch Telescope

The current 8-inch CCD detector is a thinned, back-illuminated, buried channel, Texas Instruments chip which has 800 x 800 pixels, a pixel size of 15 microns, and is cooled with liquid air. The current filter is the Hubble telescope W606 "Wide V" which has a passband of 4800 - 7200Å. In spite of the broad passband, the expected color equation (using the simulations discussed by Stone 1984) is no larger than that of the traditional V-passband. All processing is controlled with a DEC micro-VAX II and associated periperals.

The detector system has no moving parts. The CCD is mounted in the focal plane of the objective, and the tracking of stars is achieved by shifting the charge (column by column) along the chip at the diurnal rate. Once a column is shifted into the serial register, it is read and then transferred to a data file for further processing. Thus, star images can be scanned and read in a continuous process which can extend for many hours. Instrumental motions can be monitored in declination with readings of the Heidenhahn circle and in azimuth and level with four Hewlett Packard laser interferometric systems. The monitoring will be done in real-time. The old mechanical circle scanners of the telescope have been replaced with CCD scanners which are very accurate and stable. There are six scanners altogether, and the internal accuracy of a single scan made by one of the units is around ±0.01 arcsec.

Even if the telescope is very stable, the apparent zenith distances of stars can be altered with changes in the value of refraction occurring in the course of the scanning. This can be controlled by applying a correction for refraction to the observations as determined from the changing ambient observing conditions. The refraction appropriate for the passband is determined with the numerical technique discussed by Stone (1984). Environmental monitoring of temperature, pressure, and dew point is done at high accuracy and with triple redundancy. Accurate time is needed for the determination of right ascensions. The clock system consists of two Hewlett Packard high precision Cesium beam clocks (which are monitored with LORAN-C receivers) and converters for sidereal time.

3. References

de Vegt, C. 1986, in IAU Sympos. No. 109, *Astrometric Techniques*, (Reidel, Dordrecht), p173.
Koo, D.C., Kron, R.G., and Cudworth, K.M. 1986, *P.A.S.P.* **98**, 285.
Stone, R.C. 1984, *Astron. Astrophys.* **138**, 275.

THE FK5: PRESENT STATUS AND SOME DERIVED RESULTS

Heiner Schwan
Astronomisches Rechen-Institut
Mönchhofstr. 12-14
D-6900 Heidelberg
Federal Republic of Germany

ABSTRACT

Work on all parts of the FK5 is near to completion. The Basic FK5 has been published in the course of 1989 so that this catalogue is now officially released; the data have been made available to the various data centres. Work on the bright part of the FK5 Extension is finished since the end of 1988 and a tape version was sent to various institutions on request. Work on the Faint Extension is also nearly finished and the complete FK5 Extension will presumably be available on magnetic tape in a few months.

   Some results obtained from new observations of FK5 stars are presented. Comparisons of the FK5 with FC, NFK, FK3, FK4, GC, N30 were performed and are discussed with respect to systematic and individual accuracy. The parameters of galactic rotation and precession, and the distance to the Hyades cluster have been derived using proper motions which were obtained within the work on the FK5.

1. INTRODUCTION

One of the main purposes of a fundamental catalogue is to represent the conventional celestial reference frame to which the positions of other objects in the sky can be referred. In this sence the Fifth Fundamental Catalogue, FK5 (Fricke et al., 1988) represents the practical materialization of the conventional celestial reference system as defined by the Earth's equator and the ecliptic including also the theories of their motion and involved parameters. For a deeper discussion reference is made to the textbook by Kovalevsky, Mueller and Kolaczek (1989).

   The proper motions in a fundamental catalogue are of particular importance since they allow not only the transformation of the reference frame from one epoch to another epoch, but they can also be directly used for investigating the kinematics of the Galaxy or for determining the distances to nearby star clusters.

   In order to fulfill these requirements a fundamental catalogue has to be improved from time to time by including new observations and

J. H. Lieske and V. K. Abalakin (eds.), Inertial Coordinate System on the Sky, 371–381.
© 1990 IAU. Printed in the Netherlands.

using new theories and constants. In the following we want to report on the construction and the present status of the work on the FK5. By comparing the FK5 with all its four predecessors the progress shall be demonstrated which has been made with respect to internal precision and systematic accuracy since the first fundamental catalogue, the FC by Auwers (1879). New observations have recently become available providing information on the accuracy of the FK5; some of these results are presented.

Finally some results for the parameters of precession and galactic rotation and for the distance to the Hyades cluster are given. These results were derived from new proper motions obtained within the work on the FK5.

2. PRESENT STATUS OF THE WORK ON THE FK5

The FK5 will consist of two parts: first the Basic FK5 which is represented by the improved positions and motions of the classical fundamental stars given already in the FK4 (Fricke, Kopff, 1963), and second the FK5 Extension represented by about 3,000 new fundamental stars extending the fundamental system to apparent magnitude 9.5. In recent years Prof. Fricke and myself have continuously reported on the progress of this work. An extensive list of references for these reports can be found in the paper by Schwan(1988a).

2.1. The "Basic FK5"

The "Basic FK5" is the direct result of a revision of the FK4. This revision has been described by Schwan (1987) and, in some more detail, also in the introductory part to the FK5. It is therefore sufficient to give here only a very brief summary of that work. The major parts in the revision of the FK4 were the elimination of the regional errors in the FK4 system, the elimination of the error in the FK4 equinox and of its fictitious motion, the improvement of the individual accuracy of each star, and the transition to the IAU (1976) System of Astronomical Constants.

We have based the FK5 system on about 85 catalogues giving absolute or quasi-absolute observations made after 1900. The improvement of the internal precision of the FK4 could be performed by including new observations given in about 90 catalogues into the FK4 positions and proper motions without re-discussing the old observations already used in the FK4. It seems to be important to mention that this procedure was only possible because the FK4 gives the mean errors for the central epoch of each star. Mean errors of position and proper motion in compiled catalogues should always be given in this way since these quantities are uncorrelated.

In addition the FK4 positions and proper motions were rotated to the dynamical equinox by applying Fricke's (1982) correction to the FK4 equinox (including also the elimination of its fictitious motion), and by introducing the new expressions for general precession (Lieske et al., 1977) which are based on Fricke's (1977) determination of the con-

stant of luni-solar precession and on improved values for the planetary masses.

The FK5 represents therefore, as far as possible, an inertial
reference frame related to the dynamical equinox as the origin.

A magnetic tape version of the Basic FK5 has been made available by
the Astronomisches Rechen-Institut to various institutes since the end
of 1986. With the publication of the printed version of the catalogue
(Fricke et al., 1988) the Basic FK5 is officially released and the data
have been made available to the various data centres.

2.2. The "FK5 Extension"

The FK5 Extension will consist first of about 1,000 stars in the
magnitude range five to seven selected from the FK4 Supplement (Fricke,
1963), and second of about 2,000 stars selected from the list of International Reference Stars in the magnitude range 6.5 to 9.5. This part
of the work is being performed in collaboration with US Naval Observatory, where Dr. Corbin is working on the faint part of the FK5 Extension. The mean positions and proper motions are being derived from
relevant catalogues of stellar positions observed in the present century. All reductions are made in the FK4 system and still adopting
Newcomb's values for the precessional quantities. The final mean positions and proper motions will be transformed to the IAU standards.

The basic material for deriving the Bright Extension consists of
more than 100 catalogues with a sufficiently large number of fundamental stars allowing the determination of the systematic relation to
the FK4, and with also a non-negligible number of FK4 Sup stars. About
1,000 FK4 Sup stars were selected as new fundamental stars on the basis
of the precision of their positions and motions and also in trying to
achieve an even distribution on the sky as well as with respect to
apparent magnitude. Work on the Bright Extension was completed in 1988
and a magnetic tape version was sent by the Astronomisches Rechen-Institut to various institutes.

Table 1. Average mean epochs and mean errors of positions and
proper motions in right ascension (My-A) and declination (My-D)
for the Basic FK5, the Bright and Faint Extension of the FK5,
and the FK4. Right ascensions are multiplied with cos δ. The
proper motions are per century.

Catalogue	Average mean epoch		Average mean errors			
	RA	Dec	RA	Dec	My-A	My-D
Basic FK5	1955	1944	$0\overset{s}{.}001$	$0\overset{"}{.}02$	$0\overset{s}{.}005$	$0\overset{"}{.}07$
Bright Extension	56	49	.002	.04	.010	.18
Faint Extension	42	39	.004	.07	.019	.30
FK4	1917	1915	0.002	0.04	0.011	0.17

The Bright and Faint Extension will be published in one volume. The combining of the two subsets may still cause a few changes in the final star list. Work on the Faint Extension will be presumably be finished within this year, so that a tape version of the complete FK5 Extension (Bright and Faint) may be expected early in 1990.

The errors of mean positions and proper motions of the Basic FK5, of the Bright and Faint Extension and, for comparison, also of the FK4 are given in Table 1. The numbers are averages over the whole sky. An inspection of these numbers shows that the average proper motion errors of the Bright Extension are about twice as large as the errors of the Basic FK5, and the proper motion errors of the Faint Extension are about twice the errors of the Bright Extension. This is a consequence of the poor history of observation of the fainter stars. The precision of the Bright Extension of the FK5 is comparable to that of the FK4.

3. COMPARISON OF THE FK5 WITH NEW OBSERVATIONS

Comparisons of the FK5 catalogue with new observations were first presented by L. Morrison (1987). The Carlsberg Meridian Circle observations made at La Palma and observations made at Bordeaux near 1986 showed similar systematic differences Cat-FK5 in declination in the region near to δ = 55 degrees of the order of 0.1 arc sec. These differences were confirmed by observations made with the new Tokyo Photoelectric Meridian Circle in 1986 (Yoshizawa, Suzuki, 1988). This deviation must therefore be regarded as a systematic error in the FK5. The mean epoch of the FK5 declination system in that region is 1935, half a century from the epoch of the new observations; proper motion errors in the FK5 system give therefore a significant contribution. The early mean epoch in that region of the sky indicates also the considerable contribution of old observations which are likely responsible for this systematic error. It may also be worth mentioning that systematic differences between the FK5 and modern observations can now be detected and determined with much more confidence because of the significant smaller random errors in the FK5 (with respect to the FK4) as well as in the more precise modern observations.

New astrolabe observations made at Santiago de Chile in the zone -5 to -60 degrees in 1976 will be discussed by Noël and Débarbat (1989). The differences $\Delta \alpha_{\delta} \cos \delta$ between these observations and the FK5 are considerably smaller than those with respect to the FK4 demonstrating the higher systematic accuracy of the FK5 in that region. The mean standard deviations of the systematic differences in right ascension are 0.^{s}0035 and 0.^{s}0016 for the FK4 and FK5, respectively, demonstrating the increase of internal precision in the FK5 by a factor of about two. The systematic differences in declination are very similar for the FK4 and FK5. This could be expected since the corresponding changes in the transition from the FK4 to the FK5 were small. The mean errors of the systematic differences $\Delta \delta_{\delta}$ are reduced from 0.$''$048 for the FK4 to 0.$''$026 for the FK5 demonstrating again the considerable increase of internal precision in the FK5.

4. COMPARISON OF THE FK5 WITH FC, NFK, FK3, FK4, GC, N30

4.1. The catalogue comparisons

The Basic FK5 was compared with all its predecessors: the FC (Auwers, 1879), the NFK (Peters, 1907), the FK3 (Kopff, 1937) and the FK4 (Fricke, Kopff, 1963), and in addition with the N30 (Morgan, 1952) and the GC (Boss, 1937). The purpose of these comparisons is first to provide means for transforming observations from one of these catalogue systems to the FK5 system, and second to get information on the systematic and internal accuracy of the various catalogues.

Table 2. Mean epoch and equinox of the catalogues, the precession used and the differences to the IAU(1976) values at the catalogue epoch, the treatment of elliptic aberration in the respective catalogue and the number of stars; Δm, Δn are per century.

Catalogue	Equinox and Epoch	Precession	Prec. difference IAU(1976)-Cat		Elliptic Aberr.	Number of stars
			Δm	Δn		
Basic FK5	J2000	IAU(1976)	–	–	elim.	1535
FK4	B1950	Newcomb	+1.0376	+0.4360	incl.	1535
FK3	B1950	Newcomb	+1.0376	+0.4360	incl.	1535
NFK	B1900	Newcomb	+1.0444	+0.4412	incl.	925(+662)
FC	B1875	Struve	+0.4790	-0.0886	incl.	539(+83)
N30	B1950	Newcomb	+1.0376	+0.4360	incl.	5268
GC	B1950	Newcomb	+1.0376	+0.4360	incl.	33342

Some data of general interest or of importance for the catalogue comparisons are presented in Table 2. Given are the epoch and equinox of the comparison, the precession used in the catalogue, the differences Δm and Δn between the catalogue and IAU values at the epoch of the catalogue comparison (in the sence IAU(1976) minus Cat), the treatment of elliptic aberration in the catalogue positions, and the number of stars. The two numbers for FK3 indicate the addition of the 662 "Zusatzsterne" and for FC its southern extension from -10 to -32 degrees (Auwers, 1883).

The comparisons were made at the mean epoch and equinox given in Table 2 and by eliminating the effects of different treatment of elliptic aberration and of different precessional values. In principle these terms could have been included in the comparison, but they are more rigourously taken into account by using the conventional analytical expressions. In the transformation of observations from the catalogue system to the FK5 system one has therefore to consider these effects separately in addition to the systematic differences obtained from the catalogue comparisons.

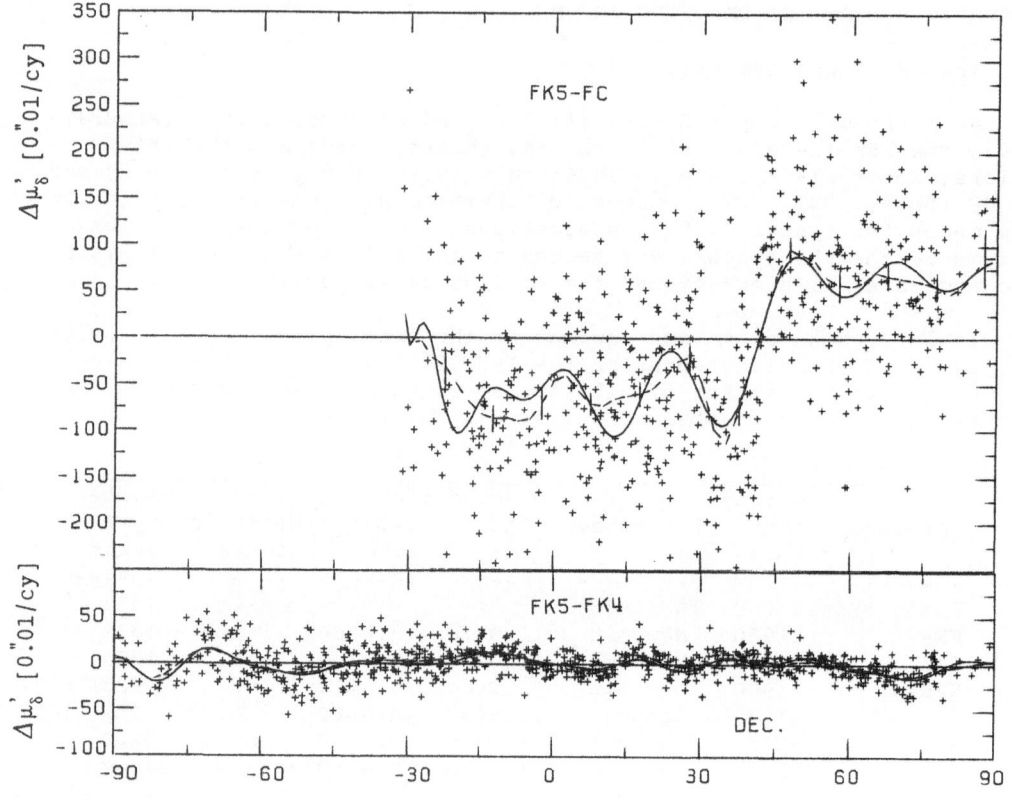

Fig. 1. Systematic differences $\Delta\mu_\delta'$ between the FK5 and FC (upper part) and FK4 (lower part); units are 0".01/cy.

As an example we present in the upper part of Fig. 1 the systematic differences $\Delta\mu_\delta'$ (according to the method by Bien et al., 1978) for FK5-FC and in the lower part for FK5-FK4 (Auwers stars were used only, i.e. star numbers 1-925). Each cross is the difference $\Delta\mu'$ for one common star. The figure illustrates impressively the large improvement made with respect to systematic and individual accuracy from FC to FK4. The FK5 proper motion errors are once more about half the dispersion of the residuals Cat-FK4.

Before discussing the catalogue comparisons in detail we want to give a few general remarks. Comparisons of the FK5 with its predecessors were performed by using only the Auwers stars because these are the only stars in FC and NFK. In addition comparisons with the whole set of FK5 stars were performed. Since the FC and NFK give no mean epochs all comparisons were made at the respective standard epochs (cf. Table 3). The positions at that epoch contain part of the proper motions and these differences are not directly comparable. We restrict the following discussion therefore on the proper motions.

4.2. Catalogue equator and equinox

A significant part of the differences arises from different zero points
(equinox and equator) in the various catalogues. The systematic
differences are more easily compared by first reducing all catalogues
to the same zero point. With the mean equatorial deviations FK5-Cat
(third to sixth column in Table 3) all catalogues were rotated to the
FK5 equinox, and the declinations were adjusted according to

$$\Delta_{red} = \Delta_{orig} - <\!\Delta\!> (1 - |\delta|/90)$$

where Δ stand for $\Delta\delta$ or $\Delta\mu'$, respectively. This reduction brings
the catalogue equator to the FK5 equator and leaves the pole unchanged.

Table 3. Mean equatorial differences in position (at the catalogue
epoch) and centennial proper motion between the FK5 and the
various catalogues, the dispersions $\sigma_{\mu}\cos\delta$, σ_{μ}, of the residuals FK5-Cat resulting from the catalogue comparisons and the
corresponding average quadratic deviation $s_{\mu}\cos\delta$, s_{μ}, of the
systematic differences (for the region north of -40 degrees).

FK5 - Cat		Mean equat. deviation at the Cat-epoch				Disp. of residuals		Mean quadr. syst. dev.	
Catalogue	Epoch	RA	Dec	My-A	My-D	$\sigma_{\mu}\cos\delta$	σ_{μ}	$s_{\mu}\cos\delta$	s_{μ}
FC	1875	$-^{s}\!071$	$+^{''}\!42$	$+^{s}\!179$	$-^{''}\!64$	$^{s}\!077$	$^{''}\!86$	$^{s}\!035$	$^{''}\!56$
NFK	1900	-.055	+.13	+.079	+.19	.034	.47	.021	.23
FK3(Auwers)	1950	+.035	-.03	+.083	-.05	.017	.22	.008	.08
FK3(Zusatz)	1950	+.039	.00	+.090	+.01	.027	.37	.009	.09
FK4(Auwers)	1950	+.035	-.02	+.085	+.03	.010	13	.007	.05
FK4(Zusatz)	1950	+.036	-.01	+.086	+.05	.031	.17	.008	.08
GC	1950	+.028	+.13	+.084	+.17	.023	.29	.008	.15
N30	1950	+.026	+.02	+.081	+.05	.020	.27	.006	.11
FK3(all **)	1950	+.037	-.01	+.086	-.02	.022	.29	.007	.08
FK4(all **)	1950	+.036	-.01	+.085	+.04	.011	.15	.007	.06

An inspection of the mean equatorial deviations in Table 3 shows
that the equators of FK3, FK4 and N30 agree practically with FK5, the
NFK and GC equators agree approximately with each other but differ from
the FK5, and the FC deviates extremely. The mean differences of the
right ascension proper motions are also very similar (with the exception of FC) and arise from the equinox correction E = 0^{s}085/cy as determined by Fricke (1982). The proper motion systems in right ascension of all catalogues (FC excluded) show therefore the same rotation
with respect to the FK5. For 1950 one derives the deviation E = 0^{s}035
between the FK5 equinox and those of FK3 and FK4, respectively; this
difference is the correction applied in the FK5. At 1950 we have

378

E = -0$.^S$016 for NFK corresponding to a difference of 0$.^S$053 with respect
to the FK3 and in accordance with the correction to the NFK equinox
(Kopff, 1937, p. 106).   The equinoxes of GC and N30 are identical
(Morgan, 1952, p. X) and deviate from FK3 and FK4 by 0$.^S$010 (Kopff,
1937); this is numerically confirmed by the corresponding mean
differences in Table 3.

4.3. Systematic deviations Cat-FK5

The systematic proper motion differences as a function of the declination and reduced to the FK5 zero point are presented in Fig. 2. The

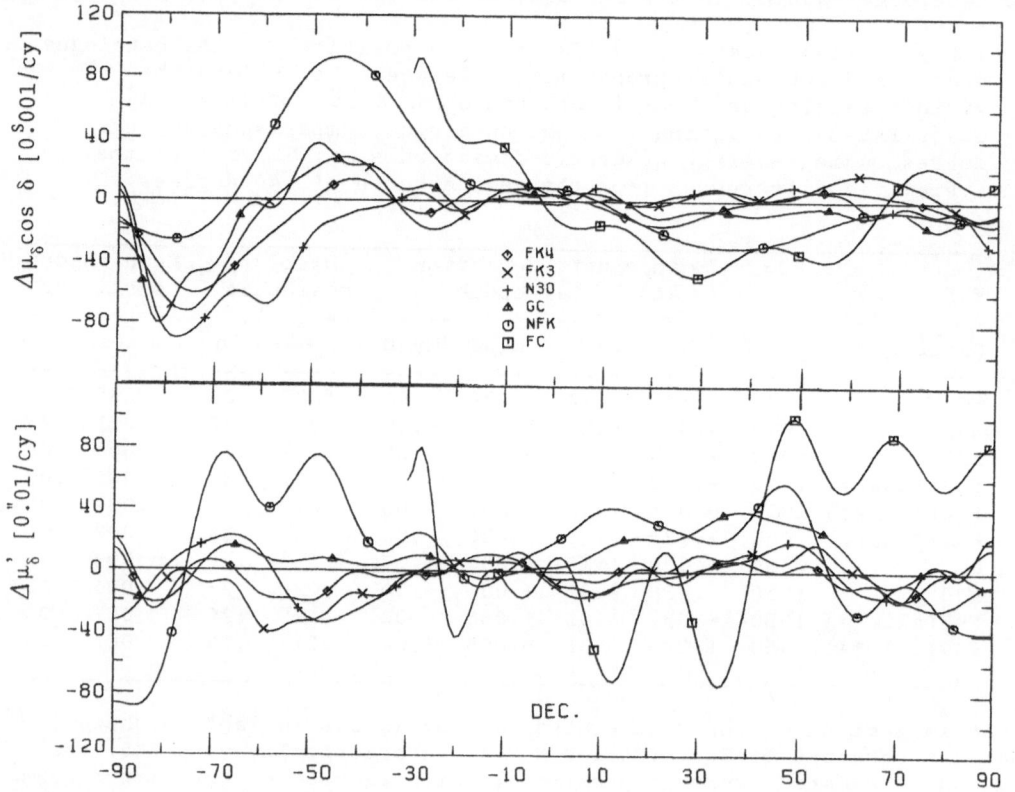

Fig. 2. Systematic differences of the proper motions in right
ascension (upper part) and declination (lower part) between
the FK5 and the catalogues FC, NFK, FK3, FK4, N30, GC. The
differences are reduced to the FK5 equinox and equator.
Units are 0$.^S$001/cy and 0$.''$01/cy, respectively.

FC deviates extremely from all other catalogues and the NFK shows also large deviations in the southern sky. The large errors in the FC reflect directly the systematic errors in Bradley's observations made at about 1755 and re-reduced by Auwers (1888); these are the only old observation used for deriving the FC proper motions. A comparison (not given) of the FK5 with Bradley's observations has reproduced the systematic differences FK5-FC in great detail. All catalogues show for $\Delta\mu_\delta\cos\delta$ a similar trend in the very southern sky. Excluding the old systems of FC and NFK one finds in general only moderate differences between the various systems, and the deviations from the FK5 take only in the southern sky considerable values. For $\Delta\mu_\delta\cos\delta$ there is some similarity between the systematic differences for GC, N30, FK3 and FK4. The mean quadratic deviations between the various catalogue systems and the FK5 are given in the last two columns of Table 3. We have restricted the averaging to the zone north of -40° since the large differences $\Delta\mu_\delta\cos\delta$ in the very southern sky would have dominated the mean. From these numbers one must conclude that the right ascension systems of GC, N30, FK3 and FK4 have, on the average, the same deviation from the FK5, whereas in declination there is a decrease in the series GC, N30, FK3, and FK4.

4.4. Internal catalogue precision

Information on the internal catalogue precision can be found in columns seven and eight of Table 3 where the dispersion of the residuals is given as resulting from the catalogue comparison. The contribution of the FK5 errors are in practice negligible. A comparison of the dispersions for the Auwers stars in the series FC, NFK, FK3(Auwers), FK4(Auwers) reveals that with each new fundamental catalogue their average precision in proper motion was increased by a factor of about two. This holds also for the transition from FK4 to FK5; the estimated precision for the Auwers stars in the FK5 is $0\overset{s}{.}004$/cy and $0\overset{"}{.}07$/cy for the proper motions in right ascension and declination, respectively.

5. PRECESSION, GALACTIC ROTATION AND THE HYADES DISTANCE IN THE SYSTEM OF THE FK5

Fundamental proper motions are the basic source for the derivation of the parameters for general precession and galactic rotation. In an extensive work Fricke (1977) has determined these quantities in the system of the FK4 and partly including results obtained in the N30 system. The transition from FK4 to FK5 system involves regional systematic corrections to the FK4 proper motions as a function of the right ascension and declination. In a recent paper Schwan (1988b) has shown that these corrections do no significantly change Fricke's values for precession and galactic rotation.

The Hyades distance is one of the corner stones for establishing the extragalactic distance scale. Although van Bueren's (1952) distance modulus m-M = $3\overset{m}{.}03$ can now definitely be ruled out there still exists a considerable discrepancy between various recent determinations. The

mean value for the Hyades distance modulus resulting from investigations using new proper motions is m-M = 3.^{m}31 (Hanson, 1980).

From work on the FK5 we have obtained improved proper motions for
all FK4, FK4 Sup and N30 stars. Among these stars there were 44
Hyades stars which could be used for determining their distances by
applying the convergence point method (Schwan, 1989). The distance modulus for the centre of these 44 Hyades stars is m-M = 3.^{m}37 ± 0.^{m}07.

REFERENCES

Auwers,A.: 1879, Fundamental-Catalog für die Zonenbeobachtungen am
    nördlichen Himmel. Publ. der Astron. Gesellschaft, IV, Leizig
Auwers,A.: 1883, Publ. der Astron. Gesellschaft, XVII
Auwers,A.: 1888, Neue Reduktion der Bradleyschen Beobachtungen aus den
    Jahren 1750 bis 1762 von A. Auwers. Dritter Band, p. 82, Petersburg
Bien,R., Fricke, Lederle,T., Schwan,H.: 1978, Veröff. Astron. Rechen-
    Inst., Heidelberg, No. 29
Boss,B.: 1937, "General Catalogue of 33342 stars for the epoch 1950",
    (GC). Publication of the Carnegie Institution of Washington, 468
Fricke,W.: 1963, Veröff. Astron. Rechen-Inst., Heidelberg, No. 10
Fricke,W.: 1963, Veröff. Astron. Rechen-Inst., Heidelberg, No. 11
Fricke,W.: 1977, Veröff. Astron. Rechen-Inst., Heidelberg, No. 28
Fricke,W.: 1982, Astron. Astrophys. 107, L13
Fricke, W., Schwan,H., Lederle,T., in collab. with Bastian,U., Bien,R.,
    Burkhardt,G., du Mont,B., Hering,R., Jährling,R.,Jahreiß,H.,
    Röser,S., Schwerdtfeger,H.M., Walter,H.G.: 1988, Veröff. Astron.
    Rechen-Inst., Heidelberg, No. 32, Verlag G.Braun, Karlsruhe
Kovalevsky,J., Mueller,F., Kolaczek,B.: 1989, Reference Frames in
    Astronomy and Geophysics, Kluwer Academic Press
Kopff,A.: 1937, Veröff. Astron. Rechen-Inst. Berlin Dahlem, Nr.54
Lieske,J.H., Lederle,T., Fricke,W., Morando,B.: 1977, Astron. Astro-
    phys. 58, 1
Morrison,L.: 1987, in IAU Coll. No. 100, held in Belgrade, 8-11 Sept.
    1987, eds. Eichhorn,H.K. et al. (in press)
Morgan,H.R.: 1952, Astron. Papers prepared for the use of the Ameri-
    can Ephemeris and Nautical Almanac, Vol. XII, Pt. III
Noël,F., Débarbat,S.: 1989, Astron. Astrophys. (in press)
Peters,J.: 1907, Veröff. des Königlichen Astron. Rechen-Inst. Berlin,
    No. 33
Schwan,H.: 1987, in IAU Coll. No. 100, held in Belgrade, 8-11 Sept.
    1987, eds. H.K. Eichhorn et al. (in press)
Schwan,H.: 1988a, in IAU Symposium No. 133, held in Paris, 1-5 June
    1987, p. 151, eds. S. Débarbat, J.A. Eddy, H.K. Eichhorn, A.R Upgren
Schwan,H.: 1988b, Astron. Astrophys. 198, 116
Schwan,H.: 1989, Astron. Astrophys. (in press)
van Bueren,H.G.: 1952, Bull. Astron. Inst. Netherlands, XI, 385
Yoshizawa,M., Suzuki,S.: 1988, paper presented at XXth General
    Assembly of the IAU at Baltimore, U.S.A., 2-11 Aug. 1988

Discussion

MORRISON: Yesterday we heard the result from radio astronomy of a correction of –0.2 arcsec/century to Fricke's value of luni-solar precession. In your recent analysis of the FK5 proper motions, you find no change to Fricke's value. Can you comment on the discrepancy between the optical and radio results?

SCHWAN: It is difficult to say how the optical proper motions can produce effects of the quoted size. But there is a considerable correlation between the precession and the motion of the equinox. It may be that these two effects cannot be well enough separated in the FK5.

MURRAY: How does the precession derived *only* from FK5 declinations compare with the new VLBI values?

SCHWAN: There are differences between the values determined from proper motions in right ascension and from declination-only, but unfortunately I cannot remember the exact numbers obtained from the individual solutions. The precessional correction from μ_δ was indeed in the direction of the VLBI results with a value of about –0.1 arcsec/century and a mean error of comparable size.

NEMIRO: The system of GC has the errors of the type $\Delta\alpha_\alpha$, $\Delta\mu_\alpha$, $\Delta\delta_\delta$, $\Delta\mu'_\delta$. Those errors must be carefully investigated and then the comparision of FK5–GC becomes much more valuable.

SCHWAN: The comparison was made with the aid of the analytical method as published in *Veröffentlichungen des Astronomischen Rechen-Instituts*, No. 29. This method determines the systematic effects depending on the right ascension, declination and the apparent magnitude simultaneously. In the viewgraph I have presented only the delta-dependent systematic differences which describe the most prominent deviations between the catalogue systems.

CELESTIAL COORDINATE SYSTEM AND THE FK5 CATALOGUE

A. P. Gulyaev
Sternberg Astronomical Institute
Universitetskij Prospekt 13
119899 Moscow
USSR

To study the motion of the celestial objects in a reference frame one should know the evolution of the frame itself. In our case it the motion of the origin of coordinates and the proper motions of the objects, the system is based upon.

It is evident, that the inertial coordinate system has just limited field of applications. In many cases the condition of the inertiality is either redundant either insufficient.

The system with well-known motion sufficient for many practical problems.

In 1980th the author of this paper had noticed that the traditional methodies of the construction of the fundamental catalogues put the limit to its precision: the systematical errors on northern hemisphere of the series of fundamental catalogues oscilates, but does not show any further decrease. This general suggestion is bieng illustrated by us with the comparison of the "FK5 - FK4" (1986.5) and "FK4 - FK3" (1963.5) systematic differences. It seems, that the comparison does confirm our idea. Therefore it is necessary to apply the new method of approach to the construction of the future Fundamental System. In other words it needs "perestroyka".

DISCUSSION

Schwan: I agree with what Dr Gulyaev has said in many respects in particular that the systems of FK3, FK4, FK5 are rather similar. But I doubt that the catalogues of the 19th century are valuable for improving the system because of the large systematic errors in those catalogues. It is therefore important to continue ground-based absolute observations with modern or new deviced instruments, at least until space astrometry has proved to provide an improved system.

Gulyaev: It is necessary to add: "absolute observations at high-altitude observatories".

Vityazev: Can you formulate the principal features of your "new thinking" policy in "perestroyka" of the FK compiling procedure?

Gulyaev: The main thing is the use of wide range of observational results including optical ground-based observations (with meri-

J. H. Lieske and V. K. Abalakin (eds.), Inertial Coordinate System on the Sky, 383–384.
© *1990 IAU. Printed in the Netherlands.*

384

dian instruments, astrographs, astrolabes), space observations (Hipparcos, Lomonosov, HST), VLBI, as well as observations in IR, UV, X-ray,
gamma-ray. The future Fundamental System must present "Grand Unifica-
tion" of the all-waves systems wich will be obtained by various
technics.

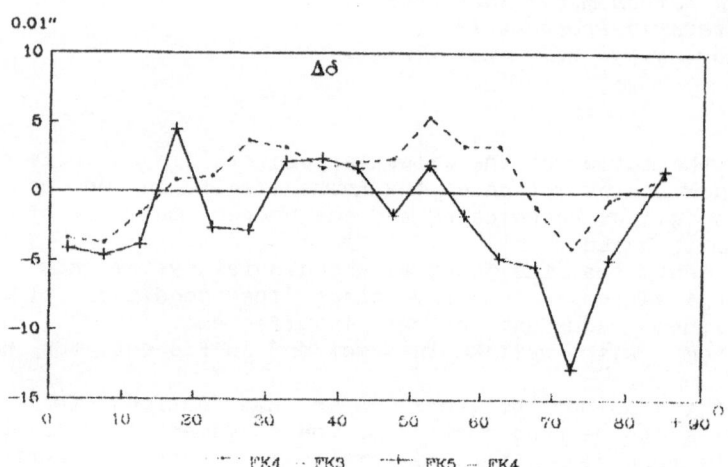

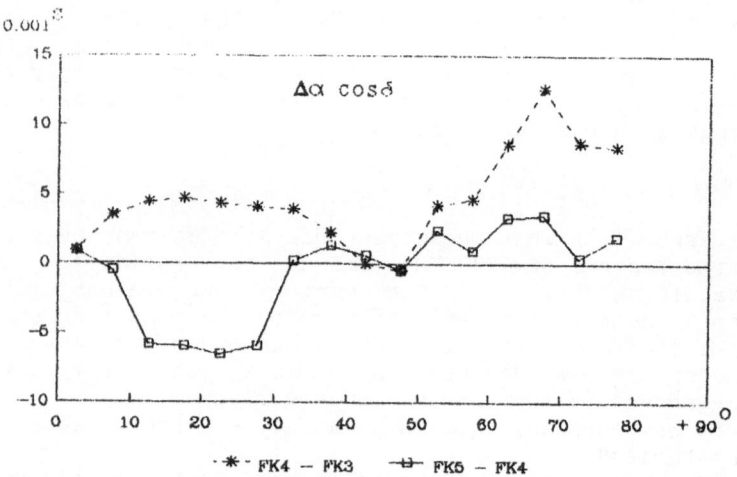

SYSTEMATIC DIFFERENCES INSTRUMENT MINUS FK5 IN THE SOUTHERN HEMISPHERE

G. CARRASCO AND P. LOYOLA
Observatorio Astronómico Nacional, Cerro Calán
Departamento de Astronomia, Universidad de Chile
Casilla 36-D
Santiago
Chile

1. Introduction

Observations of Fundamental Faint Star Catalogue (FKSZ) stars, made with the Repsold Meridian Circle at Cerro Calán National Astronomical Observatory, began in 1979 and finished in 1988. Today International Reference Star (IRS) observations are in progress. These observations correspond to the second epoch of the Santiago 67 Catalogue (Carrasco and Loyola 1981) and they are going to be used for determining the proper motions of these stars.

During this period, a series of fundamental FK5 stars along the whole meridian arc were periodically observed, in order to determine the instrumental system. The stars were observed differentially and the reductions were made according to the Zverev quasi-absolute method (1965, 1969).

The results of the observations in right ascension, of the series of fundamental stars, are given. The systematic differences $\Delta\alpha \cos \delta$, in the sense *Instrument – FK4* and *Instrument – FK5*, are presented. The mean epoch of the observations is 1982.2.

2. Observations and Reductions

From January 1979 to June 1989, three observers carried out observations of a series of 237 FK5 stars, with a total number of 9768 individual observations, in the zone $+41° > \delta > -90°$ in upper culmination and $-90° < \delta < -68°$ in lower culmination.

In order to investigate the behavior of the parameter n of the Bessel formula, a preliminary reduction of the observations was made. These values of n were corrected for the rate with declination and a second approximation was calculated (Anguita *et al.* 1975).

The instrumental coordinate system was obtained using the quasi-absolute method, which imposes the following conditions: that the instrumental system must be forced to coincide with the fundamental system at the equator and at the pole. An instrumental system was obtained in the FK4 and FK5 systems and the differences $\Delta\alpha \cos \delta$ for each observed star was computed.

The differences $\Delta\alpha \cos \delta$ obtained for each star were grouped in 5° declination zones. All residuals larger than 3σ ($\sigma_\alpha = 0\overset{s}{.}015 \cos \delta$) were eliminated after a careful investigation of the observational

J. H. Lieske and V. K. Abalakin (eds.), Inertial Coordinate System on the Sky, 385–386.
© 1990 *IAU. Printed in the Netherlands.*

386

data. The mean values for each zone, for the two positions of the instrument and in the FK4 and FK5 systems are plotted in Figure 1. For these differences, the mean standard deviations are 0ˢ022 for the FK4 and 0ˢ017 for the FK5. No clamp corrections were applied.

From Figure 1 it is possible to see that, in general, the FK5 system shows a better agreement than the FK4 system with the Repsold Meridian Circle system, especially in the zone –40° to –70°. This agreement is broken in the zone –70° to –85°.

Acknowledgements. This project was supported by the Fondo Nacional de Desarrollo Cientifico y Tecnológico, FONDECYT (Projects No. 337/87, 490/88 and 976/89) and Universidad de Chile, Departamento Técnico de Investigación (DTI Project E–1935).

References

Anguita, C., Carrasco, G., Loyola, P., Bedin, V.N., Naumova, A.A., Polojentsev, D.D., Polojentseva, T.A., Tavastsherna, K.N., and Zverev, M.S.: 1975, *Publ. Dep. Astron. Univ. Chile* **2**, 181.
Carrasco, G., Loyola, P.: 1981, *Publ. Dep. Astron. Univ. Chile* **4**.
Zverev, M.S.: 1965, *Astron. Zh.* **42**, 823.
Zverev, M.S.: 1969, *Astron. Zh.* **46**, 129 (*A.J.* **13**, 1008).

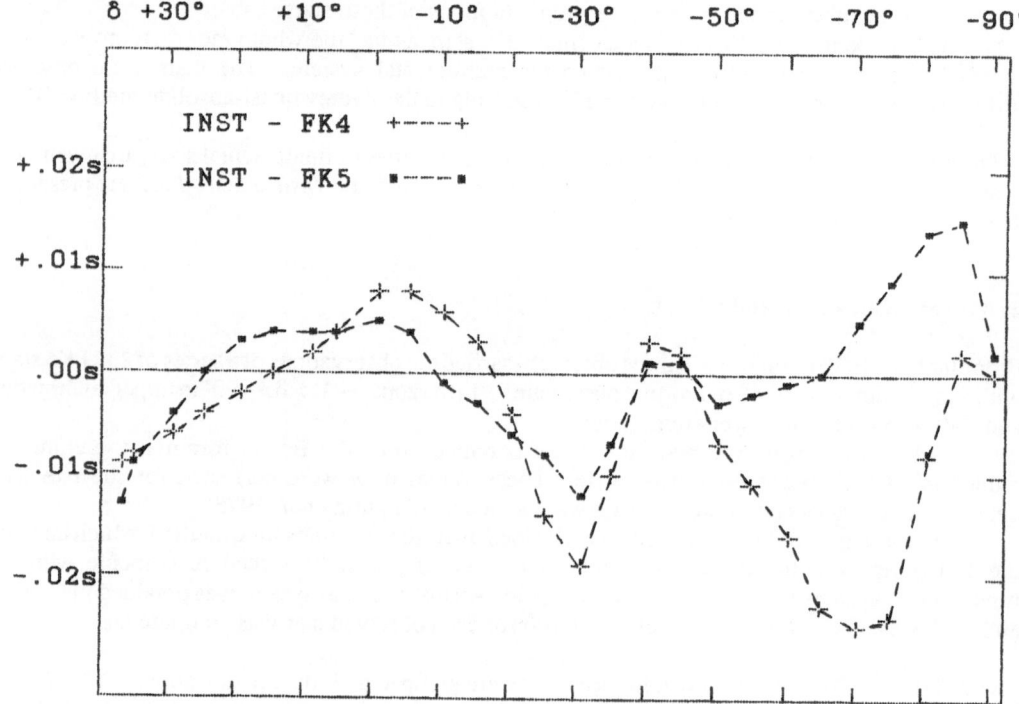

Figure 1. Systematic differences $\Delta\alpha \cos\delta$ in the sense *Instrument – FK4* and *Instrument – FK5*.

EFFECT OF THE FK5–FK4 CATALOGUE CORRECTIONS FOR THE OBSERVATIONS OF THE PHOTOELECTRIC ASTROLABE AT SHANGHAI OBSERVATORY

T. Q. Xu , Z. X. Li
Shanghai Observatory, Academia Sinica
200030 Shanghai
People's Republic of China

ABSTRACT.   On the basis of the observational materials of the photoelectric astrolabe from 1976 to 1977 at Shanghai Observatory, the possible increase improvement in observational precision after adopting the FK5 catalogue is estimated. The calculation proves that the improvements are about 6 per cent and 9 per cent in the right ascension and declination respectively.

The publication of FK5 catalogue will contribute to the improvement of precision in the case of optical astrometric observations.   Using the observational materials of the photoelectric astrolabe ZPA   from 1976 to 1977 at Shanghai Observatory, the possible improvement   in precision is estimated after adding the catalogue corrections   FK5 – FK4.
There are 12 groups in the observational programme of ZPA, mean 35 FK4 stars per group. Using the FK5 Fundamental Stars Catalogue , IAU(1976) astronomical constants, equanox correction and its motion, the catalogue corrections $\Delta\alpha_{FK5-FK4}$, $\Delta\delta_{FK5-FK4}$ of these twelve groups at 1977.0 are calculated. The mean catalogue corrections per group are given in Table 1.
After using the corrections FK5 – FK4 of Table 1   to revise the observational results of ZPA, the mean rms of a group observa- tion for time, latitude measurement decrease about 6 per cent, 9 per cent respectively.
The unbalance of the improvements between time and latitude measurements might caused by weight difference (1.5:1) in compiling the observational programme for the astrolabe ZPA.
It is quite possible that the further improvement of star catalogue can contribute not as much as one has been expecting. In order to improve the accuracy of astrometric observation, attention should also be paid for the other sources of errors: the errors of the astronomical constants, atmospheric refraction and local effects etc.

J. H. Lieske and V. K. Abalakin (eds.), Inertial Coordinate System on the Sky, 387–388.
© 1990 *IAU. Printed in the Netherlands.*

Table 1. Mean catalogue corrections for
ZPA programme

Group	$\overline{\Delta\alpha}_{FK5-FK4}$	$\overline{\Delta\delta}_{FK5-FK4}$	R.A. in the middle per group
1	$- 0.^{s}0038$	$- 0.^{''}047$	1^h
2	$-$ 21	$-$ 14	3
3	$-$ 6	$-$ 12	5
4	$-$ 10	8	7
5	5	9	9
6	5	$-$ 2	11
7	0	18	13
8	$-$ 6	17	15
9	$-$ 5	$-$ 13	17
10	$-$ 26	$-$ 14	19
11	$-$ 38	$-$ 33	21
12	$-$ 28	$-$ 14	23
mean	$- 0.^{s}0014$	$- 0.^{''}008$	

Table 2. Decreament of rms for time and latitude
measurement after using the
FK5-FK4 corrections

	rms for a measurement	
	Time	Latitude
No correction	$0.^{s}0078$	$0.^{''}082$
Correction by FK5-FK4 corrections	0. 0074	0. 075

Acknowledgement. We are grateful to Dr. Schwan who helped to provide the tape material of the FK5 Fundamental Stars Catalogue.

REFERENCE

FK5 Fundamental Stars Catalogue(tape), 1988, Astronomisches Rechen Institut, Heidelberg.

DIFFERENCES IN CHANGES OF BOROWIEC GEOGRAPHICAL COORDINATES CAUSED BY THE REPLACEMENT OF FK4 CATALOGUE BY FK5

M. LEHMANN
Space Research Center, Polish Academy of Sciences
Astronomical Latitude Observatory
Borowiec
62-035 Kornik
Poland

ABSTRACT. Astronomical observations of time and of geographical latitude carried out by means of Danjon's astrolabe OPL No. 31 in the years 1982–1988 have been reduced to the system FK4. After the introduction of the FK5 catalogue in 1989, the earlier observations were re-reduced. The geographical latitude results, before and after introducing the FK5 catalogue, are given.

At the turn of 1988/89, the new fundamental catalogue FK5 (Schwan 1989) was disseminated. In the case of the astrolabe, when the equal altitudes method of the observation reduction theory (Débarbat *et al*. 1970) was applied, the corrections to the star coordinates do not transpose directly to the final result. Logically, the character of these corrections with regard to right ascension does not have any correlating effect on geographical coordinate changes observed during the year.

The following questions regarding the effects of differences in changes of geographical coordinates were brought up for discussion:
  1) how will the "mean" coordinates change;
  2) how should their periodic components, particiularly in annual terms, change;
  3) how will the dispersion of derived results change.

Taken into consideration were the geographical Borowiec latitude observations, carried out by means of Danjon's astrolabe, which were grouped in 5-day intervals. The corrections for the polar motion were deduced from BIH-IERS (1982-1989), and so the results were computed in the "BIH 1979" system. The statistical analysis is given in Table 1 (*Before*). The basic results of Pearson's χ^2 test (Konys *et al*. 1975; Fisz 1967) proved the existence of a characteristically systematic error in the series of received latitude observations. In all cases $\chi^2_{0.05}$ is 13.8.

The series of received observations underwent spectral analysis (Jaks *et al*. 1980; Andersen 1974), and using the least suqares method the amplitudes and phases of periodical components were computed. The analysis was based on the model of observed changes of the geographical Borowiec latitude which considered 3 periodic terms: 432 days, annual and semiannual terms. Since all results were computed in the "BIH 1979" system, the global amplitudes of the 3 periodic terms mentioned above were eliminated. The estimated results of the model, using the 3 aforementioned terms, are

389

J. H. Lieske and V. K. Abalakin (eds.), Inertial Coordinate System on the Sky, 389–390.
© 1990 *IAU. Printed in the Netherlands.*

Table 1. Basic statistical analysis results of received Borowiec latitude observations, before and after estimation of 3 periodic terms.

Characteristic	FK4 (Before)	FK5 (Before)	FK4 (After)	FK5 (After)
ϕ (52°16' +)	38''740	38''700	38''738	38''690
max-min	0''526	0''447	0''475	0''455
σ	0''093	0''078	0''079	0''073
χ^2_{calc}	11.7	22.4	8.6	7.7

shown in Table 1 (*After*). In the table one can observe a significant decresase of parameters within the annual term in the FK5 system while the other components remain relatively unchanged.

The "mean" latitude change which exceed the error estimate is worth noting. The determined terms treated as the model of systematic changes have been successively removed from the observed latitude values. The statistical testing of the goodness of fit of the normal distribution of residuals has been performed. The derived descriptive characterization of the Pearson χ^2 test are shown in Table 2.

Table 2. The parameters of periodic terms in the model of observed Borowiec geographical latitude.

Term		FK4	FK5
ϕ = 52°16' +		38''748 ± 0''006	38''703 ± 0''006
432-day	Amplitude	0''026 ± 0''007	0''023 ± 0''007
	Phase	255° ± 17°	246° ± 19°
365.24-day	Amplitude	0''048 ± 0''008	0''016 ± 0''007
	Phase	252° ± 10°	238° ± 23°
182.6-day	Amplitude	0''031± 0''007	0''030 ± 0''007
	Phase	6° ± 15°	27° ± 15°

In the series of latitude observations discussed, the introduction of the FK5 caused a significant decrease of the amplitude of the annual term, while the other components remain unchanged. The results presented here show the predominance of the FK5 system over the FK4 system; however, this system still is not free of systematic error sources.

References

H. Schwan (1989), "Catalogue FK5", on magnetic tape, Astronomisches Rechen-Institut, Heidelberg.

BIH (1982–1987), IERS (1988–1989), Circulars "D", "B".

Débarbat, S., Guinot, B. (1970), "La Methode des Hauteurs Egales en Astronomie", Gordon & Breach, N.Y.

Konys, L. *et al.* (1975), "Testing the normality of a distribution in the case of one sample", *Roczniki Akademii Rolniczej w Poznaniu*, **LXXX**, z. 4 (in Polish).

Fisz, M. (1967), "Rachunek prawdopodobienstwa i statystyka matematyczna", *PWN Warszawa* (in Polish).

Jaks, W. *et al.* (1980) "Analysis of periodical variations of the Ottawa latitude", *Publ. Inst. Geophys. Pol. Acad. Sc.*, F-6 (137), 71–79.

Optical reference frame defined by Carlsberg Meridian Catalogue La Palma Number 4

LV Morrison & RW Argyle, *Royal Greenwich Observatory, UK*
L Helmer, C Fabricius & OH Einicke, *Copenhagen University Observatory, DK*
L Quijano & JL Muiños, *Real Instituto y Observatorio de la Armada en San Fernando, SP*

ABSTRACT. The contribution of the Carlsberg Meridian Catalogue La Palma Number 4 to the determination of the optical reference frame is discussed. This catalogue of almost 51000 stars provides one of the most accurate optical reference frames at the present epoch, having a density of 1 star per square degree, and an average accuracy of $0\overset{''}{.}12$ in position and $0\overset{''}{.}003$ per year in proper motion for stars with V<9. The catalogue also contains positions of reference stars with V>11 in the fields of benchmark extragalactic radio sources which can be used in linking the optical reference frame defined by the FK5 to the extragalactic frame.

1. Carlsberg Automatic Meridian Circle

The Carlsberg Automatic Meridian Circle (CAMC) is operated jointly by Copenhagen University Observatory, the Royal Greenwich Observatory and the Real Instituto y Observatorio de la Armada en San Fernando at the Spanish Observatorio del Roque de los Muchachos of the Instituto de Astrofisica de Canarias situated on the island of La Palma. The astronomical coordinates are:
17° 53′ 07″8 West,   28° 45′ 52″4 North,   2326m altitude.
The telescope has an aperture of 18cm. It has been in operation continuously since May 1984, except for short periods due to technical problems and maintenance.

The same moving-slit micrometer has been used throughout the period of observation of this catalogue, and the photoelectric circle-reading system has also remained unchanged. All the processes, from setting the telescope to the scanning of the star during meridian passage and the reading of the circle, were carried out completely automatically under the control of two minicomputers. A general description of the system is given by Helmer & Morrison (1985).

The major changes to the instrumental system during the observation of the catalogue were as follows. An air circulation system was installed in the tube of the

J. H. Lieske and V. K. Abalakin (eds.), Inertial Coordinate System on the Sky, 391–401.
© 1990 IAU. Printed in the Netherlands.

telescope in August 1985 in order to remove the internal refraction caused by the thermal stratification of the air (Høg & Fabricius 1988). An azimuth mark 130m to the south was observed from January 1986 onwards, and another 50m to the north from December 1986. These were automatically observed hourly together with the line of collimation, the horizontal flexure, and the level and zenith point of the circle which were obtained by observing the reflected image in a nadir pool of mercury.

2. Observational programme

The observing list consisted of the following programmes (limited to declinations above $-45°$, apart from the FK5 which was limited to $-55°$):

TABLE 1. Carlsberg Meridian Catalogue No.4: observational programmes

Reference frame programmes	Total size	No. stars CMC No.4
FK5	1500	
International Reference Stars	32524	21328
General Catalogue	24297	8957
Reference stars in radio fields	11397	6829
PZT stars	1957	797
Reference stars in selected Schmidt fields	1014	822
Geodetic stars (Dec +7 to +10)	374	356
Radio stars	313	203
Hipparcos non-astrometric (Dec 0 to -25)	3708	1687
Hipparcos poor SAO positions	1352	1043
Gliese catalogue of nearby stars	1114	412
Proper motion programmes		
F-type stars within 100pc	10550	9209
G-type d&g and K-type g within 300pc	9243	5665
Nearby O-B associations	5629	2548
Sample of A5-G0 stars in NGP	3098	399

Solar system	No. obs. CMC No.4
Mars	137
Callisto	92
Saturn	131
Titan	70
Uranus	284
Neptune	294
Minor planets (62)	6940

3. Real-time selection

The selection of celestial and calibration objects was performed in real-time according to a pre-assigned priority. About 7% of the time was spent on calibration objects, 73% on programme stars, 18% on FK5 stars and 2% on planets. The scanning time was a function of the magnitudes of the objects: stars brighter than 9.5 took 16s, those between 9.5 and 11.0 took 32s, and those fainter than 11.0 took 64s. FK5 stars at zenith distances greater than $60°$ were observed for 32s, and those greater than $75°$ for 64s, irrespective of their magnitude. In order to improve the efficiency of the observing, stars were observed up to 40s from the meridian after November 1985.

The same observing slits $(33''\times4'')$ were used for all objects except for Mars and Saturn which were observed in a pair of large slits $(83''\times4'')$. Very long slits $(2'\times4'')$ were used in a special search mode to locate stars with very poor input positions. Observations of these stars were, however, carried out in the normal slits after resetting the telescope.

4. Differential reduction to FK5

The real-time reduction to micrometer coordinates of the photon counts was carried out as described in Helmer *et al.*(1984) and Helmer & Morrison (1985). A differential reduction for each night was carried out the following day by the observer using interactive computer programs. When available and reliable, the hourly values of the instrumental calibrations were subtracted from the FK5 residuals before the least-squares adjustment was made for the whole night, which comprised about 9 FK5 stars per hour down to a zenith distance of $84°$. Stars above declination $+67°$ were observed at either upper or lower culmination. Effectively, on most nights the least-squares adjustment amounted to a single rotation for the whole night in right ascension and declination to bring the instrumental system into agreement with the average position of the FK5 stars. The resulting positions in the catalogue are therefore not subject to possible zonal systematic errors in the FK5 frame.

The atmospheric extinction and instrumental absorption was determined nightly by least-squares adjustment to about 50 photoelectric standards spread through the night.

The least-squares adjustment in position and magnitude was then applied to the programme stars and solar system objects. The positions and magnitudes were compiled daily into a catalogue, and stars with the requisite number of observations (normally 6) were automatically deleted from the observing list.

5. Formation of the catalogue

The scanning slit micrometer was replaced by a new one in March 1988; so it was decided to combine all the observations made in the period May 1984 to February 1988 into one catalogue, superseding the three annual catalogues published for

1984, 1985 and 1986. In total, observations carried out in 8872 hours of operation on 955 nights were compiled into this catalogue. This represents 68% efficiency in observing.

Before forming the unweighted means of the positions of programme stars, about 6% were rejected because their stellar images were badly defined, being either too faint or not properly centered in the slits. A further 1% were rejected because they had large residuals. One night was completely rejected in declination due to problems with the circle-reading system. In magnitude, 111 nights of observations were rejected because their photometric quality was too poor.

6. Accuracy of the catalogue

The internal standard deviation (mean error) of a *single* observation was calculated from the scatter of the observations about their means. These were averaged in $5°$ ranges of declination and the resultant values for stars with V<10 are plotted in Figure 1(a). The average error curves in right ascension, declination and magnitude are:

$$\sigma_\alpha = 0.''187\sec^{0.6}z \qquad \sigma_\delta = 0.''181\sec^{0.9}z \qquad \sigma_m = 0^m.055\sec z,$$

where z is the zenith distance. The mean error as a function of magnitude, reduced to the zenith, is plotted in Figure 1(b). It follows from this Figure that the average curves in right ascension and declination are only valid for stars with V<11, and the average curve in magnitude for stars with V<9.

The standard error of a catalogue position (formed from the mean of n single observations) has been investigated using stars observed on both clamps, assuming that the standard error follows the relation;

$$\epsilon_n = \sqrt{a^2 + \sigma^2/n},$$

where a is the asymptotic mean error and σ is the standard deviation. Evaluating this for $a=0.''068$ in right ascension and $a=0.''076$ in declination we derive the values in Table 2.

TABLE 2. Standard error of a catalogue position

n	z=0° δ= +29°		z=30° δ= −1°, +59°		z=60° δ= −31°, +89°		z=73.8° δ= −45°	
	ϵ_α	ϵ_δ	ϵ_α	ϵ_δ	ϵ_α	ϵ_δ	ϵ_α	ϵ_δ
3	".13	".13	".14	".14	".18	".21	".24	".34
4	".12	".12	".12	".13	".16	".19	".21	".30
5	".11	".11	".11	".12	".14	".17	".19	".27
6	".10	".11	".11	".11	".13	".16	".18	".25
7	".10	".10	".10	".11	".13	".15	".17	".23

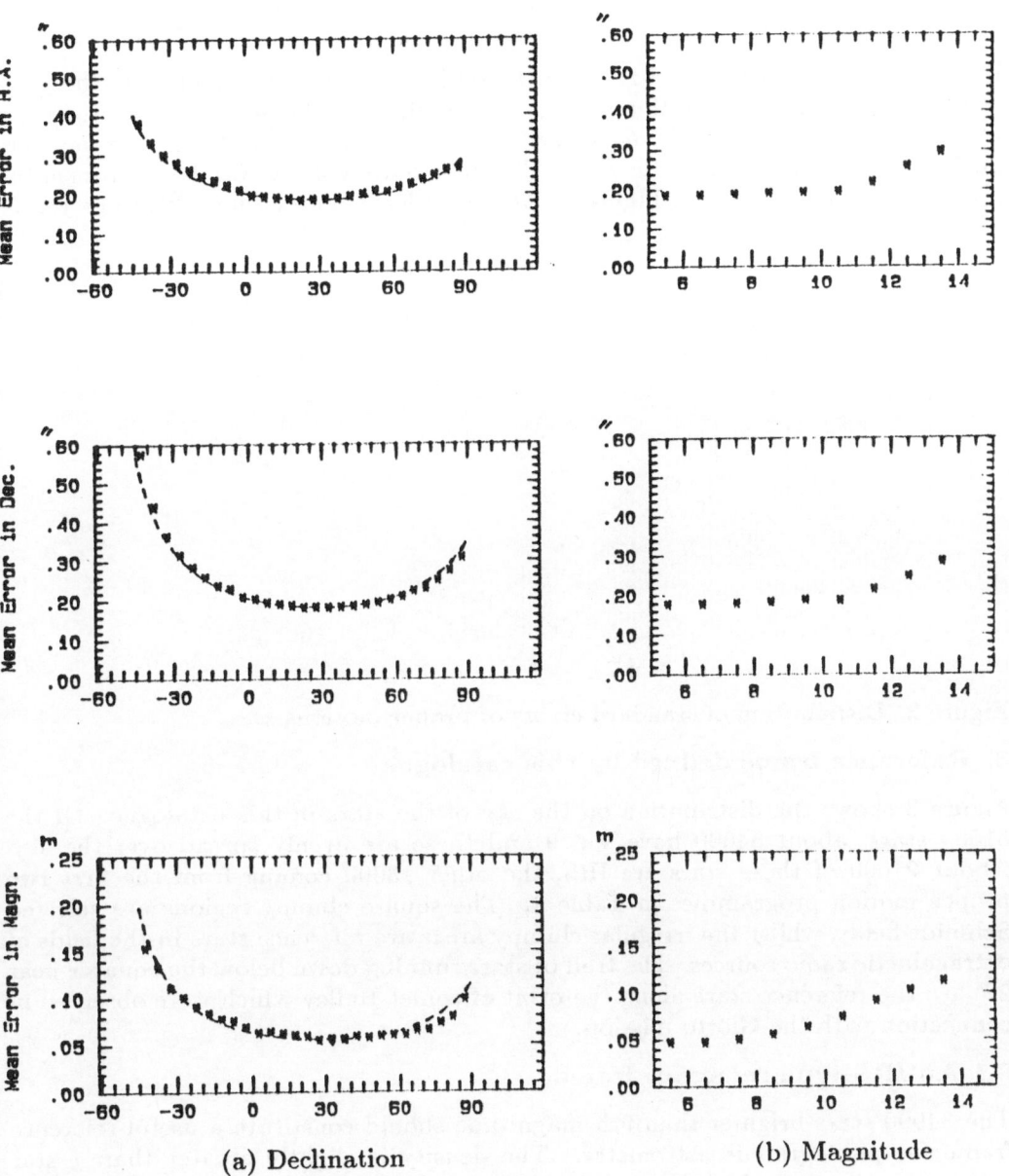

Figure 1. Mean error of a *single* observation as a function of
(a) declination, and (b) magnitude.

7. Proper motions

Proper motions were derived for almost 42000 of the stars by combining the position from this catalogue with positions from earlier epochs, using mainly the large photographic surveys of Yale, AGK2, AGK3 and Cape. Systematic corrections were applied to bring them on to the FK5. The distribution of the standard errors of the proper motions is shown in Figure 2. About 80% are better than 0"004 per year.

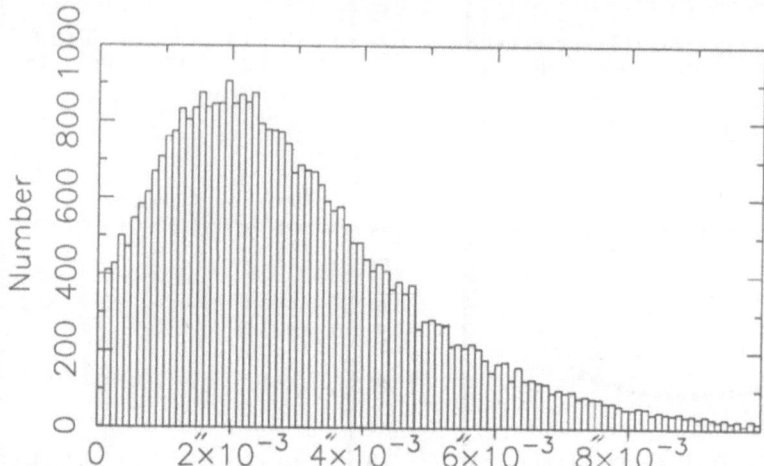

Figure 2. Distribution of standard errors of proper motions.

8. Reference frame defined by this catalogue

Figure 3 shows the distribution on the sky of the stars in this catalogue. Of the 51000 stars, about 34000 have V< 9 and these are evenly spread over the sky. About 21000 of these stars are IRS, the other 13000 coming from the first two proper motion programmes in Table 1. The square clumpy regions are selected Schmidt fields, whilst the irregular clumpy areas are reference stars in the fields of extragalactic radio sources. The trail of stars running down below the equator near 22^h are the reference stars along the orbit of comet Halley which were observed in connection with the Giotto mission.

8.1 An IRS-type reference frame

The 34000 stars brighter than 9th magnitude should constitute a useful reference frame for photographic astrometry. The density is slightly greater than 1 star per square degree, which is comparable to the IRS; but, of course, it lacks the homogenity of spectral types, as there is a concentration towards F, G and K-types originating from the first two proper motion programmes of Table 1. Also, the streakiness in Figure 3 shows that there are a few areas, particularly around 7^h and 8^h, where the density is less than 1 star per square degree.

Carlsberg Meridian Catalogue No.4

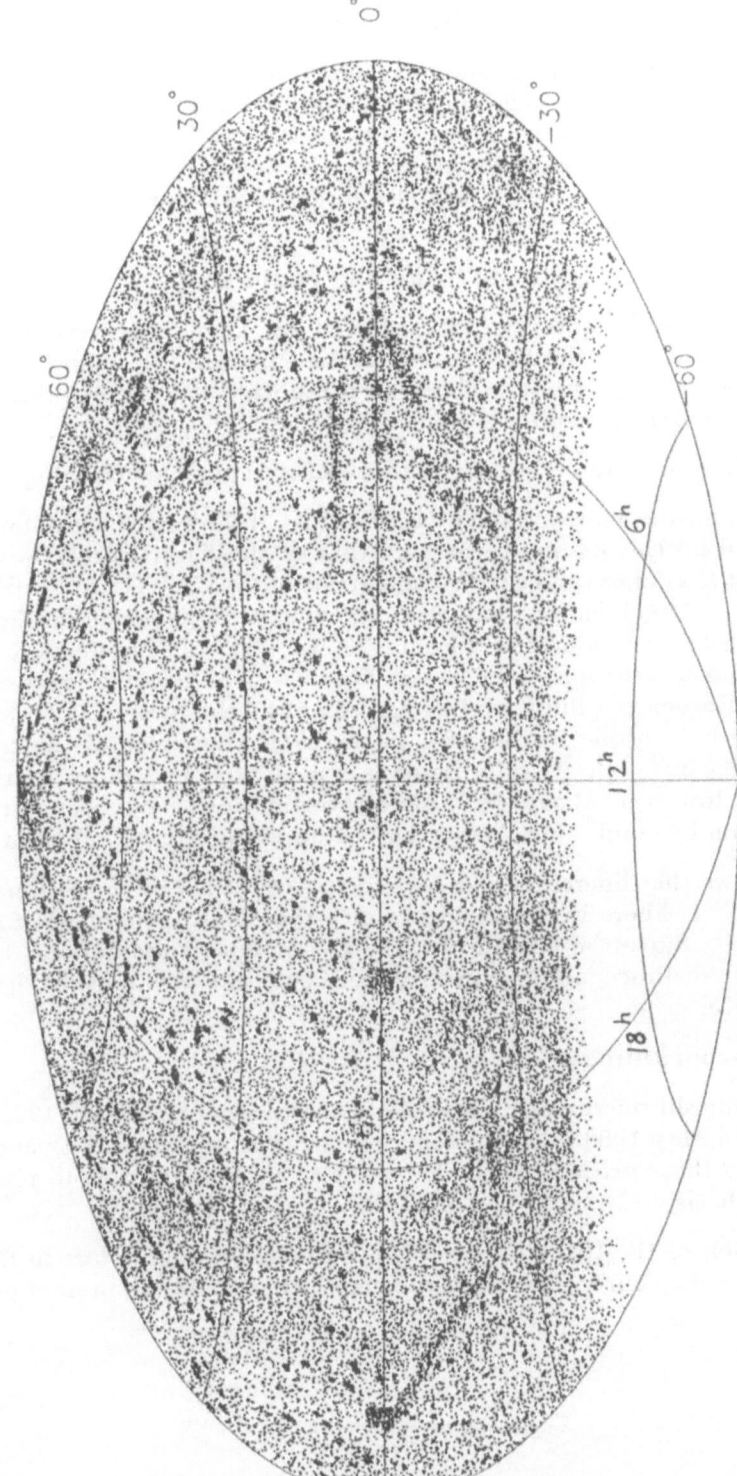

Figure 3. Equal-area plot in right ascension and declination of Carlsberg Meridian Catalogue No.4.

8.2 Link to the extragalactic frame via quasars

As can be seen from Table 1, the programme of observation of reference stars in the fields of extragalactic radio sources, mainly with V>11, is slightly more than half way towards completion. Many fields do not yet have enough observed stars to reduce photographic plates satisfactorily. However, some fields that were given higher priority have been nearly completed, and work is proceeding in taking and measuring plates of these fields.

8.3 Link to the extragalactic frame via radio stars

Observations of radio stars have been carried out, and these have been combined with observations from the Bordeaux meridian circle in a comparative study with VLBI and VLA radio positions. This work is reported elsewhere in these conference proceedings (Morrison *et al.*).

8.4 Reliability of the reference frame defined by this catalogue

Comparisons have been carried out with the FK5 in order to check how closely the frame defined by this catalogue matches that of the FK5. Similar work was also carried out at Bordeaux, and a cross-comparison of the CAMC and Bordeaux frames (Morrison *et al.*1987) showed very good agreement over the the declination range of the comparison which extended from $-20°$ to $+80°$. The systematic differences, $\Delta\alpha_\delta cos\delta$ and $\Delta\delta_\delta$, are reproduced here in Figure 4. The reference frames defined by CAMC and Bordeaux exhibit mutual agreement at the level of $0.''03$, and indicate warps in the FK5 frame reaching $0.''1$ in places at the epoch of the comparisons, 1986. Below $-30°$, the CAMC frame is not so well tied down, particularly in declination. However, above $+80°$ more investigations have been carried out in right ascension by combining observations at upper and lower culmination.

Figure 5 shows the differences $\Delta\alpha_\delta cos\delta$ for CAMC-FK5 in the declination range $+68°$ to $+90°$. There is clear evidence of a systematic difference of $0.''1$ around $+80°$ to $+85°$. Since the CAMC observations at both culminations, and thus at varying zenith distance, are in agreement, this suggests that most of the distortion lies in the FK5.

9. Future programmes

A new scanning slit micrometer was fitted to the telescope in March 1988, and came into full use in May 1988. This micrometer has improved sensitivity and accuracy, as reported in these proceedings by Fabricius *et al.* It is possible to reach V=14.5 on a routine basis.

The completion of the IRS (to $\delta = -45°$) and the reference stars in the fields of extragalactic radio sources are two of the main priorities of the present programme.

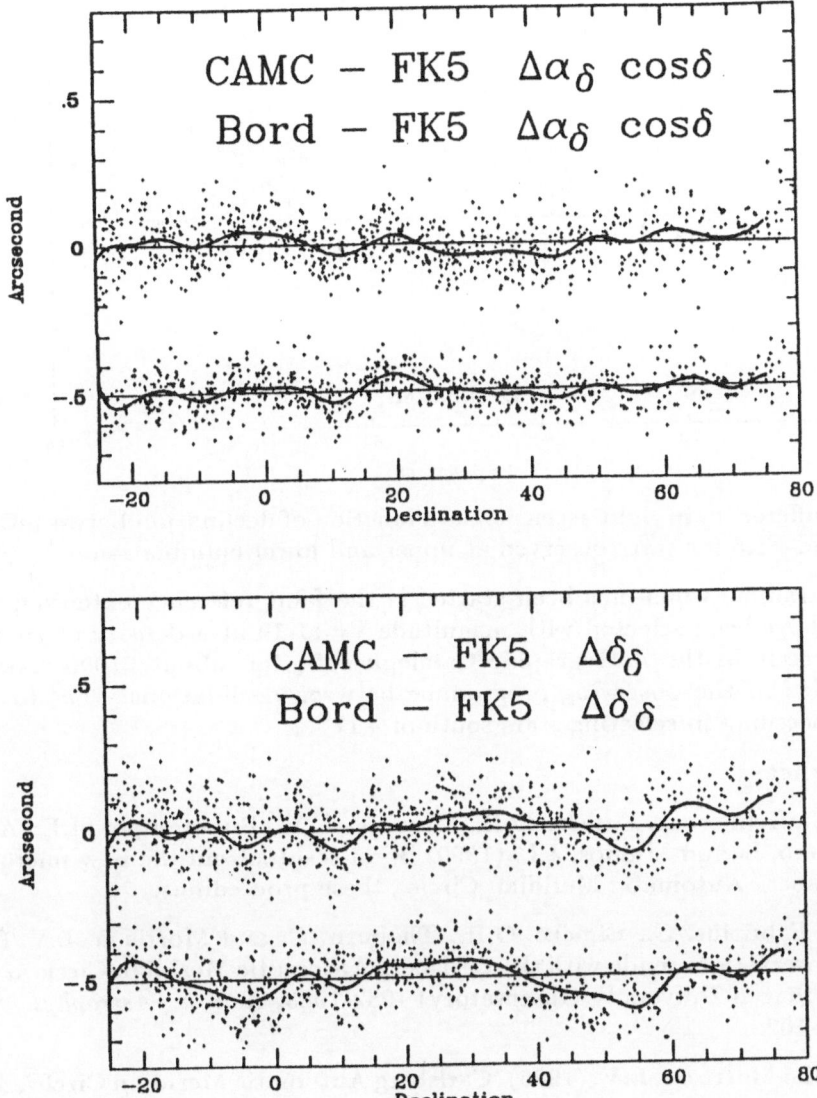

Figure 4. Systematic differences in right ascension and declination as a function of declination between CAMC and Bordeaux and FK5. The Bordeaux differences have been displaced by 0.″5 in order to make the comparison clearer.

400

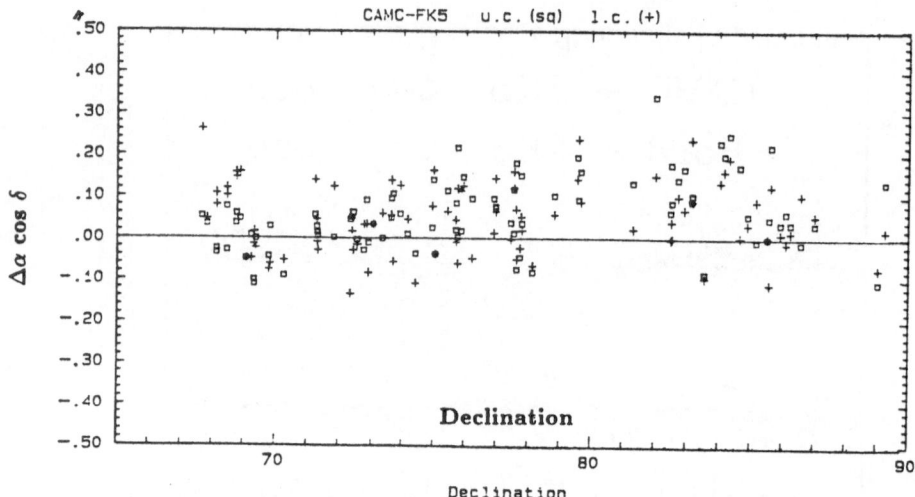

Figure 5. Differences in right ascension as a function of declination between CAMC and FK5 for stars observed at upper and lower culmination.

A new programme which has been started is the faint reference extension to the IRS. Stars have been selected with magnitude V=11-12 at a density of 1 star per square degree from the Astrographic Catalogue. So far, about 21000 stars have been included in the observing programme between declinations −17° to +90°. Work is proceeding in selecting stars south of −17°.

10. References

Fabricius, C., Helmer, L., Einicke, O.H., Morrison, L.V., Buontempo, M.E., Argyle, R.W., Quijano, L. and Muiños, J.L. (1990) 'First results from the new micrometer on the Carlsberg Automatic Meridian Circle', these proceedings.

Helmer, L., Fabricius, C., Einicke, O.H., Thoburn, C. and Morrison, L.V. (1984) 'Meridian observations made with the Carlsberg Automatic Meridian Circle at Brorfelde (Copenhagen University Observatory) 1982-1983, *Astron. Astrophys. Suppl. Ser.*, **55**, 87-102.

Helmer, L. and Morrison, L.V. (1985) 'Carlsberg Automatic Meridian Circle', *Vistas in Astronomy*, **28**, 505-518.

Høg, E. and Fabricius, C. (1988) 'Atmospheric and internal refraction in meridian observations', *Astron. Astrophys.*, **196**, 301-312.

Morrison, L.V., Gibbs, P., Helmer, L., Fabricius, C., Einicke, O.H., Requième, Y. and Rapaport, M. (1990), 'Evidence of systematic errors in FK5', in *IAU Coll. No.100* 'Fundamentals of Astrometry' [Ed. Eichhorn], Belgrade, 1987. Kluwer Acad. Publ., Dordrecht.

Morrison, L.V., Argyle, R.W., Requième, Y. and Mazurier, J.M. (1990) 'Comparison of optical and radio positions of stars', submitted to *Astron. Astrophys.*

Discussion

MIYAMOTO: Concerning the proper motion accuracy to be attained by Carlsberg Meridian Catalogues, could you tell me what kind of data for the first epoch observations will be used?

MORRISON: I have used mainly the large photographic catalogues: Yale, AGK2 and AGK3.

PULYAEV: How many polarissimae stars do you observe every night?

MORRISON: We observe 3 polar stars per hour, so it's approximately 30 per night.

HUGHES: What criterion or criteria do you use for rejecting observations? The genesis of my question is that 7% seems to be a high rejection rate and one might assume that some of the rejects are actually within a reasonable distribution.

MORRISON: The rejection figure of 7% largely comprises unsuccessful observations, such as those that were not properly scanned because of poor input positions. About 1% are rejected from criteria based on the size of the residual.

NEMIRO: Are there plans to use the Carlsberg instrument for determination of absolute positions? If not, it's a pity.

MORRISON: We do carry out observations which can eventually be used in an absolute discussion of the observations. For example, we observe three polarissimae each hour which can be used to define the absolute azimuth.

FIRST RESULTS FROM THE NEW MICROMETER ON THE CARLSBERG AUTOMATIC MERIDIAN CIRCLE

C. Fabricius, L. Helmer, O.H. Einicke
(Copenhagen University Observatory,
Brorfeldevej 23, DK-4340 Tølløse, Denmark)
L.V. Morrison, M.E. Buontempo, R.W. Argyle
(Royal Greenwich Observatory. U.K.)
L. Quijano, J.L. Muiños (Real Instituto y
Observatorio de la Armada en San Fernando. Spain)

ABSTRACT. A new micrometer has been in operation on the Carlsberg Automatic Meridian Circle since May 1988. The zenith mean error for one observation has improved from 0."19 to 0."14 and the limiting magnitude from 13^m2 to 14^m8. The first meridian observations of Pluto and observations of nine extragalactic objects are briefly discussed.

1. INTRODUCTION

The Carlsberg Automatic Meridian Circle has been in operation on La Palma since May 1984. A new photoelectric slit-micrometer has been in use since May 1988, introducing several new ideas. The slit plate, which includes the moving part of a linear encoder, is polished underneath and slides on two fixed pads. A diaphragm uncovers shorter or longer sections of either of the two slits. The photomultiplier tube is cooled to -30°C, and the double window in the cooling box comprises a Fabry lense and a filter. The small number of optical surfaces, all coated, ensures the maximum transmission of light.

2. PERFORMANCE OF THE MICROMETER

Significant improvements have been obtained in limiting magnitude as well as in mean error. The limiting magnitude is now about 14^m8 as compared with 13^m2 for the previous micrometer. The mean error for one observation of a star brighter than $m_v = 11$ and near the zenith is now 0."143 for right ascension and declination and 0^m042 for the magnitude. The old micrometer gave 0."188 in position and 0^m073 in magnitude. Further improvements of the micrometer are planned and it is hoped to reach a limiting magnitude of 15^m5 in 1991.

J. H. Lieske and V. K. Abalakin (eds.), Inertial Coordinate System on the Sky, 403–405.
© 1990 *IAU. Printed in the Netherlands.*

3. MERIDIAN OBSERVATIONS OF PLUTO

Between February and June 1989, 32 observations of Pluto were carried out. Pluto was about $m_v = 14^m.0$ and was observed 30° from the zenith. A *preliminary* comparison with the JPL integration gives residuals, CAMC-DE200, of +1".7 in R.A. and -0".4 in DEC. Observations made in 1987 by Barbieri et al. (1988), using AGK3 reference stars, yield residuals of +0".79 and -0".17 (there is an error in their printed residuals). A new ephemeris (Standish 1989), DE202, shows a difference, DE202-DE200, for 1989 of +1".2 and -0".5 for R.A. and DEC. Standish estimates that DE202 could be off by 0".5 due to systematic errors in the astrometric catalogues. This vindicates the CAMC results and demonstrates the need to get positions directly in the FK5 system.

4. OBSERVATIONS OF EXTRAGALACTIC OBJECTS

In the program for the new micrometer a list of QSO's and BL Lac's was included. Nearly all objects in the list of Argue et al.(1984) are too faint for meridian circles and most alternative objects lack accurate radio astrometry. Table 1 below gives the nine objects observed so far, and a preliminary comparison with radio astrometry from Walter (1989). Five of the objects are known to be radio sources.

TABLE 1. QSO's and BL Lac's observed with the CAMC

Name	m_v	$\Delta\alpha$	$\Delta\delta$	Remarks
0754+394	14.39			Radio
0844+349	14.71			
1101+384	13.55			Radio, BL Lac
1211+143	14.11			
3C273B	12.97	+".09	-".16	CMC4-Walter
3C273B	13.01	+".20	+".01	CMC5-Walter
1254+571	13.88			Radio
1440+356	14.63			
1652+398	13.86	+".21	+".22	CMC5-Walter
1718+481	14.45			

5. REFERENCES

Argue,A.N. et al.:1984, *Astron Astrophys* **130**,191
Barbieri,C., Benacchio,L., Capaccioli,M. and Gemmo,A.G.:
    1988, Astron. J. **96**,396
Carlsberg Meridian Catalogue, La Palma, No.4 and No.5: 1989-
    1990, CUO, RGO and ROA, (in prep.)
Standish,E.M.:1989, (private communication)
Walter,H.G.:1989, *Astron Astrophys Suppl Ser* **79**,283

Discussion

STANDISH: What determines which asteroids are observed? That is, is it possible to request observations of specific asteroids?

FABRICIUS:: Yes, proposals are welcome.

MORRISON: To supplement the reply by Fabricius, there are at present 62 of the brighter asteroids in the CAMC programme which are included in the HIPPARCOS mission.

MORANDO: Actually, the number of asteroids in the HIPPARCOS mission has been reduced from 62 to 48.

YATSKIV: What is your limiting magnitude?

FABRICIUS: The limiting magnitude is 14.8.

PINIGIN: How do you guide CAMC for such weak objects?

FABRICIUS: It is done automatically with pre-calculated ephemerides for the telescope pointing positions.

THE CONSTRUCTION OF 4 DENSE ASTROMETRIC STANDARD AREAS WITH THE CARLSBERG AUTOMATIC MERIDIAN CIRCLE

G. Chiumiento[1], C. Fabricius[2], M.G. Lattanzi[1,3], G. Massone[1]

[1] *Torino Astronomical Observatory, 10025 Pino Torinese (TO), Italy*
[2] *Copenhagen University Observatory, Brorfelde, Denmark*
[3] *Space Telescope Science Institute, Baltimore MD 21218, USA*

High precision work in photographic astrometry requires accurate monitoring of the telescope+filter+emulsion system being used.

Usually, this is done via calibration stars in standard regions, i.e., by assuming the geometry of the field as perfectly known during the reduction of plates specially taken for this purpose. Moreover, such areas can serve as testing regions for plate modelling improvements. The available material is either quite old or only covers small areas of the sky.

The high accuracy required in linking the Extragalactic Reference Frame and the Optical Reference Frame, the recent acquisition of fully digitized deep Schmidt sky surveys and the HIPPARCOS mission are demanding a decisive improvement of the present situation.

Our project is the realization of 4 astrometric regions by using the Carlsberg Automatic Meridian Circle (CAMC) located at La Palma, Canary Islands (Spain). The selected fields and their main characteristics are given in the table below. These areas are located near the equator and uniformly distributed in R.A. The star density varies across the areas (denser toward the centre) because we also require different concentric fields inside the areas (i.e., the fields of long-focus and wide-field telescopes) to be uniformly covered with adequate density. Approximate positions and magnitudes (close to the V band) are from the Guide Star Catalog (Space Telescope Science Institute, 1989). The magnitude range of the selected stars is between 10.0 and 14.5^m.

Proper motions will be determined for stars found in the Astrographic Catalogue. Precise determination of B and V magnitudes is also planned.

STANDARD AREAS	FIELD CENTRE R.A. (J1989.5) DEC.		SIZE (DEG)	NUMBER OF STARS
QSO 0111+021	$1^h 13.2^m$	$2° \ 18'$	7x7	316
NGC 2244	6 31.8	$4° \ 52'$	10x8	388
QSO 1148−001	11 50.2	$-0° \ 20'$	7x7	≈310
IC 4756	18 38.4	$5° \ 26'$	8x8	324

J. H. Lieske and V. K. Abalakin (eds.), Inertial Coordinate System on the Sky, 406.
© *1990 IAU. Printed in the Netherlands.*

PROPER MOTIONS WITH RESPECT TO THE EXTRAGALACTIC REFERENCE FRAME

A. R. Klemola
Lick Observatory, Board of Studies in Astronomy and Astrophysics
University of California, Santa Cruz, California 95064 U.S.A.

ABSTRACT. The Lick proper motion program, one of several using galaxies as a reference frame, is summarized with a statement of work accomplished for the non-Milky Way sky. The problem of identifying relatively transparent regions at low galactic latitudes is discussed, with tabular results presented for 41 windows from the literature having observable galaxies. These fields may be helpful for attaching stellar proper motions directly to the extragalactic frame.

1. Introduction

The measurement of proper motions for galactic stars with respect to an inertial frame of reference is a long-sought goal of astrometry. The absence of readily observable representatives of an inertial system led early astrometrists to devise various well-known indirect approaches with results of various degrees of success. By the end of the second decade of this century, the extragalactic nature of the nebulae was finally becoming clear, as well as their value as essentially inertial reference points for astrometry.

During the 1930's two major undertakings were begun based on photography with wide-field astrographs. One was at the Lick Observatory, with a new 51-cm astrograph completed in 1941 but not fully operational until 1947 (Wright 1950). The other was the Pulkovo program based on existing normal astrographs at several institutions (Zverev 1940), following the first successful use of photography to measure stellar proper motions by Deutsch (1937). In 1964 a third entry to this group was the Yale-Columbia astrograph at Leoncito, Argentina, now functioning as a cooperative Yale and San Juan effort for the southern sky. The details of the various programs have been presented in numerous reports, including the broad summary by Vasilevskis (1973). Other participants involved with photographic astrometry of extragalactic frame include Kiev and groups using the Tautenberg Schmidt telescope. Details of each program are described best by members of each group.

A brief description of the Lick program is given, followed by a consideration of the low-galactic latitude zone, nominally too deficient in galaxies for direct attachment of proper motions to the extragalactic frame with astrographs. A survey of the literature for regions of low transparency (windows) is presented, many of which contain faint galaxies potentially useful for astrometry.

407

J. H. Lieske and V. K. Abalakin (eds.), Inertial Coordinate System on the Sky, 407–417.
© 1990 *IAU. Printed in the Netherlands.*

2. Lick Northern Proper Motion (NPM) Program

The Lick NPM program has been described on numerous occasions over some 40 years, starting with the announcement of the commencement of work (Shane 1947) to the first major results (Klemola et al. 1987; Hanson 1987) and the most recent progress report (Klemola 1989). Its goal is the construction of a catalog of stellar positions and absolute proper motions, using galaxies as the reference frame, for the sky north of declination $-23°$ (Wright 1950).

The NPM program, with 1246 fields each 6° x 6° photographed with the 51-cm double astrograph, consists of two parts (Fig. 1 in Klemola et al. 1987). The first part consists of the 903 fields outside the Milky Way, where measurements have been completed for 741 fields. Measurements are in progress for the remaining 162 fields. The second part, consisting of the Milky Way with 343 more fields, remains for future work.

Details of the observations, input catalogs, star selection, plate measurement and reductions are given in Klemola et al. (1987). The core of the NPM program lies in its two types of input catalogs, as these will govern the uses of the derived proper motions. One is the *Input Catalog of Catalog Stars (ICCS)*, a subset from the AGK3 for the sky north of declination $-2.5°$ and a subset from the SAO south to $-23°$. These catalogs are used as reference stars for obtaining equatorial coordinates, with the AGK3 used also for intercomparison of proper motions for finding corrections to precession.

The other is the *Input Catalog of Special Stars (ICSS)*, with 109000 entries (including many duplicates) from 608 references located from systematic searches of the astronomical literature. The classes of selected objects cover the entire range encounterd in astrophysical studies (Table II in Klemola et al. 1987). The very strong concentration of many classes of stars towards the galactic plane means that proper motions for a major part of the *ICSS* will *not* become available until the completion of the Milky Way phase some years hence. The completion of the NPM program will also allow intercomparison of the Lick motions with numerous other astrometric catalogs for error studies (Klemola 1989).

First application of Lick proper motions to galactic dynamics was performed following completion of measurements and astrometric reductions for 617 fields for the sky from declination $-3°$ to $+68°$ (Hanson 1987). The results for solar motion and galactic motion, though still provisional until the completion of the non-Milky Way phase, clearly demonstrate the value of wide-field photographic astrometry for resolving central problems of galactic motions. Hanson (1987) found for the Oort constants: $A = +11.31\pm1.06$ and $B = -13.91\pm0.92$ km s^{-1}kpc^{-1}, interpreted as a nearly-flat galactic-rotation curve with local circular velocity near 200 km s^{-1}. Solar motion displayed a well-defined trend, starting with an apex near the standard position, using low-latitude stars, to apexes trending towards the direction of galactic rotation, using stars in increasingly higher latitude zones. For a *single* NPM field the absolute zero-point error in attaching proper motions to the extragalactic frame is 0.''2 cent^{-1}. The final overall systematic zero-point error from the entire NPM program for 903 fields with galaxies for almost 2/3 of the sky will be some small fraction of this.

3. Astrometry at Low Galactic Latitude

The great abundance of astrophysically interesting stars in the Milky Way makes this band extremely valuable for galactic studies. It provides the astrometrist with the strongest motivation for measuring accurate absolute proper motions to the faintest levels of apparent magnitude. An impediment to this goal is caused by the intervening galactic interstellar absorption. The present work is the systematic survey for "windows" of reduced absorption, which may be of use for astrometry.

3.1 PROBLEM OF BRIDGING THE MILKY WAY

Various options exist for obtaining stellar proper motions in the Milky Way zone, approaching an inertial reference frame to varying degrees of success. These include the use of calibrated statistical parallaxes (e.g. Stone 1978), second-order catalogs nominally on a fundamental system, like the AGK3 and SAO (on FK4) and new Heidelberg PPM (on FK5), and soon the Cape CPC2 (on FK5). The frame provided by stars from the fundamental catalog FK5 itself, and by low-latitude radio stars attached to the extragalactic frame, is normally too sparse for direct use in photography. For the future there is promise of optical interferometry linking wide intervals of the sky to high precision, a potentially valuable means of bridging the Milky Way with a suitably dense network of reference stars. A future *Hipparcos*-type program would be valuable. Without going into the relative merits of these methods, we consider here the possibilities of *direct* attachment of low-latitude stars to the extragalactic frame.

Proper motion programs with wide-field astrographs are in progress at several observatories (e.g. Kiev, Lick, Pulkovo, Tautenberg, Yale-San Juan). These make explicit use of faint galaxies for fixing proper motions to the inertial reference frame for the 70–80% of the sky lying outside the Milky Way. However, an important limitation to these programs is the reduced surface density of reference galaxies, even being totally absent over extensive areas at low galactic latitudes with strong interstellar absorption. This limitation is an artifact, arising from observations made in the usual optical domain, a limitation not shared for observations made in certain far-infrared wavelength windows and the radio region, which are relatively unaffected by the intervening galactic absorbing medium.

A study of the wavelength dependence of interstellar reddening demonstrates the advantage of observations made in the infrared domain, compared to usual blue- and visual-band photography. With *relative* absorption set at $A_V = 1.00$ mag. for the V-band (0.553 μm), we note $A_I = 0.46$ mag. in the I-band (0.90 μm), and $A_K = 0.11$ mag. in the K-band (2 μm) (Allen 1976). Since much precision astrometry currently is based on detectors, like the photographic plate and CCD devices operating in the optical to near infrared domain (up to 1 μm), this report will be oriented to this band with its possibilities and limitations

Bridging the Milky Way should be tied to the construction of a network of reference stars, spanning the entire observed range of apparent magnitude and colors. This catalog should have a surface density suitable for small-field detectors (e.g. CCD's). For $|b| \leq 10°$ there are almost 7200 deg^2, so that the size of such catalog would be numerous multiples of this.

3.2 BORDERS OF THE ZONE OF AVOIDANCE

The definition of the *border* of the zone of avoidance, a somewhat subjective concept, is operationally taken as the boundary where suitable reference galaxies for photographic astrometry become insufficient in surface density to permit proper motions to be attached to the extragalactic frame. This contrasts greatly with observations made with CCD's in essentially transparent high-latitude fields, where by magnitude B ~ 23 the surface density of galaxies predominates over the stars. The borders lie between galactic latitudes 8° to 20°, as shown from galaxy counts by Hubble (1934), Mayall (1934), Shapley (1957), Steinlin (1962), and Shane and Wirtanen (1967). The so-called boundary depends on numerous factors, including aperture, field size, wavelength region, and others, contributing to the magnitude limit reached. The problem of image crowding by galactic stars is another factor. Deep photography with the large northern and southern Schmidt telescopes, reaching 2–4 magnitudes fainter than astrographs, would produce a more reduced boundary for the zone of avoidance. A still more radical picture emerges for observations made in the infrared and finally radio regions, where absorption effects finally become non-significant. This illustrates the qualifications needed in defining the edge of the zone of avoidance.

3.3. LOW-LATITUDE OPTICAL WINDOWS

Certain low-latitude regions are sufficiently transparent to reveal galaxies on deep photographic surveys. Other so-called windows are simply directions of low absorption only to some intermediate heliocentric distance, beyond which they become more or less opaque to galaxies lying beyond. A listing of 41 windows with observable galaxies is presented in Table 1 and shown in Fig. 1 as solid points. Published coordinates are converted to equinox 1950 and new galactic coordinates as needed. Some windows are poorly defined on the sky and need re-examination.

An additional 49 regions of relatively low absorption with no information about background galaxies are also plotted in Fig. 1 as open circles. Many of these deserve furthur study for possible detection of faint galaxies suitable for fixing the astrometric frame. Some cited regions overlap with others or are essentially identical. All are retained in this *provisional* compilation. References contributing to this listing of low-absorption regions include Blanco (1988), Blanco and Terndrup (1988), Ichikawa et al. (1982), Terndrup (1988), and others. This compilation of windows of both varieties (with and without detected galaxies) will be discussed in greater detail in a paper now in preparation.

The descriptive names of individual windows in Table 1 are usually those used in publications. Some new names were assigned in the absence of clear and unique designation. In addition, a systematic designation based on the prefix LLW (low latitude window) and new galactic coordinates is assigned to each region. The quantity ρ is the great circle distance (degrees) of the window from the galactic center, while r_o is the galactocentric distance ($= 8.0 \tan \rho$ kpc) computed for fields within 30° of the galactic center. The final column gives a reference code.

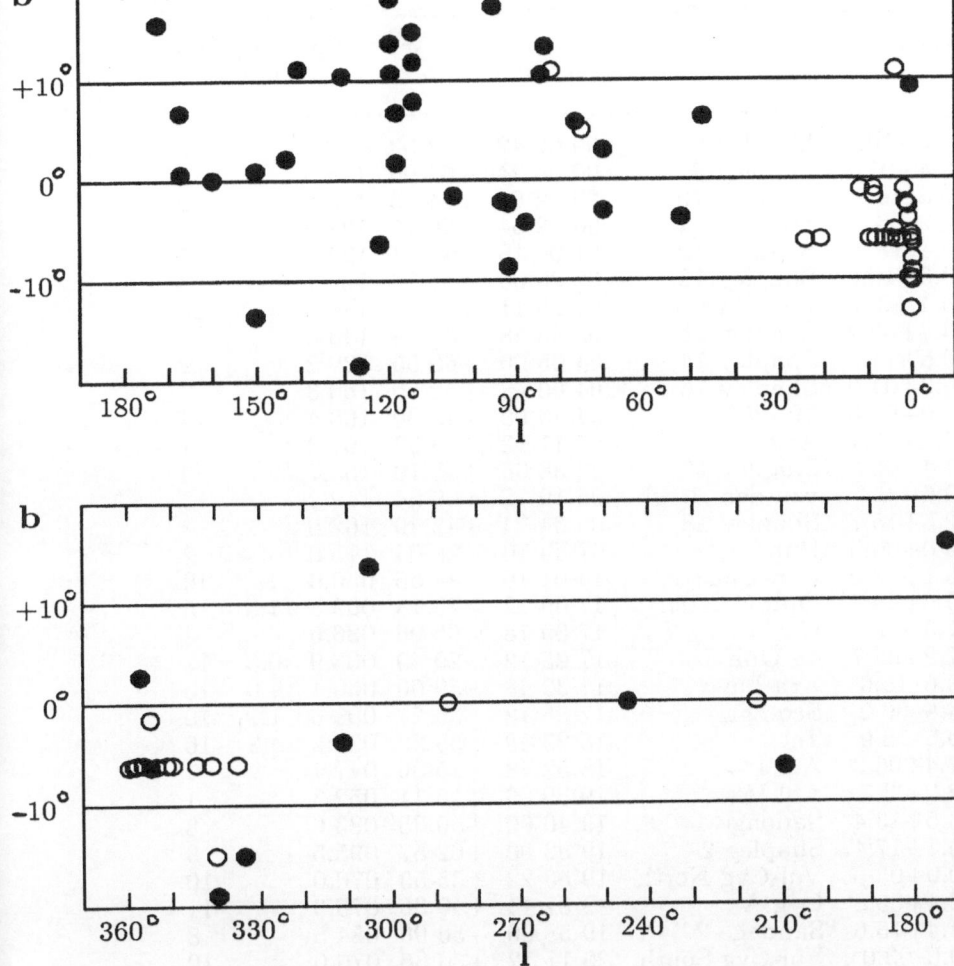

Fig. 1. Distribution of galactic windows (o) with galaxies and low-absorption regions (o) with no reported galaxies. Galactic coordinates (new system).

Table 1. Low-Latitude Windows (Eq. 1950)

Designation	Name	α h m s	δ o ,	ρ o	r_0 kpc	Ref
LLW119.6+10.8	Shapley 8	00 02 42	+73 03	119.0		9
LLW118.4+01.8	Shapley 6	00 06 24	+64 00	118.3		9
LLW119.8−18.2	Shapley 10	00 32 06	+44 15	118.2		9
LLW122.2−06.3	Shapley 11	00 42 54	+56 16	121.9		9
LLW126.8−18.4	Shapley 12	01 09 18	+44 01	124.7		9
LLW130.6+10.5	Shapley 13	02 26 00	+71 39	129.7		9
LLW150.4−13.5	Per Cluster	03 15 11	+41 15	147.8		10
LLW143.4+02.2	Shapley 15	03 33 48	+58 08	143.3		9
LLW140.6+11.2	Shapley 14	04 05 00	+66 55	139.2		9
LLW150.3+01.0	Shapley 16	04 06 48	+52 50	150.3		9
LLW160.0+00.0	3C129	04 46 00	+45 00	160.4		14
LLW167.3+00.7	Aur	05 11 12	+39 53	167.3		1
LLW167.4+06.8	Shapley 17	05 38 06	+43 10	165.7		9
LLW209.6−06.4	van den Bergh	06 19 12	+00 24	149.7		13
LLW172.5+15.7	Shapley 18	06 34 24	+42 49	162.6		9
LLW245.0+00.0	Pup	07 53 10	−28 04	115.0		2
LLW305.1+13.6	Cen Cluster	13 01 16	−48 55	056.0		10
LLW000.6+09.3	Oph Cluster	17 09 12	−23 18	009.3	1.3	5
LLW311.3−03.8	Cir	17 09 18	−65 06	036.6		3
LLW357.2+02.7	45 Oph	17 25 12	−29 49	003.9	0.5	12
LLW333.6−15.0	Ara-Pavo	17 39 48	−59 00	030.1	4.6	10
LLW354.9−06.2	Sco-Sgr	17 55 12	−36 27	008.0	1.1	10
LLW339.6−18.9	Tel	18 23 48	−55 22	027.5	4.2	10
LLW047.4+06.3	Aql I	18 52 48	+15 39	047.7		1
LLW052.2−03.7	Aql II	19 39 06	+15 11	052.3		1
LLW083.5+13.4	Sandage 1	19 40 00	+50 35	083.6		8
LLW095.7+17.4	Shapley 2	19 53 00	+62 57	095.5		9
LLW070.0+03.0	Vul-Cyg North	19 53 22	+33 53	070.0		10
LLW076.4+05.8	Cyg A	19 57 44	+40 36	076.3		11
LLW084.4+10.6	Sandage 2	19 58 00	+50 00	084.5		8
LLW070.0−03.0	Vul-Cyg South	20 17 12	+30 38	070.0		10
LLW088.1−04.2	Cyg Cl	21 19 51	+43 51	088.1		4
LLW092.2−02.3	Markkanen	21 30 24	+48 13	092.5		7
LLW093.7−02.1	Lindgren-Bern	21 35 00	+49 10	093.7		6
LLW092.1−08.6	Shapley 1	21 52 12	+43 13	092.1		9
LLW105.0−01.5	Cep	22 33 36	+56 18	105.0		1
LLW114.7+14.9	Shapley 5	22 35 36	+75 18	113.8		9
LLW114.6+11.9	Shapley 4	22 55 00	+72 36	114.1		9
LLW114.5+07.9	Shapley 3	23 13 12	+68 53	114.3		9
LLW119.7+13.8	Shapley 9	23 54 54	+76 01	118.7		9
LLW118.5+06.8	Shapley 7	23 58 36	+68 57	118.3		9

Table 1 (continued)

References: 1 = Bok (1944); 2 = Fitzgerald (1968); 3 = Freeman et al. (1977); 4 = Huchra et al. (1977); 4 = Johnston et al. (1981), Wakamatsu and Malkan (1981); 6 = Lindgren and Bern (1980); 7 = Markkanen (1978); 8 = Sandage (1976); 9 = Shapley (1935); 10 = Shapley (1957); 11 = Shapley (1957), Spinrad et al. (1982); 12 = Terzan and Ounnas (1988); 13 = van den Bergh (1976); 14 = Weinberger (1980).

3.4 SOME LOW-LATITUDE OPTICAL SURVEYS FOR GALAXIES

Böhm-Vitense (Lick) Low-Latitude Optical Survey. Deep photographs of 52 low-latitude non-stellar objects from the Shane and Wirtanen (1967) galaxy survey show that 44 are galaxies (Böhm-Vitense 1956).

Dodd and Brand (Edinburgh) Low-Latitude Optical Survey: A survey to mag. 22 yielded 29 galaxies in longitude interval l = 245° to 255° in Puppis and Pyxis (Dodd and Brand 1976).

Weinberger (Innesbruck) Very-Low-Latitude Optical Survey. A survey for l = 33° to 213° and b $\leq 2°$ yielded 207 galaxies (Weinberger 1980). This valuable survey reveals possibilities for astrometry. See Pfleiderer et al. (1981).

Hauschildt (Hamburg) Low-Latitude Optical/Radio Survey: Palomar Sky Survey red prints yielded about 260 galaxies, of which 65 form part of the study of a possible extension of the Perseus Supercluster across the Milky Way at l = 140° to 165° (Hauschildt (1987).

3.5 LOW-LATITUDE GALAXIES FOUND AT LONGER WAVELENGTHS

Low-Latitude Galaxies from Radio Surveys. Large numbers of radio sources are known within the Milky Way band ($|b| < 10°$), while systematic surveys now in progress will provide many more, as part of a current drive to obtain the whole-sky distribution of galaxies. Radio surveys have the potential of adding many galaxies for low-latitude astrometry.

Low-Latitude Galaxies from Far-Infrared Surveys. The advantage of working in the far infrared in the low-latitude zone is clearly demonstrated from the absorption penetrating ability of the *Infrared Astronomy Satellite (IRAS)*, observing at the four bands 12, 25 60 and 100 μm. Examination of sources at latitudes $|b| \geq 10°$ yielded many galaxies (Dow et al. 1988), while there is some promise for locating optically hidden galaxies in in lowest latitudes ($|b| < 10°$) despite the great confusion of sources there.

There are unique applications for precision astrometry in high-absorption, low-latitude areas, once infrared detector technology reaches the necessary level, emulating the success of the CCD. There are interesting recent developements with detectors working in the K-band at 2 μm (Koo 1988), where extinction, as noted earlier, is only one-tenth that in the optical domain . The possible role of infrared astrometry to defining the low-latitude reference frame deserves study.

3.6. LOW-LATITUDE INDICATORS OF OPACITY OR TRANSPARENCY

Dark Cloud Surveys as Markers of Regions to Avoid. Dark clouds mark lines of sight where galaxies and QSO's are unlikely to be observable in optical domain. These are compiled for the northern (Lynds (1962) and southern (Hartley et al. 1986) skies, with catalogs of 1802 and 1101 clouds, respectively. Together these catalogs permit delineation of parts of the the low-latitude zone into areas of graded levels of opacity on a scale of 1 (absorption about 1 mag.) to 6 (highly opaque).

Molecular Gas Distribution as Markers of Transparency. Observations of mm-wavelength emissions from the CO molecule is a useful tracer of molecular gas concentrations around the galaxy. Conversely *low* measures of CO emission are useful for pointing out directions with *low* total absorption. A partial survey in the galactic plane in the longitude range l = 4.3° to 90° revealed over 100 small regions with no molecular clouds, which corresponds to total visual absorption A_v under than 1.5 mag. (Verter et al. 1983). These may be directions useful for galaxy surveys.

Galactic H I Reddening Maps as Markers of Transparency. Reddening measurements provide total galactic absorption towards the region. Likewise, neutral-hydrogen column-density measurements relate reasonably well to reddening and, hence, the corresponding absorption. Contour maps of galactic reddening, derived from H I column-density measurements and spatially smoothed using the Shane-Wirtanen galaxy counts, have been constructed by Burstein and Heiles (1982). These H I reddening maps cover the sky for galactic latitudes $10° \leq |b| \leq 65°$. They may be useful for pointing out useful directions for deep galaxy surveys.

3.7. SOME STRATEGIES FOR LOW-LATITUDE ASTROMETRY

Several methods are available, or possibly becoming available, for providing reference objects close to an inertial system within the Milky Way. These include various imaging and interferometric methods in the optical and radio domains relating positions to the global system, as well as the enhanced capabilities of modern meridian astrometry. The present survey supports the option for direct attachment of faint Milky Way stars to the low-latitude extragalactic frame. The role of astrometry conducted in the absorption-penetrating infrared domain remains open to exploitation for the future. Observations from space will likely exert great impact when balanced against many of the ground-based approaches. For now it appears premature exclude any particular promising approach for *current* astrometric observations.

Partial support of this work was provided by National Science Foundation grant AST 88−03326.

REFERENCES

Allen, C. W. (1976) *Astrophysical Quantities* (3rd Ed), (The Athlone Press: University of London), p. 263.

Blanco, V. M. (1988) Astron. J. **95**, 1400.

Blanco, V. M., and Terndrup, D. M. (1989) Astron. J. **98**, 843.

Bok, B. J. (1944) Pop. Astron. **52**, 318.

Böhm-Vitense, E. (1956) Pub. Astron. Soc. Pacific **68**, 430.

Burstein, D., and Heiles, C. (1982) Astron. J. **87**, 1165.

Deutsch, A. N. (1937) Bul. Obs. Central Pulkovo **15**, Pt. 5, 8.

Dodd, R. J., and Brand, P. W. J. L. (1976) Astron. Astrophys. J. Sup. **25**, 519.

Dow, M. W., Lu, N. Y., Salpeter, E. E., and Lewis B. M. (1988) Astrophys. J. **324**, L51.

FitzGerald, M. P. (1968) Astron. J. **73**, 983.

Freeman, K. C., Karlsson, B., Lyngå, G., Burrell, J. F., van Woerden, H., Goss, W .M., and Mebold, U. (1977) **55**, 445.

Hanson, R. B. (1987) Astron. J. **94**, 409.

Hauschildt, M. (1987) Astron. Astrophys. **184**, 43.

Hubble, E. P. (1934) Astrophys. J. **79**, 49.

Huchra, J., Hoessel, J., and Elias, J. (1977). Astron. J. **82**, 674.

Hubble ,E. P. (1934) Astrophys. J. **79**, 49.

Ichikawa, T., Hamajima, K., Ishida, K., Hidayat, B., and Raharto, M. (1982) Pub. Astron. Soc. Japan **34**, 231.

Johnston, M. D., Bradt, H. V., Doxey, R.E., Margon, B., Marshall, F. E., and Schwartz, D. A. (1981) Astrophys. J. **245**, 799.

Klemola, A. R. (1989) IAU Colloq. No. 100, *Fundamentals of Astrometry* eds. H. Eichhorn, C.A. Murray, and A. Upgren (Kluyver Academic Pub., Dordrecht). In press.

Klemola, A. R., Jones, B. F., and Hanson, R. B. (1987) Astron. J. **94**, 501.

Koo, D. (1988) Bul. Am. Astron. Soc. **21**, 114.

Lindgren, H., and Bern, K. (1980) Astron. Astrophys. Sup. **42**, 335.

Lynds, B. T. (1962) Astrophys. J. Sup. **7**, 1.

Markkanen, T. (1978) Astron. Astrophys. Sup. **34**, 181.

Mayall, N. U. (1934) Bul. Lick Obs. (1934) **16**, 177.

Pfleiderer, J., Gruber, M. D., Gruber, G. M., and Velden, L. (1981) Astron. Astrophys. **102**, L21.

Sandage, A. (1976) Pub. Astron. Soc. Pacific **88**, 367.

Shane, C. D. (1947) Pub. Astron. Soc. Pacific **59**, 182.

Shane, C. D., and Wirtanen, C. A. (1967) Pub. Lick Obs. **22**, pt. 1.

Shapley, H. (1935) Harvard Bull. No. 899, 17.

Shapley, H (1957) The Inner Metagalaxy (Yale University Press: New Haven).

Spinrad. H., and Stauffer, J. (1982) Mon. Not. Roy. Astron. Soc. (1982). **200**, 153.

Steinlin, U. W. (1962) Astron. J. **67**, 370.

Stone, R. C. (1978) Astron. J. **83**, 393.

Terndrup, D. M. (1988) Astron. J. **96**, 884.

Terzan, A., and Ounnas, Ch. (1988) Astron. Astrophys. Sup. **76**, 205.

van den Bergh, S. (1976) Astron. J. **81**, 104.

Vasilevskis, S. (1973) Vistas in Astron. **15**, 145.

Verter, F., Knapp, G.R., Stark, A. A., and Wilson, R. W. (1983) Astrophys. J. Sup. **52**, 289.

Wakamatsu, K., and Malkan, M. A. (1981) Pub. Astron. Soc. Japan **33**, 57.

Weinberger, R. (1980) Astron. Astrophys. Sup. **40**, 123.

Wright, W. H. (1950) Proc. Am. Philos. Soc. **94**, 1.

Zverev, M. S. (1940) Astron. J. USSR **17**, 54.

Discussion

BASTIAN: From your graph on the galaxy density distribution it seems to me that the width of the zone of avoidance is generally 20°. Did you ever think of bridging this gap by plate overlap methods? The distance to be bridged amounts to only 3 to 4 of your plate-edge lengths, so you would not need an excessive number of overlap "steps." I do not say it would be easy, but it *might* work.

KLEMOLA: The reply is in two parts. First, the overlap is only 1° for a full 6°x6° plate. This is insufficient for the standard overlap technique. Second, the actual width of the Zone of Avoidance is wider than displayed here because the innermost galaxies are often too weak on astrograph plates. Therefore, we stop at a slightly higher latitude from the equator than the innermost isopleth of 5 galaxies per square degree.

Bronnikova: If you observed in red regions, could you not avoid the problems with the zone of avoidance?

Klemola: I have included the K-band at 2 microns, where the absorption is only one-tenth of the value in the yellow region. Infrared technology is progressing, so that it is becoming useful to consider interesting applications of infra-red astrometry to low latitude problems. There are still important questions as to what will be visible for extragalactic objects at this wavelength.

MURRAY: Have you any plans to obtain colours for your faint anonymous stars?

KLEMOLA: Yes, we do obtain colors from the blue and yellow plates. These are only approximate, owing to inadequate photometric standard stars. Our colors for faint stars are mainly useful for statistical applications.

THE YALE-SAN JUAN SOUTHERN PROPER MOTION PROGRAM (SPM)

W.F. VAN ALTENA, T. GIRARD, AND C.E. LÓPEZ
Yale University Observatory, P.O. Box 6666
New Haven, CT 06511 USA
    and
J.A. LÓPEZ AND E. MOLINA
Observatório Astronómico Felix Aguilar, Universidad Nacional de San Juan
5400 San Juan, Argentina

ABSTRACT. The Southern Proper Motion Program (SPM) is described and progress in the execution of the second-epoch is outlined, as are the reduction methods. Recent changes in the instrumentation, including the addition of a computer control room to the astrograph building and the construction and operation of a new survey machine are discussed.

1. Introduction

The Southern Proper Motion Program (SPM) is the extension of the Lick Observatory Northern Proper Motion Program (NPM) into the southern hemisphere. The scientific goals of the two projects are very similar and have been discussed extensively in the literature by Vasilevskis (1973), Klemola, *et al.* (1987), and Klemola (1988) for the NPM and by Wesselink (1974) and van Altena, *et al.* (1986,1987) for the SPM. The primary objectives are the determination of absolute proper motions of stars with respect to faint galaxies. Given these absolute proper motions, it is then possible to study the errors in the fundamental system of proper motions, by comparing the FK5 proper motions (or the International Reference System motions) with the absolute proper motions for the same stars. While the absolute proper motions will, no doubt, have systematic errors of their own, they will be different from those of the fundamental motions and should therefore help us to understand and correct the errors in the fundamental system of motions. In addition, the new absolute proper motions will provide an enormous data base for the study of galactic structure and solar motion problems, as done recently with the NPM data by Hanson (1987), and for the determination of statistical and secular parallaxes of various classes of stars too distant for the trigonometric parallax method.

2. Instrumentation

The double astrograph is located at El Leoncito, Argentina in the foothills of the Andes mountains at an elevation of 2400 meters and consists of two f/7 51-cm (20-inch) astrographs with plate scales of 55.1"/mm corrected for the blue and yellow light, respectively, mounted together inside a 1.5 meter fork mounted tube. The passband in the blue is defined by the lens transmission and the sensitivity

419

J. H. Lieske and V. K. Abalakin (eds.), Inertial Coordinate System on the Sky, 419–426.
© 1990 *IAU. Printed in the Netherlands.*

of the 17x17 inch 103a-O plate, while the yellow passband is defined by the OG4 (now catalogued as OG515) filter and the 17x17 inch 103a-G plate. The telescope is controlled by an IBM PC computer that sets the telescope on the guide star, while it is guided by an image dissector guider similar to the one used for the Lick astrograph. The IBM PC and all of the control electronics for the telescope are located in a newly constructed control room that connects to the astrograph enclosure. Precision encoders for the right ascension and declination axes have been recently constructed and installed that enable us to accurately set on the guide stars and avoid the time consuming search for the field center in the finder telescope. The computer interfaces, encoders and auxiliary electronics were designed and built by Molina, while the control room was designed by Arq. Maria Rosa Ridl of the Felix Aguilar Observatory and constructed by our workers in residence at El Leoncito.

The limiting magnitude of the astrograph is approximately 19 and 18 on the blue and yellow plates, respectively, with a 4 magnitude objective grating and a two hour exposure. Each plate contains the two hour exposure and a two minute exposure offset from the longer one by approximately two mm. The two hour and two minute exposure durations yield image diameters for the two first-order images of the long exposure that are essentially the same as that for the zero-order grating image on the short exposure. By comparing the average position of the first-order grating images on the long exposure with the zero-order image on the short exposure, we are able to extend the magnitude range of measurable stars from about the 4th magnitude to the 18th magnitude where our reference frame of galaxies is located. We can therefore connect our absolute system of faint star proper motions with respect to galaxies to the bright SRS stars.

It is necessary to survey a plate at low magnification prior to its measurement with the PDS microdensitometer in order to locate and identify the target objects, to evaluate the quality of the images, and to determine appropriate parameters for the PDS scan. The large volume of plate material to be processed in the Southern Proper Motion program has warranted the construction of a high-speed, low resolution measuring machine which may perform as both a wide-field plate surveyor and two-plate blink comparator for 17" x 17" photographic plates. The heart of the survey machine is the Mann measuring engine, which was previously used for the Yale Zone Catalogues. The Mann's precision screw drives have been replaced by high-speed ball screws driven by stepping motors controlled by programmable indexers which produce steps of 2.54 microns and accurate large slews at a speed of 20 mm/sec. A video camera and optical assembly rides on the carriage slung beneath the bridge and provides motion relative to the plate in the x-direction. One photographic plate is mounted below the optical assembly on a carriage table which rides along rails providing motion in the y-direction. The second plate is mounted above the optical assembly on a platform supported by a rigid frame attached to the lower plate's carriage table. An upper plate holder provides rotational and translational fine adjustment for alignment with the lower plate. The optical assembly consists of two plane mirrors to erect the lower plate's image, two objective lenses for adjusting the focus of each plate, a pellicle beam splitter to combine the two plates' images, a fixed collimating lens, a series of three interchangeable camera lenses, and a Panasonic b/w CCD video camera which produces images on a 19" monitor. The operator has a choice of three magnifications which allow fields of view of size 11.5 mm x 8.5 mm, 37 mm x 27 mm, or 93 mm x 66 mm; the three different camera lenses are mounted on pneumatically actuated slides which are controlled from the operator's console.

Movement of the camera and plates is controlled by an IBM PC which is also used to read plate coordinates and to create and manipulate catalogue files of stored object positions. It also provides a means for fast, efficient graphic interaction with the video image by displaying a mouse-controlled electronic crosshair directly onto the video image of the plate. The design and construction of the electronics needed to generate the crosshair are due to Molina. A general purpose survey program has been written by C. López and used to test and calibrate the survey machine. The calibration tests indicate that in high magnification mode a stellar image position can be determined with an accuracy of ±10 microns, (single coordinate standard error), by moving the stellar image until it aligns with the crosshair. The alternate method of determining a position, by moving the crosshair with the mouse until it coincides with the stellar image, is much faster but yields a somewhat lower positional accuracy of ±15 microns in high magnification. C. López is now using the survey machine to determine improved coordinates for all of the southern stars in the variable star catalogue and has obtained an average positional accuracy of 0."7 in both coordinates, which corresponds to 13 microns on the SPM plates. The services of the Gibbs Instrumentation Lab and the Yale Center for Electronic services have been employed to construct this survey machine under the supervision of Girard and van Altena.

3. Research Programs

3.1 Galactic Structure

There are numerous galactic structure problems that can be studied with the absolute proper motions that we will be determining. The most obvious area of investigation is in the determination of Oort's constants of galactic rotation A and B, as has recently been done by Hanson (1987) for the NPM. Another area of study is the current controversy over the existence of the thick disk postulated by Gilmore and Reid (1983) in contrast to the two component disk/spheroid model of Bahcall and Soniera (1980). In the following Table we list the fraction with respect to the disk of thick disk and spheroid stars at visual magnitudes 12.25, 16.25, and 18.25 for galactic latitudes 20, 30, 45 and 90 degrees using the "standard normalization" of the thick disk and spheroid, 2.0% and 0.125%, respectively, from Table 2 in Bahcall (1986).

Table. Fraction of Thick Disk and Spheroid stars to the Disk

b^{II}	V=12.25		V=16.25		V=18.25	
	F_{TD}	F_S	F_{TD}	F_S	F_{TD}	F_S
20°	0.12	0.01	0.26	0.03	0.47	0.05
30	0.16	0.02	0.32	0.05	0.65	0.13
45	0.20	0.03	0.43	0.11	0.77	0.30
90	0.22	0.05	0.56	0.22	0.76	0.61

From the table we can see that at the lowest latitude where there are still a reasonable number of galaxies to set our absolute proper motion zero point, namely 20 degrees, thick disk stars amount to 12%, 26% and 47% of the disk stars in the sample at visual magnitudes 12, 16 and 18, respectively; at high latitudes the fraction is even higher. The significance of this is that the solution for Oort's rotation constants A and B will be biased by the presence of a component that may not have the same rotational component about the galactic center as the disk. In particular, Gilmore and Reid (1983) have suggested that the thick disk lags the disk by about 100 km/sec, while several more recent determinations suggest that the lag may be closer to 30 km/sec, *e.g.* Ratnatunga and Freeman (1989). While the above figures depend on the adopted normalizations, which are still very poorly known, they highlight the point that the derived values of Oort's constants must be interpreted in terms of a galactic model that allows for the presence of some fraction of stars from other components. We will have a distribution of anonymous, randomly picked, stars ranging from the 12th to the 18th magnitudes with absolute proper motions from which we will attempt to determine the rotation constants of the different galactic components. In fact, we should be able to determine the relative fractions and kinematic characteristics of the different components from our absolute proper motions and B,V photometry.

3.2 A New Catalogue of Right Ascensions and Declinations for Faint Stars

Due to the increasing need for a faint reference frame to aid in the determination of the fiber positions in the new fiber optic spectrographs used with some of the large telescopes, we plan to construct a catalogue of faint stars with magnitudes in the range 16 - 18. The requirement is for equatorial positions of the target objects accurate to much better than the fiber size, which is usually $\leq 1"$. At present, there are two principal problems: first, a reference frame from the SRS or the SAO catalogue has a magnitude limit of approximately 9th mag., while the target objects are often 16 and fainter; and second, the density of reference objects is low, one/deg^2 for the SRS and perhaps six/deg^2 for the SAO. In addition, the SAO is not satisfactory in general since the positional accuracy at the current date is usually rather poor due to the old mean epoch of the positions. An additional source is the Guide Star Selection System Catalogue of the Hubble Space Telescope. Unfortunately, there are problems with the GSSS equatorial coordinates near the plate boundaries where the poorly determined high order plate constants on the Schmidt plates can result in systematic errors in the positions that exceed one arcsecond. This problem is a direct result of the lack of a good dense reference system at faint magnitudes. L. Taff (private communication) is exploring ways to modify the reduction procedures with the GSSSC that may yield significant improvements to the systematic accuracy of the GSSSC.

Two projects are currently underway in the southern hemisphere that will improve the reference system substantially, but unfortunately not extend the magnitude limit to the required level. The CPC2 will provide positions accurate to ±0"06 down to the 10th magnitude (de Vegt, 1988 and de Vegt, *et al.* 1988), and a new survey by the U.S. Naval Observatory will extend that magnitude limit to approximately 12 -14 (Routly, 1983 and de Vegt, 1988) at an accuracy of ±0"05. We plan to establish a relatively dense network of equatorial coordinate secondary standards in the southern hemisphere down to the 18th magnitude in the course of the second epoch of the SPM. The density of the stars in the range of 16 - 18 magnitudes would be about 10 - 15/deg^2 with an accuracy of about ±0"10 (s.e.) at a mean epoch ~ 1980. This density of faint secondary standards with absolute proper motions should then be adequate to provide a reference frame in the fields of view of most large reflectors.

4. Star Selection, Measurements and Reductions

We have been selecting stars from the SIMBAD data base for measurement on the PDS. Since SIMBAD includes virtually every object referenced in the astronomical literature, it is an ideal list of astrophysically and astrometrically interesting objects for which absolute proper motions may be of value. We have therefore adopted SIMBAD as our primary source of objects to be measured in the SPM program. Dr. Daniel Egret of the CDS in Strasbourg kindly sent us listings by BITNET of the SIMBAD objects within specific plate boundaries. Using a program written by T.-g. Yang, we extract the data of interest to us from the listing and produce a file that is easily read by other programs. This object file is then analyzed to obtain statistics on the numbers and types of objects present, it also assigns default PDS scan codes and object type codes, and the file is transformed to Survey Machine coordinates. In addition, we are in the process of acquiring a CD reader to read the GSSS catalogue on compact disks and will use it to automatically select an incremental sample of stars to measure down to the GSSS limit of about 15.

The SIMBAD output (and the GSSS supplement) are then downloaded from the VAX to the IBM PC computer that controls the Survey Machine. The PC control program written by C. López, reads the file and the survey machine moves to the first region on the old-new plate pair and the software and mouse controlled cursor on the television screen moves to the first object. Since SIMBAD contains both accurate and poor coordinates (e.g. contrast the SRS stars with variable stars known to an arcminute or worse!) we must correct the coordinates before they can be scanned on the PDS. In addition, we must delete unmeasurably faint or blended objects and objects with defects such as scratches, and also add numerous objects for the astrometric solutions and for our galactic structure research. Each object is therefore examined on both the old and new blue plate in the blink mode; the positions are updated, and the default codes for the object type, PDS scan type and image quality corrected, if necessary, and the updated record is written to an output file. At this time, supplementary objects are added to the list by visual inspection. These additional objects include: a) 80 stars at around the 12th magnitude that will be used for the "Bridge" solutions which relate the short offset exposure to the long main exposure on each plate; b) about 300 stars around the 16th magnitude that will be used as our proper motion reference frame; c) an additional 500 faint stars in the magnitude range 12 - 18 that will be used for galactic structure studies and also form, along with the stars in item b), the faint secondary reference system of equatorial coordinates that we are planning to establish; and d) about 100 galaxies that will define the zero point of our absolute proper motions. Including the objects in SIMBAD, we expect to have a list of approximately 1200 objects to measure on each of our SPM plates. Once the survey of the blue plates has been completed, the yellow plates are inserted and the output list from the blue plates is examined and unacceptable objects are deleted and the coding revised if necessary for the yellow output list. The final blue and yellow lists are then uploaded from the PC to the VAX and reformatted to the PDS input file structure.

The plates are then scanned on the Yale laser interferometer PDS microdensitometer and reduced to coordinates by the digital image centering routines described by J.-F. Lee and van Altena (1983) and further improved by Girard. The accuracy of the derived PDS x,y coordinates is approximately ±0.3 μm (s.e.) according to J.-F. Lee, et. al. (1986), while the intrinsic accuracy of the emulsions is about 1 μm, therefore we are not limited by the measuring machine.

The plate reduction procedures begin with a detailed examination of the image centering error analysis along with the image shape parameters for each image. Images that have poorly determined centers or whose "image shape" deviates significantly from the average star, or galaxy, are deleted from the lists at this point. We have found that this procedure for culling out the bad data greatly improves the solutions and does not bias our results. Our photographic photometry in B and V is obtained from the pseudo-magnitudes produced by the image centering program. While this data is internally quite accurate, there is a problem in relating the pseudo-magnitudes to the photoelectric calibration stars due to the variations in the photometric response over the plate. Our photometry is locally accurate to about 0.10 mag., but regional variations over the plate can create systematic errors of 0.5 mag. We are exploring methods to reduce those systematic errors.

Since all SPM exposures are taken with the objective grating in place, each of the brighter stars consists of the zero-order image flanked by symmetrically located grating images reduced by approximately 4 magnitudes. Due to the design of the grating, and the quality of the lenses, it is sometimes possible to measure the zero-order image and the grating images out to the third order, yielding a total of seven images for each exposure. In most cases however, only the first- or first- and second-order images can be measured, but this still yields two or four images instead of a single image. In order to take advantage of the possible use of the multiple images, the plates are scanned in an orientation such that the scan direction is parallel to the grating dispersion and with a scan length sufficient to encompass all grating images out to the third-order. The loss in time due to the long scan length is only about one and a half times the normal scan time, since most of the time in scanning an image is spent in reversing the carriage at the end of each line. The seven grating images are then split up in the image centering software and a center determined for each image separately. The symmetrically located grating images are then averaged to yield one position for each order, and the orders are averaged to yield one position with a "weight" equal to the number of orders averaged. We also archive the individual order positions for the study of systematic errors between the orders and as a function of the image diameter on the plate.

At this point we correct for the effects of atmospheric refraction and then transform the short exposure to the system of the long exposure using the "Bridge Stars" with magnitudes in the range of 10 to 13. The exact magnitude is unimportant, only that the transformation be made without any loss in accuracy and without the introduction of a systematic error, dependent for example on the brightness of the star. We therefore select stars which have image diameters in the long exposure first-order that are equal to the image diameter of the zero-order short exposure. We choose about two bridge star/deg^2 well distributed over the plate, or about 80 stars, which then introduces a zero-point error into the transformed final proper motions of the bright stars relative to the faint stars of approximately 0.4 mas/yr, which is about ten times smaller than the accuracy of our proper motions.

The determination of the relative proper motions can in principal be done at any magnitude, however since we must use the reflex proper motion of the galaxies to set the zero point of the motions, this should be done at the magnitude of the galaxies to minimize the chance of introducing a magnitude equation into the absolute proper motions. In contrast to the bridge solution, we must compromise here since as we go to fainter magnitudes the distribution and selection of galaxies improves, but the accuracy of measurement decreases rapidly as we approach the plate limit. The compromise magnitude range used by the Lick astronomers is around 15 to 17 and ours is similar. We plan to select about 300 faint reference stars for the proper motion solutions, which will require plate constants up to at least the second order in the coordinates. In addition, the real tangential velocities of the stars introduce a "cosmic dispersion" or noise that requires numerous stars to achieve a reliable average motion. We have adopted 300 faint reference stars, which is more that needed, but it will add to the density of faint stars in our final secondary equatorial coordinate system. We select all measurable galaxies that we can find in the survey process, most of which will not be in SIMBAD. The plate constants derived for the faint stars are then used to compute the relative proper motions for all stars and galaxies, and the average of the latter is used to correct the relative proper motions to absolute. The final step in the reductions is to transform the rectangular coordinates into the system of the SRS and calculate Right Ascensions and Declinations for all of the objects. In general, we find about 40 measurable SRS stars on a plate, with an average unit weight error of $0\rlap{.}''12$ in the transformations.

The repetition of the second-epoch SPM plates is being done at a rate that will enable us to maintain a 20 year epoch difference and keep the accuracy of the proper motions uniform over the southern sky. To date, approximately 60 regions have been repeated and we are just now starting the measurements and reductions. As mentioned earlier, the error introduced by the Bridge transformation appears to be about 0.4 mas/yr, which is about one-tenth of the accuracy of our proper motions. The zero-point error in our faint star reference system depends on the cosmic dispersion of the proper motions, which for the first regions at mid galactic latitude is about 1.0 mas/yr, while the accuracy of the relative proper motions at $V = 14$ is approximately 3 mas/yr. On the other hand, the dominant error is that introduced by the galaxy reflex proper motion, which depends critically on the number of measurable galaxies found in each region. That number can be depressingly small and will limit the zero-point of the absolute proper motions in each region to an accuracy of 1 to 3 mas/yr, depending on the number of galaxies.

5. Acknowledgements

We are indebted to a large number of individuals for their contributions to the SPM project. In particular we would like to acknowledge the observers at El Leoncito, past and present, the staff members of the Universidad Nacional de San Juan for their invaluable help, both financial and moral, the electronic and mechanical instrument shops at both the UNSJ and Yale, and Ting-gao Yang for his contributions to the extraction of data from the SIMBAD data files. Finally, we gratefully acknowledge the financial support of the U.S. National Science Foundation and the UNSJ.

6. References

Bahcall, J.N. (1986). *Ann. Rev. Astron. Astrophys.* 24, 577.

Bahcall, J.N., and Soneira, R.M. (1980). *Astrophys. J. Suppl.* 44, 73.

de Vegt, Chr. 1988, "Status of photographic catalogs: available material and future developments", in S. Débarbet, J.A. Eddy, H.K. Eichhorn and A.R. Upgren (Eds.), *IAU Symposium 133, Mapping the Sky*, Kluwer Academic Publ., Dordrecht, p. 211.

de Vegt, Chr., Zacharias, N., Penston, M.J., and Murray, C.A. 1988, "Current status of the Second Cape Photographic Catalogue", in S. Débarbet, J.A. Eddy, H.K. Eichhorn and A.R. Upgren (Eds.), *IAU Symposium 133, Mapping the Sky*, Kluwer Academic Publ., Dordrecht, p. 415.

Gilmore, G., and Reid, I.N. (1983). *Monthly Notices Royal Astron. Soc.* 202, 1025.

Hanson, R.B. 1987, *Astron. J.* 94, 409.

Klemola, A.R. 1988, In *Fundamentals of Astrometry*, IAU Colloquium No. 100, edited by H.K. Eichhorn, C.A. Murray, and A.R. Upgren, *Celestial Mech.* (in press).

Klemola, A.R., Jones, B.F. and Hanson, R.B. 1987, *Astron. J.* 94, 501.

Lee, J.-F., Tsay, W.S., and van Altena, W.F. 1986, In *Astrometric Techniques*, IAU Symposium No. 109, edited by H.K. Eichhorn and R.J. Leacock (Reidel, Dordrecht), p. 237.

Lee, J.-F., and van Altena, W.F. 1983, *Astron. J.* 88, 1683.

Ratnatunga, K., and Freeman, K., 1989, *Astrophys. J.* 339, 126.

Routly, P.M. 1983, in *Sky with Ocean Joined*, (S.J. Dick and L.E.Doggett, Eds.), p. 145.

van Altena, W.F., Girard, T., and López, C.E. 1987, In *Fundamentals of Astrometry*, IAU Colloquium No. 100, edited by H.K. Eichhorn, C.A. Murray, and A.R. Upgren, *Celestial Mech.* (in press).

van Altena, W.F., Girard, T., López, C.E., Klemola, A.R., Jones, B.F., and Hanson, R.B. 1986, *Highlights in Astron.* 6, 89.

Vasilevskis, S. 1973, *Vistas Astron.* 15, 145.

Wesselink, A.J. 1974, In *New Problems in Astrometry*, IAU Symposium No. 61, edited by W. Gliese, C.A. Murray, and R.H. Tucker (D. Reidel, Dordrecht), p. 201

Discussion

SMITH: Is the error quoted for the SRS at 1989 in fact the error of the Perth 70 catalog at 1989?

VAN ALTENA: Two HST calibration regions at $\delta \approx -60°$ yielded an error of ±0.25 arcsec, while the first of the SPM regions at $\delta = -35°$ gave ±0.38 arcsec for the 1989 error of the SRS.

RÖSER: You showed the plot "Peak density of image" versus "radius of image." How is "radius" defined in this plot?

VAN ALTENA: The radius is the "Gaussian radius" from a bivariate fit to the measured densities.

CORRECTION TO ABSOLUTE PROPER MOTION USING THE IAS-GALAXY MODEL

KAVAN U. RATNATUNGA
NASA Goddard Space Flight Center
Greenbelt, Maryland 20771
U. S. A.

ABSTRACT. The IAS-Galaxy model (Ratnatunga, Bahcall and Casertano 1989) is a software interface between theoretical models of the Galaxy and observed kinematic distributions. It has been developed for analysis of many kinematic catalogs to study global galactic structure. In addition, the IASG model can be used to estimate corrections needed to derive absolute parallax and absolute proper motion by evaluating, on a star-by-star basis, the expected mean motion of the reference stars.

A theoretical Galaxy model is defined on an inertial coordinate frame.* Proper motions are measured in a reference frame defined by a fundamental catalog. The observed distribution of proper motions in star catalogs can be directly compared with the expected distributions evaluated using IASG to check the accuracy of the adopted reference frame in realizing the inertial coordinate frame in the sky.

1. Introduction

Standard astrometric reduction of an observation series over a number of years in a small region of sky gives a relative parallax and a relative proper motion, with associated errors of measurement, for each of the stars in the field of view.

The reference frame is defined using quasars/galactic nuclei and/or FK5 stars. It is rare to find a sufficient number of very distant reference objects in the same field of view as the program star whose parallax and/or proper motion is being measured, to estimate the absolute motion directly in the adopted reference frame.

Correction to derive absolute proper motion often needs to be based on the expected proper motions for the stars in the neighboring region of sky. With improvements in precision of the measurements reducing random errors, the accuracy of the correction to absolute is now important to avoid systematic errors.

2. IASG Galaxy Model

2.1. GOALS

The IAS-Galaxy model is an attempt to build our current picture of the Galaxy in a computer. The program is basically a "software telescope" that can project an assumed empirical or theoretical model of the distribution function to derive expected distributions

---

* The terminology used in this paper was edited to be consistent with that adopted by G. A. Wilkins in the concluding paper of this conference.

J. H. Lieske and V. K. Abalakin (eds.), Inertial Coordinate System on the Sky, 427–429.
© 1990 *IAU. Printed in the Netherlands.*

of all directly observed quantities within any combination of selection limits. Currently, the model is, to a large extent, empirical with a gradual change to include more theoretical constraints.

The final goal is to converge on a model that fits many different sets of observations, cataloged within various selection criteria, and to use self consistently, the same model, to estimate the corrections needed to evaluate absolute proper motion and parallax.

2.2. MODEL

The model for the Galaxy is based on assumed density and kinematic distributions. The density model is derived from the observed colors and apparent magnitudes using a spectrophotometric calibration, which in turn is based on the parallax of the nearest stars. The calibration is extended to both fainter and brighter absolute magnitudes using color-magnitude diagrams of star clusters. The kinematic model is based on line-of-sight velocities that are distance independent, as well as independent of the constant of precession. The observed proper-motion distribution needs to be consistent with the density model as well as the kinematic model.

The reference system used to construct the reference frame for proper motions, includes an estimate for the constant of precession to remove the drift caused by the slow movement of the Earth's axis of rotation. This estimate must also be based on a kinematic Galaxy model. In most analyses the velocity distribution function was assumed isotropic. Eichhorn (1974) warns of the error that the observed anisotropy in velocity dispersions could introduce to the estimated precession constant.

Use of many different Galactic models for different aspects of the data processing, does not assure self consistency. The model assumed to evaluate the corrections needs to be consistent with the observed distributions. Simplifications are no longer needed to do the required computations. As our knowledge of the distribution function of our Galaxy is improved, we must feedback these improvements in estimating corrections to derive absolute parallax and absolute proper motion and to evaluate the constant of precession.

2.3. EVALUATIONS

Proper motions are measured in a reference frame, that is constructed using a reference system, which assumes a Galaxy model, defined on an inertial coordinate frame in the sky. IASG can be used to check the accuracy of the reference frame in realizing the inertial coordinate frame. For example, to check for any residual rotation of the adopted reference frame, relative to the inertial coordinate frame, about any axis, we first select stars that are located on an equatorial band in a spherical coordinate system with the pole along that axis. We then compare the observed distribution of the component of proper motion for stars perpendicular to this axis, with the expected distribution evaluated using IASG. A preliminary comparison using proper motions from the AGK3 (corrected using improved IAU 1976 constant of precession) is seen to fit with a zero point error less than one millisecond of arc per year, for the three principle axes of the Galactic coordinate system.

To estimate the correction required to derive absolute parallax and/or absolute proper motion, IASG evaluates independently for each star, the expected parallax and proper motion distributions. IASG uses all of the available data, such as the galactic coordinates and color-apparent magnitudes of the reference stars, and integrates over the error distribution of each observable and over the total expected range of any observable that is unavailable.

The expected distributions are evaluated by direct projection of the assumed distribution function to the observed plane. This is essential to estimate the optimum correction using maximum likelihood. Monte Carlo estimates can not be used within an iteration loop used to maximize the likelihood function.

2.4. LIKELIHOOD

Numerical testing shows that maximum likelihood give unbiased estimates for the correction needed to derive absolute parallax, even when the program star has a parallax comparable to those of the reference stars. Conventional estimates are significantly biased in such a case, which could arise when the limiting distances of parallax measurements are extended by observations from space.

For example, at high galactic latitude, consequent to the volume sampled and the disk density law, most disk stars are contributed at a distance of order 1 kpc at all apparent magnitudes. Therefore when measuring stars with a parallax smaller than a few milliseconds of arc, the bias caused by not using maximum likelihood is numerically found to be of order a few tenths of a millisecond of arc. The bias also depends on the uncertainty in the absolute magnitude of the reference stars. The bias being larger when only apparent magnitudes or just a mean apparent magnitude is available for the reference stars.

Re-evaluation of the corrections used to derive absolute proper motion and absolute parallax requires all available information regarding the reference stars and should be published along with each proper motion or parallax measurement. The corrections can then be re-evaluated to be consistent with any revised distribution function or improved when more information is available on the reference stars.

3. Summary

The kinematic Galaxy model assumed to estimate the corrections used to derive absolute proper motion and absolute parallax, as well as the model used to estimate the constant of precession, needs to be self consistent with the observed distributions.

The IASG Galaxy model is useful to evaluate expected distributions of all directly observed quantities in any star catalog, as well as the expected probability distributions of proper motions and parallax for each reference star, within the available information.

When the proper motions and parallaxes of the reference stars are comparable to those of the program stars, and measurement errors are not dominant, a detailed statistical analysis is important. Maximum likelihood gives an optimum way to use all available information, including individual error distributions of the observables. Using the full expected distribution and not just a tabulated mean, this procedure gives unbiased estimates of the correction required to derive absolute proper motion and parallax.

4. Reference

Ratnatunga, K. U., Bahcall, J. N., and Casertano, S. 1989, *Astrophys. J.* **339**, 106.

Eichhorn, H. 1974, *Astronomy of Star Positions* (New York: Frederick Ungar), 92.

Discussion

MURRAY: Can you describe the main components of your assumed model of galactic kinematics; in particular, did you allow for a shear in rotation velocity with height above the plane?

RATNATUNGA: The Galaxy is modeled as a sum of discrete density components, each of which is assumed to be isothermal with a mean assymetric drift. The change in the rotation velocity with height from the plane will arise naturally from change in the mixing proportions of the density components with height.

PHYSICAL DATA OF THE FUNDAMENTAL STARS

Luo Dingjiang and Zhang Baocai
Beijing Astronomical Observatory
Academica Sinica
Beijing, China P.R.

ABSTRACT. The precise positions and proper motions (J2000.0) of 1535 Basic FK5 stars in the FK5 system have been used in the reduction of the local vertical monitoring and the catalog observations at Beijing Astronomical Observatory. In addition, similar data for 1987 FK4 Sup stars, in which 980 stars indicated by "F" will be included in the FK5, were made available recently by Heidelberg via Prof. Tong Fu, director of Purple Mountain Observatory. The positions and proper motions of the other 1007 FK4 Sup stars in the FK5 system are provided by the Astronomisches Rechen-Institut with relatively low accuracies. However, the physical data, such as the visual magnitudes and the spectral types of these rs with rather large uncertainties have their origin in the Henry Draper Catalogue since the compilation of the FK4 and the FK4 Sup. More accurate visual magnitudes available in the well-defined photoelectric system and the spectrum types in the MK-system may be found, for example in the Bright Star Catalogue and its supplement.

For the convenience of users of astrometric observations and for geodetic purposes, visual magnitudes (V), color indices (V–B), spectral types (Sp), tritonometric parallaxes (p in mas), and radial velocities (RV in km/sec) for the 1535 FK4 stars and 1987 FK4 Sup stars are available.

These data are available from the authors in machine-readable form which includes the mean positions and proper motions of J2000.0 in the FK5 system and the cross-references to the GC of FK4 stars and FK4 Sup stars. Printed or magnetic versions of the data may be requested from the authors.

REALIZATION OF THE LOCAL INERTIAL GEOCENTRIC FRAME IN RELATIVITY

He Miao-Fu, and Huang Cheng
Shanghai Observatory
Academia Sinica
Shanghai, China, P.R.

ABSTRACT. There are two kinds of geocentric frames: local inertial and non-inertial geocentric frames. Ashby et al successfully constructed a local inertial geocentric frame in the neighborhood of the gravitating Earth. In the frame with origin at the Earth's center, the gravitational effects of the sun and of planets other than the Earth are basically reduced to their tidal forces, with very small relativistic corrections.

However, the spatial base vectors of the local inertial frame essentially experience the geodesic (or deSitter) precession with respect to the solar system barycentric frame. Hence the realization of the local inertial frame requires that the general precession should exclude the geodesic precession. This requirement is inconsistent with the convention that the amount of geodesic precession is included in that of the general precession given by Lieske *et al.*

THE GENERAL CATALOGUE OF STELLAR PROPER MOTION WITH RESPECT TO GALAXIES WITH ASTROPHYSIC SUPPLEMENT

N.V. KHARCHENKO
Main Astronomical Observatory
Ukrainian Academy of Sciences
252127 Kiev
USSR

ABSTRACT. Within the limits of the programme of studying of the Main Meridional Section of the Galaxy the General Catalogue of astrometric data of 26500 stars has been compiled. Corrections to the precession constants and stellar secular parallaxes have been determined.

The programme of studying of the Main Meridional Section of the Galaxy (MEGA) (Einasto *et al.*, 1985) provides for the creation of catalogues of proper motions with respect to galaxies, equatorial coordinates, stellar magnitudes in the UBVR system, absolute stellar magnitudes, effective temperatures, metal contents of all stars down to 12^m in 47 areas of the sky, each of size $5° × 5°$, and astrometric and photometric data only for stars down to 15^m-16^m in areas whose diameters are on the order of $1° \ 20'$.

The General Catalogue of astrometric data was compiled. Next basic principles were used for the compilation of the Catalogue: the application of original observations and all of bibliographic data; the analysis of systematic and accidental errors of stellar characteristics; the determination of optimal level of taking the average of these errors; the reduction of various stellar characteristics to their systems; the choice of priorities for further using of data, if these data were determined by means of various methods or with various errors.

The observations were carried using the long-focus double astrograph in Kiev. In order to improve the precision of proper motions, reference stars were selected as kinematically homogeneous group of distant stars. For the consideration of the magnitude equation in proper motions a statistical technique using the composed catalogue data only (Kharchenko, 1984) has been developed. Proper motions in reference to galaxies were converted to the General Catalogue system. This one was obtained from comparison of absolute proper motions from catalogues received on the Faint Stars Plan and AGK3. The General Catalogue contains 26500 stars.

The maximum of the distribution of r.m.s.errors of absolute proper motions is $0''\!.6$ per century. Equatorial coordinates were calculated with respect to positions of AGK3 and SAO stars with r.m.s. errors of $±(0''\!.3-0''\!.6)$. Astrophysical data of some stars were found in bibliographic sources.

Using the General Catalogue and the Pulkovo Catalogue, which was reduced to the system of General Catalogue in the sense of the magnitude equation consideration, corrections to the precession constant were determined: $\Delta k = -0.24±0.17$ and $-0.03±0.14$; $\Delta n = +0.24±0.16$ and $+0.29±0.14$ arcsec/century respectively.

J. H. Lieske and V. K. Abalakin (eds.), Inertial Coordinate System on the Sky, 431–432.

© 1990 *IAU. Printed in the Netherlands.*

Table 1. Stellar characteristics of GC

Stellar characteristic	Number of stars	Limiting stellar magnitude B (completeness until B)	
No. BD	12950	12^m5	(10^m)
RA, DEC, PM	26436	16	(13.5)
B photographic	23495	16	(13.5)
V photovisual	7777	11	(10)
U photoelectric	1339	11	
B photoelectric	1719	11	
V photoelectric	1720	11	
Spectral class	12649	12.5	(10)
Luminocity class	1405	11	
Radial velocity	1266	11	

On the basis of absolute proper motions, taken from the General Catalogue, AGK3 and papers of Deutsch (1947), and Fatchikhin (1970), stellar secular parallaxes were determined on high, middle and low galactic latitudes.

Table 2. Secular parallaxes ($\mu < 5$ "/century)

b	B =	9^m5	10^m5	11^m5	12^m5	13^m5	14^m5	15^m5
				(h/ρ) in $0''0001$				
+75°		180	160	145	130	112	100	95
+45°		142	122	110	98	87	80	78
+20° to −20°		160	135	116	100	83	72	60

References

Deutsch, A.N. (1947), *Izv. Glavnoj astron. observ. v Pulkovo*, **17**, No. 36, pp. 2–59.
Einasto, J., Maluto, V.D., Kharchenko, N.V. (1985), *Astron.circular*, No. 1394, pp. 1–6.
Fatchikhin, N.V. (1970), *Astron. Zhurn.*, **47**, 619–632.
Kharchenko, N.V. (1984), *Astrometrija i astrofizika*, **52**, 3–8.

Discussion

MURRAY: Do your values of secular parallax in different latitudes depend on different directions to the solar apex?

KHARCHENKO: We had determined the secular parallaxes by means of the equation of Kovalsky-Ery. Then the direction to the solar apex was obtained together with secular parallaxes and this direction varies for different galactic latitudes.

BROSCHE: What is the typical epoch difference on which your proper motions are based?

KHARCHENKO: The mean epoch difference of pairs of plates is 24.6 years.

FAINT REFERENCE STARS

Thomas E. Corbin and Sean E. Urban

U.S. Naval Observatory

ABSTRACT: Three major reference star projects have been completed recently at the USNO: the Southern International Reference Stars (SIRS - 19,827 stars) the southern part of the Faint Fundamentals (1,169 stars) and the Astrographic Catalog Reference Stars (ACRS - 325,416 stars). The compilation of the mean positions and proper motions of each is discussed. Reports on the progress of the USNO's Pole-to-Pole Fundamental Program and the Working Group on Star Lists are also presented.

I.   Introduction

For many years the U. S. Naval Observatory has contributed to the stellar positions and motions of the reference frame in two ways: Absolute observations, primarily of the fundamental stars, have been made on the USNO's transit circles, in particular the Six-Inch, since the beginning of this century. Both absolute and differential observations of large numbers of reference stars in the sixth through ninth magnitude range have been made on the Six-Inch, Seven-Inch and Nine-Inch. Since 1940 the reference star observations have been directed toward lists of evenly distributed stars with an average density of about one star/square degree. In the 1950's the USNO joined the Bergedorf, Bordeaux, Herstmonceux, Heidelberg, Nicolaiev, Ottawa, Paris and Pulkovo observatories in a cooperative program to observe the AGK3R (Scott 1963) in order to provide reference stars for the AGK3 (Dieckvoss 1975). This work was carried to the Southern Hemisphere in the 1960's and early 1970's in order to provide the Southern Reference Stars (SRS - Scott 1962) for the Cape astrograph plates. In this endeavor the USNO collaborated with Abbadia, Bordeaux, Bucharest, Nicolaiev, San Fernando and Tokyo in the north and Bergedorf (at Perth), Cape, Pulkovo (at Santiago), San Juan and Santiago in the south.

J. H. Lieske and V. K. Abalakin (eds.), Inertial Coordinate System on the Sky, 433–442.
© 1990 IAU. Printed in the Netherlands.

The AGK3R was compiled on the FK4 system by the USNO (Scott and Smith 1971) and resulted in positions for 21,499 reference stars north of -5°. It had been realized by Scott (1962) that the AGK3R - SRS lists were more than just reference stars for particular astrographic programs. These lists had been chosen with care to ensure even distribution over the sky, and to give preference to the stars with the best observational histories. A large contribution to both AGK3R and SRS was made by the KSZ stars, compiled by Zverev (Scott 1955). The KSZ lists were originally compiled at Sternberg Astronomical Institute. Revisions to the original KSZ stars were made by Zverev at Pulkovo. At Pulkovo, Zverev replaced stars fainter than magnitude 9.1, and deleted stars with inappropriately large proper motions (Zverev 1956). The AGK3R and SRS were selected by international agreement. Thus Scott referred to the combined lists as the International Reference Stars (IRS) and envisioned a wide range of applications.

If these applications were to be realized, then the IRS needed good proper motions. Combining 97,624 meridian circle positions from 64 catalogs, Corbin (1978) compiled mean positions and proper motions for 20,194 of the AGK3R stars in the IRS. The average mean errors of the proper motions are 0$\overset{\prime\prime}{.}$45/c in right ascension and 0$\overset{\prime\prime}{.}$46/c in declination. Proper motions for the SRS portion of the list had to await the completion of the SRS catalog.

Fricke (1973) realized that the IRS could also be used to select stars for an extension of the fundamental catalog in the seventh to ninth magnitude range. Kopff (1954,1956) had first put forth a list of stars with which to extend the fundamental list in the FK3 Supp. These stars have been well observed in the intervening years, and about 1000 will be used for the extension to the FK5. However, very few of these stars are fainter than magnitude 7.0, and thus to extend the fundamental system to the ninth magnitude, it was necessary to select from the IRS. Using the results of the NIRS and preliminary results for SIRS, Corbin (1985) selected 2072 stars for the Faint Fundamental list. A small revision changed this to 2158 stars, and this list was transmitted to Fricke in 1986. However, the positions and motions in this list were final only for the northern portion since the final southern values had to await the completion of the southern IRS (SIRS).

In recent months there has been good progress toward improving the system of faint reference stars. Work on the SIRS and southern Faint Fundamentals has been completed. In addition, the IRS has been used to compile a high density reference star catalog, the Astrographic Catalog Reference Stars (ACRS), that contains 325,416 positions and motions of stars to the tenth magnitude and covering the whole sky. Finally, the Working Group on Star Lists (IAU Commissions 8 and 24) has been considering the question of how to extend the system of reference stars to even fainter magnitudes.

II.   The Southern IRS

With the completion of the Southern Reference Star catalog (Smith et. al. 1989), it was possible to begin work on the proper motions of these stars. While the observational histories of the stars in the SRS are somewhat poorer than those in the AGK3R (Corbin 1986), this is partly offset by the epoch of the SRS being about ten years later than AGK3R.   As in the north, the problem was to reduce the stars fainter than the sixth magnitude in each catalog to the system of the FK4.

A.   The Southern Base System (SBAS):   In the Northern Hemisphere it was possible to compare the FK4 directly with ten catalogs for which screens had been used in making the observations.   This permitted a system of 5590 faint star positions and motions to be compiled that provided the basis for the reduction of the 54 other catalogs.   This was called the Base System, or BAS (Corbin 1974).   Unfortunately, the observational data in the Southern Hemisphere are such that this can be done only north of -30°. Table 1 shows the southern catalogs that can be compared directly with the FK4.   Combining these catalogs with the SRS produced a system of 3801 stars with proper motion mean errors of ±0".38/century in each coordinate. Table 2 shows the the resulting numbers of SBAS stars by declination. Quite clearly, no systematic reductions in the far south are possible with these stars.   Something additional is required.

TABLE 1. Catalogs used for the SBAS

Catalog	Number of Stars in SRS
AGK2A	1039
AGK3R	3360
Albany 10	1801
Bonn 00	283
Cape II-50	1092
Katalog von 3,356 Schwachen Sternen	234
San Luis 1910	2082
SRS	20488
Tokyo Zodiacal 1950	681
Washington 00	1363
Washington 20	1130
Washington 25	784
Washington 40	741

TABLE 2. Numbers of SBAS by 10° Zones of Declination

Declination	-80	-70	-60	-50	-40	-30	-20	-10	0	+10
no. of stars	105	12	17	91	330	281	416	534	1091	924

B. Removal of Magnitude Equation South of -30°:   The situation described above required that the FK4 be included in determining the systematic differences south of -30°.   Schwan's Analytical Method (1986) of making systematic reductions gives very good results when the reference stars cover the magnitude and declination ranges of the catalog to be reduced. Unfortunately, such is not the case in this instance.   The faint stars (SBAS) do not cover the declination range and the FK4 does not cover the magnitude range.   The solution to this dilemma was found in the characteristics of the catalogs themselves.   Using the Greenwich catalogs: Catalog of Stars for 1910.0 (Gr 10), Second Nine-Year Catalog of Stars (9Y2), First Greenwich Catalog of Stars for 1925.0 (Gr I-25), Second Greenwich Catalog of Stars for 1925.0 (Gr II-25) and First Greenwich Catalog of Stars for 1950.0 (GR I-50) in combination with the NIRS and the FK4, it was determined that those catalogs observed with fixed-wire micrometers show magnitude equations that vary with declination.   Those observed with moving-wire micrometers, however, generally do not.   Thus it was possible to select southern catalogs for which a magnitude equation can be determined between -5° and -30° and for which the magnitude equation can be used farther south.   These catalogs are listed in Table 3.

TABLE 3. Catalogs added to SBAS to form SBAS2

Catalog	Number of Stars in SRS
Cape 1 - 25	240
Cape 2 - 25	1912
Cape Standard Stars -76° to -82°	145
Cordoba Cat. of 6249 Reference Stars	801
La Plata Cat. of 3710 Galactic Stars	332
La Plata General Catalog for 1950	885
Sydney Cat. of 1499 Intermediate Stars	236

Since H. Schwan had very kindly provided the USNO with a copy of his improved positions and motions of the FK4 SUP stars (ISUP), it was possible to use these, transformed to B1950.0, in combination with the FK4 and SBAS to determine the magnitude equations for each catalog.   The procedure for each catalog was:

1. Tabular systematic differences (FK4-catalog), similar to those used for the NIRS, were computed using the FK4, ISUP and SBAS.   An average magnitude was determined for each tabular value.

2. The differences were interpolated for each star in the catalog that was used in step 1 and subtracted from that star's individual difference. The result of this was combined with the difference between the star's magnitude and the average tabular magnitude at the star's coordinates.

3. Analysis of the differences from step 2 then gave the corrections in each coordinate as a function of magnitude.   Some catalogs show magnitude equation in both coordinates.

4. The final step was to interpolate a tabular correction and a magnitude-based correction for each star in the catalog, and apply the corrections to the observed coordinates.

Combining the catalogs in Tables 1 and 3 produced a set of positions and motions (SBAS2) that are much stronger south of -30° than the SBAS, as Table 4 shows. The average mean errors of the proper motions are ±0."36/c in RA and ±0."38/c in DEC, about the same as the SBAS.

TABLE 4. Numbers of SBAS2 by 10° Zones of Declination

Declination	-80	-70	-60	-50	-40	-30	-20	-10	0	+10
no. of stars	108	160	238	355	368	322	571	561	1101	924

C. The Final SIRS: With the compilation of the SBAS2 it was possible to reduce most of the remaining catalogs. These catalogs contain a large amount of data for they include the Cordoba and La Plata zone catalogs. Since analysis of these catalogs showed that many have large magnitude equations, it was decided that the procedure outlined above would be used rather than making reductions with the faint stars alone, as was done in the NIRS. There are two advantages to this: First, the combination of SBAS2 with FK4 and ISUP gives a much larger range of magnitudes upon which to determine the magnitude equation than using the SBAS alone. Second, the resulting higher density of reference stars makes a stronger determination of the systematic differences. Using this approach, 52 more catalogs were systematically reduced, and when combined with the 20 listed above a system of 13,375 stars resulted. There were, however, 24 catalogs remaining that have too low densities of SBAS2, FK4 and ISUP to give good reductions. These last few were reduced by combining the 13,375 stars just mentioned with the FK4 and ISUP. The final body of SIRS obtained by combining these catalogs with the other data consists of 19,827 mean positions and proper motions.

The SIRS will be made available in two parts: Part I consists of 15,088 positions and motions formed from three or more catalog positions. Part II contains 4739 positions and motions that either result from only two catalog positions or have high mean errors. The average mean errors of the proper motions in Part I are ±0."43/c in RA and ±0."44/c in DEC, and their distribution is shown in Figure 1. The SIRS positions and motions have also been transformed to FK5 J2000.0 and will be available from the U. S. Naval Observatory either on this system or FK4 B1950.0.

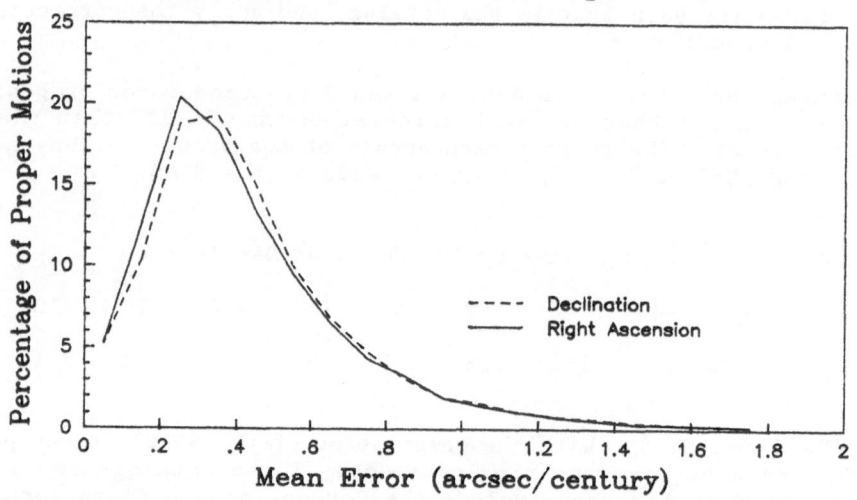

Figure 1. Frequency Distribution of the centennial mean errors of the SIRS proper motions.

III. The Southern Faint Fundamentals

There are 1169 Faint Fundamentals south of +5°, which is the northern limit of the SRS. Since the SRS added considerable weight to the determination of the positions and motions of all IRS in the SRS-AGK3R overlap zone, -5° to +5°, the final values of the Faint Fundamentals south of +5° are all taken from the SIRS. No change was made in the list of selected stars, and the details of the selection have been published (Corbin 1985). The final values have now been sent to Schwan at the Astronomisches Rechen-Institut for inclusion in the FK5, Part II. The distributions of the mean errors of the proper motions are given in Table 5, and they average ±0."28/c in RA and ±0."29/c in DEC.

TABLE 5. Faint Fundamentals - Numbers of stars by intervals of mean error of proper motion (arcsec/century)

	0.0/0.1	0.1/0.2	0.2/0.3	0.3/0.4	0.4/0.5	0.5/0.6
RA	20	228	453	303	163	2
DEC	27	169	408	387	177	1

IV. The Astrographic Catalog Reference Stars

For many years the USNO has participated in the effort to convert the published volumes of x-y measures of the Astrographic Catalog to a usable catalog. This involves first converting the measures to machine readable form and then using a suitable reference system to reduce the plates. The first of these tasks was begun by Lacroute and Valbousquet (Valbousquet 1977) when they prepared the region from -2° to +31°. Thus far the USNO effort has focused on the Perth and Cape zones, -32° to -52° and the San Fernando and Tacubaya zones, -3° to -16°.

The question of the required reference system was addressed by Corbin and Urban (1988). Due to the small size of the AC plates, 2° x 2°, a reference catalog with about 8 stars per square degree is required. The AGK3 has such a density, but as has been pointed out (Corbin and Urban 1989) there are problems with using the AGK3 at such early epochs as the AC plates, which are mostly earlier than 1915. The IRS has the required accuracy, but the density is too low, about one star per square degree.

The IRS is ideal, however, to make systematic corrections to the existing photographic catalogs so that they can be combined to yield a high density reference system. Since the NIRS was already available, work was begun at the USNO in 1987 on the Northern Hemisphere. The successful completion of this work (Corbin and Urban 1989) led to the conclusion that the ACRS could be compiled for the whole sky.

The compilation of the SIRS, described above, has allowed the northern and southern parts of the IRS to be combined, and now the IRS can be used as a whole system. In order to avoid any discontinuity in the system the IRS has been used to make new reductions of all catalogs, meridian circle and photographic, that contain any stars between -5° and +5°. In all, a total of 1.59 million catalog positions from 124 meridian circle catalogs, the Yale Photographic zones, the first Cape Photographic Catalog (CPC1), the AGK2, the Sydney Southern Star catalog, the Sydney zone catalog from -48° to -54°, and the Paris zone catalog from +17° to +25° were reduced to the FK4 system. The AGK3R, SRS, AGK3 and Second Cape Photographic Catalog (de Vegt 1989) already conform well to the FK4 and need no further reduction.

Weights and residual limits were determined in the manner described by Corbin (1977), and the data combined to give the mean positions and proper motions. There was a large variation in the numbers of catalog positions per star. Table 6 summarizes this quantity.

TABLE 6. Distribution of the Numbers of Catalog Positions

no. of cat. pos.	2	3	4	5	6	7	8	9	10 & more
no. of stars (thousands)	60	76	66	38	24	18	13	9	21
avg. epoch (+1940)	7.6	7.1	7.0	6.1	7.6	9.5	10.9	11.7	9.9

All stars are either in the AGK3 or the CPC2. Since the epoch of the CPC2 is about seven years later than the AGK3, this partly compensated for the generally poorer observational histories in the south. The results of the computation of the proper motions are summarized by zone in Table 7 and globally in Figures 2 and 3. The total number of mean positions and proper motions is 325,416 and the average mean errors for the stars with three or more catalog positions are ±0.″49/c in RA and ±0.″47/c in DEC. Like the IRS the system will be converted to FK5 on J2000.0 and will be available on either this system or FK4 B1950.0 from the USNO. When the IRS is rediscussed on FK5 and the final CPC2 is available a definitive version on FK5 will be produced and made available.

Mean Errors of ACRS Proper Motions

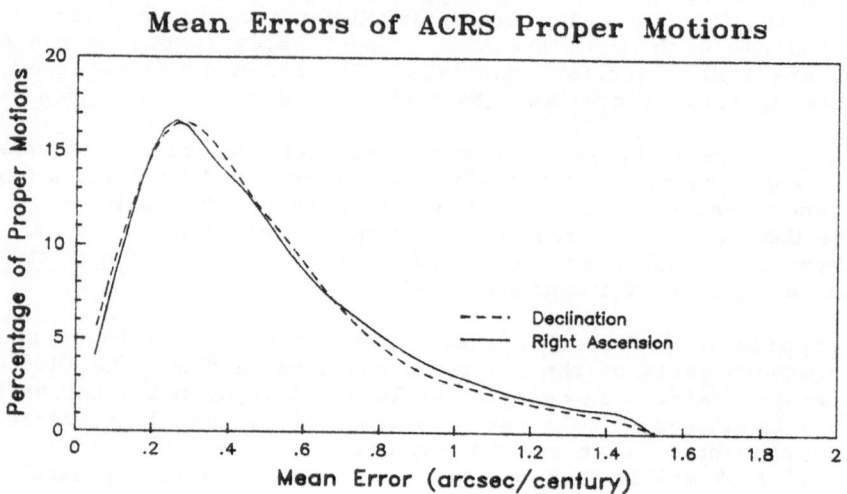

Figure 2. Mean errors of ACRS proper motions in Right Ascension and Declination.

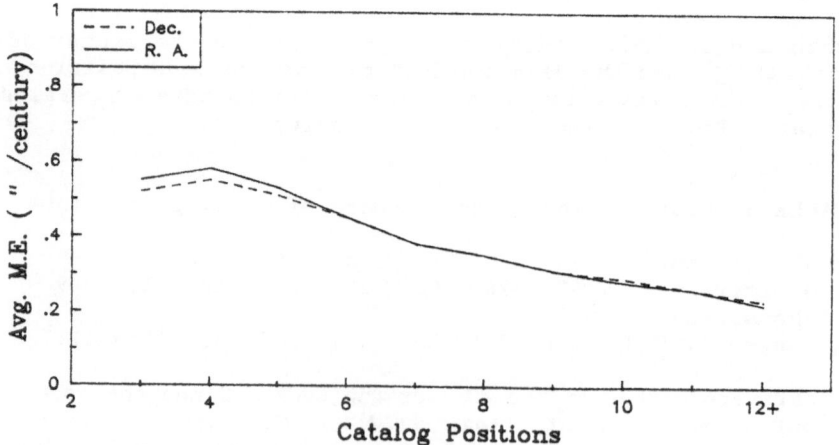

Figure 3. Mean errors of proper motions as a function of number of catalog positions.

TABLE 7. Characteristics of the ACRS by Declination

Declination	Number of Stars	Average Epoch	Mean Errors of PM	
			RA	DEC
+70 to +90	11723	1942.9	0.44	0.43
+50 to +70	30578	1944.6	0.52	0.52
+30 to +50	47282	1943.5	0.52	0.50
+10 to +30	54187	1941.9	0.50	0.46
-10 to +10	53169	1947.5	0.46	0.44
-30 to -10	45086	1950.0	0.45	0.42
-50 to -30	42380	1951.5	0.46	0.45
-70 to -50	30519	1958.1	0.43	0.42
-90 to -70	10492	1958.5	0.53	0.53

V. Current Observations

The USNO continues with its effort to make absolute observations of the
FK5, IRS and Radio Stars and differential observations of the IRS. This
is the Pole-to-Pole Fundamental Program involving the Six-Inch Transit
Circle in Washington and the Seven-Inch in Black Birch, New Zealand. In
April of this year the first circle of the Six-Inch program was completed
with a total of 122,054 observations. The second circle will begin shortly
and will require a similar number. The first circle of the Seven-Inch
should be completed in late 1990 with a higher total than the Six-Inch
because of the larger number of daytime observations. It is anticipated
that both programs will include third circles.

VI. Report from the Working Group on Star Lists

There is a general consensus among the Working Group that fundamental
stars fainter than the Faint Fundamentals should be selected so that the
observational efforts are coordinated. The question of reference stars
fainter than the IRS has produced several responses. Some feel that
reference stars to at least the 13th magnitude should be selected while
others worry about the observing load that such a list, in combination
with the IRS, would generate. Still others feel that reference stars of
this brightness should be selected, but only in selected areas, generally
around objects of special interest. The group is in general agreement
that the question of fundamental and reference stars fainter than 15th
magnitude should be deferred for the time being. Since there is good
general agreement on the identification of fainter fundamental stars, work
will begin in this area. The Hipparcos Input Catalog will provide the
basic list for the selection of these stars. Morrison (RGO) will identify
candidate stars in the 9th to 11th magnitude range, and Corbin will work
on the 11th to 13th magnitude portion. About 2000 stars in all will be
selected and submitted to the WG members for review. Stars that have been
observed at La Palma and Bordeaux will be given highest priority. Double
stars and large proper motion stars will be avoided as much as possible.
It is hoped that the WG will have agreed on a final list by the time of
the next General Assembly of the IAU.

442

Bibliography

Corbin, T.:   1974, Astron. J. 79, p. 885.
Corbin, T.:   1977, The Proper Motion System of the AGK3R, University
    Microfilms, Ann Arbor.
Corbin, T.:   1978, in IAU Symposium 109, Modern Astrometry,
    ed. F. Prochazaka and R. Tucker, Vienna, p. 505.
Corbin, T.:   1985, Celestial Mechanics 37, p. 285.
Corbin, T.:   1986, in IAU Symposium 109, Astrometric Techniques,
    ed. H. Eichhorn, Reidel, Dortrecht, p. 75.
Corbin, T. and Urban, S.:   1988, in IAU Symposium 133, Mapping the Sky,
    ed. S. Debarbat, Kluwer, Dortrecht, p. 287.
Corbin, T. and Urban, S.:   1989s, in Star Catalogues: A Centennial
    Tribute to A. N. Vyssotsky, L. Davis, Schenectady, p. 59.
de Vegt, C., Zacharias, N., Murray, C. A. and Penston, M. J.:   1989,
    in Star Catalogues: A Centennial Tribute to A. An Vyssotsky,
    L. Davis, Schenectady, p. 45.
Dieckvoss, W.:   1975, AGK3, Star Catalogue of Positions and Proper
    Motions North of -2°5, Hamburger Sternwarte, Hamburg-Bergedorf.
Fricke, W.:   1974, in IAU Symposium 61, New Problems in Astrometry,
    ed. Gliese, Murray and Tucker, Reidel, Dortrecht, p. 23.
Kopff, A.:   1954, 1956, Supplement-Katalog des FK3, Astronomisch-
    Geodatischen Jahrbuch (1954, 1956), G. Braun.
Schwan, H.:   1986, in IAU Symposium 109, Astrometric Techniques,
    ed. H. Eichhorn, Reidel, Dortrecht, p. 63.
Scott, F.P.:   1955, Conference d'Astrometrie, Bruxelles, 28-30 mars 1955,
    in Communications de l'Observatoire Royal De Belgique. No.85
Scott, F. P.:   1962, Astron. J. 67, p. 690.
Scott, F. P.:   1963, The System of Fundamental Proper Motions, in Basic
    Astronomical Data, ed. Strand, Univ. Chicago Press, Chicago, p. 11.
Smith, C., Corbin, T., Hughes, J., Jackson, E., Khrutskaya, A.,
    Polozhentsev, D., Polozhentsev, L., Yagudin, M., and Zverev, M.:
    1989, in IAU Symposium 141, Inertial Coordinate System on the Sky,
    ed. J. Lieske and V. Abalakin, in press.
Smith, C., Scott, F. P., Scott, D.K.,: in preparation,
    AGK3R, Pulb. USNO, Second Series, Vol XXVII
Valbousquet, A.:   1977, Bull. Inform. CDS, 13, p. 2.
Zverev, M. S.: 1956, Katalog Slabykh Zvezd, Moscow

Discussion

RÖSER: Could you comment on the accuracy for proper motions of about 0.48 arcsec/century on the northern sky? Our experience with PPM suggests that only by including the Astrographic Catalogue with its high accuracy could we achieve such precision.

CORBIN: First, 190,000 of the ACRS proper motions come from four or more catalog positions. Secondly, the AC has many problems and requires a rigorous reduction of its plates to yield the best results possible. That is one reason that we have compiled the ACRS.

WALTER: (1) How many of the "High Priority Radio Stars" are already among the "Astrographic Catalogue Reference Stars" or in any other of your observation programmes? (2) What kind of identifiers are provided for the ACRS in order to facilitate data retrieval of selected stars?

CORBIN: (1) All objects brighter than magnitude 10 should be in ACRS. We are observing 154 of these stars in our 6-inch/7-inch program. (2) Most of the ACRS have Durchmusterung numbers.

ASTROMETRY WITH THE TAUTENBURG SCHMIDT TELESCOPE

K.-G. STEINERT
Dresden University of Technology
Section of Geodesy and Cartography
Lohrmann Observatory
Mommsenstraße 13
DDR-8027 Dresden
Germany, Democratic Republic

ABSTRACT. The realization of an optimum astronomical reference frame depends on well-defined connection between the systems of optical and radio positions of objects. For this the physics of radio objects are to be studied carefully. Besides this it is necessary to investigate how to get accurate positions of the optical counterparts of these objects as benchmarks for establishing a radio-optical reference frame. As the optical sources are very faint, photographic telescopes with robust light-gathering power are to be used.

For 15 years research work has been done using plates of the 134/200/400 Tautenburg Schmidt telescope for determinations of photographic positions of stars and extragalactic objects. An overview of some special results obtained at Dresden Lohrmann Observatory and Potsdam Central Institute of Astrophysics of the GDR Academy of Sciences will be given.

1. Introduction

At the Second Astrometric Conference sponsored by the National Science Foundation and the University of Cincinnati, held in Cincinnati, Ohio, 17–21 May 1959, there was a paper by W. Dieckvoss (1960) concerning photographic astrometry with the Bergedorf 32-inch Schmidt telescope.

He presented satisfying results. During the discussion of Dieckvoss' paper there arose some questions connected with the geometry of Schmidt plates which may be disturbed by bending the plate in the plate holder.

In the discussion Herget remarked that in astrometric experiments which he had made with Cleveland and Michigan Schmidt telescopes the residuals of the reduction procedure showed no normal distribution. He found significant systematic effects in various parts of the field.

It seems that because of the contradictory results at the conference there was not adopted any resolution recommending the use of Schmidt telescopes for astrometric purposes.

At the conference on *The role of Schmidt telescopes in Astronomy* held in Hamburg, 21–23 March 1972, Luyten and La Bonte (1972, p. 35) from Minneapolis spoke about difficulties in the accurate determination of positions for bright stars, using an automatic measuring machine. Moreover, they stated a further improvement of plate reduction could probably be obtained by including more higher-order terms in the solution, but these terms differ from plate to plate. This had been confirmed by Dieckvoss (1972, p. 43). He recommended to use Schmidt telescopes only for small and limited

443

J. H. Lieske and V. K. Abalakin (eds.), Inertial Coordinate System on the Sky, 443–448.
© 1990 *IAU. Printed in the Netherlands.*

fields near the optical axis.

Andersen (1971) came to more optimistic conclusions concerning the Schmidt telescope in astrometry. In his opinion the form of the plate holder (quadratic or circular) plays an important role concerning the mean errors of star positions determined on the plate.

2. The Lohrmann program

Sandig (1974) proposed an astrometric program for the Tautenburg Schmidt telescope of the GDR Academy of Sciences, Central Institute of Astrophysics, taking into account:

- some optimistic results of previous astrometric experiences with Schmidt telescopes and the possibility to use the world's largest instrument of this kind;
- that it was necessary to find a way by means of a large telescope to connect the positions of very faint galaxies (18^m to 20^m)—among them radio galaxies with optical counterparts—with stars of usual catalogues (9^m to 11^m) and any faint stars;
- that these observations provide the possibility to determine absolute proper motion in a more perfect way than the classical Lick and Pulkovo programs could do with astrographs;
- that it should be possible to improve, for instance, the system of AGK3 and give therefore a contribution to the realization of an optimum astronomical reference frame;
- the possibility to solve problems of stellar astronomy in a fundamental way.

The main idea of Sandig's proposal was to take overlapping plates of belts in declination circles and then later on also in right ascension circles. So he hoped to get a network over the northern sky, including the equatorial zone.

Since astrophysical observations take the main time of the Tautenburg Schmidt it was quite obvious that such a large program could not become realized in a short time. Therefore in 1975 the "Lohrmann Program" was started as a pilot project. The photographic observations of 90 plates with about 20% overlapping in the $52°5$ declination zones were finished in 1978.

3. Preliminary results of the Lohrmann Program

Apart from the advantages of Schmidt type instruments for astrometry mentioned above, there were some serious questions to resolve. Böhme and Sandig (1978) published initial results concerning the examination of the instrument.

Having an aperture of 134 cm of the correction plate and common optical data, the 134/200/400 Tautenburg Schmidt is best fitted for extreme large limiting magnitudes and excellent resolution. The field of 24 x 24 square centimeters on the plate equals $3°1$ x $3°1$ on the sky. It was stated that the bending effect should not depend upon the size of the field only, but mainly upon the focal distance. Since the focal distance is 400 cm, the bending of 1.2 mm is comparatively small.

The authors found a star's internal root mean square error of $\pm0°19$ (9^m), $\pm0°09$ (13^m-15^m), and $\pm0°13$ (17^m) in the center of the plate and $\pm0°11$ (13^m-15^m), $\pm0°15$ (17^m) in the outer parts of the plate. On the other hand, the internal root mean square error of positions of galaxies derived from double observations on one plate for 48 to 75 measured objects as a mean value of four plates was obtained as $\pm0°24$ and $\pm0°36$ depending on the quality of the images.

To avoid the effect of magnitude equations between bright stars (9^m-10^m) and faint galaxies (17^m-

18^m) in the Lohrmann program there was used a chromatic prism in eccentric position mounted in front of the correction plate. So an object on the sky yields one image A at the plate produced from the whole aperture of the telescope, and a deflected one B by the prism about 4 magnitudes fainter and about 30" transversally displaced, corresponding to the eccentric position of the prism and its refracting angle. The long exposure (A, B images) was 20 minutes. A second, shorter exposure (20-30 sec) on the same plate after displacement of the optical axis about 1 arcmin in declination yielded an image C, in magnitude corresponding to B with full aperture, and D about 4^m fainter than B and C through the prism.

That means that a bright object leaves on the plate four images: A brighter than B and C by 4 magnitudes and D fainter by 4 magnitudes than the latter ones. Presuming a correct mutual geometry of the images $A...D$ it should be possible to reduce all images to the position of, for instance, A. At that time (Böhme and Sandig 1978) no problem seemed to exist in reduction of the plates. The closed adjustment of the whole declination circle was thought to be done with AGK3 as a reference catalogue.

The totality of selected stars was to cover the zenith zone of Potsdam PZT stars and AGK3 stars as reference stars. Moreover, faint stars of 16^m-18^m should be selected to establish direct connection with extragalactic objects.

Since the time of designing the program there arose a multitude of questions and problems of very different kind. They were investigated in Dresden Lohrmann Observatory as well as in Potsdam-Babelsberg Central Institute of Astrophysics.

Some results will be presented in the following sections.

4. Special investigations of some astrometric properties of the Tautenburg Schmidt

4.1. THE PLATE HOLDER

Andersen (1971) pointed out that a circular plate holder guarantees smaller forced deformations than a quadratic one. Böhme (1983) showed, after a thorough investigation of both types of plate holders, that there are no significant systematic effects in the positions for either the circular or the quadratic plate holders of the Tautenburg instrument. So the second epoch of the Lohrmann program will be taken with the quadratic plate holder, was was done for the first epoch.

4.2. INFLUENCE OF EXCHANGING THE MIRROR OF THE TELESCOPE

After 25 years successful work of the Karl-Schwarzschild-Observatorium, founded in 1960, the glass mirror of the Schmidt telescope was replaced by a mirror consisting of the glass-ceramic material Sitall (Marx 1986).

The glass mirror has had an excellent optical quality, but the glass-mass showed a temperature dependency. A temperature difference of 2°C between the outer and inner surfaces of the mirror yielded an image size of 1".5-2".0, which is on the order of the seeing disc. For Sitall this occurs only for a difference of 100°C.

The astrometric qualities of the new Sitall objective were tested in the course of a student's research work. Two plates, taken in 1984 July with the glass objective, were compared with two plates of the same field in the sky, taken in 1985 July with the Sitall mirror. The result was a very good agreement of the positions in both groups of exposures (1984/1985) for stars fainter than about 11 mag. Only brighter stars on plates with the Sitall mirror show larger errors and significant variations of positions, because the images of these objects at the plate show noncentric aureolas.

4.3. Plate reduction models

Numerous calculations have been carried out using different models for the plate reduction. At first, all plates of the 90 fields of the Lohrmann Program were reduced using the orthogonal transformation of the measured plate coordinates x, y into the tangential coordinates X, Y with four plate constants:

$$X - x = \ Ax + By + E$$
$$Y - y = -Bx + Ay + F \ .$$

This simple model was used because a non-rectangularity of the guide rulers of the measuring machine (ASCORECORD 3 DP/ASCOREMAT) is considered by the software.

In the course of time at Dresden Lohrmann Observatory it was understood (Witschas 1987) that for a routine program it is more useful to reduce the data with an affine model of the kind

$$X - x = \ Ax + By + E$$
$$Y - y = \ Cx + Dy + F$$

with 6 plate constants.

Dick (1989) advocates on the basis of a development by Hirte (1988) a reduction method which uses the mathematical procedure of stepwise regression. That means, in a polynomial of second or third order, that all plate constants became tested for significance. If significance is probable, the respective plate constants are used for reduction of the plates. It is known since 1972 (Luyten and La Bonte) that the constants differ from plate to plate—and therefore the significances do also.

The author is of the opinion that the method of stepwise regression is a good means to get the best mathematical model for adapting the positions on a plate to the system of standard coordinates. But for the necessity of uniformity and homogeneity in a program with a large number of plates (*e.g.*, the Lohrmann Program) it seems to be more useful to use a unique reduction model such as, for instance, the affine one under the conditions of a well-adjusted instrument and taking into account the effect of bending.

4.4. Geometry of the Schmidt plates

As was confirmed in the previous section, the method of stepwise regression in reduction of plates is a very good mathematical procedure. An imperfection of this method is that physical arguments for the selection of plate constants depending on the results of statistical tests are not known.

Witschas (1983) proposed to use the method of least squares collocation for the analysis of irregular systematic errors on Tautenburg plates. By this method it should be possible to avoid smearing effects which are introduced by any number of plate constants handling the plate as a geometrical unit.

Witschas (1987) found the root mean square errors of unit weight σ_o as shown in Table 1. The better modelling of the plate geometry by about 20% with collocation is obvious. (The large values in Table 1 arise from the catalogue errors of the AGK3). Unfortunately, the collocation analysis must be done for every single plate. In this way, the collocation has the same imperfection as the stepwise regression.

Another serious, not fully answered problem, is how to refer the A, B, C and D images of an object to a single position, so that these reduced positions represent a geometric uniform model of the photographed field.

Böhme (1989) found a general possibility to reduce the images A (exposure 20 min) and C (exposure 30 sec) of objects over the plate into one homogeneous system. He did not use equatorial coordinates, but only the differences of positions A versus C images on the plate. Thus the differences could be declared using only components of translation, rotation and inclination. The transformation into equatorial coordinates was then done in a common manner.

The mean values of the calculated and measured differences of C images after transformation to the A positions are on the order of measuring accuracy, namely 1.1 to 2.6 micrometers.

Table 1. Comparison of adjustments with plate constants and with collocation

Plate No	plate constants		collocation	
	σ_{ox}	σ_{oy}	σ_{ox}	σ_{oy}
4501	±0.''393	±0.''452	±0.''252	±0.''367
5182	±0.''376	±0.''483	±0.''289	±0.''370

5. Some results and future contributions

The main fields for application of the Tautenburg Schmidt telescope in astrometry are

- the determinatin of relative positions of objects on the sky
- the determination of absolute proper motions of the stars with respect to galaxies
- orientation of a future HIPPARCOS system with optical and radio sources.

In these fields a series of contributions were written mainly by Potsdam authors.

The determination of proper motions of different objects was done by Schilbach (1982), and Scholz and Rybka (1988). Positions of extragalactic objects were determined by Dick et al (1986). A proposal for orientation of the HIPPARCOS system was made by Dick *et al.* (1987) and by Yatsenko *et al.* (1987).

An essential task for the future is the connection between optical and radio reference frames. The success of this depends on a well defined relation between the physically very different objects in radio and optical radiation. Hitherto there is no definitive evidence that the positions of extragalactic objects, relevant to solve this problem, are independent from the frequency of their radiation.

Without any doubt the compiled radio positions catalogue CC by Walter (1989) will be the basis of future celestial reference frames. However, until now it is not satisfactorily resolved that the origin of the radio right ascensions is either 3C273, a radio galaxy with variations of its structure, or Beta Persei (Algol), a variable star.

It is quite certain that after the observations of the second epoch of the Lohrmann Program there will be a good foundation to determine exact absolute proper motions of stars of any magnitude in the declination belt of 52.°5.

References

Anderson, J.: 1971, "The Schmidt telescope as an astrometric instrument", *Astron. Astrophys.* **13**, 40–45.

Böhme, D. and Sandig, H.U: 1978, "Present status of Lohrmann Program", in F.V. Prochazka and R.H. Tucker (eds.) *Modern Astrometry*, University Observatory, Vienna, pp. 538–542.

Böhme, D.: 1983, "Einfluß des Kassettentyps beim 2-m-Schmidt-Spiegel auf die Genauigkeit astrometrischer Positionen", in K.-G. Steinert (ed.), VI. Internationales Lohrmann-Kolloquium Geodätische Astrometrie, *Mitt. Lohrmann-Obs. No. 51*, pp. 112–114 (Wiss. Zeitschrift TU Dresden 33.6).

Böhme, M.: 1989, "Geometrie von Sternpositionen auf Tautenburger Schmidtplatten mit unterschiedlicher Belichtungszeit (A/C-Bilder)", Diploma thesis, Tech. Univ. Dresden, Lohrmann Obs.

Dick, W.R., *et al.*: 1986, "Positions of extragalactic objects in the vicinity of M3 and M31 (5C3 radio field) and in the 5C1 radio field", *Astron. Nachr.* **307**, 85–88.

Dick, W.R. *et al.*: 1987, "A programme to derive from terrestrial observations the rotation of the HIPPARCOS system with reference to the inertial system", *Astron. Nachr.* **308**, 211–216.

Dick, W.R.: 1989, "Die astronomischen Eigenschaften des Tautenburger Schmidt-Teleskops und seine möglichen Beiträge zur Realisierung eines Inertialsystems", Doctor thesis, GDR Acad. Sciences, Forschungsbereich Geo- und Kosmowissenschaften.

Dieckvoss, W.: 1960, "The Bergedorf 32-inch conventional Schmidt telescope as an astrometric instrument", *Astron. J.* **65**, 214–217.

Dieckvoss, W.: 1972, "Computational solution for positions on whole Schmidt plates", in U. Haug (ed.) *The role of Schmidt telescopes in Astronomy*, ESO/SRC/Hamburger Sternwarte Bergedorf, pp. 39–43.

Hirte, S.: 1989, "Reduction methods for the astrometric analysis of Schmidt plates", in K.-G. Steinert (ed.) VII. Internationales Lohrmann-Kolloquium Geodätische Astrometrie, *Mitt. Lohrmann-Obs. No. 56*, pp. 15–16 (Wiss. Zeitschrift TU Dresden 38.2)

Luyten, W.J. and La Bonte, A.E.: 1972, "Astrometry with Schmidt telescopes" *vide* Dieckvoss (1972) pp. 33–38.

Marx, S.: 1986, "Jenaer 2-m-Spiegelteleskop in Tautenburg erhielt neuen Spiegel", *Jenaer Rundschau* **31**, 132–133.

Sandig, H.U.: 1974, "Zur systematischen Verbesserung von Positionen und zur Bestimmung absoluter Eigenbewegungen mit der Tautenburger Schmidt-Kamera", in H. Kautzleben and E. Buschmann (eds.) *2nd International Symposium Geodesy and Physics of the Earth, Proceedings* Part 1, 105–111.

Schilbach, E.: 1982, "Absolute Eigenbewegungen ausgewählter blauer Objekte in der Nähe des galaktischen Nordpols", *Astron. Nachr.* **303**, 335–340.

Scholz, R.D.: 1988, "Comparison of three catalogues of absolute proper motions of stars in the M33 field", *Astron. Nachr.* **309**, 47–52.

Walter, H.G.: 1989, "A compilation catalogue of positions of extragalactic radio sources", *Astron. Astrophys.* **210**, 455–461.

Witschas, Ch.: 1983, "Kollokation auf Schmidtplatten", *vide* Böhme (1983), pp. 75–76.

Witschas, Ch.: 1987, "Untersuchungen zum astrometrischen Informationsgehalt von Tautenburger Schmidtplatten", Doctor thesis, Tech. Univ. Dresden, Fakultät BWF.

Yatsenko, A.I. *et al.*: 1987, "The connection of the HIPPARCOS reference system to extragalactic objects by photographic astrometry", *Astron. Nachr.* **308**, 319–322.

COMPARISON OF ASTROMETRICAL ACCURACY OF BIG SCHMIDT TELESCOPES

E. SCHILBACH [1], S. HIRTE [1], AND J.-L. HEUDIER [2]
[1] *Zentralinstitut für Astrophysik*
Rosa-Luxemburg-Strasse 17A
1591 Potsdam
GERMANY, Dem. Rep.

[2] *Observatoire de la Côte d' Azur,*
Caussols, France

ABSTRACT. The measurements of AGK3 stars carried out with the automated measuring machine MAMA in Paris were used for the estimation of the accuracy of stellar positions on Tautenburg and CERGA plates. The results show good coincidence of stellar positions derived with both Schmidt telescopes. The achieved accuracy is high enough to use combined observations for improving present positions in the catalog AGK3.

1. Plate Material

The use of big Schmidt telescopes and automated measuring machines enables us to derive positions of great numbers of objects. The accuracy of measurements depends on the quality of the images and is, generally, best for faint ($15^m - 17^m$) stars. However, to get equatorial coordinates of interesting objects we have to measure also bright ($7^m - 12^m$) reference stars.

In order to estimate the accuracy of stellar positions to be achieved in this situation we used plates taken with the Schmidt telescopes at Tautenburg (134/203/401) and at Caussols, CERGA (90/152/316). In the description of the telescopes, the first number denotes the diameter of the correction plate, the second denotes the diameter of the mirror, and the third denotes the focal length.

Three plates centered at η Tau (Pleiades) were taken at nearly the same time and in the same color system, but with exposure times of 5 minutes for CERGA and 20 minutes for Tautenburg plates. The CERGA plates (30×30 cm²) and the Tautenburg plates (24×24 cm²) cover sky areas of $5^\circ\!.2 \times 5^\circ\!.2$ and $3^\circ\!.2 \times 3^\circ\!.2$, respectively. In a field of $3^\circ \times 3^\circ$ 133 AGK3 stars fainter than the $7th$ magnitude were measured with the automated measuring machine MAMA in Paris. The AGK3 coordinates were reduced to the epoch of observation by applying AGK3 proper motions and thereafter transformed into a plate-centered zenith-azimuth system.

2. Results

The form of the reduction polynomials was established by means of the method of stepwise regression (Hirte *et al.* 1989) for each plate separately. The starting model was fixed by the complete third-order polynomial in powers of coordinates. In this case the stepwise regression allows us to

J. H. Lieske and V. K. Abalakin (eds.), Inertial Coordinate System on the Sky, 449–450.
© 1990 *IAU. Printed in the Netherlands.*

find the "best" model with only significant terms among all polynomials up to the third order.

In order to estimate the accuracy of positions, the differences between the catalog coordinates of 133 AGK3 stars and their measured coordinates transformed into the system of the catalog were analysed. Under the assumption that all systematic effects are taken into account by the reduction model, these differences DX and DY may be considered as random values.

First, we compared the differences for each star measured on the Tautenburg and CERGA plates separately. For three Tautenburg plates, the errors of a stellar position range from $0\overset{''}{.}04$ to $0\overset{''}{.}10$ with the mean values of $0\overset{''}{.}05$ (1.0 μm) in X (RA) and $0\overset{''}{.}07$ (1.5 μm) in Y (declination). For the CERGA plates, a positional accuracy of $0\overset{''}{.}05$ to $0\overset{''}{.}13$ with the mean values of $0\overset{''}{.}09$ (1.4 μm) in X and $0\overset{''}{.}08$ (1.2 μm) in Y was achieved.

These results were used to estimate the error R which may arise if positions are determined on plates taken with different telescopes:

$$m_{TC}^2 = m_T^2 + m_C^2 + R^2$$

where m_T and m_C are mean errors of a stellar position on a Tautenburg and on a CERGA plate, respectively, and m_{TC} is a mean error of the corresponding difference of coordinates. Considering all nine combinations of plates we obtained m_{TC} varying from $0\overset{''}{.}10$ to $0\overset{''}{.}16$ with the mean values of $0\overset{''}{.}12$ and $0\overset{''}{.}13$ in X and Y, respectively. The error R ranges from $0\overset{''}{.}0$ to $0\overset{''}{.}11$. The mean value of R was found to be $0\overset{''}{.}07$ in X and $0\overset{''}{.}05$ in Y.

Finally, the mean error of a position in the AGK3 was estimated as

$$m_K^2 = 0.5\left(m_{KT}^2 + m_{KC}^2 - m_{TC}^2\right)$$

where m_K is a mean error of a catalog position and m_{KC} and m_{KT} are mean errors of the differences between catalog and plate coordinates(C – CERGA, T – Tautenburg). For nine plate pairs we obtained an accuracy of a catalog position in the Pleiades region between $0\overset{''}{.}47$ and $0\overset{''}{.}48$ with the mean value of $0\overset{''}{.}47$ in X and between $0\overset{''}{.}50$ and $0\overset{''}{.}52$ with the mean value of $0\overset{''}{.}50$ in Y.

These results coincide well with the value of $0\overset{''}{.}45$ found by Bastian (1989) as the average of the mean errors of position at epoch 1990.0 for the catalog AGK3. The achieved accuracy is high enough to use combined observations with the Tautenburg and CERGA telescopes for improving the proper motions in the catalog AGK3. The fact that nearly the same measuring accuracy was achieved on the plates independently from exposure time exhibits the good quality of the measuring machine MAMA. An analogous investigation based on these plates and on the catalog by Eichhorn et al. (1970) as reference catalog is now in preparation. Preliminary tests show a good agreement between catalog and measured coordinates. We intend to use the measurements of the plates to derive the presently missing proper motions for ca. 400 stars in Eichhorn's catalog.

References

Bastian, U.: 1989, "PPM: a tool for astronomers", in *Inertial Coordinate System on the Sky*, IAU Symposium No. 141 (this volume).

Eichhorn, H., Googe, W.D., Lukac, C.F., Murphy, J.K.: 1970, "Accurate positions of 502 stars in the region of the Pleiades", *Mem. Roy Astron. Society* **73**, 125.

Hirte, S., Dick, W.R., Schilbach, E., Scholz, R.-D.: 1989, In: *Errors, Uncertainties and Bias in Astronomy*, C. Jaschek and F. Murtagh (Eds.), Cambridge University Press, in press.

DETERMINATION OF ABSOLUTE PROPER MOTIONS BY USE OF AUTOMATED MEASUREMENTS OF TAUTENBURG PLATES

R.-D. SCHOLZ
Central Institute of Astrophysics
DDR - 1561 Potsdam
Germany, Dem. Rep.

ABSTRACT. From measurements of Tautenburg Schmidt plates with the APM in Cambridge positional accuracies per plate of 0."05 for stars and of 0."10 for galaxies were achieved. With 0."3/100a accuracy in a single stellar proper motion we obtained the absolute proper motion of the M3 globular cluster in good agreement between the two pairs of plates used.

1. Introduction

Proper motions of stars obtained with respect to distant galaxies are of great importance for the implementation of a nonrotating coordinate system. In the Potsdam program of determining absolute proper motions in selected fields of the Northern sky Tautenburg Schmidt plates with epoch differences of more than 20 years are used. This program is part of international efforts being made to connect existing reference frames like FK5 and forthcoming reference frames of space astrometry with an extragalactic reference frame. For a description of the program see Dick *et al.* (1987) and Yatsenko *et al.* (1987).

The development of high speed scanning microdensitometers for accurate coordinate measurements of the images of all objects on a Schmidt plate has brought some progress in proper motion work, especially concerning the tangential motions of groups of objects against the background of large numbers of galaxies. Kibblewhite *et al.* (1982) reported on very promising first results of proper motion studies from measurements on Palomar and UK Schmidt plates with the Automated Photographic Measuring (APM) facility at Cambridge/UK. This work is now being continued by Evans (1988). Owing to the assistance of Prof. Argue and the APM group in Cambridge it was possible to measure some Tautenburg Schmidt plates on the APM. First results of the reduction of these measurements are briefly described here.

2. Results

In each of 4 fields (one with the M3 globular cluster in the centre, one with M3 in the plate corner and two fields including parts of he Virgo galaxy cluster) two pairs of plates were measured. The

451

J. H. Lieske and V. K. Abalakin (eds.), Inertial Coordinate System on the Sky, 451–452.
© 1990 *IAU. Printed in the Netherlands.*

24×24 cm² plate area corresponds to 3.2×3.2 square degrees. The number of objects measured on one plate varied from 16000 to 70000 .

Prior to the determination of proper motions we compared plates of one epoch and eliminated most of noise images. The positional accuracy with different magnitude classes of stars and galaxies was investigated in plate-to-plate solutions. In 7 of 8 cases an accuracy from 0.8 to 1.5 microns ($0\rlap{.}''04$ to $0\rlap{.}''08$) was obtained for faint stars. For galaxies we achieved a positional accuracy of 1.5 to 2.0 microns ($0\rlap{.}''08$ to $0\rlap{.}''10$) except the second epoch M3 plates with only 4 microns ($0\rlap{.}''2$). For the determination of proper motions we used complete third and second order polynomials, a linear model and the best model of stepwise regression (see Hirte *et al.* 1989). An iterative procedure considering larger coordinate shifts for brighter stars was used in the plate matching of different epochs. For the first time we obtained an accuracy of $0\rlap{.}''3/100a$ for faint stars, whereas the accuracy for AGK3 stars was comparable with former results (*e.g.* Scholz and Rybka 1988).

Webbink (1988) stressed the need of determining absolute proper motions of globular clusters directly with respect to galaxies. For the most part former investigations have been relative proper motion studies affected by serious uncertainties in the reduction to absolute proper motions. Tucholke *et al.* (1988) used stars from the Lick program with known absolute proper motions as reference stars and obtained mean cluster motions with an error of $0\rlap{.}''05$ to $0\rlap{.}''10$ /100a. Our preliminary results in an attempt to derive the absolute proper motion of the M3 cluster are described here.

All stars within a circle of 8.2 arc minutes radius around the cluster centre were taken as M3 stars. In order to minimize possible magnitude effects only one magnitude interval with 790 reference galaxies and 680 M3 stars was selected. The mean proper motion of the cluster from two pairs of plates was obtained as $0\rlap{.}''11/100a$ in x and $-0\rlap{.}''20/100a$ in y. The variation in the mean cluster motion with different reduction models was less than $0\rlap{.}''02/100a$. The difference of the cluster motion between the two pairs of plates was about $0\rlap{.}''05/100a$. Using more pairs of plates available in the Tautenburg plate archive we could minimize possible systematic effects.

References

Dick,W.R., Ruben,G., Schilbach,E., Scholz,R.-D. (1987) *Astron. Nachr.* **308**, 211–216.

Evans,D.W. (1988), Thesis, Cambridge.

Hirte,S., Dick,W.R., Schilbach,E., Scholz,R.-D. (1989) In: *Errors, bias and uncertainties in astronomy*, Proc. IAU Colloq. Strasbourg, in press.

Kibblewhite,E.J., Irwin,M.J., Bridgeland,M.T., Bunclark,P.S. (1982), *Occ. Rep. R. Obs. Edinburgh* **10**, 79–89.

Scholz,R.-D., Rybka,S.P. (1988), *Astron. Nachr.* **309**, 47–52.

Tucholke,H.-J., Brosche,P., Geffert,M. (1988) in: J.E.Grindlay, A.G.Davis Philip (eds.), *The Harlow-Shapley Symposium on Globular Cluster Systems in Galaxies*, IAU, 525–526.

Webbink,R.F. (1988) *ibidem*, 49–60.

Yatsenko,A.I., Rybka,S.P., Scholz,R.-D. (1987) *Astron. Nachr.* **308**, 319.

OPTICAL ASTROMETRY OF EXTRAGALACTIC RADIO SOURCES WITH THE TAUTENBURG SCHMIDT TELESCOPE

W.R. DICK
Zentralinstitut für Astrophysik
Rosa-Luxemburg-Strasse 17a
DDR-1561 Potsdam
Germany, Democratic Republic

I.I. KUMKOVA
Institute of Applied Astronomy
8 Zhdanovskaya ul.
197042 Leningrad
USSR

ABSTRACT. Optical positions of objects from the IAU Commission 24 Working Group list of benchmark radio sources have been derived which will contribute to the link of the radio and optical reference frames. Results for 11 objects with an r.m.s. position error of 0$\overset{''}{.}$2 are presented and discussed.

1. Introduction

Since 1978, when a working group of IAU Commission 24 was established, a continuing effort has been made to obtain precise optical positions of extragalactic radio objects. Most of the published positions are for objects brighter than 18 mag, whereas the majority of recommended benchmark radio sources (Argue *et al.* 1984) have visual magnitudes beyond 18 mag. For statistics on the Northern sky see Dick and Kumkova (1989).

At the end of 1986 a program was started with the Tautenburg Schmidt Telescope with special emphasis on objects of 18 to 19 mag. The first measurements concentrated on those objects for which precise optical positions had already been published. Some details of the observations and measurements as well as a discussion of the first results were reported by Dick (1989). We used 50 to 80 reference stars from the AGK3 distributed over the whole plate.

2. Results

Table 1 presents results for the 11 objects. Each position is the mean of two values from a pair of plates. From the differences between these two values an "internal" standard error for the mean values of less than 0$\overset{''}{.}$15 (r.m.s.) was derived. A comparison with optical positions published by other authors shows that some of the derived coordinates seem to have systematic offsets from these values. The unweighted mean of the offsets is given in columns (5) and (6). Only those published positions with r.m.s. errors less than or equal to 0$\overset{''}{.}$2 (as given by the authors) have been taken into consideration. Their number is listed in column (7).

J. H. Lieske and V. K. Abalakin (eds.), Inertial Coordinate System on the Sky, 453–454.
© 1990 *IAU. Printed in the Netherlands.*

For 8 objects a comparison is possible with the radio reference frame (Walter 1989). In most cases the residuals are identical with those calculated from a comparison with the compilation catalogue of Argue *et al.* (1984) at the 0″.01 level. Neglecting the errors in the radio positions, which are considerably smaller than the errors in the optical positions, an "external" standard error of 0″.2 in $\Delta\alpha \cos \delta$ and 0″.3 in $\Delta\delta$ was derived. The mean offsets indicate that in the optical positions a systematic error is probably inherent due to a magnitude equation on Tautenburg Schmidt plates.

Some additional results (*cf.* Dick 1990) which may be of general interest for photographic astrometry are published elsewhere.

Table 1. Optical positions of extragalactic radio sources

(1) Source	(2) m_v	(3) R.A. (B1950.0) h m s	(4) Dec. ° ′ ″	(5) $\Delta\alpha \cos \delta$ ″	(6) $\Delta\delta$ ″	(7) n	(8) Epoch 1900+
0106+013	18.5	1 6 4.496	+ 1 19 1.27	+0.02	−0.06	3	87.75
0440−003	18.5	4 40 5.270	− 0 23 21.02	+0.27	−0.16	3	86.85
0552+398	18	5 52 1.392	+39 48 22.11				87.07
0642+449	19	6 42 52.984	+44 54 30.90	+0.36	+0.02	2	86.85
0736+017	18	7 36 42.486	+ 1 44 0.04	+0.34	+0.00	6	87.07
0952+179	18.0	9 52 11.822	+17 57 44.58	−0.06	+0.27	3	87.23
1055+018	18.0	10 55 55.288	+ 1 50 2.87	+0.19	+0.86	1	87.22
1328+254	18.0	13 28 15.904	+25 24 37.64	+0.33	−0.24	1	87.30
1442+101	18.4	14 42 50.460	+10 11 11.96	+0.36	+0.18	3	87.22
1638+398	18.5	16 38 48.177	+39 52 29.22	−0.08	+0.84	1	87.64
1641+399	16.3	16 41 17.596	+39 54 10.36	+0.13	+0.28	5	87.64

References

Argue, A.N. *et al.* (1984) "A catalog of selected compact radio sources for the construction of an extragalactic radio/optical reference frame", *Astron. Astrophys.* **130**, 191–199.

Dick, W.R. (1989) "Determining optical positions of benchmark radio sources with the Tautenburg Schmidt telescope (first results)", *Mitt. Lohrmann-Obs.* TU Dresden Nr. 56, 20–22.

Dick, W.R. (1990) "The astrometric properties of the Tautenburg Schmidt Telescope and its possible contributions to the realization of an inertial reference frame (Dissertation Abstract)", *Astron. Nachr.* **311**, No. 2, in press.

Dick, W.R., and Kumkova, I.I. (1989) "The connection of the optical and radio reference systems by means of photographic astrometry", 6th Intern. Symp. Geodesy and Physics of the Earth, Proceed. Part I, *Veröff. Zentralinst. Physik der Erde* Potsdam No. 102, 70–74.

Walter, H.G. (1989) "A celestial reference frame based on extragalactic radio sources", *Astron. Astrophys. Suppl. Ser.* **79**, 283–289.

Discussion

MORRISON: Have you considered using reference stars from the Carlsberg Meridian Catalogues?

W. DICK: When the CMC containing reference stars in the fields of extragalactic radio sources will be available, we will examine the possibility of using it. The question is whether the number of about 20 reference stars situated in the centre of the Schmidt plates is sufficient for modelling the plates. This has to be tested.

CCD PARALLAXES FOR FAINT SOUTHERN HIGH PROPER MOTION STARS

C.A. ANGUITA, M.T. RUIZ
Obs. Astronomico Nacional
Universidad de Chile
Casilla 36-D
Santiago, Chile

ABSTRACT. In April 1985, we started a program to measure trigonometric parallaxes for faint southern high proper motion stars, using a CCD at the Cassegrain focus of the Cerro Tololo Interamerican Observatory (CTIO) 1.5-m telescope. The program stars ($m_R > 16$; $\mu \geq 1$ arcsec/year) were selected from the LHS Catalogue and the University of Chile proper motion program.

The X and Y positions of the stellar image centroids were obtained using the algorithms of DAOPHOT program packages. The precision of the measurement of one stellar image is about 6 milliarcseconds (0.02 pix). For stars with $16 < m_R < 19.5$ a precision of 2 milliarcseconds in the parallax determination can be obtained in a one year period. Trigonometric parallaxes for some stars common to other parallax programs are given, showing an agreement with those results within the quoted mean errors.

Several technical aspects of the present program are also discussed.

Discussion

VAN ALTENA: Is there a difference between the short term repeatability and the long term precision of the CCD observations?

ANGUITA: No, we have found no significant differences in short term and long term precision of our observations. I can say this, since we have studied our CCD results very carefully in relation to these "could-be" differences because during the four years of this ongoing project we have used three different CCD chips from RCA.

SHAKHT: It is known that the proper motions of VB8 and VB10 might have perturbations due to suspected invisible companions. Could you say something about these perturbations?

ANGUITA: The time-bases of our VB8 and VB10 CCD observations are not large enough to say anything reliable concerning perturbations due to invisible companions. In the future— some years from now—certainly we will be able to say if there are or are not such perturbations. If I recall correctly, the perturbation period is about 5 years for VB10 and a little longer for VB8.

CHEREPASHCHUK: What value of seeing do you have during your observations?

ANGUITA: Better than 1.6 arcsec at the 50% intensity level.

J. H. Lieske and V. K. Abalakin (eds.), Inertial Coordinate System on the Sky, 455.
© 1990 *IAU. Printed in the Netherlands.*

A COMPARISON OF THE ACCURACY OF THE DETERMINATION OF RIGHT ASCENSION BY MEANS OF TRANSIT INSTRUMENTS OF DIFFERENT TYPES

S.P. Izmailov, N.G. Litkevich, S.N. Sadzakov, V.D. Simonenko, T.I. Suchkova, S.A. Tolchelnikova-Murri, V.I. Turenko
Kharkov State University Observatory
310022 Kharkov, USSR

ABSTRACT. A special series of observations of the groups of stars performed in Pulkovo, Chile, Kharkov and Belgrade were used to compare the mean errors of the right ascension by means of the method eliminating the influence of the errors of the source catalogue. The results show the advantage of the small transit instruments over transit circles of the classical type.

INDIVIDUAL STAR CATALOGUES OF THE UNIFIED TIME SERVICE OF KHARKOV SCIENTIFIC AND RESEARCH INSTITUTE AND KHARKOV UNIVERSITY ASTRONOMICAL OBSERVATORY

S.P. Izmailov, N.G. Litkevich, V.D. Simonenko and V.I. Turenko
Kharkov State University Observatory
310022 Kharkov, USSR

ABSTRACT. The Time Service Catalogue (TSC), which is obligatory for use in all USSR Time Services, was created for the epoch 1958 in the 1970s. Research done in the succeeding years detected a seasonal variation of TSC errors. Thus, a new Time Service Catalogue (TSC2) was required, in order to remove the seasonal errors as well as to satisfy the increasing demands for accuracy in the determination of universal time. During 1986–88 the data of Kharkov Common Time Service for 1980–1987 were reanalyzed using the new system of astronomical constants and two new catalogues were produced.

THE KHARKOV RIGHT ASCENSION CATALOG OF DOUBLE STARS AND HIGH LUMINOSITY STARS FOR THE DECLINATION ZONE +30° TO +90°

V.M. Kirpatovskij
Kharkov State University Observatory
310022 Kharkov, USSR

Paper not available.

THE SRS CATALOG OF 20,488 STAR POSITIONS
CULMINATION OF AN INTERNATIONAL COOPERATIVE EFFORT

C.A. Smith[1], T.E. Corbin[1], J.A. Hughes[1], E.S. Jackson[1], E.V. Khrutskaya[2],
A.D.Polozhentsev[3], D.D. Polozhentsev[2], L.I. Yagudin[2], M.S. Zverev[2]

[1] U.S. Naval Observatory, Washington, DC, U.S.A.
[2] Pulkovo Observatory, Leningrad, U.S.S.R.
[3] Leningrad State University, Leningrad, U.S.S.R.

ABSTRACT

A major international effort to observe and compile the results of
observations from many transit circle programs into a single catalog of
positions referred to the FK4 system came to a conclusion with the completion
and distribution of the Southern Reference Star (SRS) catalog of 20,488
stars. Previous discussions focussed on the adjustments to the observational
material to refer it to the FK4 system and on the random errors as estimated
from residual differences. In the present discussion, we give the results
of internal comparisons which have been made between the individual
contributing catalogs and the final combined SRS catalog. Also, results
of a comparison between the SRS catalog and the AGK3R catalog are given
where they overlap in the declination zone from +5 to -5 degrees. The
possibility of magnitude equation and color error in the SRS catalog is
discussed.
The reduction procedure used to transform the version of the SRS
catalog based on the FK4/B1950.0 system to the version based on the
FK5/J2000.0 system is given.

INTRODUCTION

In 1961, a program of observations was started which, by the time it
ended in 1973, was to involve six transit circle telescopes in the southern
hemisphere and seven in the north. In the southern hemisphere, observations
were made from Argentina, Australia, Chile and South Africa. Observations
in the northern hemisphere were made from France, Japan, Rumania, Spain,
the USA and the USSR.
A list of stars from -90 to -30 degrees declination was chosen at
the Cape Observatory. At the U.S. Naval Observatory a list of stars from
-30 to -5 degrees declination was chosen. The combination of those two
lists and the AGK3R stars in the declination zone from -5 to +5 degrees
comprised the list of about 20,500 stars called the Southern Reference
Star (SRS) catalog. The selection of stars was made in accordance with
guidelines coordinated through the SRS Committee of Commission 8 of the
International Astronomical Union. The principal criteria for selection
were that the visual magnitudes of the stars should lie mostly in the
range from 7.5 to 9.5 with excursions outside of that range as necessary
to achieve a uniform distribution on the celestial sphere of about 1 star

457

J. H. Lieske and V. K. Abalakin (eds.), Inertial Coordinate System on the Sky, 457–463.
© 1990 *IAU. Printed in the Netherlands.*

per square degree. Stars with good observational histories were given preference in order to improve the proper motions.

The compilation of the observations into a single catalog of positions referred to the fundamental system of the FK4 catalog was carried out jointly at the Pulkovo Observatory and at the U.S. Naval Observatory. Results of observations from the participants were received over an eight year period from 1974 to 1982. Work on the compilation of the SRS catalog was completed in December 1987. Distribution of the catalog was made in April 1988 after a review of procedures and comparisons with other catalogs had been completed. Preliminary proper motions were distributed with the SRS catalog, but should not be regarded as a part of it. See Corbin (1974,1978) for a discussion of the work on the proper motions of the SRS stars.

There was general agreement that a version of the SRS catalog referred to the FK5 system should also be made available. The question of the proper technique for accomplishing the transformation is still under discussion, but an approach was adopted and justified as closely paralleling the technique used at the Astronomisches Rechen-Institut for the reduction of the FK4 to the FK5 system.

The SRS catalog is the reference catalog for the 2nd Cape Photographic Catalog (CPC2), the first example of a modern photographic catalog taken in the visual spectral range (5300-6400 Å). The plates were measured on the automatic measuring machine, GALAXY. The measurement program is described in Nicholson(1978). Reductions at Hamburg Observatory for the individual plates using the SRS catalog referred to the FK5/J2000 system are nearly finished, and preparations for a complete plate overlap solution are underway. The accuracy of the CPC2 is estimated at $0\overset{''}{.}06$. For further details see Nicholson et al.(1984), de Vegt(1988) and de Vegt(1989).

COMPARISON OF THE SRS CATALOG WITH THE INDIVIDUAL CONTRIBUTORS

Evidence of a small but significant magnitude equation in the declination differences between a preliminary version of the CPC2 realized in early 1989 and depending on a single plate solution and the SRS in the declination zone from -5 to +3 degrees has been noted by de Vegt (personal communication). From those differences alone, it is not possible to conclude with certainty from which catalog the magnitude equation comes.

However an analysis of the differences between the declinations of the individual SRS contributors and the SRS catalog could be expected to indicate something about the extent to which magnitude related differences exist among them. The following table (Table I) shows the differences in declination collected in half-magnitude intervals for the seven catalogs which contributed to the SRS catalog in the -5 to +3 degree zone of declination.

It is quite clear from the table that each of the participants, with the exception of Perth and Washington, show some systematic behavior at the 0.1 to 0.2 arcsec level over the magnitude range from 6.5 to 9.0. Unfortunately, there being no reliable standard known to be free of magnitude equation, it is not possible to free the SRS catalog of the residual effects of the magnitude equation of the individual contributors, which, it appears, can be rather substantial.

Table I
Magnitude Dependent Systematic Differences in Declination
Individual Catalog Minus SRS Catalog
in the Declination Zone -5 to +3 Degrees

	6.5	7.0	7.5	8.0	8.5	9.0 mag
						unit = arcsec
Abbadia	0.198+	0.198+	0.032+	0.079+	0.020+	0.040+
(me)	57	22	7	2	3	3
(N)	16	45	80	197	260	147
Bordeaux	0.151+	0.100+	0.047+	0.054+	0.042+	0.037+
	30	6	3	2	4	3
	21	63	89	186	281	132
Bucharest	0.234+	0.167+	0.110+	0.048+	0.075+	0.034+
	31	14	6	4	1	6
	17	55	93	187	272	134
Leoncito	0.153-	0.040-	0.037-	0.047-	0.014-	0.007+
	8	3	2	1	1	1
	59	216	330	738	1063	526
Nicolaiev	0.172-	0.003-	0.027+	0.012-	0.054-	0.133-
	22	8	7	1	2	2
	25	93	134	318	495	250
Perth	0.030-	0.036-	0.011+	0.015+	0.034+	0.041+
	13	2	2	0	0	1
	58	207	328	735	1064	540
Washington	0.002+	0.020-	0.021-	0.004-	0.009-	0.027-
	8	1	1	0	0	1
	59	222	333	757	1100	555

Although no significant magnitude equation in right ascension between
the preliminary CPC2 and SRS catalogs was noticed, an internal analysis
of the SRS catalog differences with respect to the SRS contributors was
made. Only the Nicolaiev right ascensions showed significant systematic
behaviour, being rather strongly positive at the 0.1 arcsec level in the
brightest and faintest SRS catalog magnitude limits, and zero or slightly
negative in the middle of the magnitude range.

When the comparisons are extended over the full declination range of
the SRS catalog, from -90 to +5 degrees, vestiges of the behavior in the
-5 to +3 degree zone persist, but are greatly diminished. It is clear that
observations in the equatorial zone are the most seriously affected.

Residual differences in right ascension and declination were examined
for a color equation correlated with spectral type. No significant variation
was found.

COMPARISON OF THE SRS CATALOG WITH THE AGK3R CATALOG

A comparison of the SRS and the AGK3R catalogs in the zone of overlap between -5 and +5 degrees in declination was made. There were about 3,300 stars in common or more than 300 per one degree zone of declination. The AGK3R position was brought to the mean epoch of the SRS position using the NIRS (also known as AGK3RN) catalog proper motions of T. Corbin. One outlying difference in right ascension was rejected. No differences in declination were rejected. The average difference in epoch between the SRS and AGK3R catalogs is about 10 years.

Taking the differences globally and forming the mean differences in right ascension and declination over the entire zone in the sense (SRS minus (AGK3R + NIRS proper motion)) gave a systematic difference of +0.02 arcsec in each coordinate. The greatest regional difference in the right ascension coordinate occurred in the declination zone from -4 to -5 degrees, where the difference was +0.06 arcsec with a mean error of 0.02 arcsec for 307 differences. The greatest regional difference in the declination coordinate occurred between +3 and +4 degrees where the difference was +0.07 arcsec with a mean error of 0.01 arcsec for 347 differences.

The following table gives differences collected by magnitude interval.

Table II. SRS-(AGK3R+NIRS Prop. Motions) in the Decl. Zone -5 to +5 Deg.

Avg. Mag.	Avg. RA Diff.	ME	N	Avg. DEC Diff.	ME	N
6.25	-0".053	0".029	34	0".036	0".028	34
6.75	-0.028	9	217	37	14	217
7.25	-0.015	7	330	22	10	330
7.75	-0.007	7	471	21	9	471
8.25	0.003	5	932	12	7	932
8.75	0.051	5	986	25	7	987
9.25	0.080	10	322	19	13	322

ME denotes the mean error, N denotes the number of stars.

In right ascension, Table II shows a small, but well determined magnitude equation. On the other hand, the comparison of the SRS with the preliminary CPC2 by deVegt shows no significant magnitude equation. In declination, Table II shows no magnitude equation, but de Vegt's comparison of SRS with the preliminary CPC2 shows a well determined magnitude equation.

Our interpretation of these conflicting results is that despite the results given in Table I, where a magnitude equation is found in the individual transit circle catalogs in declination, neither the SRS nor AGK3R has a serious magnitude equation in declination since it is not evident in their differences. The potential for a serious magnitude equation in the SRS declinations from a few northern hemisphere catalogs has been avoided by a strong contribution from the southern hemisphere catalogs.

We restate the result that no significant magnitude equation in right ascension is found among the SRS contributing catalogs. We may also conclude that, since no magnitude equation is found between the CPC2 and SRS right ascensions, the SRS right ascensions are essentially free of magnitude equation. This implies that the magnitude equation found in the SRS-AGK3R differences comes mainly from the AGK3R catalog.

It is significant that problems of magnitude equation occur in the equatorial zone. This is where all of the transit circles participating in the SRS (and AGK3R) programs observed at extremes of zenith distance. Image quality as affected by seeing and the consequent increase in the diameter of the seeing disk could produce a magnitude equation. It is not surprising that everywhere else in the southern hemisphere, at more modest zenith distances, no significant magnitude equation is found, either among the catalogs contributing to the SRS catalog, or between the SRS and preliminary CPC2.

An analysis using NIRS positions instead of AGK3R positions was also tried, but as expected, no significant change was noted.

TRANSFORMATION OF THE SRS CATALOG TO THE FK5/J2000.0 SYSTEM

The SRS positions are the combined results of observations of all participants in the SRS observing campaign from 1961 to 1973. Preliminary catalogs were compiled at both the U.S. Naval Observatory (Washington) and the Pulkovo Observatory (Leningrad). The two compilations were compared, reconciled and combined to give the SRS catalog. The proper motions are given principally as a means to take into account the small differences in epoch of observation between the SRS and the CPC2, which are at most about two years.

Because of the inhomogeneous nature of the proper motion sources, and the likely presence of magnitude-related systematic errors in the proper motions, they may not safely be used to extend the SRS system of positions very far from the mean epoch of observation, which is about 1968. A discussion by T. Corbin of all the SRS observations, including those made before and after the 1961 to 1973 campaign, and their reduction to the FK4 system is in an advanced stage of preparation and will yield the desired high quality proper motions required for transferring the system to more distant epochs.

Two sets of positions and proper motions are given. One set is referred to the equator and equinox of B1950.0 and the other to the equator and equinox of J2000.0. The positions are referred to the mean epoch of observation in both cases. The B1950.0 positions were linked directly to the FK4 system by observational techniques.

The SRS catalog has been referred to the FK5 system and the epoch and equinox of J2000.0 in accordance with the IAU 1976 conventions described in Resolution No. 1 of the IAU Sixteenth General Assembly (Trans. IAU XVIB, 56, 58).

The precise SRS positions and approximate proper motions are given for the equator and equinox of B1950.0 and are referred to the FK4 system. The SRS was reduced to the FK5 system by using coefficients of H. Schwan (Astronomisches Rechen-Institut, Heidelberg) communicated to us as part of a computer subroutine package described in Schwan (1988). The coefficients represent the systematic differences between the FK4 and FK5 catalog positions and proper motions for the mean epoch and equinox of B1950.0 and are given in the system of the FK4 catalog.

The arguments needed as input to the computer subroutine are the right ascension, declination and magnitude. The subroutine returns systematic corrections to the FK4 positions and proper motions. In general, the magnitudes of the SRS lie outside the range of magnitudes of the stars in the FK4 catalog and are therefore inappropriate to use as arguments, because the magnitude-related error of the FK4 system is not defined outside the magnitude range of the FK4 stars. The problem was resolved by adopting as the magnitude argument the mean magnitude of the FK4 stars within 10 degrees of declination and two hours of right ascension of a given SRS position. For stars within 10 degrees of either pole, ALL of the FK4 stars within 20 degrees of the pole and within two hours of right ascension of the SRS star were used to compute the mean FK4 magnitude used as the input argument.

After the SRS positions and proper motions had been referred to the FK5 system by using the procedure described above, they were transformed from the equinox B1950.0 basis to J2000.0 using the algorithm given in the 1988 edition of the Astronomical Almanac, p. B42. The algorithm was applied with the foreshortening terms set equal to zero. By the use of that algorithm, elliptic terms in aberration are removed, the IAU 1976 precession is introduced, the FK4 equinox error in the right ascension position and proper motion system is corrected, and the time scale for the proper motions is shifted from units of tropical centuries to Julian centuries of exactly 36,525 days.

This transformation procedure gives a first order reduction from the FK4 to the FK5 system. A more precise reduction of many of the SRS catalogs observed from 1961 to 1973 is possible from a re-discussion of the FK4 star observations made concurrently with the SRS observations on a nightly basis. A program to reduce first the SRS and then the AGK3R to the FK5 system as directly as possible, going back to the observations of FK4/FK5 stars when they are available, is now in progress at the observatory in Washington.

REFERENCES

Corbin, T.E. 1974 *Astron. J.* **79**, p. 885.

Corbin, T.E. 1978 Proc. IAU Coll. 48, *Modern Astrometry*, ed. Prochazka, F.V. and Tucker, R.H., Inst. of Astronomy (University Observatory) Vienna, p. 505.

de Vegt, C., Zacharias, N., Penston, M.J. and Murray, C.A. 1988 Proc. IAU Symp. 133, *Mapping the Sky*, ed. Debarbat, S., Eddy, J.A., Eichhorn, H.K. and Upgren, A.R., Kluwer Acad. Publ., Dordrecht, p. 415.

de Vegt, C., Zacharias, N., Murray, C.A. and Penston, M.J. 1989 *Star Catalogues: A Centennial Tribute to A.N. Vyssotsky*, Contr. of the Van Vleck Observatory No. 8, ed. Philip, A.G.D. and Upgren, A.R., L. Davis Press, Schenectady, N.Y., p. 45.

Nicholson, W. 1978 Proc. IAU Coll. 48, *Modern Astrometry*, ed. Prochazka, F.V. and Tucker, R.H., Inst. of Astronomy (University Observatory) Vienna, p. 515.

Nicholson, W., Penston, M.J., Murray, C.A., and de Vegt, C. 1984 *Mon. Not. R. Ast. Soc.* **208**, p. 911.

Schwan, H. 1988 *Astron. and Astrophys.* **198**, p. 363.

Discussion

RÖSER: You made the comparison SRS – NIRS and SRS – (AGK3R + NIRS pm) in the zone –5° to +5° in order to study the possible magnitude equation in SRS. May I ask Tom Corbin, how large is the weight of AGK3R in NIRS at epoch 1958? If it's rather large, then both comparisons are almost the same and it is not surprising that the results are the same.

SMITH: We were not surprised that the results are the same.

CORBIN: AGK3R had high weight in the proper motions. There are other late epoch catalogs of high weight in this zone also: PFKSZ, Larink's Schwacher Sternen, Semirot's Bordeaux 50, and Bucharest KSZ +11° to –11°. However, only AGK3R contains all of the AGK3R stars.

CORBIN: Did Bucharest observe with screens?

[*Several people respond that they do not know*]

CORBIN: It should be noted that the six-inch Washington instrument makes rigorous use of screens and could be used as a standard as well as the Perth results.

SMITH: The Seven-inch transit circle at Leoncito also used screens and yet it shows a significant magnitude equation. Which one should we believe?

DÉBARBAT: I do not know the answer for Bucharest regarding the use of a screen, but I know that in Paris they used screens for their meridian observations for developing catalogues. Have you looked at the spectral type for a magnitude effect?

SMITH: We find no systematic effects correlated with spectral type.

BRONNIKOVA: The biggest contribution to magnitude equation in declination is from the Bucharest catalog. If you did not consider it when you put together the initial catalog, then do you still have that dependency left?

Smith: All of the catalogs were included with weights. Bucharest was not excluded, but contributed about 15% of the total weight.

HØG: The magnitude equation of the Perth catalogue was determined by means of some observations made through a neutral filter. The equation was found to be zero within certain limits and was published by Nikoloff and Høg in the Perth 75 catalogue, pages 14 and 15. Have you made use of this result in your discussion?

SMITH: No. This discussion is confined to the problem of SRS observations made within a few degrees of the equator.

PPM: A TOOL FOR ASTRONOMERS

U. Bastian
Astronomisches Rechen-Institut
Mönchhofstr. 14
D-6900 Heidelberg
Federal Republic of Germany

ABSTRACT. The new catalogue of Positions and Proper Motions ("PPM")
was compiled to replace the AGK3 and SAO catalogues as an astrometric
reference on the northern celestial hemisphere. It provides more than
180 000 reference stars, is on the system of FK5 (J2000.0) and has a
higher accuracy than the two older catalogues.
Some properties of PPM and its High-Precision Subset are presented.

1. INTRODUCTION

PPM gives positions and proper motions of 181731 stars north of -2.5
degrees declination. Its main purpose is to provide a convenient, dense
and accurate net of astrometric reference stars on the northern
celestial hemisphere. This net is designed to represent as closely as
possible the new IAU (1976) coordinate system on the sky, as defined by
the FK5 catalogue. In other words, it is a representation of the FK5
system at higher star densities and fainter magnitudes.

Two older catalogues of similar character have served the same
purpose in the past decades: AGK3 and the SAO Catalogue. There are
three major reasons to replace these now:

1) SAOC and AGK3 are representations of the now obsolete FK4
system of positions and proper motions. Astronomers should have a
direct access to the FK5 system.

2) The accuracy of positions and proper motions in AGK3 and SAOC
is no longer satisfactory. Astronomers should have a more accurate
tool. Over more than a century astrometrists have accumulated a vast
treasure of measured star positions. The power of present-day
computers makes it easy to analyse and combine this amount of data.

3) Proper motions in AGK3 and SAOC were derived from only two
separate source positions per star. This lack of redundancy lead to a
large number of coarse errors in those catalogues. With more than two
measurements per star such errors can be largely avoided. In this way
astronomers now get a more reliable astrometric tool.

PPM is available on magnetic tape from the astronomical data centers CDS at Strasbourg, France and NSSDC at Greenbelt, Maryland - or
from the authors. A printed version is aimed at.

J. H. Lieske and V. K. Abalakin (eds.), Inertial Coordinate System on the Sky, 465–467.
© 1990 IAU. Printed in the Netherlands.

2. PROPERTIES OF PPM

The table below shows a summary of the accuracy budget of PPM. Each line of the table gives the following data for the particular set of stars indicated: The number of stars in the set, the average number of source positions per star, the average of the mean epochs (for right ascension and declination), the average of the mean errors of proper motion (for right ascension and declination) and the average of the mean errors of position at epoch 1990 (again for right ascension and declination). Units are seconds of arc and seconds of arc per century, respectively. At the bottom of the table the corresponding values for AGK3 and SAOC are given for comparison.

Table 1: Accuracy budget of PPM

set of stars	No. stars	No. obs.	mean epochs		mean err. prop. mot.		mean err. pos. 1990	
PPM, all stars	181731	6.2	1931.5	1930.7	0.43	0.42	0.27	0.27
PPM, HPS stars	31841	7.8	1950.3	1948.0	0.24	0.25	0.12	0.12
PPM, FK5 stars	1365	---	1954.2	1945.2	0.08	0.10	0.04	0.05
AGK3	181581	2.0	1945	1945	0.95	0.95	0.45	0.45
SAOC, north	133000	2.0	1930	1930	1.5	1.5	0.9	0.9

Three subsets of PPM are shown in the table: The first line refers to the entire catalogue, the third line to the FK5 stars and the second line to the High-Precision Subset (HPS) of PPM. This subset is defined as the set of PPM stars for which either Carlsberg Meridian Catalogue or AGK3R observations are available. Its superior precision in present-day positions is mainly due to the work done at the Carlsberg Automatic Meridian Circle. Applications of PPM demanding utmost accuracy rather than high star density should use HPS stars only.

Note that (on average) more than 6 measured positions were used per star. This redundancy allowed to discover (and eliminate) a large number of coarse errors in the source catalogues. The positions and proper motions given in PPM should be highly reliable, therefore.

3. PRESENTATION OF DATA IN PPM

The physical units used and the arrangement of stars are the same as in the printed SAOC. The stars are arranged in belts of 10 degrees width. In addition to the positions and proper motions and their respective mean errors PPM gives the following data for each star: Spectral type and magnitude (both copied from AGK3), SAOC number, HD number, AGK3 number, DM number and a set of flags. The astrometric data are given for equator and equinox J2000.0, on the system of FK5. The flags indicate double stars, members of the HPS, peculiar object designations and some uncertain astrometric data, for instance.

As PPM presently does not cover the southern celestial hemisphere the PPM magnetic tape contains also a J2000.0 (FK5) version of the southern portion of SAOC. In addition all data given for J2000.0 are repeated for equator and equinox B1950.0 on the system of FK4.

Discussion

WARREN: Prof. Jaschek (*see editorial remark below*) used his usual insight when he made the comments that you cited. We recently performed statistics on archived catalogs disseminated from the data centers and we found that the two most frequently requested catalogs are the SAOC and the Bright Star Catalogue. The SAO is often used for applications such as spacecraft tracking and guidance and for general reference because it includes the whole sky; however, it is well known that the SAO catalog has its real problems in the south. The PPM is a valuable piece of work and will help the situation, but clearly there is a lot to do yet because such a catalog must cover the whole sky in order to be useful for many present-day applications.

EDITOR'S NOTE: The remark by Jaschek is contained in *Mapping the Sky*, Proceedings IAU Symposium 133 (S. Débarbat *et al*. eds.) page 381: *"Now it is unnecessary to ask your opinion about SAO, because there seems to exist a certain consensus that it is not the best catalogue which could have conceivably been constructed. But the general use made of it shows that for non-astrometrists it is the catalog "par excellence." I think this underlines very clearly the needs of the non-specialist for astrometric data, and it tells also that the astrometric community has not been very active in responding to these needs. If astrometrists feel that one could do better than SAO, then DO IT—but please do not tell that one has to wait until HIPPARCOS is reduced, or give some other date from here in twenty years. Users wish to have data NOW."*

BASTIAN: I fully agree. It is clear that a southern-hemisphere equivalent has to be provided as quickly as possible. We shall do this within about a year. The improvement in the south will indeed be much larger, simply because the SAOC has such an utterly low accuracy in the south.

CORBIN: (1) You have over 400,000 AC positions in your list. Are they of different stars? You have only 182,000 in your catalog.

(2) What zones do you cover with your AC data?

(3) How do you determine the mean errors of your proper motions? From propagation of errors?

BASTIAN: (1) The Astrographic Catalogue was designed to be a two-fold coverage of the sky. In fact it is even 2.5-fold. So we have 2.5 independent positions for each star.

(2) We used all AC zones from $-2°$ to $+90°$ declination. We were lucky to find punched cards with the necessary data on them in the Strasbourg and Hamburg observatories. At Strasbourg we found *all* AC data complete from $-2°$ to $+32°$ on 6 tons of punched cards. Luckily we did not need to read all of these cards, because most of the data had already been read onto a tape. You had this tape in your possession and kindly gave it to us. At Hamburg we found punched cards from $+32°$ to the pole. But these did not contain *all* AC data for these zones. Dieckvoss 20 years ago had selected only the data for the AGK3 stars from the printed AC volumes. These data were kindly given to us by Drs de Vegt and Steinbach.

(3) Yes, error propagation using separately determined weights for each catalogue. The weights were determined in different ways for different catalogues.

THE SYSTEM OF THE PPM CATALOGUE

SIEGFRIED ROESER
Astronomisches Rechen-Institut
Mönchhofstr. 12 – 14
D-6900 Heidelberg
Federal Republic of Germany

ABSTRACT. The coordinate system of the Catalogue of Positions and Proper Motions (PPM) was constructed as an extension of the FK5 sytem to higher star densities and fainter magnitudes. Zonal deviations from the FK5 were minimised on scales of 7×7 square degrees via AGK3R and CMC catalogues. Compared with these catalogues, no magnitude equation was found for stars brighter than $m(\text{pg}) = 10.5$.

1. Introduction

PPM (Roeser and Bastian, 1989) is a new catalogue of reference stars on the northern celestial hemisphere. It contains the positions and proper motions of 181731 stars for equinox and epoch J2000.0 on the system of FK5.

The Fifth Fundamental Catalogue (FK5) (Fricke *et al.*, 1988) is the presently adopted official coordinate system of the IAU. This system is defined by the positions and proper motions of the 1535 basic fundamental stars evenly distributed over the sky. The FK5 defines the system, but how shall an astronomer refer his observations to FK5? The spatial stellar density of FK5 is so low (1 star per 27 square degrees) that only instruments able to perform large-scale astrometry like meridian circles or astrolabes can make direct use of FK5. So, the FK5 stars serve as the knots in a network which spans over the whole sky, but secondary knots have to be introduced in order to make the net denser. This densification of the net can be achieved with a single instrument provided that this instrument measures large angles free of all kinds of systematic errors. The HIPPARCOS satellite, according to its design and measuring principle, would have been the ideal instrument if it had worked as intended. Under the present circumstances, however, HIPPARCOS will not be able to supply positions *and* proper motions with the required systematic accuracy. Of course, PPM cannot be a substitute to the desired HIPPARCOS system, it should only be considered as a practical tool to refer observations to the system of FK5.

This paper describes the measures, U. Bastian and myself have taken to construct the system of PPM from catalogues which were produced by conventional earthbound observing techniques.

469

J. H. Lieske and V. K. Abalakin (eds.), Inertial Coordinate System on the Sky, 469–478.
© 1990 *IAU. Printed in the Netherlands.*

2. The Problem

The problem of the construction of the PPM system is twofold. First, the FK5-system has to be extended to higher star density and fainter magnitudes. Second, all the observational catalogues which are used for PPM must be rigourously reduced to this system. We will concentrate on the first topic now and cover the second part at the end of the paper. For the northern hemisphere — only this is covered by PPM at present — a "natural" extension of FK5 is AGK3R (Scott, 1968), and for even higher star densities AGK3 (Heckmann *et al.*, 1975), if properly transformed to FK5. Table 1 shows a comparison of the spatial stellar densities between these three catalogues. Loosely speaking, it can be interpreted that one FK5 star determines the system for about 225 AGK3 stars. The mean photographic magnitude in FK5 is 5.5 mag, it increases to 8.8 mag in AGK3R and to 9.7 mag in AGK3/PPM.

Table 1. Spatial stellar density and mean photographic magnitudes in the catalogues FK5, AGK3R and AGK3/PPM.

Catalogue	Stars/sq. deg.	mean magnitude (phot.)
FK5	0.04	5.5
AGK3R	1	8.8
AGK3/PPM	9	9.7

In conclusion from Table 1 we are faced with the problem of extending a coordinate system defined for a few bright stars to much fainter stars without making a systematic error usually called magnitude equation. In a similar way we must verify that the final system has no colour equation, *i.e.* red stars must not be rotated with respect to blue stars.

So, the construction of the system of positions and proper motions of PPM was carried out in a "natural" way. Starting from FK4/5 we went via AGK3R and CMC to AGK3/PPM applying all the necessary corrections found from catalogue comparisons.

3. Catalogue Comparisons

In this section we describe the methods used for the comparisons of catalogues and the reduction of individual catalogues to our system. Systematic differences between catalogues were investigated depending on the position on the sky, on magnitude and on spectral type. Position-dependent systematic differences are determined in the following way: In right ascension and declination they are given by a spatial moving average filter whose grid-points are the stars common to both catalogues. The spatial frequency of this filter is chosen such that about 30 to 40 neighbouring stars contribute to each star's systematic correction.

For comparison in magnitude and spectral type simple step functions are determined on magnitude intervals of 1 mag and 1 spectral class. In our case, where the primary catalogue to compare with is AGK3, a finer subdivision would overinterpret the effects, because of the uncertainties in AGK3 magnitudes and spectral types. Usually global magnitude and colour corrections are applied, but dependence of the colour equation on declination, for instance, was always investigated.

4. The Construction of the System

For practical reasons we started with FK4 as reference system. Systematic comparisons of some of the source catalogues with FK4 were already available. AGK3R has been compared with FK4 by Schwan (1985 and private communication) with his method of representing spatial systematic differences and magnitude equations by series developments. Schwan found that AGK3R is on the system of FK4. As AGK3R is the reference star catalogue for AGK3, we can say that reducing rigourously AGK3 onto AGK3R means AGK3 is on the system of FK4 for the mean epoch of 1958. In the same paper, Schwan (1985) also compared AGK3 with AGK3R and found no significant deviation of AGK3 from AGK3R at the observational epoch of AGK3. To study the proper motion system of AGK3, Schwan compared it with NIRS (Corbin, 1976) and detected small deviations. We added these corrections to AGK3 proper motions, and in all comparisons below the term AGK3 should always be understood as AGK3 including Schwan's corrections. In the construction of the PPM system we repeated this comparison using the methods described above with the intention to analyse the difference between AGK3 and AGK3R on spatial scales smaller than those investigated by Schwan. The results of this comparison are shown in Figure 1. From bottom to top in this figure, the systematic differences between AGK3R and AGK3 as functions of magnitude, spectral type, right ascension and declination are plotted. Plots on the left hold for differences in r.a. times cos δ, those on the right for differences in declination. No magnitude equation and no spatial systematic differences between AGK3 and AGK3R were found even on scales of 7×7 square degrees, a scale which was much smaller than that used by Schwan in his comparisons. This figure documents that the authors of AGK3 successfully reduced their measurements onto the system of AGK3R with one exception.

This exception shows up in the comparison between AGK3 and AGK3R declinations as a function of spectral type. A small systematic slope from blue stars towards red stars is noted, a so-called colour equation. The effect seems to be very small, but Figure 1 is a global curve, *i.e.*, all 20 500 stars on the northern hemisphere common to AGK3 and AGK3R enter this comparison. To further investigate this effect we repeated this plot for different declination zones. This is shown in Figure 2 and gives the hint to the solution. The effect is stronger at lower declinations, and reverses its sign near the pole. So, one must suspect that this colour equation is caused by differential colour refraction. This effect has not been considered in the construction of AGK3, at least it is not mentioned in the AGK3 introduction (Heckmann *et al.*, 1975; see also de Vegt, 1988). To correct AGK3 for this effect, all AGK3 declinations were corrected by $\Delta\delta$ given by

$$\Delta\delta = A + B \tan{(\phi - \delta)},$$

where ϕ is the geographic latitude of Hamburg, and the coefficients A and B are determined from the actual AGK3R − AGK3 differences for each spectral type bin. The coefficients B — so derived — agree to within 10 per cent with theoretical values derived from atmospheric dispersion and the spectral energy distribution of the stars and photographic plates. Today this effect is well known in photographic astrometry, and one tries to minimise it by using yellow-sensitive plates.

The proper motion system of AGK3 is simply given by the difference AGK3 − AGK2, where modifications to the original AGK2 were introduced by Heckmann *et al.* (1975) in order to reduce AGK2 to FK4. It must be suspected, however, that differential colour refraction was not considered in the reduction of AGK2 also, meaning that the proper motion system is essentially unaffected by

472

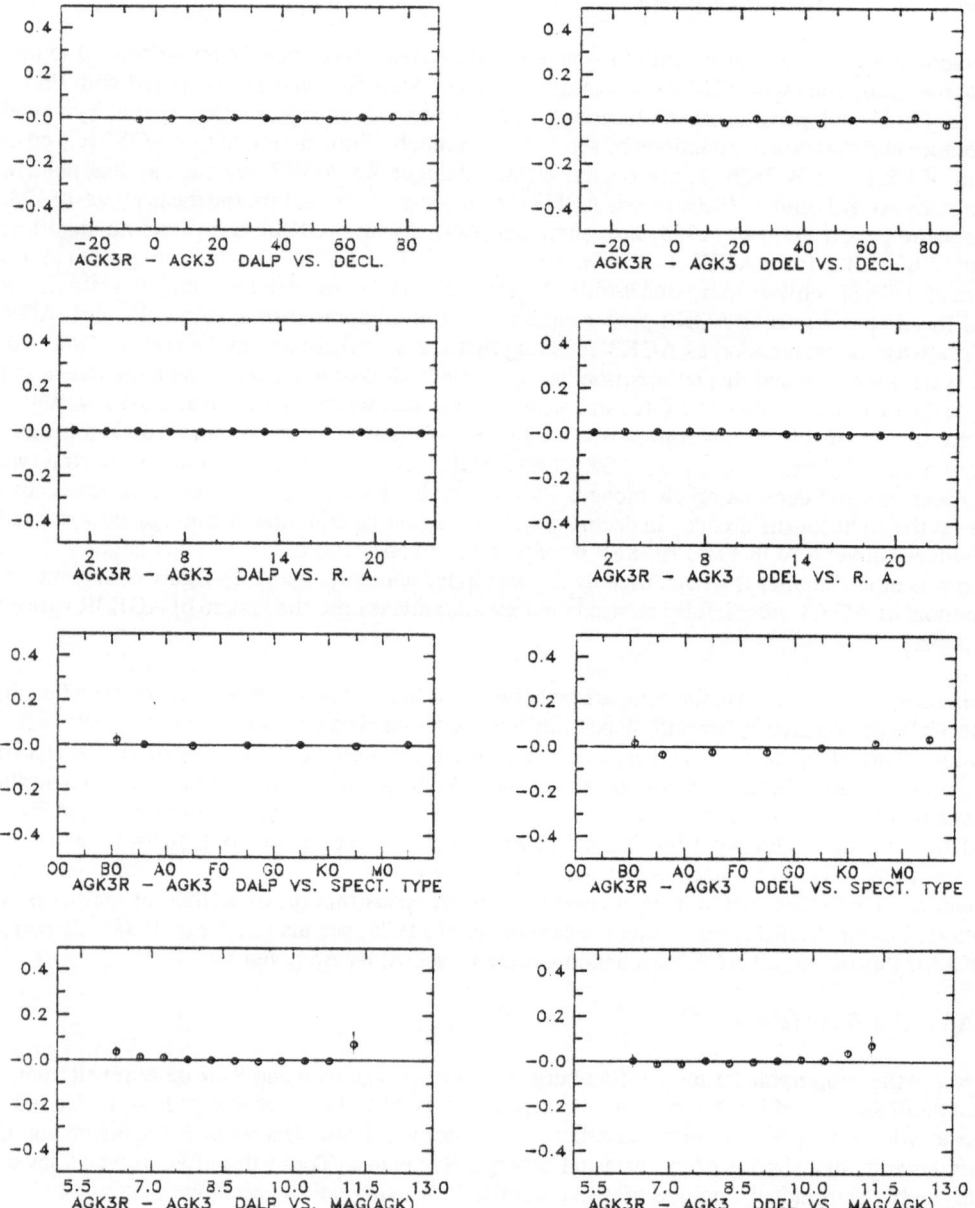

Figure 1. Systematic differences AGK3R – AGK3 (after Schwan's corrections) in right ascension times cos δ (*left*) and declination (*right*). From top to bottom the differences depending on declination, right ascension, spectral type and magnitude are shown. Units are sec of arc.

this effect. This problem is further discussed below, where we compare AGK3 with CMC.

In constructing the proper motion system of PPM we were guided by the idea that only the most reliable catalogues should be used and, especially, the system should represent FK5 as precisely as possible at present epoch. This led to the following conclusion: the system of PPM is determined by AGK3R at epoch 1958 and by CMC (1985-88) at epoch 1986.5.

The comparison with the CMC catalogues is described below. Although the systems of the individual CMC catalogues differ slightly from each other, we discuss them as one entity after reducing them individually to the FK4 system using Schwan's (1988) results. Then AGK3 was compared with CMC (on FK4). This comparison is shown in Fig. 3, which has the same scheme as Fig. 1. Let us first consider the magnitude dependence. In right ascension we note a flat region of almost zero systematic difference between magnitudes 6 and 10.5, where the bulk of PPM stars lies. All stars brighter than 6 mag fall in the first bin in Fig. 3. There the observed r.a. in CMC are systematically larger than those computed from AGK3 for this epoch. But only 770 stars or 0.4 per cent of our ensemble lie in this magnitude range. In the majority these stars are bright blue stars, which are difficult objects in photographic astrometry, as can also be seen from the plots showing the dependence on spectral type in this Figure. This should be remembered if one uses data for these stars from PPM. For stars fainter than 10.5 a systematic difference between CMC and AGK3 of about 0".1 can be found. It is quite possible, that this effect is due to errors in the AGK3 proper motion system, but there is no final proof. For this magnitude range, the number of stars in CMC which can

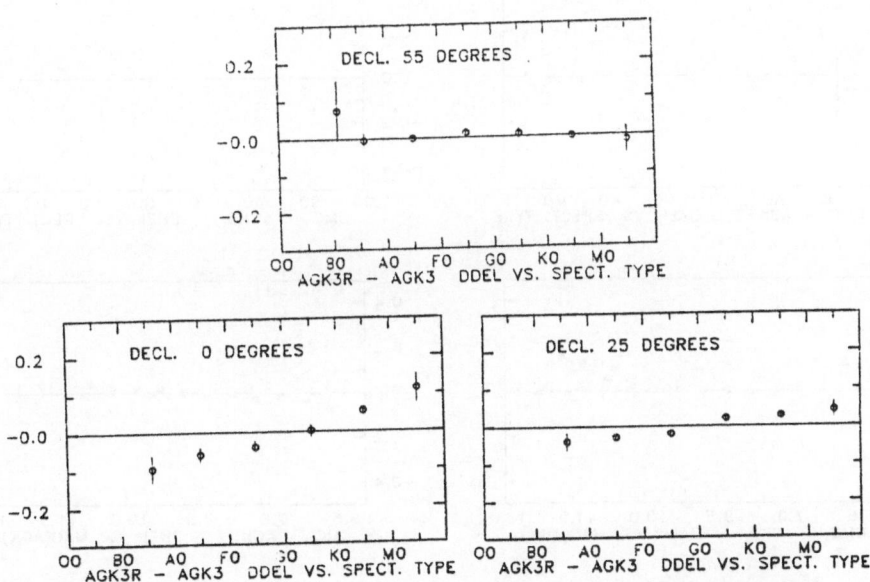

Figure 2. Systematic differences $\Delta\delta$ AGK3R – AGK3 depending on spectral type for declinations 0, 25 and 55 degrees. Units are sec of arc.

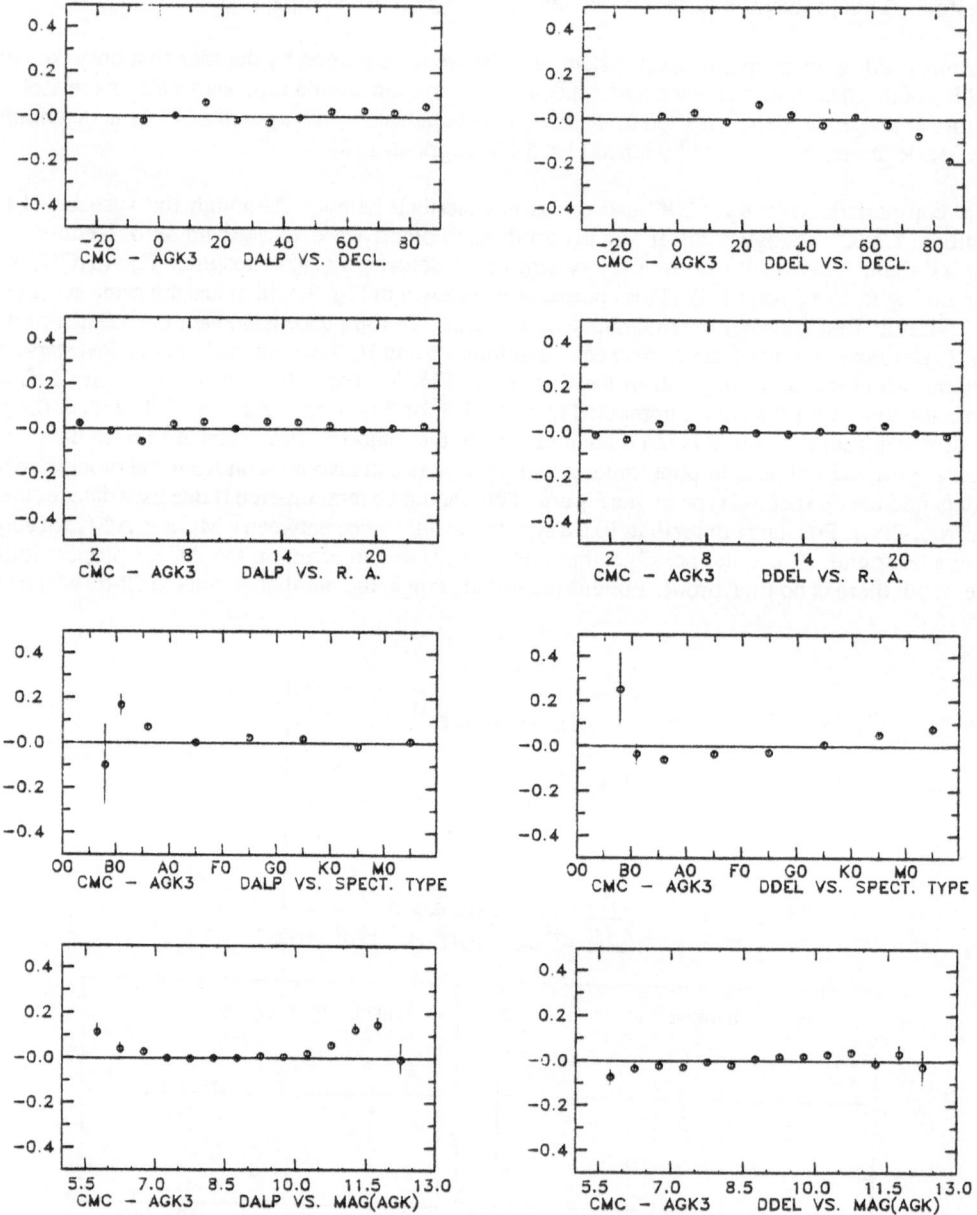

Figure 3. Systematic differences CMC (on FK4) – AGK3 (after Schwan's corrections) in right ascension times cos δ (*left*) and declination (*right*). From top to bottom the differences depending on declination, right ascension, spectral type and magnitude are shown. Units are sec of arc.

be used for a comparison strongly decreases. For stars brighter than 10 mag, the ratio CMC/PPM is about 0.1; it decreases to 0.05 between $m = 10$ and $m = 10.5$ and to 0.03 for stars fainter than 10.5 mag. This means that the systematic corrections would be determined from a relatively small number of comparison stars. We did therefore not correct the PPM system for this effect. But any user of PPM is free to do so, of course, using Fig. 3 for this purpose.

In declination we note a very smooth slope from the bright to the faint end. The size of the slope is 16 milli-arcsec/mag, which is so small that we did not apply this correction.

The results shown in Fig. 3 are rather remarkable. It is by no means straightforward, that in comparing the proper motions of a photographic catalogue AGK3/2 with a present-day photoelectric meridian circle catalogue magnitude-dependent effects are so minute. As the differences in epoch AGK3 – AGK2 and CMC – AGK3 are almost the same, the comparisons above are essentially comparisons between CMC and AGK2 modulo the sign. The smallness of all the effects in Fig. 3 is a proof for the good work done in the compilation of AGK2.

The investigation of colour-dependent effects shows similar features as in the comparisons with AGK3R. The differences $\Delta\alpha \cos \delta$ show no colour dependence for spectral types from late B to M, and we already mentioned that stars earlier than B2 are problematic in PPM. As in the comparison with AGK3R, the influence of neglecting the colour refraction is inherent in the declination differences. It is larger than in the case of AGK3R suggesting that AGK3 proper motions also carry this defect. This would indicate that in AGK2 the effect of atmospheric dispersion differs from that in AGK3, contrary to our expectation above. This could be studied in a new reduction of AGK2 using a suitable reference catalogue. We did not do this and therefore left the PPM proper motion system unchanged in this respect.

The remaining plots in Fig. 3 show the position-dependent systematic differences between AGK3 and CMC on FK4. These are determined and removed with the moving-average-filter method explained in section 3. Again, it is noteworthy that these systematic differences are rather small with only one minor exception. This refers to $\Delta\delta$ for declinations larger than about +75 degrees. There is a difference between CMC (on FK4) and AGK3 of about –0''2. From the comparison CMC – NIRS (see CMC, 1986) we find $\Delta\delta$ about –0''1, and Schwan (1988) determined +0''1 for the resp. difference CMC – FK4. This explains this relatively large discrepancy between CMC (on FK4) and AGK3.

5. Systematic Reductions of Individual Catalogues

PPM is constructed from the following source catalogues (see Roeser and Bastian, 1989): The Astrographic Catalogue (AC), AGK1, Yale zone catalogues, AGK3, AGK2, AGK3R and CMC. The reduction of AC, AGK1 and Yale is briefly described below.

For the reduction of the 12070 AC plates a single-plate, second-order polynomial fit algorithm was selected. Each observatory's AC zone was then investigated with respect to higher-order distortions, "coma", magnitude and/or spectral type dependent systematic deviations from AGK3, and corrected.

The reductions of the Yale and AGK1 zones are performed with the methods of catalogue comparison described in section 3.

6. Conclusion

The individual steps for the derivation of the PPM system can be summarized as follows: As starting hypothesis the system of AGK3 positions and proper motions on equinox 1950.0 was selected. First, the corrections found by Schwan (1985) were applied. This is an intermediate system, nominally on FK4, from which systems of position at epoch 1958.0 and 1986.5 can be computed.

The 1958.0 positional system was compared with AGK3R, the only correction applied to this system is the correction in declination due to the effect of differential colour refraction (see Figures 1 and 2).

The 1986.5 positional system was compared with CMC. Spatial systematic differences in right ascension and declination as found from the comparison were applied, and, also at this epoch, the same declination correction due to colour refraction as above must be performed. No correction for magnitude effects has been made as explained above.

The system of positions and proper motions on FK4, equinox 1950.0, is then given by the 1986.5 and the 1958.0 systems and their difference.

The final transformation from FK4 system at equinox and epoch 1950.0 to the FK5 system at equinox and epoch J2000.0 was performed in exactly the same way as in the construction of FK5.

It is a rather difficult undertaking to estimate the systematic errors of the PPM system of position and proper motions. But, it seems appropriate to give here at least rough figures even if they are uncertain. Let us assume that both AGK3R and CMC have a systematic accuracy of their positional system of 0".04. It can be concluded from Figure 1, that the systematic difference between AGK3R and PPM is smaller than 0".04, and after applying the systematic differences to CMC derived from Figure 3, the deviation of PPM from CMC (on FK4) is also less than 0".04. Then we immediately find the mean epoch of the PPM system to be 1972, and we may expect typical systematic errors of position at mean epoch of 0".03, and a typical error of the proper motion system of 0.20 arcsec/ century.

References

CMC (1985-1987): *Carlsberg Meridian Catalogue La Palma*, Numbers 1 to 3. Copenhagen University Observatory, Royal Greenwich Observatory and Real Instituto y Observatorio de la Armada, San Fernando.

CMC (1988): *Carlsberg Meridian Catalogue La Palma, Number 4*. Copenhagen University Observatory, Royal Greenwich Observatory and Real Instituto y Observatorio de la Armada, San Fernando. Private communication.

Corbin, T. (1978): "The proper motions of the AGK3R and SRS stars". *IAU Colloquium No. 48*, Eds. F.V Prochazka and R.H. Tucker, 515.

Fricke, W., Schwan, H., Lederle, T. (1988): *Fifth Fundamental Catalogue (FK5)*. Part I. The Basic Fundamental Stars. *Veröff. Astron. Rechen-Institut Heidelberg* Nr.32.

Heckmann, O., Dieckvoss, W., Kox, H., Günther, A. and Brosterhus, E. (1975): *AGK3. Star catalogue of positions and proper motions north of –2.5 declination.* Hamburg-Bergedorf 1975.

Roeser, S. and Bastian, U. (1989): *PPM - Positions and Proper Motions of 181731 stars north of –2.5 degrees declination.* Astronomisches Rechen-Institut, Heidelberg, 1989.

Scott, F. P. (1968): "The AGK3R, SRS and related projects", In: *Highlights of Astronomy*, ed. L. Perek. International Astronomical Union. Reidel, Dordrecht, 279.

Schwan, H. (1985): "The systems of the positions and proper motions in star catalogues AGK3, AGK3RN, and N30", *Astron. Astrophys.* **149**, 50.

Schwan, H. (1988): Private communication.

de Vegt, Ch. (1988): "Status of Photographic Catalogs: Available Material and Future Developments", *Proc. IAU-Symp. 133 Mapping the Sky.* Eds. S. Débarbat, J.A. Eddy, H.K. Eichhorn and A.R. Upgren. Kluwer, Dordrecht.

Discussion

COLE: (1) The transformation from FK4 to FK5 is magnitude dependent. How did you handle this problem? (2) I note that a slightly different approach was used for the SRS. The average magnitude of FK4 stars in the neighborhood of the SRS star was used. I do not know if there is any significant difference in those two approaches.

RÖSER: (1) We have to perform this transformation for stars fainter than those in the FK5. Nothing is really known about such a transformation for these faint stars. So, we made this transformation for the fainter stars using Schwan's transformation functions at $m_v = 6.5$, which is the faintest realistic point. (2) I have not in mind the size of the magnitude dependent term in this transformation. The average magnitude in the FK5 is $m_v = 4.7$, so we can calculate the size of this effect in comparing the transformation functions at 4.7 and at 6.5 magnitudes.

SMITH: In applying Schwan's analytic functions to reduce a catalog of faint stars which was referred to the FK4 system (such as the AGK3R catalog) to the FK5 system, to what extent do you think the average magnitude of the FK4 stars in the vicinity of the particular AGK3R star should be used as an input argument to the analytic function — in preference to some constant value far from the average magnitude of the FK4 stars?

RÖSER: In order to study the size of the effect, I will go home and look into the transformation function. I will send my response to the editor.

[Written Response]

First, the FK4 system is only well-defined for $m_v < 6.5$ mag. At present the FK5 system given by the basic FK5 (Fricke *et al.*, 1988) is also not defined for faint magnitudes. So the transition between FK4 and FK5 at magnitudes fainter than $m_v = 6.5$ is, in principle, undefined.

Second, in constructing the PPM on FK5 we made the transition at $m_v = 6.5$ for all stars fainter than that. Your question refers to magnitude dependent effects correlated with the position on the sky. Checking these effects on Schwan's tables for the transition, the only noticable effect occurs in proper motions in right ascension. There the position dependent difference between $m_v = 5$ and $m_v = 7$ is at its maximum 0.015 arcsec/cy, which is so small, that it does not matter at which magnitude the transition is made.

RATNATUNGA: After all the corrections for systematic effects which reduce the catalogs to a single proper motion system of FK5, how do you ensure that this is an inertial frame of reference?

RÖSER: We do not know how "inertial" the FK5 is. We tried the best we could do to reduce our catalogue to the FK5. It can, of course, not be "more inertial" than the FK5. It would be highly desirable to have an instrument able to measure large angles between a great number of stars free of systematic errors like the HIPPARCOS mission as it was originally conceived.

BASTIAN: PPM, aiming to be a representation of the FK5 at higher star density and fainter magnitude — aiming to be on the system of the FK5 as closely as possible — it cannot be better than the FK5. This would be the ultimate limit. And from the discussions today we know that the FK5 is not perfectly inertial.

THE MACHINE-READABLE DURCHMUSTERUNGEN: CLASSICAL CATALOGS IN CONTEMPORARY FORM

Wayne H. Warren Jr.
Astronomical Data Center (ADC)
National Space Science Data Center (NSSDC) /
World Data Center A for Rockets and Satellites
NASA Goddard Space Flight Center
Greenbelt, Maryland 20771
U. S. A.

François Ochsenbein
Centre de Données Astronomiques de Strasbourg
11, rue de l'Université
F-67000 Strasbourg
France

Barry N. Rappaport
Jet Propulsion Laboratory
4800 Oak Grove Drive
Pasadena, California 91009
U. S. A.

ABSTRACT. The entire series of *Durchmusterung* (DM) catalogs (*Bonner, Southern, Córdoba, Cape Photographic*) has been computerized through a collaborative effort among institutions and individuals in France and the United States of America. Complete verification of the data, both manually and by computer, the inclusion of all supplemental stars (represented by lower case letters), complete representation of all numerical data, and a consistent format for all catalogs, should make this collection of machine-readable data a valuable addition to digitized astronomical archives.

1. Introduction

The *Durchmusterung* catalogs need little introduction to most astronomers, particularly those working in the area of postional astronomy and the identification of stars. The *Bonner DM* (BD) and its southern extension, the *Southern DM* (SD) were completed and published by Argelander (1859-62) and his assistant Schönfeld (1886) following many years of painstaking observations. The visual techniques used for these surveys were identical, except that the SD observations were made with a larger telescope and, thus, the resulting catalog extends to fainter magnitudes. The *Córdoba DM* (CD, Thome 1892-1932) extends the German visual survey to the south celestial pole, while the *Cape Photographic DM* (CPD, Gill and Kapteyn 1895-1900) repeats the southern zones using photographic techniques.

The present paper briefly outlines the procedures used for the computerization of the entire set of DM catalogs. This work has been accomplished over the last 15 years and results in the availability of all DM catalogs in a uniformly-formatted, machine-readable edition.

479

J. H. Lieske and V. K. Abalakin (eds.), Inertial Coordinate System on the Sky, 479–481.
© 1990 *IAU. Printed in the Netherlands.*

2. The Machine-Readable Catalogs

2.1. PROCEDURE

All data were keyed directly from the published catalogs to disk storage, but the procedures differed somewhat depending upon local circumstances and available software to partially automate the work. The following sections briefly describe the procedures used for the four catalogs.

2.1.1. *Córdoba Durchmusterung.* All data were keyed at the NSSDC with the help of a preprocessing program to automatically insert data that remained the same throughout each zone. A computer program was written to process individual zones and to create hard copy in the exact format and structure as in the published catalog. This allowed the matching of last stars on each page to detect missing records. Individual zones were then supplied to a number of volunteers for proofreading. Many zones were also proofread at the ADC, where all error correction, incorporation of errata and corrigenda, and final assembly of the catalog were done by WHW.

2.1.2. *Cape Photographic Durchmusterung.* The northern zones ($-18°$ to $-32°$ and $-35°$) were keyed and partially verified at Case Western Reserve University under the supervision of BNR. Unverifed northern and the remaining zones were keyed and verified by a commercial firm with funding by the NSSDC and supervision and checking by BNR. Final checking, incorporation of errata, and final assembly were done at the ADC by WHW.

2.1.3. *Bonner Durchmusterung.* The edition of Küstner (1903) was used, the keying of the data being divided according to the following table:

Zones	Location
$+89°$ to $+60°$	Centre de Données Astronomiques de Strasbourg (CDS)
$+59°$ to $+26°$	Observatoire de Nice
$+25°$ to $+24°$	National Space Science Data Center (also CDS)
$+23°$ to $+20°$	B. N. Rappaport ($+23°$ also done at CDS)
$+19°$ to $-01°$	National Space Science Data Center ($+14°$ also done at Nice)

Zones $+89°$ to $+60°$ were also verified at the CDS and zones $+59°$ to $+26°$ were proofread there. Software for data entry was written by FO, who also supervised the work. The remaining zones were proofread at the ADC. Redundant zones were compared by computer at the ADC, where final checking, incorporation of supplemental stars and errata, analysis and flagging of "missing" stars, and final assembly were done by WHW.

2.1.4. *Southern Durchmusterung.* Zones $-02°$ to $-21°$ were done at the CDS under the supervision of FO. The remaining zones were keyed and proofread at the ADC by WHW. All zones were carefully examined, checked for sequencing and record counts, and assembled into the final catalog by WHW.

2.2. FINAL CATALOGS

The zone coverages and final catalog statistics are given in the following table. Record counts are greater than star counts because stars that have been deleted from the catalog are flagged in the machine version, while their records and data have been left in the catalog so that zone counts and sequencing agree with the published catalogs.

DM	Zones	Number of Records	Number of Stars
BD	+89° to −01°	325037	324948
SD	−01° to −23°	134834	134832
CD	−22° to −89°	613959	613953
CPD	−18° to −89°	454877	454875

3. Summary

The entire set of DM catalogs has been computerized in a uniform format. All known corrigenda and errata, including changes made in various reprinted editions of the BD and SD, have been incorporated into the catalogs. The machine-readable catalogs will allow all DM stars not already in the SIMBAD data bank to be entered. The full catalogs are being disseminated through the worldwide network of astronomical data centers. More detailed information on each catalog, including individual zone statistics, can be found in documentation distributed with the machine-readable catalogs.

4. Acknowledgments

The authors gratefully acknowledge the efforts contributed by numerous staff members of the collaborating institutes and by volunteers who proofread DM data. This immense project would not have been possible without their help.

5. References

Argelander, F. W. A. 1859-1862, *Bonner Sternverzeichniss*, erste bis dritte Sektion, *Astronomischen Beobachtungen auf der Sternwarte der Königlichen Rhein.*, Friedrich-Wilhelms-Universität zu Bonn, Bände 3-5.

Gill, D. and Kapteyn, J. C. 1895-1900, *Cape Photographic Durchmusterung, Ann. Cape Obs.* 3 (1895, Part I: zones −18° to −37°); 4 (1897, Part II: zones −38° to −52°); 5 (1900, Part III: zones −53° to −89°).

Küstner, F. 1903, *Bonner Durchmusterung des Nördlichen Himmels*, zweite berichtigte Auflage, Bonn Universitäts Sternwarte (Bonn: A. Marcus und E. Weber's Verlag).

Schönfeld, E. 1886, *Bonner Sternverzeichniss*, vierte Sektion, *Astronomische Beobachtungen auf der Sternwarte der Königlichen Rheinischen Friedrich-Wilhelms-Universität zu Bonn* 8, Part IV (Bonn: Adolph Marcus).

Thome, J. M. 1892-1932, *Córdoba Durchmusterung, Resultados del Observatorio Nacional Argentino* 16 (1892, Part I: −22° to −32°); 17 (1894, Part II: −32° to −42°); 18 (1900, Part III: −42° to −52°); 21 (Part I) (1914, Part IV, −52° to −62°); 21 (Part II) (1932, Part V: −62° to −90°).

482

THE SECOND CAPE PHOTOGRAPHIC CATALOGUE: CPC2

C.A. MURRAY, M.J. PENSTON
Royal Greenwich Observatory
Herstmonceux Castle
Hailsham, E. Sussex BN27 1RP, U.K.

and

N. ZACHARIAS, CHR. DE VEGT
Hamburger Sternwarte
D-2050 Hamburg
Germany, Fed. Rep.

ABSTRACT. Between 1962 and 1972 the southern hemisphere has been covered by a 4-fold overlap pattern on 5820 plates, taken with a scale of 100 arcsec/mm and using a visual bandpass. More than 2 million pairs of x,y coordinates of the 2 exposures from 276259 stars have been measured at RGO with the GALAXY machine. The conventional plate adjustment, carried out at Hamburg Observatory, is now complete. About 150 000 primary stars will achieve a catalogue accuracy of about 0.06 arcsec at epoch of observation.

The rigorous FK4 and FK5 versions, using the SRS catalogue (C. Smith 1988), will be supplemented by preliminary new proper motions for those stars common with the SAOC. Finally, a rigorous block adjustment will be performed within the next year.

AN ASTROMETRIC CATALOG OF FOUR MILLION STARS

V.V. NESTEROV, V.S. KISLIUK, H.I. POTTER
Central Astronomical Observatory
Pulkovo
196140 Leningrad, USSR

ABSTRACT. The goal of the present study is

(1) Completion of a general reduction of the *Carte du Ciel* astrophotographic catalogs, measurement of the positions in them on modern astronegatives, derivation of proper motions and compilation of a catalog of stars to magnitude 12.5.

(2) Copying the data of all 272 volumes of astrophotographic catalogues onto magnetic tapes (all together 8.5 million measurements).

(3) Measurement of modern astronegatives on FON-OBZOR-S and FOKAT-Y programs.

(4) Studies of the magnitude equation of the AGK3 catalog and other catalogs which will be presumably used as reference catalogs.

(5) Development and comparative studies of several methods of determination of equatorial coordinates on the basis of full information on astronegative measurements. Application of these methods to compilation of the catalog of 4 million stars.

A COMPREHENSIVE ASTROMETRIC DATA BASE:
AN INSTRUMENT FOR COMBINING EARTH-BOUND OBSERVATIONS
WITH HIPPARCOS DATA

Roland Wielen
Astronomisches Rechen-Institut
D-6900 Heidelberg, Federal Republic of Germany

ABSTRACT: The astrometric data bank ARIGFH will contain all relevant astrometric data on stellar positions and proper motions of stars from ground-based observations and space missions. For each star in the ARIGFH, the best available position and proper motion shall be derived. We rediscuss the accuracy of proper motions and positions of fundamental stars, resulting from a combination of data in the FK5 with the expected results from a revised HIPPARCOS mission. The FK5 data could be significantly improved even by rather degraded positions from a revised HIPPARCOS mission.

1. THE ASTROMETRIC DATA BASE ARIGFH

The Astronomisches Rechen-Institut Heidelberg has started to collect 'all' relevant astrometric data on positions and proper motions of stars in a comprehensive astrometric data bank, called ARIGFH. Presently, we deal with the results from earth-bound observations, made with various types of instruments (meridian circles, astrographs etc.). In the future, the ARIGFH should also contain observations obtained from space, such as the HIPPARCOS and TYCHO data.

The ARIGFH is named in memory of the old German project 'GFH', i.e. Geschichte des Fixstern-Himmels. We do not use, however, the data published in the volumes of the GFH, since, for many reasons, it is preferable to go back to the original catalogues.

From all the data collected in the ARIGFH, we plan to derive the 'best' position and proper motion for each observed star. This would produce a comprehensive and most accurate general catalogue of stellar positions and proper motions (ARIGC). By keeping the data base always up-to-date, subsequent versions of the ARIGC, derived and improved in appropriate intervals of time, would then represent always a summary of our actual knowledge of stellar positions and proper motions.

J. H. Lieske and V. K. Abalakin (eds.), Inertial Coordinate System on the Sky, 483–488.
© 1990 *IAU. Printed in the Netherlands.*

2.   STRUCTURE AND STATUS OF THE ARIGFH

The ARIGFH will exist mainly in two versions: as an ARIGFHOBS and as an ARIGFHSYS. In the ARIGFHOBS, the 'direct' observations are listed. For this purpose, the positions will be reduced back to the original epoch and equinox of observation in order to get rid of the various constants of precession used over the centuries. The ARIGFHOBS will be already in a standard format suitable for further treatment.

Using the 'raw data' of the ARIGFHOBS and available fundamental systems, we shall then produce an ARIGFHSYS in which the positions and proper motions are, as far as possible, reduced to a common system (e.g. to the FK5 system at present, later perhaps to a HIPPARCOS or other system) and to a common equinox (e.g. J2000).

The general outline of the tasks for the ARIGFH and the ARIGC is the following: (1) All relevant astrometric catalogues have to be brought into a machine-readable form. (2) The stars in a catalogue have to be identified with the master catalogue of the ARIGFH. This master catalogue consists initially of the more than 500 000 stars of the CDA catalogue, and will be continuously supplemented by observed stars not found in the CDA. (3) Reduction of the astrometric data back to their original epochs and equinoxes of observation. (4) Determination of systematic corrections for each catalogue. Reduction of the astrometric data to a common system and equinox. (5) Determination of the position and, if possible, of the proper motion of each star of the ARIGFH within the fundamental system. This procedure corresponds basically to what is called an 'individual correction' in fundamental astrometry. The result will be the ARIGC.

At present, our main task is still to put the older astrometric catalogues into machine-readable form. The following steps have to be carried out: (1) Identification of suitable catalogues. We expect that the ARIGFH will finally contain more than 2000 individual astrometric catalogues. (2) Locating each catalogue in a library. It is astonishing how often the bibliographic information about an astrometric catalogue is very poor, causing a time-consuming search procedure. (3) Decision about which data given in the catalogue should be typed in which format (raw data or combined results ?, precession ?, cross-references ?, etc.). (4) Typing of the catalogue. Each catalogue is typed twice, by two independent persons, in order to eliminate typing errors. At present, about 1000 astrometric catalogues are available to us in machine-readable form, containing a few million observed positions.

Some subparts of the full ARIGC will probably be finished first. For example, the individual accuracy of the basic FK5 and of the bright extension of the FK5 could be checked and perhaps improved by using data from the ARIGFH. In the basic FK5, the individual observations made before about 1960 have not been included directly, but only through the use of the normal equations of the FK4 (for which in turn the normal equations of the FK3 have been used in order to represent the older catalogues ...). A complete and unified rediscussion of all the catalogues which have entered into the basic FK5 is certainly desirable. For the bright extension of the FK5, only observations after about 1900 have been used, in contrast to the basic FK5. Hence we can expect an individual improvement of the proper motions of the stars of the bright extension of the FK5 by including older catalogues. Similarly, we hope to improve the individual accuracy of the faint extension of the FK5 by using more and older catalogues.

3.   COMBINING HIPPARCOS DATA WITH EARTH-BOUND OBSERVATIONS

One of the motives for setting up the astrometric data base ARIGFH is to combine the results of the HIPPARCOS satellite mission with earth-bound observations. It has been shown in an earlier paper (Wielen 1988) that earth-bound observations would considerably improve the proper motions of the (nominal) HIPPARCOS mission. In view of the probably reduced accuracy of the revised mission with respect to the nominal HIPPARCOS mission, this seems now even more important. In such a combination of earth-bound observations with the HIPPARCOS and TYCHO data, older catalogues contribute more to the individual accuracy of the resulting proper motions than do very recent catalogues. This is due to the larger epoch differences and the often still acceptable positional accuracy of the older catalogues. Of course, systematic errors of the older catalogues have to be carefully eliminated by a reduction to a fundamental system (FK5 or HIPPARCOS system) before we can use such older catalogues for improving the individual accuracy of proper motions (and of predicted positions based on such proper motions).

In the following, we shall rediscuss the accuracy of the proper motions of stars in the FK5 resulting from a combination of earth-bound observations (represented here for simplicity by the FK5 itself) with data from the revised HIPPARCOS mission. At present, the accuracy of the HIPPARCOS data is highly uncertain, mainly due to the unknown duration of the revised mission. Hence, we shall use the typical mean error of a final HIPPARCOS position, $\varepsilon_{p,hipp}$, as a free parameter (within reasonable limits), and we shall assume (hopefully too pessimistically) that HIPPARCOS does not provide proper motions at all. The latter assumption is not very restrictive because it will turn out that proper motions based on a combination of HIPPARCOS positions alone with the earth-bound observations of FK5 stars cannot be significantly improved by corresponding HIPPARCOS proper motions.

In Table 2, we give the mean errors of a proper motion, $\varepsilon_{\mu,tot}$, derived from a combination of the data in a fundamental catalogue with a HIPPARCOS position which has an assumed mean error $\varepsilon_{p,hipp}$. The method for deriving these results has been described by Wielen (1988). The basic data for the fundamental catalogues, mainly provided kindly by H. Schwan (1989), are listed in Table 1. In order to show the importance of the older observations, I have constructed some 'rudimentary' catalogues, which contain only 'modern' observations, by eliminating older fundamental catalogues from the basic FK5 (e.g. FK5 minus FK3). This can be done by using the equations given by Wielen (1988).

Table 1.   Basic data for catalogues considered

Catalogue	Central epoch	Mean error of position	proper motion
	T	$\varepsilon_p(T)$ (mas)	ε_μ (mas/year)
FK5 Basic	1950	20	0.7
FK5 Faint Extension	1940	70	3.0
FK3	1903	44	2.7
FK5 Basic minus FK3	1962	22	1.5
NFK ("FK2")	1880	70	4.8
FK5 Basic minus NFK	1956	21	1.0
HIPPARCOS revised mission	1990	$\varepsilon_{p,\text{hipp}}$ used as a parameter	? (neglected)

Table 2. Mean error $\varepsilon_{\mu,\text{tot}}$ of a proper motion based on a combination of a HIPPARCOS position with fundamental catalogues

Catalogue combined with HIPPARCOS	Assumed mean error in position for the revised HIPPARCOS mission $\varepsilon_{p,\text{hipp}}$ (mas)					(Catalogue alone)
	2	5	10	20	50	(∞)
	$\varepsilon_{\mu,\text{tot}}$ (mas/year)					
FK5 Basic	0.41	0.42	0.44	0.50	0.62	(0.70)
FK5 Faint Extension	1.27	1.27	1.28	1.31	1.49	(3.00)
FK3	0.48	0.50	0.51	0.54	0.74	(2.70)
FK5 Basic-FK3	0.70	0.71	0.75	0.87	1.19	(1.50)
NFK	0.63	0.63	0.64	0.66	0.77	(4.80)
FK5 Basic-NFK	0.53	0.54	0.57	0.66	0.88	(1.05)

Table 3.   Mean error $\varepsilon_{p,tot}(t)$ of a position predicted for an epoch t
for $\varepsilon_{p,hipp}$ = 20 mas at T = 1990

	Catalogues and combinations			
	FK5 Basic		FK5 Faint Extension	
	alone	with HIPPARCOS	alone	with HIPPARCOS
Central epoch	1950	1970	1940	1986
$\varepsilon_{p,tot}$ (")	0.020	0.014	0.070	0.019
$\varepsilon_{\mu,tot}$ (mas/year)	0.70	0.50	3.00	1.31
Epoch t		$\varepsilon_{p,tot}(t)$ (")		
1990	0.034	0.017	0.166	0.020
1991	0.035	0.018	0.168	0.020
1992	0.036	0.018	0.171	0.021
1995	0.037	0.019	0.179	0.022
2000	0.040	0.021	0.193	0.026
2010	0.047	0.024	0.221	0.037

In Table 3, we have listed the mean error of a position predicted for an epoch t, $\varepsilon_{p,tot}(t)$, based on a combination of FK5 data with a HIPPARCOS position of $\varepsilon_{p,hipp}$ = 20 mas. The value of 20 mas is chosen as an example only, and is hopefully much too pessimistic.

The results presented in Table 2 show that the proper motions of the FK5 stars can be significantly improved by a HIPPARCOS position, even if the HIPPARCOS position is not as accurate as predicted by the nominal mission ($\varepsilon_{p,hipp}$ = 2 mas): The mean error $\varepsilon_{\mu,tot}$ of the proper motion of a star in the basic FK5 is typically decreased by a factor of about 1.5, that of a star in the faint extension of the FK5 by a factor of more than 2. The combination of the FK3, containing observations until about 1930 only, with HIPPARCOS gives nearly the same accuracy of proper motions as the basic FK5. Furthermore, the combination of the NFK, which used observations until about 1900, with HIPPARCOS produces still better proper motions than those given in the FK5 itself, if $\varepsilon_{p,hipp}$ is smaller than about 30 mas. All this proves again the special importance of older observations for deriving proper motions from a combination of ground-based observations with HIPPARCOS data.

488

The improvement in accuracy of positions predicted on the basis of a combination of a HIPPARCOS position with FK5 data, with respect to the FK5 itself, is even more impressive: Table 3 shows that the positions predicted for the next two decades are improved by a factor of about 2 for the stars in the basic FK5, and by a factor of more than 6 for the stars in the faint extension, if we could use a HIPPARCOS position with $\epsilon_{p,hipp}$ = 20 mas or better. This is partially due to the improved accuracy of the proper motion (Table 2), but also due to the fact that the HIPPARCOS position at about 1990 is (hopefully) more accurate than the FK5 position for such a recent epoch.

We conclude that even a HIPPARCOS mission with an accuracy much degraded with respect to the nominal one, would allow a great improvement in the accuracy of both the proper motions and the positions (for recent or future epochs) of fundamental (and many other) stars. Since the present system of positions of the FK5 could also be significantly improved by results from the revised HIPPARCOS mission, a combination of HIPPARCOS data with earthbound observations may result in a new catalogue (FK6) of fundamental stars.

REFERENCES

Schwan, H.: 1989, paper in this volume (IAU Symposium No. 141) and private communication.
Wielen, R.: 1988, in IAU Symposium No.133, Mapping the Sky, Eds. S. Debarbat, J.A. Eddy, H.K. Eichhorn, A.R. Upgren, Kluwer Publ. Comp. , Dordrecht, p. 239.

A PRESENTATION OF THE WORK PERFORMED WITH THE BELGRADE LARGE MERIDIAN CIRCLE DURING THE PERIOD 1968–1988

S. Sadžakov, Z. Cvetković and M. Dačić
Astronomical Observatory
Volgina 7
Yu-11050 Belgrade
Yugoslavia

ABSTRACT. The work performed with the Belgrade Large Meridian Circle and the results obtained during the last twenty years are presented.

1. Catalogue of the declinations of the latitude programme stars (KŠZ). [Sadžakov, N.S., and Šaletić, P.D.: 1972, *Publ. Obs. Astron. Beograd*, No. 17, 1]

The Belgrade Catalogue of Latitude Stars (KŠZ) covers the celestial sphere between +13° and +90°. It contains 3957 stars between magnitudes 3.0 and 9.4. Every star was observed four times on the average, in the interval 1968–1971. The mean epoch of observations is 1969.46. The rms-error of a single observation is 0".34. A comparison with the AGK3 was made by zones and yielded an rms difference of 0".28.

2. Catalogue of NPZT programme stars. [Sadžakov, S., Šaletić, D., and Dačić, M.: 1981, *Publ. Obs. Astron. Beograd*, No. 30, 1]

The catalogue of NPZT stars is obtained on the basis of the observational material compiled by use of the Meridian Circle of Belgrade Astronomical Observatory during 1973–1980. It contains 1838 stars with magnitudes $6.5 \leq m \leq 8.5$. The mean epoch of observations is 1977.02 in α and 1978.78 in δ. The rms-error of the right ascension $\varepsilon_\alpha \cos \delta$ is 0^s030 and of declination ε_δ is 0".26. Comparison with the FK4 of about 340 stars yields an error in α of 0^s015 and in δ of 0".14.

3. Belgrade catalogue of double stars. [Sadžakov, S., and Dačić, M.: 1989, *Publ. Obs. Astron. Beograd*, in press]

In the period between March 1981 and April 1987, 1576 stars were observed in Belgrade during the Double Star Programme. The measurements and the treatment of the observational material were performed by use of the relative method. Both coordinates α and δ were observed simultaneously, on the average every three minutes. The rms error of a single observation of double stars between −30° to +60° is $\varepsilon_\alpha \cos \delta = \pm 0^s026$ and $\varepsilon_\delta = \pm 0".32$.

4. A catalogue of positions of 290 stars situated in the vicinity of radio sources. [Sadžakov, S., Dačić, M., and Cvetković, Z.: 1989, *Astron. J.* in press]

J. H. Lieske and V. K. Abalakin (eds.), Inertial Coordinate System on the Sky, 489–490.
© 1990 *IAU. Printed in the Netherlands.*

Simultaneously with these observations we carried out in 1982–87 observations of 290 stars from 78 parts of the sky situated in the vicinity of radio sources. A programme star was observed on the average 5.5 in α and 5.8 times in δ. The mean observational epoch of the catalogue is 1984.60 in α and 1984.70 in δ. The rms-error of a single observation in right ascension is $\varepsilon_\alpha \cos \delta = 0\overset{s}{.}024$ and $0\overset{''}{.}30$ in δ. Comparison with fundamental stars yields $\pm 0\overset{s}{.}009$ in α and $\pm 0\overset{''}{.}13$ in δ.

5. A catalogue of right ascensions and declinations of FK4 stars. [Sadžakov, S., and Dačić, M.: 1989, *Astron. Astrophys. Suppl. Ser.* No. 77, 411]

This catalogue contains the positions of 576 FK4 stars observed during 1981–87. The mean epoch is 1983.90 for α and 1983.84 for δ with mean errors for a single observation of $0\overset{s}{.}022$ in α and $0\overset{''}{.}32$ in δ.

6. General catalogue of Latitude Stars (IKŠZ). [Sadžakov, S.: 1978, *Publ. Obs. Astron. Beograd*, No. 24, 1]

The values of declinations and proper motions in IKŠZ were derived on the basis of about 36000 star positions, the rms error of their determination ranging from $0\overset{''}{.}20$ to $0\overset{''}{.}62$ with the average being $0\overset{''}{.}35$. The rms error of the determination of a single position is $\varepsilon_\delta = 0\overset{''}{.}08$ and the proper motion $\varepsilon_\mu = 0\overset{''}{.}005$ with a mean epoch of 1969.44.

7. Declinations and proper motions of the stars of the International Latitude Service on the basis of meridian catalogues from 1929–1972 (BSKŠZ1; BSKŠZ2). [Sadžakov, S., and Šaletić, D.: 1975, *Publ. Obs. Astron. Beograd*, No. 21, 1]

The declinations in this catalogue have been obtained with an internal accuracy of $\varepsilon_\delta = 0\overset{''}{.}023$ and the proper motions are characterized by $\varepsilon_\mu = 0\overset{''}{.}0023$ with a mean epoch of 1954.0. The total number of stars is 440.

8. Investigation of systematic errors $\Delta\delta_\alpha$ in the latitude observations of different observatories from their comparision with the Belgrade General Catalogue of Latitude Stars and with the photographic catalogue AGK3. [Sadžakov, S.: 1979, *Publ. Obs. Astron. Beograd*, No. 27,1]

The systematic differences for the errors in proper motions in the AGK3 and the present catalogues are practically the same. Our analysis establishing the existence of the systematic errors of the $\Delta\delta_\alpha$ type in the latitude observations are in harmony with the results of many other authors. The non-uniform data handling, non-unified coordinate systems, varying number of stars, and the swift and frequent changes in the methods of processing make research in this field dificult and to some extent even unreliable. Our publication contains some recommendations for resolving these difficulties.

9. Observations of the Sun and inner planets with the Large Meridian Circle in Belgrade. [Sadžakov, S., and Dačić, M.: 1987, *IAU Colloquium No. 100*, Beograd, in press]

Since 1975 the Belgrade Astronomical Observatory has made regular observations of the Sun and inner planets. During the observations of the Sun, Sukharev's filter was used before 1985 and a new one from high quality glass afterwards. The errors of observation are $0\overset{s}{.}006$ and $0\overset{''}{.}02$ in α and δ for the Sun, $0\overset{s}{.}014$ and $0\overset{''}{.}06$ in α and δ for Mercury and $0\overset{s}{.}004$ and $0\overset{''}{.}03$ in α and δ for Venus.

THE FLAGSTAFF MEASURING MACHINE

Gart Westerhout and Dave Monet
U.S. Naval Observatory
Washington, DC 20392
USA

Abstract. The U.S. Naval Observatory is building a measuring machine which will hold four PSS plates simultaneously and will automatically measure them, in a highly controlled environment, in about 12 hours. The images will be recorded by two camera subsystems with a scale of 0.9 arcseconds/pixel on a 0.33 x 0.26 degree CCD frame. The original glass plates of the PSS-II and glass copies of the PSS-I will be measured. This is a joint CALTECH-USNO program.

INTRODUCTION

To our knowledge, no other papers in this Symposium describe the second Palomar Sky Survey (PSS-II). Therefore, we start this paper with a short description. The PSS-II consists of 894 fields on 5-degree centers from declination 0 to 90 degrees, each exposed in three colors:

Blue: III aJ + GG 385
Red: III aF + RG 600
Near-IR: IV N + RG 9

The plates measure 35 x 35 cm or 6.6 x 6.6 degrees, and the plate scale is 67.1 arcsec/mm. The blue and red plates go to 23 and 22 magnitude, respectively, which is 2 magnitudes fainter than PSS-I. In addition, the U.S. Naval Observatory has commissioned the taking of a short-exposure (2 minute unbaked) blue plate to allow an astrometric tie-in with the fundamental reference system (existing astrometric catalogs). This short plate will be taken within 1 month of the PSS-II blue plate.

The accuracy of relative positions and absolute proper motions measurable on the PSS plates is limited by the S/N characteristics and the plate scale to a centroiding accuracy of approximately 1 - 1.5 micrometers, corresponding to 0.1 arcseconds. Thus relative positions of 0.1" are expected. Reference catalog accuracies will limit absolute positions to the 0.2" to 0.3" level, but may be improved with future catalogs. With the PSS 30-year baseline, proper motions of 0.003"/year are expected and will be complete from 13 - 19 magnitude outside the Milky Way. It is likely that these will be absolute proper motions, as the number of QSO's per plate at V = 19 is about 300.

EXPECTED CATALOG PRODUCTS.

All processing will be done in real time, but all CCD frames will be retained for future use. The first catalog available will be a Raw Catalog, containing X and Y positions in microns, estimators of object classification, and an uncalibrated density estimate. This will be a plate-by-plate catalog, with no inter-plate correlation nor photometric calibration. Subject to manpower and computer limitations, this Raw Catalog will be available to qualified, active users. We cannot support a general distribution.

The User Catalog will be calibrated both astrometrically and photometrically, using the best existing

J. H. Lieske and V. K. Abalakin (eds.), Inertial Coordinate System on the Sky, 491–492.
© 1990 *IAU. Printed in the Netherlands.*

astrometric catalogs and photometry from the STScI guide star catalog and other existing photometric data. The detection lists for each field will be correlated and merged to provide for each object a mean position, mean magnitude, color, proper motion, and object classification. The fields will then be merged into a single catalog. It is also intended to isolate the QSO's to determine the zero point of the proper motion system.

User support documentation to be provided will be the source code for image processing, the source code for list correlation, and the calibration formulae and coefficients that convert the Raw Catalog into the User Catalog. The User Catalog will be widely distributed.

PRECISION MEASURING MICRODENSITOMETER (PMM).

The only company we have found capable of providing a 30 x 40 inch active measuring area (to mount four plates at one time) using granite and air-bearing technology is ANORAD. Metrology in X and Y will be interferometric, with 1/16 wavelength local precision and 0.5 micrometers/meter global. There will be a separate auto-focus on the upper and lower Z-axis stages. The machine will be provided with two measuring CCD cameras and a microdensitometer with a slit-viewing TV camera. It will be housed in a class 100,000 clean room enclosure and kept to within 0.5 degrees C temperature.

Each camera subsystem is a blemish-free VIDEK CCD camera with 1320 x 1035 pixels, 8-bit digitization, 10 MHz readout rate, and 6.8 micrometer square pixels. A 2:1 transfer lens will translate this to 13.6 micrometer square pixels on the plate, providing a scale of 0.9 arcseconds/pixel, and a 0.33 x 0.26 degree area per CCD frame. The microdensitometer subsystem will be supplied by Perkin-Elmer and will be substantially the same as that used on their PDS machine.

The philosophy behind the computer follows from the condition that we want to mount the four plates, leave them for a while to assume constant temperature, then measure them automatically in about 12 hours. This provides 17 seconds computer time for each frame; the computer speed then dictates how much information can be extracted from a frame in these 17 seconds. The configuration consists of three Silicon-Graphics 4D/240 "advanced workstations" boxes. Each box will have 4 RISC CPUs, 32 MByte Physical Memory, 600 MByte Disk, 1/4 Inch Tape and will use UNIX and FORTRAN-77 with VMS Extensions.

We are coding in FORTRAN, and work in a multi-thread, multi-CPU parallel computational environment. The full-frame operation co-adds four VIDEK images, flattens the field and normalizes, uses a box median filter to get the sky, and uses a star finder and primitive splitter. For each star detection it then does 0th, 1st, and 2nd density moments (total intensity, position and shape), makes gradient and separation measures, and does a non-linear least squares fit.

The PMM will be delivered in December 1990. Software development is in progress, and production is expected to start early in 1991. The production rate is 1000 plates per year and is expected to catch up with the observing in a few years. Total plate material is about 6000 (3 PSS-II, 2 PSS-I, 1 short J), and therefore the total program is expected to take 6 years to accomplish. The final product is a homogeneous catalog of astrometric positions, proper motions, colors, magnitudes, and identifiers for non-stellar objects.

CURRENT STATUS OF THE PROJECT OF PHOTOGRAPHIC FOURFOLD COVERAGE OF THE NORTHERN HEMISPHERE OF SKY

V.S. KISLYUK
Main Astronomical Observatory
Ukrainian Academy of Sciences
252127 Kiev
USSR

The project of fourfold coverage of the northern hemisphere of the sky by means of wide-angle astrographs was proposed in 1977 (Kolchinsky and Onegina, 1977). This project is known as FON *(fotograficheskij obzor neba)* — photographic survey of sky.

The distinctive pecularity of the project FON is that it is carried out by means of not one but six of the same type astrographs (K.Zeiss, Jena) which have been installed at six observatories. All instruments have the same apertures (D = 400 mm) but different focal lengths namely F = 2000mm (Kiev, Zelenchuk, Zvenigorod, Dushanbe) and F = 3000 mm (Abastumani and Kitab). Available fields of all instruments consist of 8° x 8° and their working fields were adopted as 4° x 4°.

In order to ensure the fourfold coverage of sky six observatories were combined in the four groups as it is shown in Table 1. Each group photographs once the whole northern sky.

Between the groups the plate centres are displaced on 2° in RA and/or DEC. Inside of each group the plate centres are displaced as shown in the table.

To consider the magnitude equation on each plate two exposures are made with duration of 18 min and 40 sec. It allows us to obtain the images of stars in the system B down to 15^m-16^m and 12^m-13^m correspondingly. A particular attention was paid to unique investigations of all astrographs (Ivanov et al, 1985).

Such analysis, including the characteristics of objectives, comparison of quantities of stellar images on plates, investigations of magnitude equation and other errors, allowed us to make conclusions about the excellent optics of all the instruments and to choose optimum conditions for observations.

Systematic observations on the FON project were started in 1982. At present nearly 90% of required plates have been obtained. The observations are suggested to be finished in 1991. The photographic collection will be used for determination of accurate positions and proper motions of all stars contained in the Astrographic Catalogue (*Carte du Ciel*) using the last one as the first epoch of observations.

Measurements of plates will be carried out by means of PARSEC (programme automatic radial-scanner coordinatometer) which has been created recently (Sergeev et al, 1987). The Astrographic Catalogue will be used as the input catalogue for PARSEC. Preliminary analysis showed quite enough reliability and high accuracy of measurements of plates by means of this machine.

J. H. Lieske and V. K. Abalakin (eds.), Inertial Coordinate System on the Sky, 493–494.
© 1990 *IAU. Printed in the Netherlands.*

Preliminary estimates of the accuracy of star positions from measurements of one plate give the values of errors of about ±(0".30 -0".35). It is expected that after the combining of all plates of fourfold coverage using overlapping plate techniques we will have an opportunity to obtain an accuracy of star positions ±(0".10 - 0".15). We also expect that using the data of *Carte du Ciel* the errors of proper motions of stars will be ±0.007"/year.

Table 1. Number of plates of FON project

Group	Station	From DEC	to DEC	RA	Num.of bands	Num.of plates
1	Kitab	$-4°$	$32°$	16^m	10	900
	Kiev	32	60	16	8	720
		64	76	32	4	180
		80	84	64	2	44
		88	180	1	8	
2	Zelenchuk	-6	58	16	17	1530
		62	74	32	4	180
		78	86	64	3	69
		90	360	1	4	
3	Abastumani	-6	30	16	10	900
	Zvenigorod	30	58	16	8	720
		62	74	32	4	180
		78	86	64	3	69
		90	360	1	14	
4	Dushanbe	-4	60	16	17	1530
		64	76	32	4	180
		80	84	64	2	44
		88	180	1	8	

References

Kolchinsky, I.G., Onegina, A.B. (1977), *Astrometrija i Astrofisika* (Kiev), No 33, p.11–16.

Ivanov,G.A., Rachmatov,E., Yurevitch,V.A., *et al* (1988), *Jenaer Rundschau*, No 2, pp.90–91.

Sergeev,A.V., Sergeeva,T.P., Riabokon,A.V., *et al* (1987), *Fundamentals of Astrometry*, IAU Colloq. No 100, Beograd.

THIRD PRELIMINARY CATALOGUE OF STARS OBSERVED WITH THE PHOTOELECTRIC ASTROLABE

LU LIZHI
Beijing Astronomical Observatory
Beijing, 100080
China

ABSTRACT On the basis of the data observed with the Photoelectric Astrolabe Mark II of the Beijing Astronomical Observatory during the period from 1982 to 1988, residuals of 355 fundamental stars and 2921 catalogue stars are reduced and analyzed. The relations between the residuals V and the magnitude M, and the spectral type S of FK5 stars are discussed. The position corrections (PACP3−FK5), $\Delta\alpha's$ and $\Delta\delta's$, from the double passage of the stars are determined. There are 910 in $\Delta\alpha$ and 696 in $\Delta\delta$. The mean precisions of $\Delta\alpha's$ and $\Delta\delta's$ are ±3.9 ms and ±0″.067, respectively. Under certain conditions 328 $\Delta\alpha's$ and 12 $\Delta\delta's$ from the observed single passage of the stars are determined. Finally systematic corrections of PACP3−FK5 are given.

1. Introduction

Since 1976, two preliminary catalogues of stars have been compiled[2][3]. With the positions at J2000.0 of the Fifth Fundamental Catalogue (FK5) and new astronomical constants (IAU, 1976), the data observed with the Photoelectric Astrolabe Mark II (PA II) during 1979 to 1988 are re-reduced to the FK5 system. With the data of the catalogue stars observed from 1982 to 1988, the Third Preliminary Catalogue observed with PA II is compiled with the stars whose magnitudes are from 0.1 to 7.3 and the declinations are from 11° to 69°. The mean observational epoch is 1985.9. Finally, with the position corrections of $\Delta\alpha$ and $\Delta\delta$ of the FK5 stars obtained from the double transits, the systematic corrections(PACP3−FK5) are given.

2. The Reduction of Residuals and the Instrumental System Errors

2.1 The Reduction of Residuals of Stars

There are 355 stars in the fundamental group. The corrections of astronomical time, latitude, zenith distance, and residuals v_i are obtained by the method of least squares. The residuals of the catalogue stars are calculated with the corrections of astronomical time, latitude and zenith distance of the reference group of stars. Then, the mean values of the residuals of stars are calculated by weighted average.

2.2 The Correlations of Residuals with Magnitude

The mean values of residuals of FK5 stars for every 0.5 of magnitude are made by weighted average according to the precision of residual. The results are given in Table 1.

Table 1. The Correlations of Residuals V With Magnitude M

$\bar{M}$	2.00	2.80	3.31	3.73	4.24	4.79	5.25	5.75	6.29
$V_m(0″.01)$	4.4	1.6	2.9	0.4	0.7	-1.2	-1.1	-0.2	-0.2

2.3 The Correlations of Residuals With Spectral Types

The mean values of residuals of FK5 stars are calculated for each spectral type after the system corrections on magnitude is considered. The results are shown in Table 2.

495

J. H. Lieske and V. K. Abalakin (eds.), Inertial Coordinate System on the Sky, 495–496.
© 1990 *IAU. Printed in the Netherlands.*

Table 2. The Correlations of Residuals V With Spectral Type S

S_p	B	A	F	G	K	M
$\bar{V}_s(0''.01)$	-1.1	0.2	-0.5	1.0	0.2	0.2

3. The Group Corrections

With the results of PA II from 1980 to 1988, the group difference of the astronomical time, latitude and zenith distance, Δt, $\Delta\varphi$, and Δdz are obtained. The group corrections of t, φ and dz are calculated by a chain method. The results are given in Tables 3.

Table 3. The Group Corrections and Precisions of t, φ, and dz

Group	0	2	4	6	8	10	12	14	16	18	20	22
$t(0^s.0001)$	8	15	18	38	4	-4	-24	-39	-4	-2	-12	2
$\varphi(0''.001)$	15	-1	3	10	-3	-27	50	46	7	-47	-19	-34
$dz(0''.001)$	5	3	19	0	4	6	-20	-6	-23	-11	-2	25

4. The Calculation of 2K

Strictly speaking, the calculation of 2K should be made using the stars of $\cos q = 0$, that is $q = 90°$. But these stars are very few. So we used the stars of $|\cos q| < 0.1$ (in this catalogue there are 130 stars) to calculate 2K and obtained $2K = 0''.013 \pm 0''.014$.

5. $\Delta\alpha$ and $\Delta\delta$ of Stars

There are 910 $\Delta\alpha$'s and 696 $\Delta\delta$'s obtained from the passage observations of the same stars at both east and west sides with the average precisions $\pm 3.9ms$ and $\pm 0''.067$, respectively.

There are 328 $\Delta\alpha$'s and 12 $\Delta\delta$'s obtained from single passage.

6. The Systematic Corrections of the Catalogue of Stars

By the method [4] and [5] and with the $\Delta\alpha$ and $\Delta\delta$ of FK5 stars obtained from the double passage, the systematic corrections of the preliminary catalogue of stars (PACP3−FK5) are analyzed. The system corrections are given in Tables 4.

Table 4. The Systematic Errors (PACP3−FK5) $\Delta\alpha_\delta$ and $\Delta\delta_\delta$

$\delta°$	15.0	17.5	20.0	22.5	25.0	27.5	30.0	32.5	35.0	37.5	40.0
$\Delta\alpha_\delta(0^s.0001)$	-7	27	28	18	11	8	8	6	2	-6	-10
$\Delta\delta_\delta(0''.001)$	0	23	26	18	8	-1	-5	-7	-5	-1	

$\delta°$	42.5	45.0	47.5	50.0	52.5	55.0	57.5	60.0	62.5	65.0
$\Delta\alpha_\delta(0^s.0001)$	-11	-6	6	17	26	26	18	6	-8	-10
$\Delta\delta_\delta(0''.001)$					-4	-8	-10	-5	1	15

The author wishes to thank Prof. Luo Dingjiang for his valuable discussions, Ms. Peng Yizhi for her providing some observational data, and the observers working on the Photoelectric Astrolabe.

References

[1] Luo Dingjiang et al. *Publications of the Beijing Astronomical Observatory*, No.1 (1979), 56.

[2] Lu Lizhi, Luo Dingjiang et al. *Acta Astronomica Sinica*, **21** (1980), 305.

[3] Lu Lizhi, Luo Dingjiang et al. *Publications of the Beijing Astronomical Observatory*, No.3 (1983), 70.

[4] Bien, R., Veröff. *Astron. Rechen-Institut, Heidelberg*, No.29 (1978).

[5] Working Group of GCPA, *Acta Astronomica Sinica*, **24** (1983), 267.

THE ESTABLISHMENT OF AN ASTROMETRIC STANDARD REGION IN THE NGP: STAR CATALOGUE

L.K. PAKULYAK
Main Astronomical Observatory
Ukrainian Academy of Sciences
252127 Kiev
USSR

ABSTRACT. The positions and proper motions catalogue in the area near the NGP is presented.

The astrometric standard region (AS) in the NGP is supposed to be a supplement to three standard regions in Praesepe, Pleiades and IC4756 clusters, completing the net of them on the northern sky available for observations at any station in any moment of time (Russell, 1984).

The adopted technique of AS derivation, as it exists now, allows the updating of the catalogue with new data.

AS is centered on the Coma cluster with centre coordinates of $12^h 24^m$ in RA and $+26°$ in Dec. A square area of $6° \times 6°$ is covered with 16 plates of the double wide-angle astrograph (F = 2 m, D = 40 cm). Centres of plates were displaced by $2°$ in both coordinates. The average plate epoch is 1984.2. Plates were made to the west and to the east from the astrograph column for the elimination of the instrumental magnitude equation. Two exposures on every plate were made for this purpose also. The main principle of the star choice is a homogeneous density of stars both in general and in any magnitude interval. The catalogue includes objects up to 15^m. The average density of stars is 84 stars per sq. degree. The measuring accuracy is ranging 2.3 to 3.0 μm.

The plate reduction model was chosen with the Eichhorn-Williams statistical test and optimum models for each plate were fitted together with the determination of instrumental aberrations, namely the coma coefficient and the tangential distortion terms (Pakulyak, 1989).

Each plate was reduced separately with the optimum reduction model. Final positions were derived as the average of all images, weighted with generalised Shlesinger coefficients. This technique is the same as Shlesinger one but a non-linear reduction model is used. Generalised weights were calculated for every field star in relation to the reference stars configuration (Kiselev, 1984).

First iteration mean star positions appeared to be sufficient because the second iteration gave only a little better accuracy (see Table 1) due to the "spreading" of corner plates errors on the whole star field.

Positional errors (internal convergence and reduction errors) for both coordinates which depend on the zone of overlap are given in the same table.

Proper motions were derived for 537 objects up to 13^m and improved for 525 stars from the AGK3 list. The improvement performed with 4 catalogues employed : AGK2, AGK3, AS and the reduced

J. H. Lieske and V. K. Abalakin (eds.), Inertial Coordinate System on the Sky, 497–498.
© 1990 *IAU. Printed in the Netherlands.*

Astrographic Catalogue (Oxford section) on the area 12° x 6° near the NGP (Stock, Cova, 1983). The epoch of the last one is 1901.1 and overlap with AS region is near 50%. Weighted equations for proper motion determinations together with positions were written for all catalogues and all AS plate images. The positional accuracy of these stars is of ±0ˢ.014 in RA and ±0".19 in Dec. The accuracy of proper motions is ±0.00035 s/yr and ±0.0047 "/yr in RA and Dec correspondingly. For field stars there were only AS and Oxford section positions; corresponding proper motion errors are of ±0.00039 s/yr and ±0.0032 "/yr.

The magnitude equation in proper motions in the whole star magnitude range is of 0".019 in RA and −0".011 in Dec for the AGK3 list stars and of 0".003 in RA and near zero in Dec for field stars and eliminated by three iterations.

Magnitudes for AGK3 stars were converted to the B-system by linear equations for stars, common with SAO catalogue. The mean accuracy is of 0ᵐ13. The same attempt for field stars failed because of errors worse than 0ᵐ64. That's why field stars magnitudes are given the same as in the Astrographic Catalogue.

The total number of objects catalogue contains is 13197 in magnitude range 6ᵐ to 15ᵐ with 1062 objects, which have proper motions.

Table 1. Positional accuracy of AS stars

i-fold overlap	Number of stars	1st iter.(Positional errors)		2 nd iteration	
		inter conv.	reduct.	inter.conv.	reduct.
1	425	0".35	0".35	0".27	0".27
2	329	0.21	0.18	0.19	0.14
3	321	0.27	0.19	0.22	0.14
4	1316	0.21	0.15	0.18	0.13
6	467	0.19	0.13	0.15	0.11

References

Kiselev, A.A. (1984), *Problemy astrometrii*, pp.44–56.
Pakulyak, L.K. (1989), *Kinematika i fizika neb.tel*, **251**, 2, pp.24–28.
Stock, J.S., Cova, S. (1983), *Rev.Mexicana Astron.y Astrof.*, **251**, 4, pp.233–260.
Russell, J.L. (1984), IAU Symposium 109, pp.697-703.

RESEARCH ON THE ERRORS IN DETERMINATION OF PRECISE OPTICAL POSITIONS OF
RADIO SOURCES

LU CHUN-LIN, LI DONG-MING
Purple Mountain Observatory
Academia Sinica
Nanjing, China

To link the optical reference frame to extragalactic radio frame, an observation program is in progress in Purple Mountain Observatory (PMO) and Shanghai Observatory. In this program, 70 extragalactic sources will be observed with the 1.56 m astrometric reflector in Shanghai. The secondary reference stars are determined with the twin astrograph (D = 40 cm, f = 300 cm) in PMO, and about 50 radio stars and some bright extragalactic sources will be observed with the twin astrograph.

It has been noticed that there are significant differences between optical positions of radio sources and radio stars obtained by different observers and between the optical and radio positions. Since the internal errors of these results are small, we think that these differences are mainly brought about by systematic errors. In order to obtain reliable results, research on some important systematic errors (*viz.* errors of reference catalogs, magnitude equation and distortion of telescope objective) has been carried out and some primary results have been obtained.

1. Errors of reference catalogs

To investigate the effect of errors of reference catalogs on the determination of optical positions of radio sources and radio stars, we compared AGK3RN and AGK3—which are the most extensively used reference catalogs in the northern sky—with the FK5. In 223 fields centered at extragalactic sources selected from the IAU Commission 24 Working Group catalog (Argue *et al.* 1984) and radio stars selected from the High Priority Hipparcos Radio Stars (Walter 1988), the common stars are compared. There are about 9 stars in common between FK5 and AGK3 in a 10° x 10° field on average. Because there are too few common stars between FK5 and AGK3RN, we compared them through AGK3. In every 5° x 5° field, the common stars between AGK3RN and AGK3 are compared. Then AGK3RN is compared with FK5 in the sense of (FK5 – AGK3) – (AGK3RN – AGK3). The results indicate that a systematic trend is not obvious in the difference between AGK3 and FK5, but it is significant in the difference between AGK3RN and FK5. On the other hand, there are a large number of fields with local difference of order of 0".1 or larger.

2. Distortion of telescope objective

For wide field telescopes, distortion is one of the most important aberrations which affect position determinations. We developed a new star-pair method on the basis of the classical star-pair method

499

J. H. Lieske and V. K. Abalakin (eds.), Inertial Coordinate System on the Sky, 499–500.
© 1990 *IAU. Printed in the Netherlands.*

to determine distortion (Lu *et al.* 1990). In this method, the effect of gnomonic projection on the distance between the two stars of the pair is removed using the geometry of projection. Then the coefficient of distortion can be determined with a strict reduction formula. This method eliminated the defects of the classical star-pair method. A number of star pairs on a plate can be used in the determination of distortion. With this method, we obtained distortion coefficients of the twin astrograph in PMO, which are consistent with that obtained in the solution of plate constants.

3. Magnitude equation

The magnitude equation is the most complicated error in photographic astrometry. It is especially important to take that equation into account in determination of optical positions of radio sources because of the large difference between magnitudes of the sources and reference stars.

We are trying out a method to determine the equation. In this method, a short exposure (1 minute) and a long exposure (30 minutes) are taken on the same plate. By analysing the relation of image diameters and the differences between the measured coordinates of two exposures, the behavior of the equation is detected, which will be used to remove the effects of the magnitude equation from the measured coordinates.

References

Argue, A.N., de Vegt, C., Elsmore, B., Fanselow, J., Harrington, R., Hemenway, P., Kuhr, H., Kumkova, I., Niell, A.E., Walter, H.G., Witzel, A.: 1984, *Astron. Astrophys.*, **130**, 191.
Lu Chun-lin, Li Dong-ming: 1990, *Acta Astronomica Sinica*, In press.
Walter, H.G.: 1988, private communication.

Discussion

CORBIN: I regret to inform you that, in my opinion, your comparison FK5 – AGK3RN is not valid for the following reasons:
1. The FK4 stars in AGK3 have positions and motions from FK4 not from the AGK3-2 program.
2. The AGK3 proper motions are not on FK4 due to the fact that FK4 – FK3 was applied to AGK2, whereas AGK2 is not on FK3, but rather the system of AGK2A.
3. Thus, your in your expression (FK5 – AGK3) – (AGK3RN – AGK3) the AGK3 stars included the two sets of parentheses are not on the same system, due to the proper motions.
4. What your diagrams really show is FK4 – AGK3.
I continue to see studies that assume the proper motions of AGK3 to be on FK4, and I again caution against this.

LU CHUN-LIN: 1. According to our analysis and the Introduction of the AGK3, the positions and proper motions of the FK4 stars in the AGK3 are not from the FK4.
2. The effect of the AGK3 proper motion system on our results is not significant. The same comparisons have been carried out at 1958.0 (the epoch of the AGK3, so there would be no effect of AGK3 proper motions) and they result in the same conclusions.
3. In the comparison of the FK5 and AGK3, the FK4 Supp stars, whose accuracy is 7 mas in position and 2 mas/y in proper motion in the FK5 system, are used. The results are consistent with FK5B.
The detailed results will soon be published elsewhere.

TORINO PROGRAMME ON RADIOSOURCES ASTROMETRY: PROGRESS REPORT.

G. Chiumiento[1], M.G. Lattanzi[1,2], G. Massone[1],
R. Morbidelli[1], R. Pannunzio[1], M. Sarasso[1],

[1] *Torino Astronomical Observatory, 10025 Pino Torinese (TO), Italy*
[2] *Space Telescope Science Institute, Baltimore MD 21218, USA*

ABSTRACT. Measurements of precise positions of optical counterparts of extra-galactic radio-sources, with the aim to compare them with VLBI positions, is in progress at Torino Observatory. The Carlsberg Automatic Meridian Circle Catalogue (CAMC) is used for plate reductions. Presently, about 40% of the plate material required to complete the project has been taken and some significant improvements have been made both in the material and the reduction methods. Improvements are also due to the publication of a new volume of the CAMC catalogue that significantly increases the number of stars in the fields.

I. INTRODUCTION

In a previous paper, (Chiumiento et al, 1989, hereinafter paper I) we reported some preliminary results on a programme of determination of precise optical positions of extragalactic radio-sources. Our observing list is a sample of 75 targets from the list proposed by the Working Group of the IAU Commission no. 24 (Argue et al, 1984). The selection criteria and a description of the telescopes used are described in paper I. The primary reference stars have been taken from the Carlsberg Meridian Catalogues no. 1, 2, 3 (La Palma: 1985, 1986, 1987).

During that first attempt, in spite of the small number of plates available, we obtained fairly good results. At this time, the plate material has been increased considerably and about 40% of the programme completed. Additionally, the following improvements have been introduced as standard procedure: a) 3 plates are taken for each object both at the 38 cm photographic refractor and the 105 cm astrometric reflector; b) the plates have been measured in direct and reverse directions with the digitized Ascorecord machine of Torino Astronomical Observatory; c) due to the increase in the number of CAMC stars, more sophisticated plate models than the six-constant linear one may be attempted. Second order effects, coma and magnitude terms can also be checked for; d) the secondary reference frame stars were increased from 10 to 15. They were selected within 12′ the radiosource. The magnitude of these stars is intermediate between that of the CAMC stars and the radiosources.

II. REDUCTION AND EVALUATION OF INTERNAL ERROR

The reduction procedure of our material (internal error evaluation and quality

501

J. H. Lieske and V. K. Abalakin (eds.), Inertial Coordinate System on the Sky, 501–502.
© 1990 *IAU. Printed in the Netherlands.*

control) may be summarized as follows: 1) measure the refractor plates in direct and reverse directions; 2) combine the two sets for the same plate and evaluate the inter-agreement of the measurements σ_m (column 1 in Table 1); 3) derive RA and DEC for the intermediate reference stars; their internal error is measured through the mean error of unit weight σ_0 of the adjustment with the CAMC stars (column 2); this must be consistent with the error obtained at step 2 plus the error of CAMC positions; 4) average the individual positions of the intermediate stars from different refractor plates; the dispersion of these positions is reported in column 3 of Table 1; 5) repeat steps 1 to 4 for reflector plates using the intermediate stars as reference frame and the target radiosources as test stars.

Column 4 in Table 1 was computed using the formula $\sigma_m^2 + \sigma_0^2/n$ where n is the number of reference stars. The figures in columns 3 and 4 would match. However, we can see that the match is satisfactory for refractor plates only. The poor result for reflector plates seems to be due to the low quality of faintest radiosources images. This can be inferred, on a plate basis, by comparing the value $0''.076$ in column 1 of Table 1 with the corresponding $0''.053$ obtained for the secondary reference stars images on the reflector plates. This value is very close to what expected considering the more favourable scale (S) of the reflector ($S_{reflector}/S_{refractor} \simeq 1.43$). That image quality in our measurements is strongly dependent on the magnitude of the radiosources, is clearly confirmed in Figure 1. Here, the individual values used in computing the average in column 3 of Table 1 are given.

Comparison of our preliminary optical positions with the radio ones in the catalogue by Argue et al. gave the following mean residuals:

$$< \Delta\alpha > \simeq 0''.18 \qquad < \Delta\delta > \simeq 0''.15$$

Seeking for improvements on the faintest target positions, some of the plates will be re-measured with a PDS. If necessary, new plates will be taken for the worst cases.

Table 1. Internal error table

Figure 1.

Refractor plates			
1	2	3	4
$0''.082$	$0''.185$	$0''.117$	$0''.086$
Reflector plates			
1	2	3	4
$0''.076$	$0''.094$	$0''.233$	$0''.079$

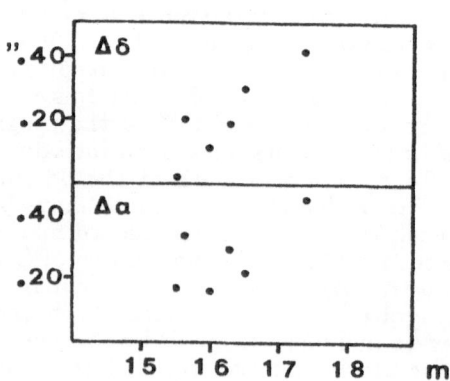

REFERENCES

Argue, A. et al.: 1984, *Astron. Astrophys.* 130, 191.
Chiumiento et al.: IAU Coll. No. 100, Belgrade 1987, *Cel. Mechanics*, in press.

OCCULTATION OF RADIO SOURCES FOR THE LINKAGE OF RADIO AND STELLAR REFERENCE FRAMES

Mitsuru Sôma, Masanori Miyamoto and Shinko Aoki
National Astronomical Observatory
Mitaka, Tokyo 181, Japan

ABSTRACT. The right ascension of the radio source 3C273B, which serves as a right ascension zero point in radio astrometric work, has been determined from lunar occultations and photographic observations.

We re-analyze here the lunar occultations of 3C273B using the recent precise lunar ephemeris and obtain its right ascension referred to the FK5 equinox at J2000.0. The obtained right ascension is $12^h 29^m 06\!\!.\!\!^s 6946 \pm 0\!\!.\!\!^s 007$ at its mean observation epoch of 1963.62.

Predictions of occultations of radio sources by the Moon and planets are also given. Observations of them are encouraged in order to improve the accuracy of the linkage between radio and stellar reference frames.

1 Introduction

In the astrometric observations of radio sources by VLBI, relative right ascensions can only be determined. The radio source 3C273B has been recommended as right ascension zero point in radio astrometric work and its right ascension $12^h 26^m 33\!\!.\!\!^s 246$ referred to the FK4 equinox at B1950.0 was based on lunar occultations and photographic observations (Hazard et al., 1971). The occultations were analyzed using the lunar ephemeris $j = 2$ based on the theory by Brown with the corrections to the ephemeris, which were deduced, with slight modifications, from an analysis of occultations of stars by Morrison and Sadler (1969).

There is a confusion about this adopted right ascension. For example, Witzel and Jonston (1982) assume that in this value the E-terms of aberration are included while Clark et al. (1976) and Elsmore (1979) assume that the terms are not included. The right ascension $12^h 29^m 06\!\!.\!\!^s 6997$ referred to the FK5 equinox at J2000.0 was converted from the B1950.0 value, but its conversion procedure leaves some problem as Aoki et al. (1986) pointed out.

In Sect. 2 we re-analyze the lunar occultations of 3C273B used by Hazard et al. using the recent precise lunar ephemeris. In order to improve the accuracy of the linkage between radio and stellar reference frames, observations of occultations of radio sources by the Moon and planets are important. In Sect. 3 prediction of occultations by the Moon is given and in Sect. 4 prediction of occultations by the planets is given.

J. H. Lieske and V. K. Abalakin (eds.), Inertial Coordinate System on the Sky, 503–511.
© 1990 IAU. Printed in the Netherlands.

2   Re-analysis of occultations

We analyze the lunar occultations of 3C273B used by Hazard *et al.* (1971), based on the lunar theory ELP2000-82 by Chapront-Touzé and Chapront (1982). The constants for the lunar theory and the correction to the Watts' (1963) charts we use are those obtained by Sôma (1985) from an analysis of lunar occultations of stars.

Fig. 1 shows the difference between the ephemerides used by Hazard *et al.* and used by us. The difference reaches about 0″.3 at 1964. Fig. 2 shows the difference between the ephemeris LE200 of JPL and the ephemeris used by us. The LE200 is the lunar ephemeris appearing in *The Astronomical Almanac* for the year 1984 onwards and it is consistent with the planetary ephemeris DE200. The constant difference in longitude of about 0″.4 shown in Fig. 2 is almost due to the difference between the center of gravity and the center of the mean limb adopted by Watts (1963). The linear trend of about 0″.004/year is due to the difference in the mean motion, which probably includes the difference in the adopted precession constant. The small periodic terms are due to the difference between the adopted values for the eccentricity, inclination, etc. The difference is less than 0″.04.

In deriving the right ascension of 3C273B, we adopt the declination $\delta = +2°03'08″.59$ (J2000.0) obtained from VLBI observations. Of 5 occultations Hazard *et al.* cited we omit that of 1963 Mar. 11 (grazing occultation) because the accuracy of the Watts' charts of that field is poor and the accuracy of determining the right ascension is not good due to the position angle being near 90°. We also omit the occultation of 1964 Sept. 7 as Hazard *et al.* did because of its poor observational accuracy.

The geodetic coordinates (longitude λ, latitude ϕ, and height h) of the observatories (Australian National Radio Astronomical Observatory at Parkes and Arecibo Observatory) are taken from *The Astronomical Almanac* for the year 1981 (p. J46 and p. J36). The data for reducing the regional geodetic coordinates to the geocentric coordinates are taken from *The Astronomical Almanac* for the year 1990 (p. K13).

The obtained right ascension of 3C273B referred to the FK5 equinox at J2000.0 (without the E-terms of aberration) is

$$\alpha = 12^{h}29^{m}06^{s}.6946 \pm 00^{s}.007$$

and its mean observational epoch is 1963.62. The difference 0″.08 between this value and the presently adopted value is within the accuracy of observation.

It should be noted that the value obtained here is dependent on the precession constant used in the analysis. The precession constant we use is the constant in the IAU (1976) System of Astronomical Constants (Duncombe *et al.*, 1977).

In order to improve the accuracy of the connection between the radio reference frame and the optical reference frame, further observations of occultations of radio sources by the Moon and planets are recommended.

3   Prediction of occultations by the Moon

Table 1 shows the radio observatories for which lunar occultations are calculated. Coordinates of these stations are taken from *The Astronomical Almanac* for the year 1990 (pp.

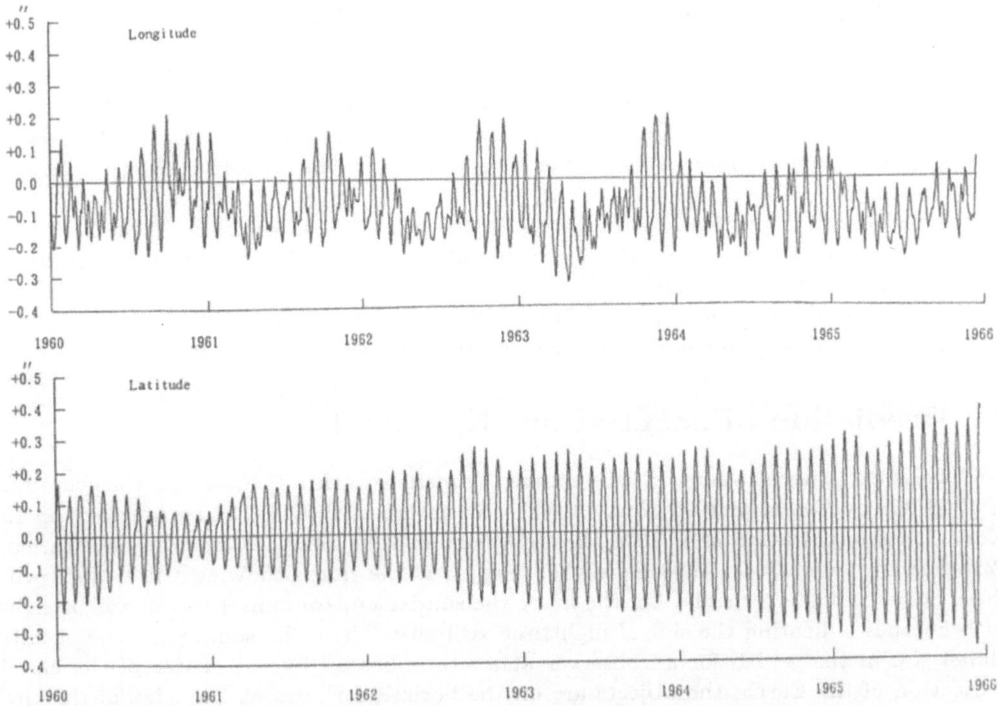

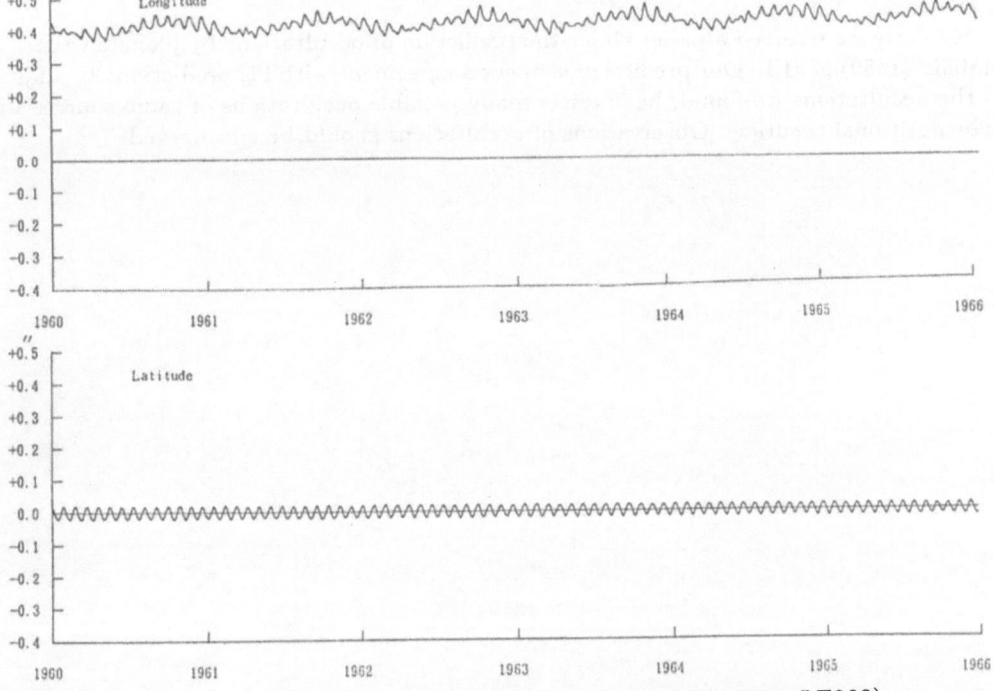

Fig. 1   Difference in lunar ephemeris (Hazard *et al.* minus ELP2000)

Fig. 2   Difference in lunar ephemeris (ELP2000 minus LE200)

J6–J16). Radio sources are taken from Argue *et al.* (1984), Morabito *et al.* (1986) and the recent JPL catalog JPL 1989-3 (Sovers, 1989).

Catalogs of stars may include systematic errors of about 0″.1. Accordingly if we rely on only one radio source 3C273B for the connection between radio and stellar reference frames, the radio reference frame should include the error of that field in the stellar reference frame. Therefore it is important that positions of many radio souces should be determined independently.

Prediction of occultations by the Moon for the year 1991 is shown in Table 2. It shows the observatories where occultation is visible at more than 20° above horizen when the Sun's altitude is less than −10°.

Occultation of the radio source 3C273B will occur during 2000–2002.

We can provide occultation predictions for any observatory on request.

4 Prediction of occultations by planets

We have found 4 occultations of radio sources by planets during 1990–2000: the occultation of 1748-253 by Venus on 1992 Nov. 11, the occultation of P 2208-137 by Mercury on 1996 Mar. 7, the occultation of P 2008-159 by Venus on 1998 Mar. 6, and the occultation of 1215-002 by Mercury on 1998 Sept. 26. Fig. 3 is the maps showing the regions where these occultations are visible. On the maps the sunrise and/or sunset terminator is shown, with hatches indicating the side of nighttime visibility. The radio source and the occulting planet are in the zenith for an observer at a site indicated by the center of the circular projection of the Earth; the objects are on the horizen for sites at the edge of the circle. The altitude above the horizen can be estimated for any site shown on the map; the cosine of the altitude is the distance of the site from the center of the circle divided by the radius of the circle.

Recently we received a paper about the prediction of occultations by planets written by Linfield (1989) of JPL. Our prediction is in good agreement with his prediction. In addition to the occultations we found, he predicts many possible occultations of radio sources with poor positional accuracy. Observations of occultations should be encouraged.

Table 1. Radio Observatories

	Name of Observatory	East Long.	North Lat.	Height
				m
Ch:	Chatanika Incoher. Scatter Fac.	−147°27ʹ1	+65°06ʹ2	235
HC:	Hat Creek Radio Ast. Obs.	−121 28.4	+40 49.1	1043
Gs:	Goldstone Complex	−116 50.9	+35 23.4	1036
Hv:	Harvard Radio Ast. Sta.	−103 56.7	+30 38.2	1603
Ri:	Richmond POLARIS Obs.	− 80 23.1	+25 36.8	11
Hs:	Haystack Obs.	− 71 29.3	+42 37.4	146
CT:	Cerro Tololo Inter-Amer. Obs.	− 70 48.9	−30 09.9	2215
Rb:	Robledo Deep Space Sta.	− 4 14.9	+40 25.8	774
On:	Onsala Space Obs.	+ 11 55.2	+57 23.6	5
SR:	Radio Space Research Sta.	+ 27 41.2	−25 53.4	1382
By:	Byurakan Astrophysical Obs.	+ 44 17.5	+40 20.1	1500
Gu:	Guaribidanur Radio Obs.	+ 77 26.1	+13 36.2	−
Bj:	Beijing Normal Univ. Obs.	+116 21.6	+39 57.4	70
Ka:	Kashima Space Commun. Center	+140 39.8	+35 57.3	32
Ti:	Tidbinbilla Deep Space Sta.	+148 58.8	−35 24.1	656

Table 2   Occultations by the Moon in 1991

Date	Radio Source	α_{2000}	δ_{2000}	Observable Places
1 7	1237−101	12 39 43.07	−10 23 28.8	Gs Hv Ri Hs
1 25	CTD 26	4 3 5.59	26 0 1.5	SR
1 29	GC 0802+21	8 5 38.60	21 6 50.6	Ti
2 1	P 1055+01	10 58 29.61	1 33 58.8	Ka
2 3	1237−101	12 39 43.07	−10 23 28.8	Bj Ka
2 5	P 1354−174	13 57 6.03	−17 44 1.3	Hv
2 8	P 1622−253	16 25 46.91	−25 27 38.3	CT
2 22	CTD 26	4 3 5.59	26 0 1.5	CT
2 22	3C 133	5 2 58.47	25 16 25.4	On
2 24	0610+260	6 13 50.12	26 4 36.9	CT
2 27	MC 0938+119	9 41 13.55	11 45 32.0	Gu Ti
3 1	P 1055+01	10 58 29.61	1 33 58.8	Rb
3 2	P 1223−074	12 26 16.33	− 7 41 6.2	SR
3 2	P 1225−083	12 28 19.84	− 8 38 17.2	Rb
3 3	1237−101	12 39 43.07	−10 23 28.8	Hs
3 22	3C 133	5 2 58.47	25 16 25.4	HC
3 30	1237−101	12 39 43.07	−10 23 28.8	Ka
4 1	P 1354−174	13 57 6.03	−17 44 1.3	Ri
4 4	P 1622−253	16 25 46.91	−25 27 38.3	SR
4 5	1748−253	17 51 51.27	−25 23 59.8	Ti
4 18	3C 133	5 2 58.47	25 16 25.4	Bj
4 24	P 1055+01	10 58 29.61	1 33 58.8	By Gu
4 26	P 1223−074	12 26 16.33	− 7 41 6.2	Ti
4 26	P 1225−083	12 28 19.84	− 8 38 17.2	Gu
4 26	1237−101	12 39 43.07	−10 23 28.8	Rb On By

Table 2   Occultations by the Moon in 1991 (Continued)

Date	Radio Source	α_{2000}	δ_{2000}	Observable Places
5 2	1748−253	17 51 51.27	−25 23 59.8	SR
5 23	P 1225−083	12 28 19.84	− 8 38 17.2	SR
5 24	1237−101	12 39 43.07	−10 23 28.8	Hv Ri
5 25	P 1354−174	13 57 6.03	−17 44 1.3	By Gu
5 29	P 1657−261	17 0 53.15	−26 10 51.7	HC Gs Hv Ri
5 30	1748−253	17 51 51.27	−25 23 59.8	CT
6 3	P 2124−12	21 27 30.49	−11 51 20.2	Ti
6 25	P 1657−261	17 0 53.15	−26 10 51.7	Ka
6 26	1748−253	17 51 51.27	−25 23 59.8	Ti
6 30	P 2124−12	21 27 30.49	−11 51 20.2	SR
7 8	GC 0322+22	3 25 35.91	22 24 12.2	Rb
7 17	P 1225−083	12 28 19.84	− 8 38 17.2	Ti
7 23	1748−253	17 51 51.27	−25 23 59.8	SR
7 28	P 2124−12	21 27 30.49	−11 51 20.2	CT
7 28	P 2128−12	21 31 35.26	−12 7 4.8	Ri Hs
8 4	GC 0322+22	3 25 35.91	22 24 12.2	HC Gs
8 6	3C 133	5 2 58.47	25 16 25.4	By
8 15	P 1354−174	13 57 6.03	−17 44 1.3	Gu
8 19	P 1657−261	17 0 53.15	−26 10 51.7	Hv Ri
8 24	P 2124−12	21 27 30.49	−11 51 20.2	Ti
8 25	3C 446	22 25 47.26	− 4 57 1.4	Ti
8 31	GC 0322+22	3 25 35.91	22 24 12.2	Bj Ka
9 2	3C 133	5 2 58.47	25 16 25.4	Ri
9 3	P 0601+24	6 4 55.27	24 29 23.2	Hs
9 15	P 1657−261	17 0 53.15	−26 10 51.7	Ka
9 18	OV−235	19 23 32.19	−21 4 33.3	CT
9 20	P 2124−12	21 27 30.49	−11 51 20.2	SR Gu
9 20	P 2128−12	21 31 35.26	−12 7 4.8	By
9 27	GC 0322+22	3 25 35.91	22 24 12.2	By
9 30	0554+242	5 57 4.56	24 13 53.7	Ch HC
9 30	0556+238	5 59 32.03	23 53 53.9	Ch
9 30	P 0601+24	6 4 55.27	24 29 23.2	HC
10 18	P 2124−12	21 27 30.49	−11 51 20.2	Ri
10 18	P 2128−12	21 31 35.26	−12 7 4.8	HC Gs
10 19	3C 446	22 25 47.26	− 4 57 1.4	Ti
10 25	GC 0322+22	3 25 35.91	22 24 12.2	Hv Ri Hs
10 26	3C 133	5 2 58.47	25 16 25.4	Gu
10 27	0554+242	5 57 4.56	24 13 53.7	Bj Ka
10 27	P 0601+24	6 4 55.27	24 29 23.2	Gu
10 28	P 0618+23	6 21 0.34	23 18 43.9	On
11 14	P 2124−12	21 27 30.49	−11 51 20.2	Bj
11 15	3C 446	22 25 47.26	− 4 57 1.4	SR
11 21	GC 0322+22	3 25 35.91	22 24 12.2	Gu Bj Ka
11 22	GC 0423+23	4 26 54.95	23 27 48.4	Ch
11 23	P 0504+23	5 7 6.41	23 51 13.7	Ch
11 24	0554+242	5 57 4.56	24 13 53.7	Rb By
11 24	0556+238	5 59 32.03	23 53 53.9	Rb On By
11 24	P 0618+23	6 21 0.34	23 18 43.9	Ch HC Hs
11 26	GC 0759+18	8 2 48.06	18 9 49.3	Rb On By
12 1	P 1223−074	12 26 16.33	− 7 41 6.2	SR
12 11	P 2124−12	21 27 30.49	−11 51 20.2	Rb
12 19	GC 0322+22	3 25 35.91	22 24 12.2	Hv Ri Hs
12 21	0554+242	5 57 4.56	24 13 53.7	Gs Hv Ka
12 21	0556+238	5 59 32.03	23 53 53.9	Ch HC Gs Bj Ka
12 21	P 0618+23	6 21 0.34	23 18 43.9	On By Bj Ka
12 23	GC 0759+18	8 2 48.06	18 9 49.3	Ch HC Gs Hv Ri Hs
12 28	P 1223−074	12 26 16.33	− 7 41 6.2	CT
12 30	P 1354−174	13 57 6.03	−17 44 1.3	By

1748-253 by Venus (1992 Nov. 11)

P 2208-137 by Mercury (1996 Mar. 7)

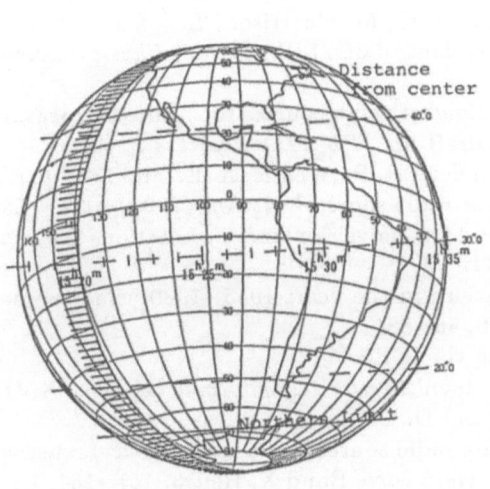

P 2008-159 by Mercury (1998 Mar. 6)

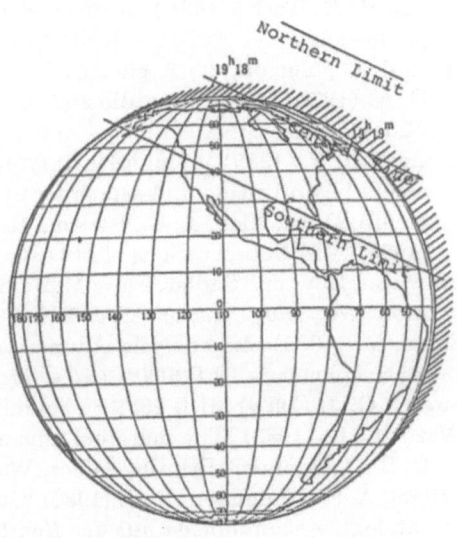

1215-002 by Mercury (1998 Sept. 26)

Fig. 3　Occultations by planets

References

Aoki, S., Sôma, M., Nakajima, K., Niimi, Y., Fujishita, M., and Takahashi, Y. (1986) 'The conversion from the B1950 FK4-based position to the J2000 position of celestial objects', in H. K. Eichhorn and R. J. Leacock (eds.), *Astrometric Techniques* (IAU Symp. No 109), D. Reidel Publishing Company, Dordrecht, pp. 123–131.

Argue, A. N., de Vegt, C., Elsmore, B., Fanselow, J., Harrington, R., Hemenway, P., Johnston, K. J., Kühr, H., Kumkova, I., Niell, A. E., Walter, H., and Witzel, A. (1984) 'A catalog of selected compact radio sources for the construction of an extragalactic radio/optical reference frame', *Astron. Astrophys.* **130**, 191–199.

Chapront-Touzé, M. and Chapront, J. (1982) 'ELP2000-82', Magnetic Tape.

Clark, T. A., Hutton, L. K., Marandino, G. E., Counselman, C. C., Robertson, D. S., Shapiro, I. I., Wittels, J. J., Hinteregger, H. F., Knight, C. A., Rogers, A. E. E., Whitney, A. R., Niell, A. E., Rönnäng, B. O., and Rydbeck, O. E. H. (1976) 'Radio source positions from very-long-baseline interferometry observations', *Astron. J.* **81**, 599–603.

Duncombe, R. L., Fricke, W., Seidelmann, P. K., and Wilkins, G. A. (1977) 'Joint report of the working groups of IAU commission 4 on precession, planetary ephemerides, units and time-scales', *Trans. IAU* **XVI B**, 56-67.

Elsmore, B. (1979) 'Review of radio astrometry I. Radio measurements', in F. V. Prochaska and R. H. Tucker (eds.), *Modern Astronomy* (IAU Coll. No. 48), Univ. Obs. Vienna, pp.93–99.

Hazard, C., Sutton, J., Argue, A. N., Kenworthy, C. M., Morrison, L. V., and Murray, C. A. (1971) 'Accurate radio and optical positions of 3C273B', *Nature Physical Science* **233**, 89–91.

Linfield, R. P. (1989) 'Using planetary occultations of radio sources for frame-tie measurements I. Motivation and search for events', draft for *JPL TDA Progress Reports*.

Morabito, D. D., Niell, A. E., Preston, R. A., Linfield, R. P., Wehrle, A. E., and Faulkner, J. (1986) 'VLBI observations of 416 extragalactic radio sources', *Astron. J.* **91**, 1038–1050.

Morrison, L. V. and Sadler, Flora McBain (1969) 'An analysis of lunar occultations 1960–66', *Mon. Not. Roy. Astron. Soc.* **144**, 129–141.

Sôma, M. (1985) 'An analysis of lunar occultations in the years 1955–1980 using the new lunar ephemeris ELP2000', *Celes. Mech.* **35**, 45–88.

Sovers, O. J. (1989) 'JPL 1989-3', submitted to the IERS.

Watts, C. B. (1963) 'The marginal zone of the Moon', *Astron. Pap. Amer. Ephem.* **XVII**, U. S. Government Printing Office, Washington, D. C.

Witzel, A. and Johnston, K. J. (1982) 'Candidate radio sources for a radio/optical reference catalog', *Abhandlungen aus der Hamburger Sternwarte* Band X, Heft 3, 151–164.

Discussion

FEISSEL: Radio source 3C273B has played a historical role in the initial definition of the right ascension origin of the kinematic reference frame, based on extragalactic objects, in agreement with that of the dynamical one based on solar system objects. In recent years, other processes have been used, *e.g.* (a) define the agreement with the FK5 system by a few tens of quasars whose positions are known in the FK5 system, or (b) maintain the directions of axes of the reference frame by a no-rotation condition based on well measured extragalactic radio sources.

SÔMA: I think occultations are still useful for checking the agreement between the radio and stellar or dynamical reference frames.

STEINERT: Radio source 3C273 has a variable structure. Is that not a disadvantage for a zero-point of the RA system?

SÔMA: Yes. I think a more stable radio source should play the role of the zero-point of right ascension.

FEISSEL: This zero-point of the radio system in RA (*i.e.* 3C273) has its history and it seems not necessary to change it.

SÔMA: The variability of the structure, for example, has not yet been established.

MORRISON: I was involved in the original reduction of the lunar occultations which led to the position published in the paper by Hazard *et al.* It seems to me somewhat fortuitous that your new position reduced with a modern lunar ephemeris should agree so closely with the original value. Do you agree?

SÔMA: Yes. The close agreement is merely fortuitous. In the present work, I also wanted to stress this point.

TREUHAFT: Is the dominant error contributing to your 7 millisecond error in the right ascension of 3C273 due to the uncertainty in the structure of the limb of the Moon?

SÔMA: Yes. I think observational timing error and the error in the limb correction are the main sources of error in the derived right ascension.

Comparison of optical and radio positions of stars

LV Morrison & RW Argyle, *Royal Greenwich Observatory, UK*
Y Requième & JM Mazurier, *Observatoire de L'Université de Bordeaux I,*
France

ABSTRACT. The positions of radio stars measured with respect to the VLBI network of extragalactic sources provides an important link between the optical and extragalactic reference frames. The establishment of this link was brought a stage nearer realization with the publication by Florkowski *et al.*(1985) of the radio positions of 20 stars measured with the Very Large Array (VLA) of the National Radio Astronomy Observatory, and the publication by Lestrade *et al.*(1985, 1988) of 10 stars measured with the VLBI network.

Several of these radio stars are known to be double, so it is necessary to check whether the optical and radio emission come from the same position in the system before these sources can be used as standard links between the reference frames.

The accuracy of the radio positions is typically 0".03, which is more accurate than the best optical positions available from catalogues at the present epoch. Even the FK5 has only an accuracy of about 0".04 at the epoch 1990; so astrometrists have started observational programmes to measure the positions of the radio stars.

Here we report on the optical positions obtained in the past few years with the automated meridian circle at Bordeaux in France (BORD) and the Carlsberg Automatic Meridian Circle (CAMC) at the Observatorio del Roque de los Muchachos, Instituto de Astrofísica de Canarias on the island of La Palma. The positions are referred to a homogeneous reference frame which is close to that of the FK5, covering the northern hemisphere and extending into the southern hemisphere as far as $-45°$. This improved optical reference frame gives the comparison with the radio reference frame advantages over previous work, such as Johnston *et al.*(1985), which was hampered by warping of the FK4 frame and the problem of extending even that inferior frame to fainter reference stars, especially in the southern hemisphere.

The new optical positions are intercompared with the radio ones and conclusions are drawn about the sizes of the offsets. The full text of this paper can be found in *Astron. Astrophys.* (1990).

J. H. Lieske and V. K. Abalakin (eds.), Inertial Coordinate System on the Sky, 513–514.
© 1990 *IAU. Printed in the Netherlands.*

514

References

Florkowski, K.J., Johnston, K.J., Wade, C.M. & de Vegt, C.: (1985), *Astron. J.* **90**, 2381.

Johnston, K.J., de Vegt, C., Florkowski, D.R. & Wade, C.M.: (1985), *Astron. J.* **90**, 2390.

Lestrade, J.-F., Preston, R.A., Requième, Y., Rapaport, M. and Mutel, R.L.: (1985), in *HIPPARCOS, Scientific Aspects of the Input Catalogue Preparation*, 251, ESA SP-234.

Lestrade, J.-F., White, G.L., Jauncey, D.L. and Preston, R.A.: (1988), in *HIPPARCOS, Scientific Aspects of the Input Catalogue Preparation II*, 481, Comissió Interdepartamental de Recerca i Innovació Tecnològica, Generalitat de Catalunya.

Discussion

BATTEN: Since RS CVn is a known spectroscopic binary, I assume that when you speak of checking it for duplicity you refer to a possible visual (or interferometric) companion. I agree that any discrepancy in the radio and optical emissions of Algol is unlikely to be explained by associating the former with C—the radio emission is almost certainly associated with AB.

RÉQUIÈME: I agree.

HUGHES: It has been shown by de Vegt that some types of radio stars are not suitable for astrometric comparisons. There are differences between H_2O and SiO masers and active or "flare" stars. We should not lump all of these together simply as "radio stars."

YE: When making very precise comparisons between radio and optical positions of objects, do you think the radio positions may have small differences due to observations at different wave bands and that the radio position may not be the same counterpart in the optical region for the same celestial object?

RÉQUIÈME: We are looking for such small differences which are already evident for some radiostars like R Aqr and Alpha Sco A.

COMPARISON OF PROPER MOTIONS OF STARS FROM AGK3 AND SPECIAL COMPILED CATALOGUE IN REGIONS WITH GALAXIES

S.P. RYBKA
Main Astronomical Observatory
Ukrainian Academy of Sciences
252127 Kiev
USSR

ABSTRACT. The procedure of construction of special compiled catalogue is presented. The compiled catalogue is based on 6 catalogues obtained following the KSZ plan. The mean error of proper motions of stars is $\pm 0''009$ per year. AGK3 proper motions compared with those in the compiled catalogue yield precession corrections $\Delta k = -0''0046 \pm 0''0010$, $\Delta n = +0''0045 \pm 0''0010$. Systematic differences between these catalogues are represented by a series development using products of Hermite polynomials, Legendre polynomials and Fourier terms.

The determination of absolute proper motions of stars with respect to galaxies plays an important part in the approximation of an inertial system. Proper motion programmes using galaxies as reference frame are in progress at several observatories.

There are comprehensive catalogues of proper motions compiled at Moscow, Tashkent, Pulkovo and Goloseevo following the KSZ plan (Deutsch, 1952). Since the Soviet programme is carried out by several observatories it is a good opportunity to combine individual catalogues in a general system (Rybka, 1980).

The Special Compiled Catalogue (SCC) is based on 6 individual catalogues. Specially selected distribution of sky areas in the compiled catalogue is optimum for the determination of corrections to the precession values. It may also reduce the residual magnitude equation errors originating from their investigation by statistical method (Rybka, 1985).

The resulting compiled catalogue for 21817 stars with magnitudes 8^m -15^m5 for 75 sky fields was completed in the declination zone from $+60°$ to $-2°$. The average rms error of an individual SCC proper motion is $\pm 0''009$. The SCC provides a data base for studies in the fields of stellar kinematics and astrometry. This paper presents first results based on SCC proper motions.

The differences between AGK3 proper motions and those in SCC were examined for corrections to the precession. There is a sample of 894 AGK3 stars in the compiled catalogue.

Results for precessional values are $\Delta k = -0''0046 \pm 0''0010$, $\Delta n = +0''0045 \pm 0''0010$. The computed quantities do not significally differ from values derived by Asteriadis from AGK3 proper motions

J. H. Lieske and V. K. Abalakin (eds.), Inertial Coordinate System on the Sky, 515–516.
© 1990 *IAU. Printed in the Netherlands.*

($\Delta k = -0\rlap{.}''0036 \pm 0\rlap{.}''0002$, $\Delta n = +0\rlap{.}''0044 \pm 0\rlap{.}''0002$) and by du Mont from the Lick Pilot proper motion programme ($\Delta k = -0\rlap{.}''0034 \pm 0\rlap{.}''0011$. $\Delta n = +0\rlap{.}''0038 \pm 0\rlap{.}''0006$).

Besides the above investigation other systematic errors were analysed in order to provide the linkage between AGK3 and SCC systems. They were performed with the new analytical method developed at Heidelberg. The spherical harmonics were replaced by products of Hermite polynomials, Legendre polynomials and Fourier terms. Using transformed declinations these types of functions are much more suited for modelling systematic differences between catalogues under considerations.

Only significant terms of series development were analysed. The F-test and γ-test with 1 % significance level were used for this purpose. The resulting systematic differences between AGK3 and SCC catalogues depend on Right Ascension and Declination and do not depend on the magnitude of stars.

The largest systematic differences occure in proper motions in Declination, especially in the declination zone from 0° to 30°. The root mean square difference in Declination is $0\rlap{.}''054 \pm 0\rlap{.}''0016$ and in Right Ascension is $0\rlap{.}''0037 \pm 0\rlap{.}''0016$.

The above results of analysis of differences of proper motions of stars given in AGK3 and SCC catalogues may be used for the connection between these systems within the sky areas covered with SCC catalogue.

References

Deutsch, A.N. (1952), *The utilization of extragalactic objects for construction of absolute proper motion system*, Academic Publishers Nauka, Moscow.

Rybka, S.P. (1980), "On the composing of the compiled absolute proper motion catalogue", *Pis'ma v Astron. Zhurn.* **6**, 55–57.

Rybka, S.P. (1985), "Corrections to precession constants from stellar proper motions in the areas with galaxies", *Kinematica i fizika neb. tel* **1**, 17–20.

AN ADVANCED ASTROMETRIC SENSOR ARRAY

David G. Monet and Harold D. Ables
U. S. Naval Observatory, Flagstaff Station
P.O. Box 1149
Flagstaff, Arizona 86002

The U.S.Naval Observatory has commissioned the fabrication of an array of six charge-coupled devices (CCDs) bonded to a single silicon substrate for precision astrometry of sources spanning a large range in apparent brightness. The array includes one CCD with rapid readout capability for imaging bright stars. We present the design parameters and the anticipated performance characteristics of this sensor.

The Electrical Engineering Department of Auburn University under contract to the U.S.Naval Observatory (USNO) has developed the technology to produce an array of six CCDs mounted on a single silicon substrate. Four of these advanced astrometric sensor arrays (AASAs), each composed of 5 Thomson- CSF CCD7882s and 1 Fairchild CCD222, are being fabricated by Auburn University for the USNO. Fabrication of these arrays required development of silicon-on- silicon bonding techniques that survive repeated thermal cycles between $+25^{\circ}$C and -140°C, electrical isolation of each CCD for independent connections to power and readout electronics subsystems, and accurate geometrical alignment of the CCDs. The AASAs will be used to measure positions of bright stars with respect to much fainter reference fields in "stare" mode operation and to make 3-color photometric observations in "scan" mode operation. A schematic representation of the AASA is shown in Figure 1.

The Fairchild CCD222 is an interline transfer device on which the signal can be electronically shuttered and read out in less than 0.05 second. With typical integration times of 10 minutes on the Thomson-CSF CCDs, the dynamic range of the AASA over which accurate astrometric measurements can be made should exceed 13 magnitudes. Based on experience with a Texas Instruments 800 x 800 CCD camera, we expect an astrometric accuracy of ±0.005 arcsec for a single observation made under favorable observing conditions and a final accuracy approaching ±0.001 arcsec for 40 or so combined observations. If this performance is realized in practice, absolute parallaxes and proper motions of stars as bright as 4th or 5th magnitude can be measured to an accuracy of ±0.001 arcsec and the positions of

J. H. Lieske and V. K. Abalakin (eds.), Inertial Coordinate System on the Sky, 517–518.
© 1990 *IAU. Printed in the Netherlands.*

these stars can be referred to an inertial extragalactic reference frame (quasi-stellar objects and galaxy nuclei) with the same accuracy.

We anticipate using one of the AASAs in the "scan" mode with appropriately placed color filters for 3-color photometric observations. In this mode the telescope focal plane image on the CCD and the charge image in the CCD are scanned along columns at the same rate. Accurate geometric alignment of the Thomson-CSF CCDs will enable sequential 3-color observations of selected fields. Auburn University employed a chemical milling process to produce a silicon mask for aligning the CCDs to within a few microns during the bonding process.

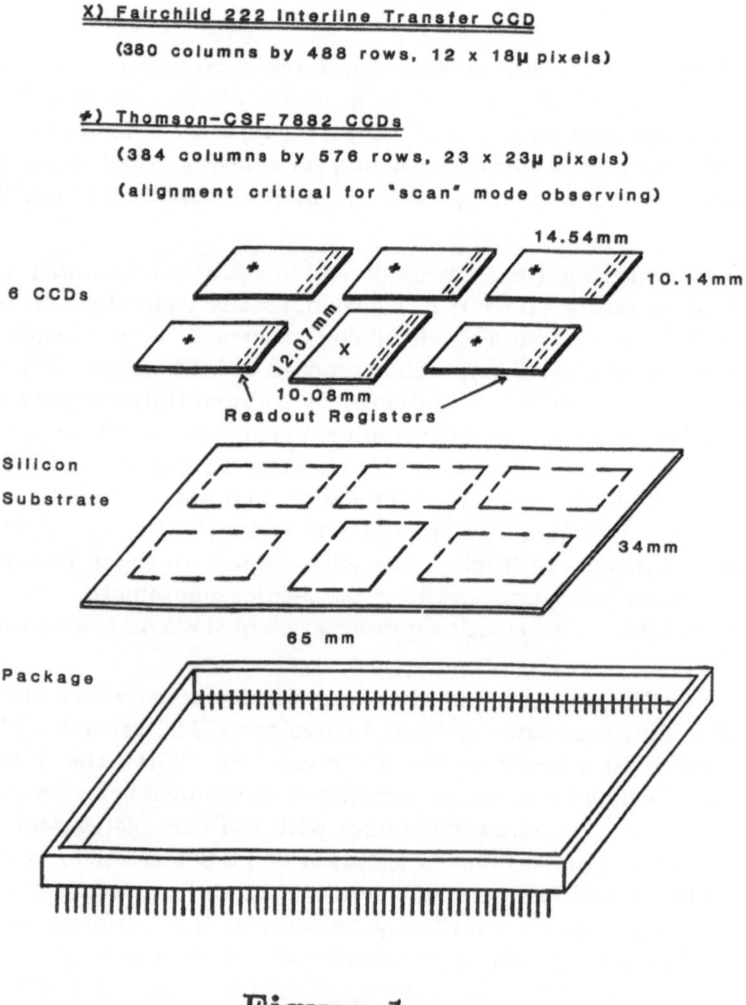
X) Fairchild 222 Interline Transfer CCD
(380 columns by 488 rows, 12 x 18μ pixels)
*) Thomson-CSF 7882 CCDs
(384 columns by 576 rows, 23 x 23μ pixels)
(alignment critical for "scan" mode observing)
14.54mm
10.14mm
6 CCDs
12.07mm
10.08mm
Readout Registers
Silicon Substrate
34mm
65 mm
Package

Figure 1.

MODERNIZATION OF CHINESE PHOTOELECTRIC ASTROLABES AND IMPROVEMENT OF OPTICAL REFERENCE SYSTEM

LI DONG-MING [1], WANG HONG-QI [2], ZHAO GANG [3], WANG ZE-ZHI [4], AND WANG RUI [5]

[1] *Purple Mountain Observatory, Nanjing, China*
[2] *Shanxi Observatory, Xian, China*
[3] *Shanghai Observatory, Shanghai, China*
[4] *Beijing Observatory, Beijing, China*
[5] *Yunnan Observatory, Kunming, China*

The features of the astrolabe are that:

(1) a span of 6 hour of R.A. of star positions can be tied in a group of observation of 2 hours, all-night observation can cover more than half of the sky, and
(2) the declination system can be set up with high precision without the need of a precise circle (Li 1987).

Observation with the astrolabe will make an important contribution to the improvement of the fundamental reference system (Fricke 1972, 1985; Guinot *et al.* 1961; Li 1987).

In the past 3 years, we have reequipped the 4 photoelectric astrolabes (1 Mark I and 3 Mark II). The modernization of Chinese photoelectric astrolabes includes the following:

(a) Automation of observational process. Now these instruments can operate automatically all night and observe about 40 stars per hour.
(b) modernization of photoelectric record technique to extend the limiting magnitude to about 10th magnitude or fainter. The later part of the work will be finished early next year.

The Chinese photoelectric astrolabes will first contribute to the improvement and expansion of the FK5 and will obtain valuable data to the linking of radio/optical, space/ground-based reference frames around 1994. Careful comparison of modern astrolabe catalogs, meridian catalogs and space catalogs will be conducive to the research on the sources and magnitude of systematic errors of ground-based catalogs from various angles, will help to explore the possibility to improve the fundamental proper motion system in the light of the revealed systematic errors and will find out the trend of development of various techniques.

The working program to improve the fundamental reference system with Chinese photoelectric astrolabes is limited to the northern sky at present. We have noticed that South American colleagues are carrying out astrolabe observations of bright stars on the southern sky (Clauzet 1986). The

J. H. Lieske and V. K. Abalakin (eds.), Inertial Coordinate System on the Sky, 519–520.
© *1990 IAU. Printed in the Netherlands.*

importance of this work will be more remarkable if the observations can be extended to SRS stars (IAU Comm.8 1986). Chinese astronomers are interested in cooperating with foreign colleagues, especially those in South America, to accomplish this task.

While carrying out the observations mentioned above, we will pay particular attention to determination of precise positions of radio stars in the fundamental optical reference system. 86 high-priority Hipparcos radio stars with $\delta > -4°$ will be included in the observation program. The observation will be extended to $\delta = -15°$ with the astrograph. It is expected that precise positions of more than 110 radio stars will be obtained. Among them about 50 will be observed with both astrolabe and astrograph. The astrolabe has strong capability of smoothing local errors. Optical positions of radio stars determined with the astrolabe refer to a fundamental position system of large range. Comparison of the results of astrolabe, meridian and photography will be conducive to revealing possible systematic differences between defferent techniques and the causes of systematic difference in radio–FK5.

Besides determination of star positions, the modernized photoelectric astrolabe can obtain photometric data and resolve binaries to a certain extent. It can be used to observe planets and astroids. At present, 4 Chinese photoelectric astrolabes are engaged in observation of 2000 bright stars. After it is accomplished in early next year, they will be put into the observation program mentioned above, including some stars of astrophysical interest.

References

Clauzet, L.B.F.: 1986, *Transactions of the IAU*, **XIXB**, 121.
Fricke, W.: 1972, *Ann. Rev. Astron. Astrophys.*, 10, 101.
Fricke, W.: 1985, *Celest. Mech.*, 36, 207.
Guinot, B. *et al.*: 1961, *Bull. Astron.*, 23, 307.
IAU Commission 8: 1986, *Transactions of the IAU*, **XIXB**, 111.
Li Qi: 1987, *Astron. Astrophys.*, 174, 307.

INDEX